DIELECTRIC RELAXATION AND DYNAMICS OF POLAR MOLECULES

World Scientific Series in Contemporary Chemical Physics – Vol. 8

DIELECTRIC RELAXATION AND DYNAMICS OF POLAR MOLECULES

Vladimir I Gaiduk

Institute of Radio Engineering & Electronics
Russian Academy of Sciences
Moscow

World Scientific
Singapore • New Jersey • London • Hong Kong

Published by

World Scientific Publishing Co. Pte. Ltd.

P O Box 128, Farrer Road, Singapore 912805

USA office: Suite 1B, 1060 Main Street, River Edge, NJ 07661

UK office: 57 Shelton Street, Covent Garden, London WC2H 9HE

British Library Cataloguing-in-Publication Data
A catalogue record for this book is available from the British Library.

DIELECTRIC RELAXATION AND DYNAMICS OF POLAR MOLECULES

ISBN 981-02-2123-1

Printed in Singapore by Uto-Print

PREFACE

Interaction of electromagnetic radiation with polar media results in absorption of radiation and in the change of its velocity of propagation. These fundamental physical phenomena may be described theoretically by employing the complex dielectric permittivity ε. This in turn is expressed in terms of real and imaginary parts ε' and ε'' by $\varepsilon = \varepsilon' - i\varepsilon''$. Dielectric properties of polar media have a great importance for practical application, since they govern the propagation of electromagnetic waves in the atmosphere, in water, in ice and in various aqueous systems. Dielectric phenomena are used also for the determination of constant eleotric moments of polar molecules and of their concentrations in various solutions and mixtures. The $\varepsilon(\nu)$ spectra, where ν is the frequency of electromagnetic radiation, supply information about molecular dynamics and the structure of fluids, particularly about time and space scales of molecular events at thermal equilibrium.

Basically the dielectric response arises from a torque $\boldsymbol{\mu} \times \boldsymbol{E}$, exerted by a polar molecule with electric moment $\boldsymbol{\mu}$ in the presence of radiation field $\boldsymbol{E}(t)$. The torque tends to turn the molecule along the direction of the field but this turning is opposed by the inertia of rotating molecules and by their interaction with neighboring molecules. The resultant $\varepsilon(\nu)$ spectra are spread from very low frequencies up to 10-30 THz, where we recall that 1 THz = 10^{12} Hz, so these spectra reach far infrared region. It is difficult to interpret these spectra without using microscopic or semi-microscopic molecular theory.

In this book the Reader will find equations, which provide a basis for dielectric theory; the description of a number of molecular models; the interpretation of dielectric spectra of polar fluids and briefly of gases (the technical details are disregarded).

The relative simplicity of the mathematical treatment provides an opportunity for studying dielectric response in polar fluids, that is, fluids which consist of molecules each of which contains a constant electric dipole moment μ. These fluids embrace nonassociated liquids, namely those whose molecules behave independently to a first approximation, and associated H-bonded liquids whose molecules are connected by semi-chemical bonds so as to form a molecular network in space. To this type of liquid one may include water, aqueous solutions, water bounded by biological molecules. The dielectric behavior of isotropic media is found as an important particular case of anisotropic polar fluids.

The systematic description of the theory and its applications follows after two introductory chapters, in which definitions of appropriate physical parameters are given, the basic ideas of the theory are listed and wideband dielectric spectra in liquids / rotational spectra in gases are briefly discussed.

This book was born on a terra bordered on physical electronics, molecular and chemical physics. Namely, an approach, which was previously used in physical electronics, was applied to polar fluids. This approach was considerably improved and was termed the *dynamic method.* The latter was elaborated by the author in collaboration with Boris Tseitlin and (on the first stage) with Yuri Kalmykov. Many calculations of dielectric spectra were performed by the author in collaboration with T. A. Novskova, V. V. Brekhovskikh, V. V. Gaiduk, and S. V. Titov.

The very idea to submit the monograph to the World Scientific was gifted to the author by Prof. Myron Evans and Prof. William Coffey. Several visits of the author to Trinity College Dublin and to the Dublin Institute of Advanced Studies helped the author to start writing the monograph. The author tried to show that the evolution of analytic models from very simple to rather complex occurred under "pressing" demand to obtain a satisfactory agreement between the theory and experimental data observed in various polar media.

The book is nearer to the textbook than to a traditional "pure" monograph; the author tried to present in the book all derivations of the formulas; the number of references is reduced to a minimum, and these are concentrated in few places in order not to distract the attention of the Reader.

V. I. G.

Fryazino,
Moscow region,
Russia

ACKNOWLEDGEMENTS

Many new results, presented in the book, were obtained by the author jointly with Boris Tseitlin; part of them remains yet unpublished. Boris Tseitlin had also read almost all the text and checked derivations of most analytical expressions.

The book was read by Prof. Myron Evans who suggested corrections related to English text. The part two of the book, devoted to the general theory (chapters 3-6), was read by Prof. B. K. P. Scaife who made many useful remarks. Prof. J. McConnell has corrected the first (introductory) chapter, where he also partially wrote sections 1.1-1.4.

The author owes a great deal to all above mentioned scientists for their valuable help, as well as to Prof. W. Coffey and Prof. J. McConnell for their hospitality in Dublin.

The author appreciates the help of Ms. Sook Cheng Lim and of Mr. Choon Yang Chiam in resolving of many problems that arose during a long period while the author corresponded with the WS Publishing Co.

The author recognizes that his English is still poor but hopes that in spite of it the material of the monograph will be clear to the Reader.

V. I. G.

CONTENTS

LIST OF SYMBOLS

AC	Alternating current (time-varying field)	ED	Extended (rotational) diffusion (model)
ACF	Autocorrelation function	EM	Electromagnetic
CCR	Cone confined rotator (model)	FIR	Far infra-red (spectral region)
CF	Constant field (model)	GT	(Gaiduk and Tseitlin 1994)
CR	Confined rotator (model)	HF	High frequency (usually in the FIR spectral region)
CS	Cosine squared (model)	HM	Hybrid model
DWP	Double well potential	SF	Spectral function

Symbol	Value	Example	Symbol	Value	Example
i	$\sqrt{-1}$		$\check{f}$	Symbol, referring to librations	
*	Complex conjugation	$\chi^* = \chi' + i\chi''$	$\mathring{f}$	Symbol referring to rotation	
$\hat{\boldsymbol{E}}$	Complex amplitude		$\in$	"is element of" sign	$x \in [a, b]$
$\langle f \rangle$	Ensemble average of quantity f(G)	$\langle X \rangle = \int XW d\Gamma$	$\propto$	Proportio-nality sign	$G \propto N$

$\hat{A} = \hat{A}(\omega)$	Complex orientational factor
$\hat{B} = \hat{B}(\omega)$	Complex angular velocity factor
$B = h/(8\pi^2 cI)$	Rotational constant (in cm^{-1})
$b = (I/I_c) - 1$	Form-factor of a symmetric top molecule
$c = 2.9979 \cdot 10^{10}$ cm/s	Velocity of light
$C(t)$, $\tilde{C}(\omega)$	Single-particle dipolar ACF and its spectrum
D	Self-diffusion coefficient
$d\Gamma$	Volume element in a phase space
$E = E_m \sin \omega t = \mathrm{Re}\,[\hat{E}\, e^{i\omega t}]$	Radiation electric field
$\hat{E}, E_m$	Complex, real amplitude of radiation field
$\mathbf{E}(k)$	Complete elliptic integral of the second kind,
$F(t), F(\gamma)$	Induced distribution functions of polar molecules
$\{F_B, F_D, F_G, F_K, F_T\}$	Induced distributions: Boltzmann, Debye, Gross, orientation-correlation, isothermal
$f = 2\beta/\pi$	Normalized libration amplitude in the CR model
$G = \mu^2 N / (3k_B T)$	Normalized concentration of molecules
g	Kirkwood correlation factor
H	Steady state energy (Hamiltonian) of a dipole
$h = 6.6262 \cdot 10^{-27}$ erg·s	Planck constant
$h = H/(k_B T)$	Normalized energy of a dipole
I	Moment of inertia of a molecule
$J = 0, 1, 2, \ldots$	Rotational quantum number
$\mathbf{K}(k)$	Complete elliptic integral of the first kind
$K(z)$	Spectral function (in an anisotropic medium)
$k = (\omega/c)\sqrt{\varepsilon} = k' - i\,k''$	Complex propagation constant
$k_B = 1.3807 \cdot 10^{-16}$ erg /·K	Boltzmann constant
k_μ	Factor correcting μ value in a liquid
$L(z)$	Spectral function (convenient to employ for an isotropic medium)
l	Projection of the normalized angular momentum on

	the symmetry axis
M, m	Molecular mass (gram·mole); mass of a molecule
$m_H = 1.6726 \cdot 10^{-24}$ g	Mass of a proton
N	Concentration
$N_A = 6.0220 \cdot 10^{23}$ (g·mole)$^{-1}$	Avogadro constant
$n^* = \mathrm{Re}\,[\sqrt{\varepsilon^*}\,] = n + i\kappa\,]$	Complex refraction index
$n_\infty = n(\omega \to \infty)$	Optical refraction index
$Q = \Omega \bullet [\mu \times \hat{\boldsymbol{E}} \exp(i\omega t)\,]$	Complex power describing interaction a dipole with EM field
$q = \cos\Delta = \mu_E/\mu$	Direction cosine (see Fig. 4.1)
$\check{p} = \eta\check{\Omega},\ \mathring{p} = \eta\,\mathring{\Omega}$	Normalized angular frequency of librators, rotators
p, p_m	The field parameter of the conservative field
$\boldsymbol{P}, P$	Polarization, the gas pressure
$s_n = \dfrac{8f^2}{\pi^2}\,\dfrac{\sin^2[\pi(n-f)/2]}{(n^2-f^2)^2}$	Statistical weight of the n th Fourier amplitude for the confined rotator model
T	Temperature
$t,\ t_0,\ t_v$	Time, initial instant after the last collision, lifetime (random quantity)
$W(\Gamma) = C\exp\,[-H/(k_B T)])$	Steady state distribution function
$x = \omega\eta = 2\pi c\nu\eta x$	Normalized field frequency
$x_n = xf/n$	
$x_D = \eta/\tau_D = 2\pi c\nu_D\eta$	Normalized frequency of the loss maximum $\varepsilon''(\omega)$
$y = \eta/\tau = \sqrt{I/(2k_B T)}/\tau$	Normalized collision frequency
$y_n = \eta f/n$	
$z = x + iy$	Normalized complex frequency
$z_n = zf/n = x_n + i\,y_n$	
Z	Statistical sum
$\alpha = \omega\varepsilon''/(cn) = 4\pi\nu\mathrm{Im}\{\sqrt{\varepsilon^*}\,\}$	Absorption coefficient
β	Libration amplitude
Γ	Phase variables
$\gamma = \omega t_0$	Initial phase

$\Delta\nu_J = 1/(2\pi c\tau_J)$	Half-width of the J th absorption line
$\varepsilon = \varepsilon' + i\varepsilon''$, ε_s	Complex, static dielectric permittivity
$\eta = \sqrt{I/(2k_B T)}$	Normalizing time parameter
λ	Wavelength of EM radiation in vacuo
λ	Precession angle (see Fig. 3.3b)
$\mu = \mu_0 k_\mu (n^2_\infty + 2)/3$	Dipole moment of a molecule in liquid
μ_0	Dipole moment of a molecule in vacuo
$\nu = \omega/(2\pi c) = 1/\lambda$	Wave number (in cm^{-1}), called «frequency»
$\Pi = \int_{-\infty}^{\infty} \omega\, \chi''(\omega)\, d\omega$	Integrated absorption
Φ	The period of the $q(t) = \mu_E(t)/\mu$ function
$\check{\Phi}$, $\mathring{\Phi}$	The periods of libration, rotation
$\Phi(t)$, $\tilde{\Phi}(\omega)$	Many-particle dipolar ACF and its spectrum
$\Phi = \omega t - kz = \omega t - k'z + ik''z$	The complex phase
τ	The mean lifetime [of the potential well $U(\vartheta)$]
τ_D	Debye relaxation time
$\chi^* = \chi' + i\chi''$	Susceptibility of an isotropic medium, effective susceptibility of an anisotropic medium
Ψ	Angle between vector $\boldsymbol{E}_m$ and the symmetry axis Z of the potential well (see Fig.4.1)
$\Psi_v = \tau^{-1} \exp(-t_v/\tau)$	The exponential distribution over lifetimes t_v
$\Omega = d\vartheta/dt$, $\check{\Omega}$, $\mathring{\Omega}$	Angular velocity, angular frequencies of librators and hindered rotators
$\omega = 2\pi c\nu$	Angular radiation frequency

Functions

$\mathrm{sn}(u,k)$, $\mathrm{cn}(u,k)$, $\mathrm{dn}(u,k)$	The Jacobian elliptic functions
$\mathrm{erf}(x) = (2/\sqrt{\pi}) \int_0^x \exp(-t^2)\, dt$	The error function
$E_1(z) = \int_z^{\infty} \exp(-t)\, t^{-1}\, dt$	The exponential integral function of complex argument z
$w(z) = \frac{i}{\pi} \int_{-\infty}^{\infty} \frac{\exp(-t^2)\, dt}{z - t}$	Probability integral of complex argument

PART ONE

INTRODUCTION TO THE DIELECTRIC SPECTROSCOPY

CHAPTER 1
BASIC TERMS, PROCESSES AND MODELS

This book is devoted to description of certain relatively simple microscopic processes in condensed and gaseous states. The main attention is given to polar liquids. As for our approach to polar gases, the key idea consists in an attempt to describe a rotational spectrum "from the viewpoint" of a liquid but with inclusion of necessary quantum corrections. Indeed, one can take the envelope of the polar gas spectrum as a basis for description of condensed-state absorption spectrum. It should be noted, that an additional high frequency shift of the absorption band appears in the latter simultaneously with the rise of Debye relaxation. Both phenomena can be explained by intermolecular interactions.

Einstein (1905) was the founder of the classical microscopic description of *translational* Brownian motion. Dielectric spectra are underlain by *reorientation* of polar molecules. In the study of dielectric behavior of polar fluids we can distinguish the following three periods.

1st period. STUDY OF THE LOW FREQUENCY RELAXATION

Debye (1929), when investigating dielectric relaxation, obtained the relation between the relaxation time τ_D, now known as Debye relaxation time, and the coefficient ζ of molecular friction. He also calculated the maximum value of the loss factor ε'' and predicted for water the angular frequency ω_D of this maximum, given by $\omega_D = \tau_D^{-1}$. It should be noted that this frequency ω_D belongs to microwave region while microwave technique was unknown at the time of prediction.

The books of Frohlich (1958), McConnell (1980), and Scaife (1989) present a comprehensive account of fundamental achievements in this field. It was then customary to neglect in the equations of motion the term $I{\cdot}d^2\theta/dt^2$, proportional to the moment of inertia I and to the second time derivation of the angular variable θ. This term describes orientation of a molecular dipole and gives rise to nonlinear dynamics. Thus, it is presumed that the moment of inertia is negligibly small. Only the low frequency part of the orientational spectrum was obtained theoretically. At $\omega >> \omega_D$ the theoretical absorption coefficient $\alpha(\omega)$ reaches the "Debye plateau", while in experiments when ω rises it goes through the maximum value and then drops in the far infrared (FIR) spectral region.

2nd period. QUALITATIVE DESCRIPTION OF WIDEBAND SPECTRA IN LIQUIDS

This discrepancy between theory and experiment gave rise to many publications in which efforts were made to resolve the problem. In these works the moment of inertia I of a molecule was considered to be nonzero. We shall give now a few examples without however claiming to present a historical description of such efforts. Thus Nora E. Hill (1963) proposed a model, in which a molecule is confined to a cage formed by the neighboring molecules. Then G. Wyllie (1971) specified that the cage is related to the smallest polyhedron whose faces are the normal bisectors of the lines joining the center of the molecule under consideration to the centers of neighboring molecules. C. Brot (1975) and his collaborator (Lassier and Brot 1968) supposed that the molecule executes perturbed angular oscillations called librations. A librating molecule has an equilibrium orientation and this orientation undergoes rotational diffusion motion during the relaxation time, which is much greater than the libration period. The sets of orientations are called potential wells and these wells are separated by potential barriers. With I nonzero a distinction appears between weak collisions. which perturb only the librational state of a molecule in its well, and strong collisions which cause a reorientation jump into another well. Moreover, on account of the finite mass of a molecule this jump is not instantaneous and the process is non-Marcovian (McConnell 1980, Sec. 4.2).

The activity of the European Molecular Liquid Group (EMLG) greatly stimulated the study of mechanisms of dielectric relaxation. Particularly, conferences arranged by EMLG concentrated the attention of many investigators on some basic unsolved problems of dielectric theory. Many references and useful information about this period the Reader may find in the monographs of M. Evans, G. Evans, Coffey, and Grigolini (1982) and of Coffey, Evans, and Grigolini (1984). In the review article of Gaiduk and Kalmykov (1987) a new approach to dielectric theory was suggested and a new confined rotator model was described, where the rectangular potential well was considered. In the first two books stochastic equations were taken as a basis of the theory, while in the above review article a simpler formalism was elaborated. The corresponding approach, termed "the dynamic method", assumes the approximation of instantaneous collisions and the introduction of induced distribution functions. In all these investigations and in many others it was found possible to avoid the difficulty, arising from the existence of the

Debye plateau, and to provide qualitative explanation of dielectric spectra of various polar fluids in all the range of orientational relaxation, including the Debye and Poley absorption regions.

3rd period. QUANTITATIVE DESCRIPTION OF DIELECTRIC BEHAVIOUR IN ISOTROPIC AND ANISOTROPIC MEDIA

Remarkable progress was achieved recently in both the theoretical approaches mentioned above, "stochastic" and "dynamic". In particular, the dynamic method was developed by Gaiduk and Tseitlin (1994) (to be referred throughout as GT), so as to provide a solution of nonlinear equations of motion for a dipole, rotating / librating in a conservative potential well, and calculation of spectrum of the orientational autocorrelation function by using the above solution. The most interesting physical problem in the study of dielectric relaxation can be formulated as follows. We need to find an adequate potential profile $U(\theta)$ in order to describe quantitatively the dielectric relaxation in various classes of fluids (we mean both low and high frequency phenomena). We think that this purpose may be achieved in near future. The main purpose of the book is related to the systematic description of the dynamic method and of its applications, having in mind the physical problem just formulated.

1.1 Displacement and Polarization Vectors for a Constant External Field

A dielectric may be defined as a substance that does not conduct electricity. We shall recall some results in electromagnetic theory related to dielectrics. Associated with static electric field $\boldsymbol{E}$ for free space there is an electric field $\boldsymbol{D}$ related to a dielectric and called the displacement vector. If the medium is isotropic, that is to say, if the medium has no directional properties, and if it is homogeneous, then $\boldsymbol{D}$ and $\boldsymbol{E}$ satisfy the relation

$$\boldsymbol{D} = c_0 \boldsymbol{E}, \tag{1.1}$$

where c_0 is a constant. The polarization vector $\boldsymbol{P}$ is defined by

$$\boldsymbol{P} = [\,(c_0 - 1) / (4\pi)\,]\,\boldsymbol{E}. \tag{1.2}$$

It follows from (1.1) and (1.2) that

$$\boldsymbol{D} = \boldsymbol{E} + 4\pi \boldsymbol{P} \tag{1.3}$$

and that $\boldsymbol{E}$, $\boldsymbol{P}$, $\boldsymbol{D}$ are in the same direction. The polarization is an additional field produced by charges / dipoles of the medium. In vacuum $\boldsymbol{P} = 0$ and $\boldsymbol{D} = \boldsymbol{E}$. If the dielectric is anisotropic, Eq. 1.3 will be replaced by a set of equations

$$D_1 = c_{11} E_1 + c_{12} E_2 + c_{13} E_3,$$

$$D_2 = c_{21} E_1 + c_{22} E_2 + c_{23} E_3,$$

$$D_3 = c_{31} E_1 + c_{32} E_2 + c_{33} E_3, \tag{1.4}$$

where

$$c_{23} = c_{32}\ , c_{31} = c_{13}\ , c_{12} = c_{21}\ , \tag{1.5}$$

and subscripts 1, 2, 3 refer to the three axes of a coordinate system. The vectors $\boldsymbol{E}$ and $\boldsymbol{D}$ are no longer in the same direction.

We shall show that using the dynamic method one can calculate *directly* the complex permittivity ε thus avoiding the tedious problem related to the preliminary calculation of the coefficients c_{ij} in Eq. 1.4 (i, j = 1, 2, 3).

Unless it is stated to the contrary, we shall in future assume that the dielectrics which we are studying are homogeneous and isotropic. This will permit us to employ (1.1) rather than (1.4).

Suppose that the electric field $\boldsymbol{E}$ having been in operation for some time is shut off. If the dielectric is fluid, the system of charges will as a result of their thermal motion revert to a random arrangement with zero dipole moment of a dielectric sample. This reversal is known as *dielectric relaxation.* But usually the latter term has more general meaning: it denotes the dielectric behavior of a fluid in a wide frequency range. It is this notion of dielectric relaxation which will be mentioned further.

1.2 Relative Complex Permittivity and Susceptibility

We next consider the situation where the external field in a fixed direction acting on the dielectric is periodic in time with angular frequency ω. Thus in any point of a medium we generalize (1.3) to

$$\boldsymbol{D}(t) = \mathrm{Re}\{\hat{\boldsymbol{D}}(\omega)\exp(i\omega t)\},\ \boldsymbol{E}(t) = \mathrm{Re}\{\hat{\boldsymbol{E}}(\omega)\exp(i\omega t)\}. \tag{1.6}$$

The *complex amplitudes* $\hat{\boldsymbol{D}}$ and $\hat{\boldsymbol{E}}$ may depend on coordinates and are related by the equation

$$\hat{\boldsymbol{D}}(\omega) = \varepsilon(\omega)\hat{\boldsymbol{E}}(\omega), \tag{1.7}$$

where $\varepsilon(\omega)$ is defined as the complex relative permittivity. In vacuum $\varepsilon = 1$ and $\hat{\boldsymbol{D}} = \hat{\boldsymbol{E}}$. It is customary to write

$$\varepsilon(\omega) = \varepsilon'(\omega) - i\varepsilon''(\omega), \tag{1.8}$$

where $\varepsilon'(\omega)$ and $\varepsilon''(\omega)$ are both real and are termed, respectively, the relative *permittivity* and the *loss factor* (or simply *loss*). The significance of the minus sign in (1.8) is related to the fact that because of the response of the medium, described by the polarization term in (1.3), the time dependence $D(t)$ lags behind in phase the external field $E(t)$. Assuming for simplicity $\hat{E}$ to be real, we have

$$D(t) = \mathrm{Re}\{\hat{E}\ (\varepsilon' - i\,\varepsilon'')\ (\cos\omega\, t - i \sin\omega\, t)\} = \hat{E}\ (\varepsilon' \cos\omega\, t + \varepsilon'' \sin\omega\, t) \quad .$$

We shall see later that $\omega\varepsilon'' > 0$ at thermal equilibrium, so the phase lag is indeed positive. For the permittivity given by (1.8) the plane electromagnetic wave, propagating in the + Z-direction,

$$E(t) = \hat{E}_0\, \mathrm{Re}\{\exp(i\omega\ t - kZ)\} \tag{1.8a}$$

will be absorbed by the medium. Indeed, the propagation constant k, given by

$$k = k' - i\, k'' = (\omega / c)\, \sqrt{\varepsilon}\,, \tag{1.8b}$$

possesses the positive loss factor k'' simultaneously with the positive ε''. The same will result for the wave propagating in the -Z direction, in this case we take $\omega t + kZ$ in (1.8a) and retain (1.8b), see also Sec. 1.3. If the field is constant, then on comparing (1.1) with (1.6)-(1.8) we have

$$D(0) = D_s\,,\ \varepsilon'(0) = \varepsilon_s\,,\ \varepsilon''(0) = 0, \tag{1.9}$$

subscript s being referred to the word "static".

Up to this point we have not considered the concrete cause of the dielectric polarization. We shall assume, that the latter is induced by reorientation of polar molecules in an external time-varying electric field $\boldsymbol{E}(t)$. The relative complex susceptibility $\chi(\omega)$ is associated with $\varepsilon(\omega)$, for which the following definitions are to be found in the literature:

$$\chi(\omega) = [\varepsilon(\omega) - 1] / (4\pi)\,, \tag{1.10}$$

$$\chi(\omega) = [\varepsilon(\omega) - n_\infty^2] / (4\pi), \tag{1.11}$$

$$\chi(\omega) = \frac{\varepsilon(\omega) - n_\infty^2}{4\pi}\, \frac{2\varepsilon(\omega) + n_\infty^2}{3\varepsilon(\omega)}\,. \tag{1.12}$$

In proposing (1.10) the influence of induced polarization, such as the response of electrons in atomic and molecular shells, has been neglected, while in (1.11) the latter mechanism of polarization is accounted for and the symbol n_∞ denote the optical refractive index. In both equations the susceptibility $\chi(\omega)$ means the response of polar medium to reorientation of molecular dipoles. The complex function $\chi(\omega)$ can be calculated theoretically on the basis of a semi-microscopic model.

In (1.12) we account also for the difference between the averaged electric field $E(t)$ in any point and the time varying internal field $E_c(t)$ at this point, so the multiplier

$$\frac{2\varepsilon(\omega) + n_\infty^2}{3\varepsilon(\omega)} \tag{1.13}$$

means the internal field correction for $\chi(\omega)$ (Gaiduk, Novskova and Brekhovskikh 1991). It must however be admitted that there is no solid theoretical basis for (1.12). Furthermore, we shall employ Eq. 1.11 for a *qualitative* description of dielectric relaxation and Eq. 1.12 for a more rigorous description. Putting $\omega = 0$ we find that for Eq. 1.11

$$\chi_s = \chi'(0) = [\varepsilon_s - n_\infty^2] / (4\pi), \quad \chi''(0) = 0 \tag{1.14a}$$

and for Eq. 1.12

$$\chi_s = \chi'(0) = gG, \tag{1.14b}$$

where g is the Kirkwood correlation factor. It is related to the experimental parameters ε_S, n_∞, G:

$$g = \frac{\varepsilon_S - n_\infty^2}{4\pi G} \frac{2\varepsilon_S + n_\infty^2}{3\varepsilon_S} \; ; \tag{1.15}$$

the normalized concentration G is given by

$$G = \mu^2 N_A / (3k_B T), \tag{1.16}$$

k_B is the Boltzmann constant, T is absolute temperature, μ is a constant dipole moment of a polar molecule in a fluid, N is the number density of molecules (the concentration). In a pure liquid, having density ρ and molecular mass M [g·mole] N is given by

$$N = N_A \rho / M, \tag{1.17}$$

where N_A is the Avogadro number.

The g-factor accounts for the correction of reorientation of polar molecules at equilibrium.

In this book the susceptibility $\chi(\omega)$ will be calculated from a number of reasonable molecular models. Thus, the permittivity $\varepsilon(\omega)$ may be expressed through $\chi(\omega)$ from the quadratic equation (1.12), while (1.11) yields a simpler linear relation:

$$\varepsilon(\omega) = 4\pi\chi(\omega) + n_{\infty}^{2}. \tag{1.18}$$

A physical quantity, proportional to the radiation frequency ω, that we shall employ especially in the discussion of experimental results, is the wave number, denoted by ν and defined by

$$\nu = \omega / (2\pi c). \tag{1.19}$$

Since $\nu = 1/\lambda$, where λ is the wavelength *in vacuo*, the dimension of ν is that of the reciprocal of a length. We shall express ν in units cm^{-1}. For brevity this quantity is usually called "frequency".

In theoretical calculations it is more convenient to use the quantity χ^* (which is the complex conjugate of the susceptibility χ), since χ^* has a positive imaginary part, the same is true for the permittivity ε^*. Thus Eq. 1.12 can be rewritten in the following way:

$$\chi^*(\omega) = \frac{\varepsilon^*(\omega) - n_{\infty}^2}{4\pi} \frac{2\varepsilon^*(\omega) + n_{\infty}^2}{3\varepsilon^*(\omega)}, \tag{1.20}$$

where

$$\varepsilon^* = \varepsilon' + i\,\varepsilon'', \chi^* = \chi' + i\chi''. \tag{1.20a}$$

The quantity χ^* will also be termed susceptibility.

1.3 Complex Refractive Index and Absorption Coefficient

The Maxwell equations for an isotropic medium with constant electric permittivity ε yield the wave equation (Scaife 1989, Sec.1.9)

$$\nabla^2 \mathbf{E}(t) = \frac{1}{c^2}\frac{\partial^2}{\partial t^2}, \quad \text{where } \nabla^2 = \frac{\partial^2}{\partial X^2} + \frac{\partial^2}{\partial Y^2} + \frac{\partial^2}{\partial Z^2}; \tag{1.21a}$$

X, Y, Z are rectangular coordinates. For a plane wave the solution of (1.21a) we may take in the form

$$\boldsymbol{E}(t) = \mathrm{Re}(\,\hat{\boldsymbol{E}}_0 \exp\{i\omega[t - \hat{n}Z/c]\}\,) = \mathrm{Re}(\,\hat{\boldsymbol{E}}_0 \exp\{i\omega t - \hat{k}Z\}\,), \quad (1.21b)$$

where $\hat{n}$ and $\hat{\kappa}$ are complex refractive index and propagation constant, respectively,

$$\hat{n}(\omega) = n(\omega) - i\kappa(\omega); \quad (\omega/c)\,\hat{n}(\omega) = k' - ik'', \qquad (1.22)$$

and Z is in the direction of propagation of a wave (the Reader should not mix the Greek $\kappa = -\,\mathrm{Im}\{\hat{n}\}$ with the complex propagation constant k !). The amplitude $\hat{E}_0$ of a wave is supposed to be constant (in space). If we take $\hat{E}_0$ to be real, then

$$E(t) = \hat{E}_0 \exp(-ik''Z) \cos(\omega t - nZ/c). \qquad (1.23)$$

We see that $n(\omega)$ is the index of refraction of the plane wave, and it depends on the properties of the medium. The phase velocity of the wave is $c/n(\omega)$ and is less than the light velocity since $n > 1$.

We notice that in small volume the mean value of the energy of radiation decreases in space by factor e^{-1} in the distance $(2k'')^{-1}$, since

$$\frac{1}{8\pi}\,\overline{[\mathrm{Re}\,E(t)]^2} = (\hat{E}_0^2/[8\pi])\left\langle \cos^2(\omega t - nZ/c)\right\rangle \exp(-2k''Z)$$

$$= (\hat{E}_0^2/[16\pi])\exp(-2k''Z).$$

Hence, $2k''$ is the *absorption coefficient* of radiation, which we shall denote $\alpha(\omega)$, so that

$$\alpha(\omega) = 2k''. \qquad (1.24)$$

To express the absorption coefficient in terms of the complex permittivity we employ (1.7) and (1.21a) to show that

$$\varepsilon(\omega) = \hat{n}^2(\omega) \qquad (1.25)$$

Then from (1.7) the real refractive index is given by

$$n(\omega) = \mathrm{Re}\{\sqrt{\varepsilon(\omega)}\} = \frac{1}{\sqrt{2}}\sqrt{\sqrt{\varepsilon'^2+\varepsilon''^2}+\varepsilon'^2}. \tag{1.26}$$

It should be noted that often the loss angle δ is introduced in the following way

$$\varepsilon'(\omega)= |\varepsilon(\omega)| \cos\delta, \quad \varepsilon''(\omega) = |\varepsilon(\omega)| \sin\delta.$$

Thus,

$$\tan\delta(\omega) = \varepsilon''(\omega) / \varepsilon'(\omega). \tag{1.27}$$

In low absorbing media δ is very small. In this book we are mainly interested in the opposite case of strongly absorbing polar fluids, for which the maximum value of δ has an order of a unity.

In many cases, where two components of permittivity, ε' and ε'', are employed, it is convenient to express them through the real (n) and imaginary (κ) parts of the complex refractive index, since n and κ can be measured in the far infra-red range. According to (1.22) and (1.25)

$$\varepsilon(\omega) = (c^2/\omega^2)\, k^2 = n^2 - \kappa^2 - 2i\,\kappa\, n,$$

from which it follows

$$\varepsilon' = n^2 - \kappa^2, \quad \varepsilon'' = 2\,\kappa\, n. \tag{1.28}$$

At last, we rewrite Eq. 1.24 in forms which will be used extensively later:

$$\alpha(\omega) = \frac{\omega\varepsilon''(\omega)}{cn(\omega)}, \tag{1.29}$$

or equivalently

$$\alpha(\omega) = 4\pi\nu\, \mathrm{Im}\sqrt{\varepsilon'(\nu) + i\,\varepsilon''(\nu)}\,, \tag{1.30}$$

that is, introducing again the symbol of complex conjugate,

$$\alpha(\omega) = 4\pi\nu\, \mathrm{Im}\,[\sqrt{\varepsilon^*(\nu)}]. \tag{1.31}$$

Since at thermal equilibrium $n(\omega)$ and $\alpha(\omega)$ are non-negative quantities, (1.29) shows that

$$\omega\varepsilon''(\omega) > 0. \tag{1.32}$$

1.4 Debye Model of Rotational Diffusion

To compare theoretical and experimental results for dielectric relaxation one may plot as functions of ω the relative permittivity $\varepsilon'(\omega)$, the loss factor $\varepsilon''(\omega)$, the refractive index $n(\omega)$ and the absorption coefficient $\alpha(\omega)$. A very useful and simple procedure, proposed by the brothers K. S. and R. H. Cole (1941), is to plot $\varepsilon''(\omega)$ as a function of $\varepsilon'(\omega)$. If we assume that

$$\omega > 0, \tag{1.33}$$

then on account of (1.32) the Cole-Cole plot lies in the upper half-plane.

Many theories have been advanced to provide the value of the complex relative permittivity $\varepsilon(\omega)$ for a prescribed physical model. For the present we shall be content to accept the Debye theory of relaxation as a reference point for other theories. Debye (1929) proposed that dielectrics contain polar molecules, each of which has the same constant electric dipole moment μ. He assumed each molecule to be spherical and he supposed that the dipole is situated at the center of the molecule. The Debye theory of relaxation is based on the consideration of rotational Brownian motion of polar molecules. We denote by I the moment of inertia of a molecule about a diameter, by $\boldsymbol{E}$ electric field that acts on the molecule, by θ the angle between the diameter of the molecular dipole moment $\boldsymbol{\mu}$ and $\boldsymbol{E}$, and by ζ the coefficient of microscopic molecular friction between the polar molecule and its environment, which is supposed to consist of *nonpolar* molecules. Then, if the microscopic system is assumed to obey the macroscopic laws of motion / electrodynamics, the equation of motion of the polar molecule is

$$I\, d^2\theta/dt^2 = -\mu E \sin\theta - \varsigma\, d\theta/dt. \tag{1.34}$$

Here and further we slightly alter the original notations of Debye to provide the agreement of this section with notations used in the theoretical part of this book.

To simplify the calculations Debye neglects in (1.34) the acceleration putting $\hat{E}_0 \exp\{i\omega\,[t - nZ/c]$ equal to zero. Thus for the average torque due to external field the following expression was obtained:

$$\varsigma\, d\theta/dt = -\mu E \sin\theta. \tag{1.35}$$

On the other hand, the angle θ of a given molecule changes also because of torque (of white noise type) due to other molecules. As a result, Debye had obtained the Smoluchowski kinetic equation. The latter is written in the configuration space for the distribution function $F(\theta, t)$ of dipoles:

$$\frac{\partial F}{\partial t} = \frac{1}{\sin\theta}\frac{\partial}{\partial t}\left[\sin\theta\left(\frac{\langle\Theta^2\rangle}{4\delta t}\right) + \frac{\mu}{\varsigma}EF\sin\theta\right]. \tag{1.36}$$

Here $F\ d\Omega$ is the number of dipoles, whose dipole moments are directed inside the small solid angle $d\Omega$; $\langle\Theta^2\rangle$ is the mean square polar angle between two electric moments μ, calculated for molecules quitting the volume centered near the angle Ω during small time interval δt. The distribution F does not depend on the other angles of spherical coordinates because of the symmetry of the system under consideration.

Three important particular cases I, II and III were investigated by Debye.

I. Electric field E is constant. In this case $F(\theta)$ is the Boltzmann distribution, constant in time and the coefficient $\langle\Theta^2\rangle/\delta t$ of rotational diffusion is related to the friction coefficient ζ as follows:

$$\langle\Theta^2\rangle/\delta t = 4k_BT/\zeta. \tag{1.37}$$

The greater the friction ζ, the less pronounced is diffusion. It is understandable that the latter increases with temperature T. The ratio (1.37) was assumed to be unaltered in other two cases.

II. Electric field E, being constant in the time interval $[-\infty, 0]$, *is instantly switched off at* $t = 0$. In this case the approximate (for small field E) solution of Eq. 1.36 has the form

$$F(\theta, t) = A\ [1 + \frac{\mu E}{k_BT}\cos\theta\,\exp(-\,t/\tau_D)\,], \tag{1.38}$$

where A is constant and τ_D is the Debye relaxation time:

$$\tau_D = \zeta\,/\,(2k_BT). \tag{1.39}$$

In the derivation of (1.39) it was supposed that

$$E << k_BT, \tag{1.40}$$

this condition usually holds for external but not for intermolecular fields. It follows from (1.38)-(1.39) that a mean dipole moment vanishes with time accordingly to

$$\langle \cos\theta \rangle = \mu \frac{\mu E}{3k_B T} \exp(-t/\tau_D), \tag{1.41}$$

where $\mu E/(3k_B T)$ is much less than unity. This means that a stochastic Brownian motion influences a given molecule much more strongly, than the electric E field does. We retain the main result (1.39) of this section also for the third case.

III. Small AC field is harmonic in time:

$$E(t) = \text{Re}\,[E_m \exp(i\omega t)] = E_m \cos \omega t. \tag{1.42}$$

In this case, which is especially interesting for us, the mean dipole moment of a small volume of a dielectric is also time-varying with the same frequency ω but with some phase shift, depending on the value of τ_D:

$$\mu \langle \cos\theta \rangle = \frac{\mu^2}{3k_B T} \text{Re}\left[\frac{E_m \exp(i\omega t)}{1 + i\omega \tau_D} \right]. \tag{1.43}$$

1.4.1 Debye Formula for the Complex Permittivity

Because of a time-lag of the polarization, the loss parameter ε'' is non-zero and its value depends on the product $\omega \tau_D$. From (1.43) one can obtain the following expression for the normalized permittivity:

$$\frac{\varepsilon(\omega) - \varepsilon_\infty}{\varepsilon_s - \varepsilon_\infty} = \frac{1}{1 + i\omega\tau_D}, \tag{1.44}$$

that is,

$$\varepsilon'(\omega) = \varepsilon_\infty + \frac{\varepsilon_S - \varepsilon_\infty}{1 + \omega^2 \tau_D^2}, \quad \frac{(\varepsilon_S - \varepsilon_\infty)\, \omega \tau_D}{1 + \omega^2 \tau_D^2}, \tag{1.45}$$

where $\varepsilon_\infty = \lim_{\omega \to \infty} \varepsilon'(\omega)$. In the case, considered by Debye (1929), we must take here

$$\varepsilon_\infty = n_\infty^2, \qquad \varepsilon_s = G = \mu^2 N / (3k_B T), \tag{1.46}$$

τ_D being the free parameter of the model of rotational diffusion.

To the great surprise of many investigators experiments had shown that Eq. 1.44 and 1.45 [but not Eq. 1.46 !] are often valid for systems quite differing from those initially considered. An example of such a system is a pure liquid, in which a polar molecule is embedded in a medium consisting of the same kind of polar molecules. If a dipole moment μ is large, molecular interactions are strong and the maximum loss factor ε″ is great.

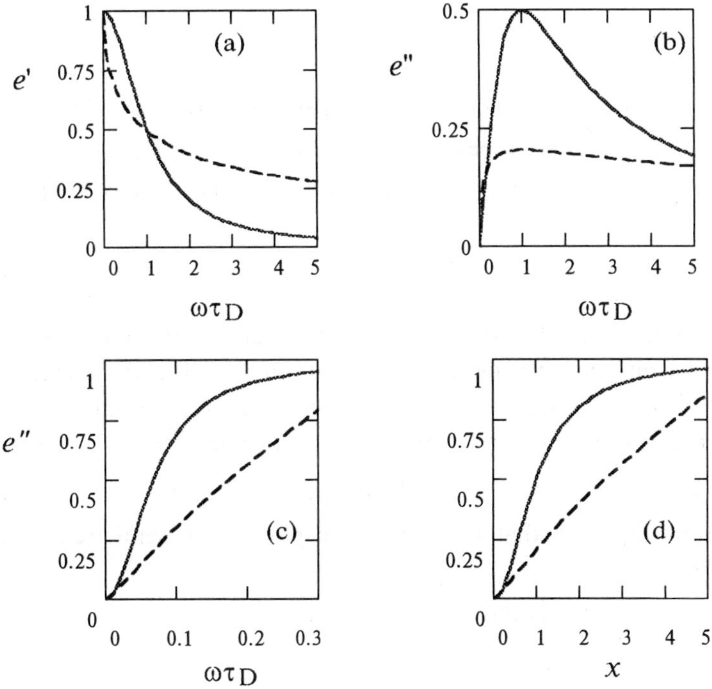

Fig. 1.1 Dielectric spectra for the Debye model of rotational diffusion. Solid lines for the Debye formula (1.44), dashed lines - for the Cole-Cole formula (1.55), $\beta = 0.5$. a: real part of $e^*(\omega\tau_D)$, b: imaginary part of $e^*(\omega\tau_D)$, c and d: the normalized absorption vs. $\omega\tau_D$ and x (the latter for the ratio $\tau_D/\eta = 15$).

Having this in mind, one can generalize the Debye formula (1.44), considering ε_s and ε_∞ to be the free parameters of the Debye theory of rotational diffusion. In this generalized case Eq. 1.46 are no longer applicable.

1.4.2 Sketch of the Debye Relaxation

In order to illustrate the frequency dependencies (1.44), (1.45) it is convenient to denote by $e^* = e'(\omega) + ie''(\omega)$. Then the Debye formula takes the form:

$$e^*(\omega\tau_D) = e' + ie'' = \frac{\varepsilon^*(\omega) - \varepsilon_\infty}{\varepsilon_s - \varepsilon_\infty} = \frac{1}{1 - i\omega\tau_D}, \qquad (1.47a)$$

so that $e^*(0) = 1,\ e^*(\infty)$.

In Fig. 1 a, b, c the frequency dependencies $e'(\omega\tau_D)$, $e''(\omega\tau_D)$, and $\omega\tau_D e''(\omega\tau_D)$ are plotted respectively by solid lines. In Fig. 1d normalized absorption $\omega\tau_D e''(\omega\tau_D)$ is plotted as function of the normalized frequency x:

$$x = \omega\eta, \text{ where } \eta = \sqrt{\frac{I}{2k_B T}}, \qquad (1.47b)$$

which is conveniently used to describe the FIR spectra of the Poley absorption. The former normalized frequency $(\omega\tau_D)$ is often used in the low frequency region. In Fig. 1.1c the ratio τ_D/η is taken equal to 15. The graph of $e''(\omega)$ has its only maximum at $\omega = \tau_D^{-1}$, the maximum value $e''(\omega)$ is given by

$$e''_D = (1/2)\,(e_s - e_\infty). \qquad (1.47c)$$

It follows from Eq. 1.45 that the Cole-Cole plot is the semicircle (see Fig. 1.2a):

$$[e'(\omega) - R]^2 + [e''(\omega)]^2 = R^2 \qquad (1.48)$$

with a radius $R = (1/2)(e_s - e_\infty)$.

We next examine the behavior of the refractive index $n(\omega)$ and the absorption coefficient $\alpha(\omega)$. According to the Debye theory, when $\omega\tau_D >> 1$, Eq. 1.45 yield

$$e'(\omega) = \varepsilon_\infty + \frac{\varepsilon_s - \varepsilon_\infty}{\omega^2\tau_D^2}, \quad \varepsilon''(\omega) = \frac{\varepsilon_s - \varepsilon_\infty}{\omega\tau_D}, \qquad (1.49)$$

so that approximately

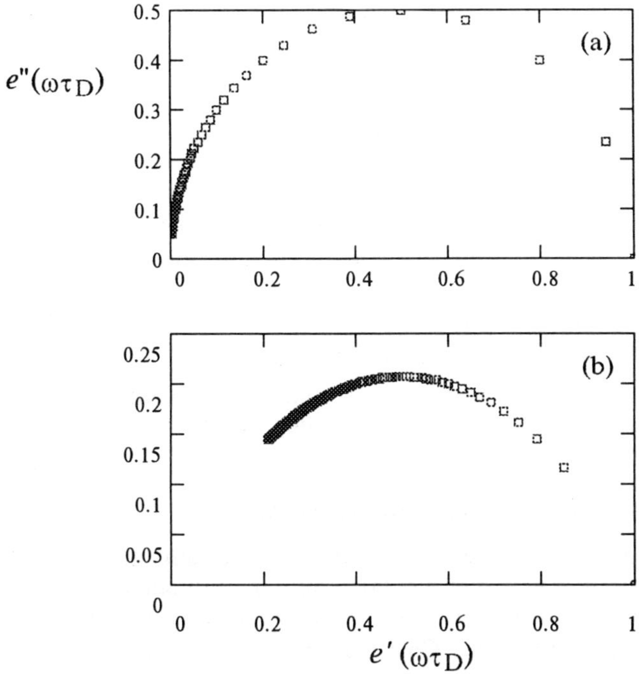

Fig. 1.2 The Cole-Cole diagram, calculated for the Debye formula (1.44) (a) and for the empirical dependence (1.55) with $\beta = 0.5$ (b). Separate points correspond to $\omega\tau_D = 0$, 0.25, 0.5... .

$$|\varepsilon(\omega)|^2 = \varepsilon_\infty{}^2 + 2\,\varepsilon_\infty \frac{\varepsilon_s - \varepsilon_\infty}{\omega^2\tau_D^2} + \frac{(\varepsilon_s - \varepsilon_\infty)^2}{\omega^2\tau_D^2} \tag{1.50}$$

and therefore

$$|e(\omega)| = \varepsilon_\infty + \frac{\varepsilon_s - \varepsilon_\infty}{\omega^2\tau_D^2} + \frac{(\varepsilon_s - \varepsilon_\infty)^2}{2\varepsilon_\infty\omega^2\tau_D^2}. \tag{1.51}$$

On substituting into (1.26) and (1.29), we have the Debye results for $\omega\tau_D \gg 1$:

$$n(\omega) = \sqrt{\varepsilon_\infty}\left[1 + \frac{(\varepsilon_S - \varepsilon_\infty)(\varepsilon_S + 3\varepsilon_\infty)}{8\varepsilon_\infty^2 \omega^2 \tau_D^2}\right], \tag{1.52}$$

$$\alpha(\omega) = \alpha_\infty = \frac{\varepsilon_S - \varepsilon_\infty}{c\tau_D \sqrt{\varepsilon_\infty}}. \tag{1.53}$$

The graph of $\alpha(\omega)$ tends at $\omega \to \infty$ to the constant value α_∞, which is termed as the "Debye plateau". That is, in terms of the reduced absorption

$$\lim_{x \to \infty} [x\varepsilon''(x)] = \frac{1}{c\tau_D \sqrt{\varepsilon_\infty}}. \tag{1.54}$$

The generalization of the Debye theory was made by McConnell (1980) and by many others, see e.g. (Scaife 1989); (Coffey, Evans, and Grigolini 1984).

1.4.3 Restrictions on Applicability of the Debye Theory

The Debye model of rotational diffusion was the first successful model based on molecular considerations. The experimental results may disagree with that of Debye at low and especially at high frequencies, as we shall now indicate.

a) In the low frequency region the Cole-Cole plot $e''(e')$ may differ from the semicircle, shown in Fig. 1.2a. To reduce such a disagreement Eq. 1.44 may been altered empirically e.g. to the formula (Scaife 1989, p.67)

$$e^*(\omega\tau_D) = \frac{\varepsilon^*(\omega)}{\varepsilon_S - \varepsilon_\infty} = \frac{1}{1 + (-i\omega\tau_D)^{1-\beta}}, \quad (0 \le \beta < 1). \tag{1.55}$$

The Cole-Cole plot $e''(e')$ is the circular arc with end points on the axis of real (e') and the center below the axis. By a suitable choice of β values it is possible to match (1.55) with some experimental results. The parameter β is sometimes interpreted as accounting for a *distribution of relaxation times.* However (1.55) lacks a microscopic basis. The example of the plot (1.55) for $\beta = 0.5$ is presented in Fig. 1.2b. Comparing with Fig. 1.2 a, we see that the difference of dielectric spectra and $e''(e')$ dependence from that described by the Debye formula (1.44) is appreciable.

More complex is a question of disagreement of frequency dependencies. The Cole-Cole plot and the absorption curve are employed frequently for comparison with experimental results, see e.g. (Chantry 1977). The agreement with experiment is generally good for moderate frequencies. At higher frequencies:

i) the e"(e') plot leaves the semicircle in the neighborhood of its left-hand extremity;

ii) the experimental absorption curve quits the "Debye plateau" and rises to a high maximum, due to what is called Poley absorption (Poley 1955), and then decreases steadily to zero. Accordingly the real part of permittivity ε' decreases from the Debye value ε_∞ to n_∞.

The contradictory behavior between the theoretical and experimental absorption curves indicates that the Debye theory is valid only in the low frequency region, where the magnitude of $\omega\tau_D$ does not exceed few unites.

1.5 About the Dynamic Method

The book presents an analytic theory of dielectric spectra. The theory relates the complex dielectric susceptibility $\chi(\omega)$ to molecular dynamics of reorienting polar molecules, constituting a conservative system of particles governed by an intermolecular potential U. We assume that the latter exists during a mean lifetime denoted by τ and that U is produced by neighbors of a given molecule. Formulas for the complex susceptibility $\chi(\omega)$ and permittivity $\varepsilon(\omega)$ will be given in a closed form. The approach used in the book is termed the *dynamic method.* It is based on a few concepts which will now be described.

1.5.1 Instantaneous Collisions

The strong collision approximation can be applied for the case of polar fluids. Each particle rotates / librates between such collisions *regularly* under influence of an intermolecular potential U during some lifetime t_v. The latter is a random exponentially distributed quantity; the mean value denoted by τ is also termed lifetime. The value τ plays the role of the coefficient of friction. The term "regular" denotes in this context that a polar particle has a known law of motion. As a rule, it is periodic in time. In result of a strong collision the particle instantly and randomly changes its *angular momenta.* It seems that the strong collision should be understood as a change in immediate surroundings of a given molecule. The specific nature of this process will not be investigated.

1.5.2 Complex Power

The complex susceptibility χ is proportional to the integrated and averaged value of complex power Q, consumed by electromagnetic field $E(t)$ in a unit volume of the medium:

$$\chi^*(\omega) = \frac{iN}{\pi \left|\hat{E}\right|^2} \left\langle \int_{t_0}^{t_0+t_v} Q(t)\,dt \right\rangle, \tag{1.56}$$

Here

$$Q(t) = e_q\,\mathbf{v}(t)\bullet\hat{\boldsymbol{E}}\exp(i\omega t) = \left(\mu \times \frac{d\vec{\theta}}{dt}\right)\bullet\hat{\boldsymbol{E}}\exp(i\omega t), \tag{1.57}$$

$\chi = \chi' + i\chi''$ is the complex conjugate of χ; $\langle\bullet\rangle$ denotes statistic averaging, e_q is the electric charge of a particle; $\mathbf{v}$ is its velocity and $d\theta/dt$ is the angular velocity of a dipole $\mu(t)$; t_v is an instant of a strong collision; $\hat{E}$ is a complex amplitude of the external AC field. Further we employ notations θ and ϑ for molecular rotation in space and on a plane, respectively. It is also convenient to introduce a real amplitude $\boldsymbol{E}_m$ of the AC field in a given point of a medium. Putting $\hat{E} = -i\boldsymbol{E}_m$, we have

$$\mathrm{Re}\,\{\hat{E}\exp(i\omega t)\} = \boldsymbol{E}_m \sin\omega t. \tag{1.57a}$$

According to (1.56) the real part of the susceptibility is related to the imaginary part of Q ("reactive power") while the loss factor χ'' does to the real one (to "active power"). To obtain Eq. 1.56 we employ Maxwell equations *in vacuo* with external “sources” (charges or dipoles) and the t_0 theorem. Using the latter, we transform the time average $\left\langle \sum_i Q_i \right\rangle$ for *ensemble* of particles to the integral $\int Q\,dt$ over time *t for one particle*. This integral is carried during the interval $t \in [t_0, t_0 + t_v]$ between the consecutive strong collisions. The t_0 theorem was applied previously in physical electronics (Gaiduk 1960a, Gaiduk,

Palatov, Petrov 1979, p.108), where the integration was carried along the trajectory of an electron between the cathode and anode of an electron device.

It should be noted that the direct use of Eq. 1.56 for the determination of $\chi(\omega)$ involves tedious calculations (Gaiduk and Kalmykov 1987, part I), since the dynamic quantity $Q(t)$ is influenced by the AC field $\boldsymbol{E}(t)$. We should make the next step in order to remove this drawback.

1.5.3 Linear Response

For the standard situation, encountered in frames of the linear response theory, after some transformation of Eq. 1.56 we may substantially simplify the calculation of the susceptibility $\chi(\omega)$. Indeed, for the low-intensity field $\boldsymbol{E}(t)$ the problem may be solved *in terms of thermal equilibrium.* In the dynamic method some basic quantity is introduced, termed the *spectral function* (SF) $K(z)$. This characterizes the dielectric response for an undamped (periodic) motion of a dipole in a conservative potential well $U(\vartheta)$ during mean lifetime τ; $K(z)$ is proportional to the spectrum of the autocorrelation function of $q(t)$. The latter is the normalized time dependent projection of the vector $\mu(t)$ on the direction $\boldsymbol{E}_{\mathrm{m}}$ of electromagnetic field, see above Eq. 1.57a:

$$q(t) = \mu_{\mathrm{E}}(t) / \mu, \tag{1.58}$$

that is to say, q is the time-varying direction cosine. *When the SF $K(z)$ is calculated, the susceptibility $\chi^*(\omega)$ is found as a rational function of K.* This is the key statement of the dynamic method, and we consider it in greater detail.

When regarding the dependence on ω of the susceptibility χ we distinguish *two mechanisms* of the wave-medium interaction, specific for molecular rotation at equilibrium. These mechanisms are conditioned by:

a) periodic rotation of a dipole in a potential well of a prescribed profile during lifetime τ;

b) stochastic (Brownian) reorientation of molecules.

In our approach, regarding the **a** factor, one should calculate the spectral function $K(z)$, which is the basic function of the problem under consideration; $K(z)$ also describes dielectric response of the medium in the FIR spectral range. Accounting for the **b** factor, we express the resultant quantity χ^* through the same function $K(\mathrm{z})$. Ii is convenient to employ dimensionless quantities. Thus for a spherical or a linear molecule with moment of inertia I we introduce time η by

$$\eta = \sqrt{I / (2k_B T)} \, . \tag{1.59a}$$

Then we write

$$x = \eta\omega, \; y = \eta/\tau \tag{1.59b}$$

and we see immediately, that x is the normalized radiation frequency and y is the normalized frequency of strong molecular collisions. Then we introduce the complex frequency

$$z = x + i\,y = \eta\,(\omega + i\,/\,\tau). \tag{1.59c}$$

We define the spectral function $K(z)$ by

$$K(z) = 3iz\left\langle \int_0^{\infty} q_0\;(q - q_0)\;\exp\;(i\,z\varphi)\;d\varphi \right\rangle, \tag{1.60}$$

where φ is the reduced time t/η and the brackets $\langle \bullet \rangle$ denote ensemble average over the phase space Γ. We note that $q(\varphi)$ is a function of time, q_0 is the value of $q(\acute{\varphi})$ at the instant $\varphi = 0$ of a "strong collision". The q_0 coordinate is regarded as a point in a phase space Γ. Thus, the $K(z)$ function is proportional to the *spectrum* of the autocorrelation function of a dipole moment.

In order to calculate the SF $K(z)$ we proceed as follows:

1) We find the periodic time function $q(\varphi)$ as a solution of nonlinear equations of motion of a dipole, librating / rotating in a prescribed potential well U.

2) Employing the steady state distribution

$$W = C\;\exp\;[-\,H\,/\,(k_B T)\,], \tag{1.61}$$

where C is a constant and H is the Hamiltonian of a particle, we find the statistical average (1.56) for a particular case of the Boltzmann induced distribution F, such that

$$F = F_B = \text{const} \bullet \exp\;[\;\mu(t_0) \times \boldsymbol{E}(t_0)\;]/(k_B T). \tag{1.62}$$

In this way we obtain the complex susceptibility

$$\chi^*{}_B\,(\omega) = G\,K(z), \tag{1.63}$$

corresponding to F_B. The basic frequency dependence $\chi^*{}_B\,(\omega)$ does *not* describe the Debye relaxation, since at F_B the polarization $P(t)$ changes *in phase* with the external EM field $\boldsymbol{E}(t)$ (Note, in the HF region this dependence $\chi^*{}_B\,(\omega)$ is rather close to that, corresponding to a more adequate induced distribution F, see the next point **3**).

3) We obtain the closed-form expression for $\chi^*(\omega)$, taking an appropriate *induced distribution F,* which corresponds to the prescribed collision model (for $F \neq F_B$). In this way we account *implicitly* for the Brownian motion. The notion "implicitly" means in this context that in the dynamic method the *stochastic equations are actually not considered.* This property of the method leads to a simple formal expression for the susceptibility $\chi^*(\omega)$.

For definiteness we consider in this subsection the modified Gross collision model, for which[XIX] the correlation-orientation induced distribution F_K was employed (GT, Sec. IVB). The latter, as well as the Gross distribution F_G, is *self-consistent,* since the induced distribution F is related to the unknown quantity χ^*. The following rational relation of χ^* to the spectral function K is found to hold:

$$\chi^* = \frac{gGzK(z)}{gx + \frac{iy}{\sigma}K(z)}, \tag{1.64}$$

where $\sigma/3$ is the dispersion of a direction cosine q, so that

$$\sigma = 3\left[\left\langle q^2\right\rangle - \left\langle q\right\rangle^2\right], \tag{1.65}$$

and the Kirkwood correlation factor g is related to the experimental quantities ε_s, n_∞ , G and to σ by

$$g = \frac{\varepsilon_s - n_\infty^2}{4\pi G\sigma}\,\frac{2\varepsilon_s + n_\infty^2}{3\varepsilon_s}. \tag{1.66}$$

The polarization $\boldsymbol{P}(t)$, corresponding to χ^* from (1.64), changes behind the field $\boldsymbol{E}(t)$. This property of the correlation-orientation collision model has the important result that:

a) the Debye relaxation appears in the low frequency region, and
b) the relaxation time τ_D and other parameters of the Debye formula (1.44) are expressed through the parameters of the employed molecular model.

The complex permittivity ε^* can be expressed through the theoretical quantity χ^* as a root of quadratic equation (1.12), which we rewrite in the form

$$(\varepsilon^* - n_\infty^2)(2\varepsilon^* + n_\infty^2) - 12\,\pi\,\varepsilon^*\chi^* = 0, \tag{1.67}$$

where $\varepsilon^* = \varepsilon' + i\,\varepsilon''$.

We conclude that according to Eq. 1.29, 1.31, 1.64 - 1.66 the complex permittivity ε^* and the absorption coefficient

$$\alpha = 4\pi\nu \,\mathrm{Im}\, \sqrt{\varepsilon^*} \tag{1.68}$$

can be calculated, if we know the SF $K(z)$ and the correlation factor g. Here the normalized frequency, expressed in cm^{-1}, is defined as

$$\nu = x \,/\, (2\pi c\eta) = \omega \,/\, (2\pi c). \tag{1.69}$$

In this monograph only relatively simple molecular models are considered. In most cases we shall use the theory, in which $K(z)$ is expressed or in the form of ordinary integrals over the energy of rotational motion, or in form of double integrals.

1.5.4 The Effective Susceptibility

The dielectric properties of an *anisotropic* medium are described in the monograph in terms of an *effective complex susceptibility* χ. It is a *scalar* quantity, related to the complex propagation constant k by

$$k^2 = (\omega^2/c^2)\,(4\pi\chi + n_\infty^2\,). \tag{1.70}$$

The two basic crossed orientations are characterized by reference to $\boldsymbol{E}_{\mathrm{m}}$ and Z, viz. longitudinal ($\boldsymbol{E}_{\mathrm{m}} \parallel Z$) and transverse ($\boldsymbol{E}_{\mathrm{m}} \perp Z$), $\boldsymbol{E}_{\mathrm{m}}$ being the amplitude of AC field in the point under consideration [see Eq. 1.57a] and Z being the symmetry axis of the potential $U(\vartheta)$. We introduce the longitudinal ($K_{\parallel}$) and transverse ($K_{\perp}$) spectral functions for these cases, respectively. Let Ψ be the angle between $\boldsymbol{E}_{\mathrm{m}}$ and Z. In this general case the spectral function is given by

$$K(z) = K_{\parallel}(z)\cos\Psi + K_{\perp}(z)\sin\Psi, \tag{1.71}$$

where according to Eq. 1.60 the two basic SF's are

$$K_{\parallel}(z) = 3iz\left\langle q_{\parallel}(0)\int_0^{\infty}\left[q_{\parallel}(\varphi) - q_{\parallel}(0)\right]\exp(iz\varphi)\,d\varphi\right\rangle, \tag{1.72a}$$

$$K_{\perp}(z) = 3iz\left\langle q_{\perp}(0)\int_0^{\infty}\left[q_{\perp}(\varphi) - q_{\perp}(0)\right]\exp(iz\varphi)\,d\varphi\right\rangle, \tag{1.72b}$$

$q_{\parallel}(\varphi)$ and $q_{\perp}(\varphi)$ are the direction cosines for the cases $\boldsymbol{E}_{\mathrm{m}} \parallel Z$ and to $\boldsymbol{E}_{\mathrm{m}} \perp Z$, respectively. Thus *we avoid the problem of calculating the dielectric permittivity tensor* χ_{ij} (see Eq. 1.4), since in (1.70) the effective susceptibility χ and k^2 are both scalars.

1.5.5 Transition from Anisotropy to Isotropy

The transition to an *isotropic medium* is made by the averaging of Eq. 1.71 over Ψ. Thus we suggest that an isotropic medium as if consists of *multiple anisotropic "islets"* with randomly directed symmetry axes Z. Each islet is characterized by its nonzero intermolecular potential $U(\vartheta, \Psi)$. The space scale of any islet should be much less, than the radiation wavelength λ, in all orientational-relaxation region. This condition is easily satisfied even on the edge of this range. The minimum wavelength λ_{min} is about $(5 \cdot 10^{-3})/n_\infty$ cm for nonassociated liquids and about $(10^{-3})/n_\infty$ cm for water. These λ_{min} values are much larger, than the supposed dimension of an islet, which possibly is not much greater than the diameter of a polar molecule (the latter is usually commensurable with 10^{-8} cm).

In an isotropic medium the mean value of a direction cosine is usually zero, $\langle q \rangle = 0$, and the dispersion $\sigma = 1$. Noting that the g-factor (1.66) reduces to the well known form (1.15) and denoting the spectral function $L(z)$, we can rewrite Eq. 1.61 and 1.64 as follows:

$$K(z) \approx L(z), \tag{1.72c}$$

$$\chi^* = G\,L(z) \qquad \text{(at } F = F_B\text{)}, \tag{1.73}$$

$$\chi^* = \frac{g\,G\,z\,L(z)}{g\,x + i\,y\,L(z)} \qquad \text{(at } F = F_K\text{)}. \tag{1.74}$$

According to Eq. 1.71 and 1.72 the spectral function L depends on two basic SF's: $K_\parallel(z)$ and $K_\perp(z)$. In this respect the susceptibility $\chi^*(\omega)$ of an isotropic medium *appears as a somewhat more complex notion* than the longitudinal and transverse susceptibilities, since $\chi^*(\omega)$ combines the properties of both quantities $\chi^*_\parallel(\omega)$ and $\chi^*_\perp(\omega)$.

1.5.6 The Low Frequency Approximation

The Debye relaxation parameters ε_s, τ_D, ε''_D and ε_∞, where $\varepsilon''_D = \max\{\varepsilon''(\omega)\}$, can readily be related to the parameters of a microscopic model. When the condition $x << 1$ is satisfied, which usually holds in the Debye spectral region, we may put $x = 0$ in Eq. 1.60. Using notation K_0 for K in this limiting case, we have

$$K_0 = K(iy) = 3\left\langle q_0 \int_0^\infty \left[q(\varphi) - q_0\right] \exp(-y\varphi)\, dy\varphi \right\rangle. \tag{1.75}$$

We see that K_0 is real. Substituting K_0 for K in (1.64), we immediately obtain the Debye formula

$$\chi^*(x) = \chi_\infty + \frac{\chi_s - \chi_\infty}{1 - ix\tau_D / \eta}, \tag{1.76}$$

where

$$\chi_s = \chi^*(0) = gG\sigma, \tag{1.77}$$

$$\chi_\infty = \chi^*(x)\,|_{x=\infty} = GK_0, \tag{1.78}$$

$$\tau_D = \frac{\eta}{\chi_s - \chi_\infty} \lim_{x \to 0} \frac{\chi''(x)}{x} = \frac{\tau g \sigma}{K_0}. \tag{1.79}$$

The specificity of the Debye relaxation in various fluids can be understood from the viewpoint of molecular rotation in a prescribed potential well $U(\vartheta)$. For example, according to (1.79), the Debye relaxation time τ_D is proportional to the lifetime τ, to dispersion σ and to K_0^{-1}. The parameter σ describes the statistical properties of a fluid, while K_0 is related to the trajectory of an individual particle in the potential well. In our impact-collisions approximation the lifetime τ *is a free parameter of the model,* which is fitted with account of the experimental relaxation time τ_D. The increase of the welldepth U_0 greatly influences the "static" response K_0 and the dispersion σ. These quantities also differ for the longitudinal and transverse susceptibilities $\chi^*_\parallel(\omega)$ and $\chi^*_\perp(\omega)$.

In an isotropic medium we have by analogy with Eq. 1.75-1.79:

$$L(z) \to L(iy) \equiv L_0; \tag{1.80}$$

$$\chi_s = g\,G;\ \chi_\infty = GL_0\,;\ \tau_D = \tau g / L_0. \tag{1.81}$$

*   *   *

Thus, we have mentioned six main peculiarities of the dynamic method (Sec.1.5.1 - 1.5.6), which will be used in the monograph. The detailed description of this method for classic ensemble of dipoles is given in Part one (Chap. 3 - 6).

1.5.7 Spectrum of a Harmonic Oscillator

We shall consider the properties of a single resonant line taking, as an example, the one-dimensional classic harmonic oscillator. For the latter the dispersion is described by the Lorentz line, which is a widely used frequency dependence. In this section we follow Gaiduk and Kalmykov (1987, p.35).

We assume that a charge e_+, having the mass m, is affected by the superposition of *elastic conservative field*, characterized by the oscillator frequency ω_0, and by *weak AC field* $\boldsymbol{E}(t) = \boldsymbol{E}_m \cos \omega t$. Let us write the equation of motion of a particle in both conservative and AC fields:

$$\frac{d^2 r}{dt^2} + \omega_0^2 r = \frac{e_+}{m} E_m \cos\Psi \cos\omega t, \tag{1.82}$$

where Ψ is the angle between the vector $\boldsymbol{E}_m$ and the fixed direction of the vibrations, $r(t)$ is the coordinate of a charge. The dielectric susceptibility is given by Eq. 1.73 or 1.74 for the Boltzmann or correlation-orientation induced distributions, respectively (F_B or F_K). Now the spectral function $L(z)$ is represented by the Lorentzian

$$L(z) = \frac{1}{1 - z^2}, \tag{1.83}$$

so that

$$\chi_B^* = G L = \frac{G}{1 - z^2} \qquad \text{(for } F_B\text{)}, \tag{1.84}$$

$$\chi^* = \frac{G z L}{x + i y L} = \frac{G}{1 - x z} \qquad \text{(for } F_G\text{)}, \tag{1.85}$$

where the normalized quantities are

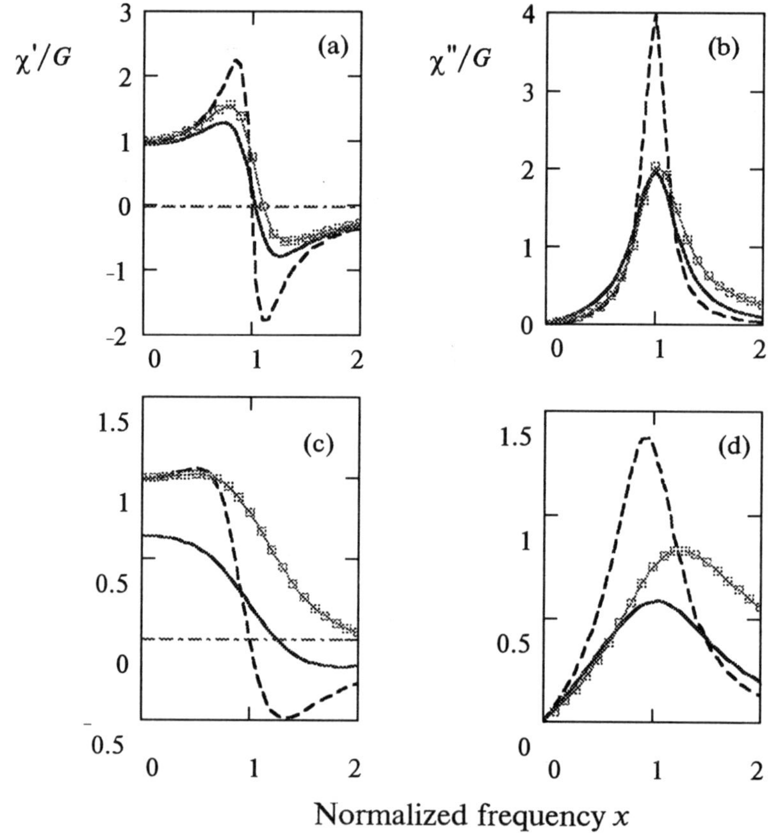

Fig. 1.3 The dielectric behaviour of ensembles of harmonic oscillators. The real (to the left) and imaginary (to the right) parts of the complex susceptibility vs. frequency x, calculated for the Boltzmann (solid lines), Gross (dashed lines) collision models and for the Van Vleck-Weisskopf line (dots). The normalised collision frequency $y = 0.25$ (on the top) and 0.75 (beneath).

$$G = \frac{e_+^2 N}{3m\omega_0^2}, \quad z = x + i\,y, \quad x = \frac{\omega}{\omega_0}, \quad y = \frac{1}{\omega_0 \tau}, \tag{1.86}$$

τ being the mean lifetime between strong collisions. Setting $g = 1$, we consider here the Gross collision model, in which $F_K \equiv F_G$.

The dielectric spectra, found from Eq. 1.84 and 1.85, we compare in Fig. 1.3 with those, calculated for the well known Van Vleck-Wesscopf line (Townes and Schawlow 1975, Brown 1956):

$$\chi^*_{\mathrm{VVW}} = G\frac{1 - iyz}{1 - z^2}. \tag{1.87}$$

In Fig. 1.3 the dielectric spectra are pictured for two collision frequencies y (0.25 and 0.75). The resonance dielectric response, described by the Lorentz-like terms 1.84, 1.85, and 1.87, have the following properties:

(a) Dielectric loss χ'' is maximal near the resonance frequency ($\omega \approx \omega_0$); when the collision frequency y decreases, the absorption peak, gaining in value, becomes more narrow and tends to infinity, if $y \to 0$, i.e. if $\tau \to \infty$. In this limit $\chi''(\omega)$ becomes proportional to δ-function.

(b) The real part χ' of the susceptibility vanishes in the neighborhood of the resonance frequency ω_0 (at F_G we have $\chi' = 0$ precisely for $\omega = \omega_0$). This result can be interpreted as follows. The reactive power tends to zero when ω approaches the resonance region. For $\omega > \omega_0$ the susceptibility χ' becomes negative and passes with the increase of ω through the point of minimum. The depth of this minimum lessens with the increase of the collision frequency y. If $x \to \infty$, the real part χ' once again tends to zero but now from below.

These features are common for all collision models. The differences between the three lines pictured in Fig. 1.3b and 1.3d, are following:

1. For the fixed y value the absorption maximum is for F_G approximately twice that found for F_B.
2. When collision frequency y rises, the loss maximum χ'' shifts to left or to right, respectively for the induced distributions F_G or F_B

For frequent collisions the Van Vleck-Wesscopf line becomes essentially asymmetric, so that the plateau appears on the absorption curve $x\chi''(x)$ just as in the Debye theory, see Fig. 1.1 c, d. The Van Vleck-Wesscopf line possesses some principal drawback, viz. the integrated absorption

$$\Pi = \int_{-\infty}^{\infty} \omega\, \chi''(\omega)\, d\omega \tag{1.88}$$

is infinite for this line, while for both lines (1.84) and (1.85) Π has the standard value:

$$\Pi = \pi N e^2_+ / (3\ m). \tag{1.89}$$

This fact makes the Gross collision model more preferable, than the Van Vleck-Wesscopf's one. It should be noted that in accord with (1.89) the integrated absorption Π does not depend on the collision frequency y. Because of this property the absorption curve widens with the increase of y, correspondingly the maximum absorption decreases with y.

The same feature is characteristic for ensembles of *polar* molecules, in which the integrated absorption of an isotropic medium is given by

$$\Pi = (2/3)\ \pi\ \mu^2 N / I. \tag{1.90}$$

We remark that the Van Vleck-Wesscopf's approach was the first, in which the Debye theory was applied to resonance phenomena in gases. As a result, a reasonable theoretical behavior of this spectrum was obtained for the low frequency limit ($\omega \to 0$). But the derivation of Eq. 1.87 *is not based on a rigorous description* of dielectric relaxation.

The applicability of the oscillator model to real media, e.g. to the study of *atomic vibrations* in molecules, deserves special consideration. On the other side, the Lorentz line is a typical line-form, that is observed in a rarefied polar gas. However, this description can be obtained only, if quantum corrections are taken into account (see Chap. 16). This spectrum is discrete, it consists of *infinite number* of similar lines. On the contrary, the classical description yields the *only* resonance line. As a first approximation, the resonant frequency ω_0 of a classic oscillator may be identified with the frequency Ω_J of the $J + 1 \leftarrow J$ rotational transition, where J is the quantum number, $J = 0, 1, 2, \dots$. We are particularly interested to mention the difference between the dispersion dependencies, that are characteristic for *oscillators* and *rotors*, since the Reader can find the widespread use (without necessary substantiation) of the oscillator model for the description of the rotational spectrum of polar gas. It is important that for a *classic ensemble* of rotors the extended diffusion model is more adequate than the above described oscillator model.

1.6 Classification of Semi-microscopic Molecular Models

In the context of this monograph the notion "model" means the totality of physical ideas and a theoretical method, which results in a closed-form analytical formula for the complex susceptibility χ^*. Using this formula one can calculate the wideband dielectric spectra that characterize the prescribed molecular model of a polar fluid.

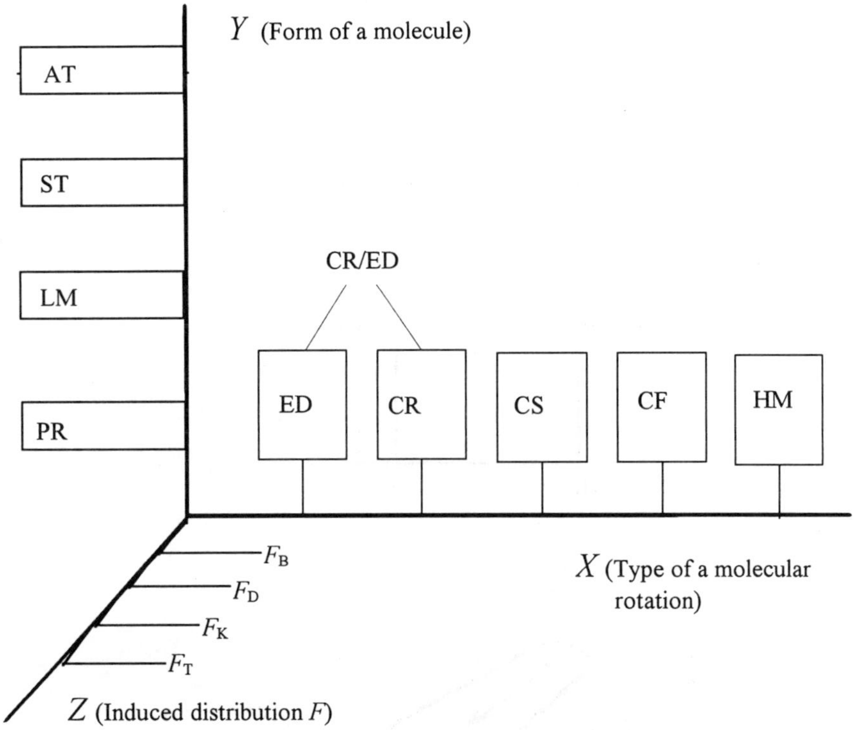

Form of a molecule		**Induced distribution**		**Type of molecular rotation**	
AT	Asymmetric top	F_B	Boltzmann	ED	Extended rotational diffusion
ST	Symmetric top	F_D	Debye	CR	Confined rotator
LM	Linear molecule	F_G	Gross	CS	Cosine squared
PR	Planar rotator	F_K	Correlation-orientation	CF	Constant field
		F_T	Isothermal	HM	Hat model

Fig. 1.4 Classification of molecular models pertaining to the type of rotational motion (X), form of a molecule (Y) and induced distribution (Z).

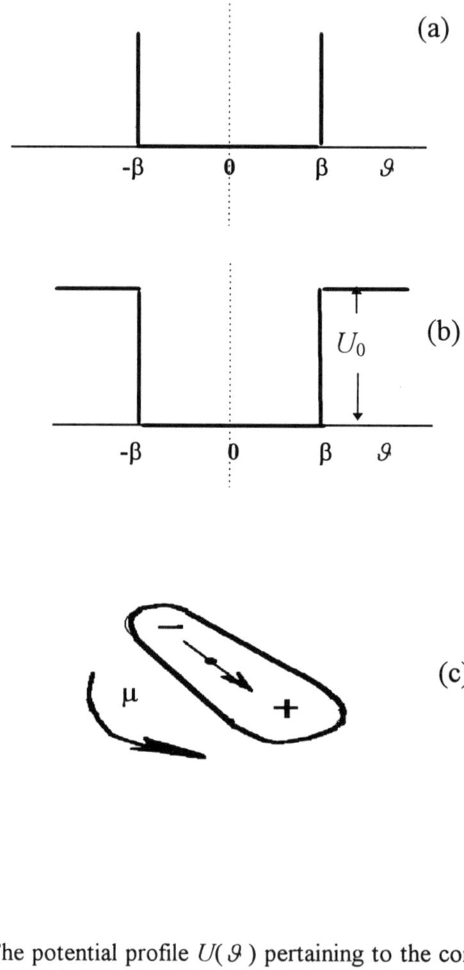

Fig. 1.5 The potential profile $U(\vartheta)$ pertaining to the confined rotator (CR) model (a), the hybrid CR/ED model (b) and the extended diffusion model (c). In the latter $U = 0$.

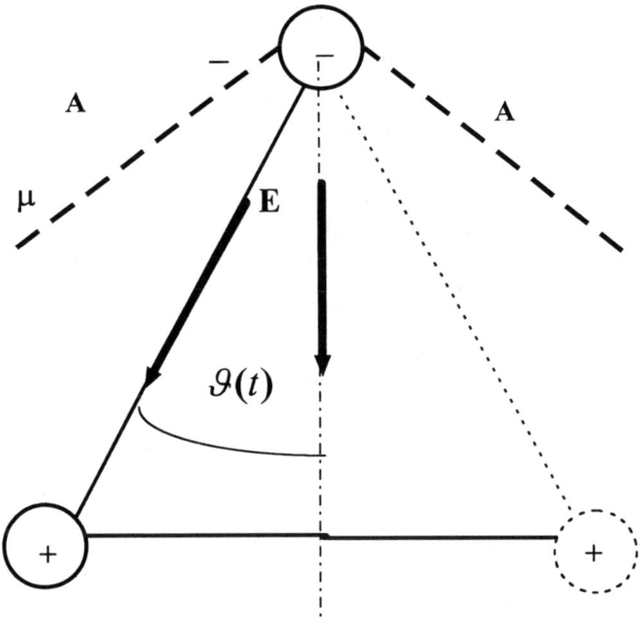

Fig. 1.6 Scheme of libration / hindered rotation in a field model. Dashed lines A show the limiting angular shift (max ϑ) for a dipole of a given energy H and the E_0 vector - the direction of a conservative field (near the bottom of a potential well). This field is constant in the constant field (CF) model and varies as cos ϑ in the cosine squared (CS) model, see Fig.1.7.

In Fig. 1.4 the conditional coordinate system X, Y, Z is shown, using which it is convenient to offer to the Reader some classification of molecular models. Here

Y axis represents the *form of a polar molecule*, that is, the type of its symmetry. A reservation should be made that for the sake of simplicity we have no appropriate theory for liquids constituting of asymmetric top molecules;

X axis represents the *type of regular (periodic in time) rotation* of a molecule in a conservative (averaged in time) intermolecular potential U;

Z axis represents the *collision models*, viz. the induced distribution F of molecules at instants t_0 just after strong instantaneous collisions.

Hence, a point $\{X, Y, Z\}$ corresponds to one model. We see that the number of possible models is not small. However, we intend to show in the monograph, that in spite of these considerations the number of "good" models, which agree with wideband experimental dielectric spectra, is *strongly restricted.* The relation of a *molecular shape* to dielectric relaxation is studied in the monograph only for molecules, free rotating during the lifetime τ, viz. for the extended diffusion (ED) model, where $U = 0$.

For other potentials $U(\vartheta)$ we shall consider ensembles of linear molecules (that rotate in space) or ensemble of planar rotors. Since the differences between various induced distributions F_D, F_G, F_K, and F_T are not great, for a rough study we may choose the simplest Gross collision model ($F = F_G$), in which the Kirkwood correlation factor $g = 1$.

The most important differences between models are related to their *type of molecular rotation* at equilibrium, to which the X axis in Fig. 1.4 corresponds. Fig. 1.5-1.7 illustrate a few simple potentials; we mention here two groups of models:

a. Models, in which *free rotation / libration* of dipoles is the characteristic feature. In such models $U = 0$ during almost all time τ between strong collisions. However, U may differ from zero at instants, when a dipole undergoes collisions with "walls" of the well. These models are considered in Chap. 7 - 9.

b. *Field models* (Chap. 10-13), where libration / hindered rotation of a dipole takes place in some prescribed potential U, depending on the angular shift ϑ of a dipole relative to the symmetry axis Z of this potential. For simplicity, these models are applied mainly to a motion on a plane.

We have no a priori knowledge, what a potential $U(\vartheta)$ is the most adequate. That is why our approach is *semi-microscopic*. Indeed, we take a prescribed potential $U(\vartheta)$, since we cannot get it from the first principles.

THE FIR SPECTRAL REGION. If the potential well U is deep ($U \gg k_B T$), then in the case of the second **b** group (of field models) a dipole performs almost harmonic librations, to which *narrow* absorption band corresponds. On the contrary, in the case of the first **a** group the absorption band is *wide*, because the rotational motion of a dipole in the rectangular well is quite *anharmonic*: Indeed, molecules with different energies have also different angular velocities, to which different resonant radiation frequencies of the electromagnetic field correspond.

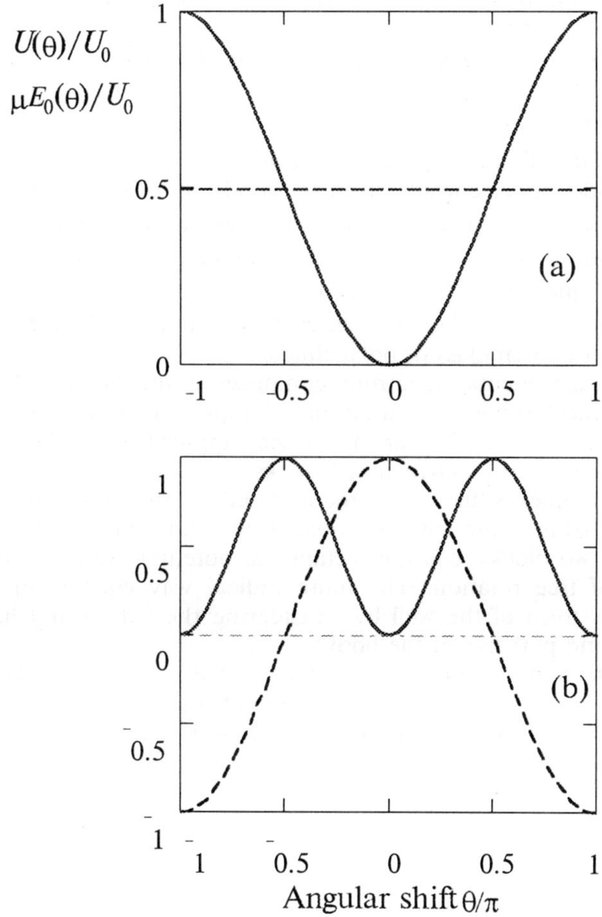

Fig. 1.7 The potential profiles used in the constant field (a) and in the cosine squared (b) potential profiles (solid lines). The corresponding θ-dependencies of the conservative field $E_0(\theta)$ are shown by dashed lines.

In the limiting case of *rare collisions* (for $\tau \to \infty$, $y \to 0$) we come to an important new notion - the *limiting line*. For such a line the absorption bandwidth is minimal and the absorption peak is maximal. This limit may be termed the *Landau damping in polar fluids.*

THE LOW FREQUENCY REGION. If the lifetime τ is rather long, we may derive a simple formula for the theoretical Debye relaxation time τ_D as a function of the parameters of a prescribed molecular model. In a liquid the ratio τ_D/η is usually much greater than unity.

The model under consideration is "good" in a first approximation, if both calculated parameters (the absorption bandwidth $\Delta\nu_L$ in FIR spectral region and the position $\nu_D = 1/\tau_D$ of the low frequency loss maximum) agree with the observed quantities. The better this agreement, the more adequate is the prescribed potential $U(\vartheta)$, the better this potential characterizes molecular dynamics in a chosen class of polar fluids.

If the experimental spectrum comprises a number of Debye relaxation regions, we meet some new situation. "Simple" models, like those shown in Fig. 1.5 a, b, c or 1.6, 1.7, are no longer applicable. In these cases we may combine two or several "elementary blocks".

More complex situation is encountered, if the absorption band exhibits two maxima (this is the case of liquid water and aqueous solutions). We can also employ two blocks, e.g. the rectangular potential well in combination with the model of free rotation. The more radical way consists in an attempt to alter the very form of the well by considering the hybrid and hat models (see Chap 9, 13 and part five of the book).

In the second introductory chapter we shall show that wideband dielectric spectra of polar fluids combine non-resonant and resonant features, that were demonstrated separately by Figures 1.1, 1.2, and 1.3.

CHAPTER 2
ROTATIONAL AND DIELECTRIC SPECTRA

In this chapter we try to picture typical dielectric spectra observed in gaseous and liquid states. The classification of pictures. presented below, is given in Fig. 2.1. Here we are interested mainly in results of *experimental observations*, although most figures are accompanied by results of calculations. The molecular models applied and the interpretation of experimental spectra are touched briefly: in the remainder of the book we shall return to these and other questions in greater detail.

2.1 Gaseous and Gas-like States

2.1.1 Rotational Spectrum in a Gas State

We consider the rotational spectrum of CO gas (Fig. 2.2 a, b, c), which is typical for the simplest (diatomic) polar molecules. The CO molecule has small dipole moment (μ = 0.1 D) and rotational constant (B = 1.93 cm^{-1}), so that the first rotational line falls into the millimeter region (see Fig. 2.2 b). If we take for the comparison HCl molecule, where μ = 1.08 D and B = 10.44 cm^{-1}, we find, that the first transition $J = 1 \leftarrow 0$ is placed in the submillimeter range, viz. the first line frequency is equal to $2B$ = 20.9 cm^{-1}.

This absorption spectrum of carbon monoxide (CO) gas was calculated for the Gross collision model and also for the system of Lorentz lines, see curves 1 and 2, respectively. The set of fitted parameters comprises the lifetimes τ_J of rotational transitions or, equivalently, the halfwidths $\Delta\nu_J = \dfrac{1}{2\pi c \tau_J}$ of absorption lines. The experimental (Gray 1974) $\Delta\nu_J$ values were employed in these calculations; these decrease with the increasing of rotational number J (see Fig. 2.2 d). Note, the absorption coefficient in Fig. 2.2 a, b, c is calculated as follows:

$$\alpha\ [\text{dB/km}] = 10 \log \{ \exp (\alpha\cdot[\text{cm}^{-1}]) \} \approx 2.38\cdot10^5\ \alpha\cdot[\text{cm}^{-1}]. \qquad (2.1)$$

NONASSOCIATED LIQUIDS WITH WEAK INTERMOLECULAR INTERACTIONS

State of matter	Gas		Gas-like liquids	Polar molecule in nonpolar solvent
Dielectric spectra	Low pressure Fig. 2.2a	High pressure Fig. 2.2b	Fig. 2.3	Fig. 2.5, 2.6
Effect of molecular rotation on vibration spectra	Fig. 2.4a, 2.4b			Fig. 2.4c

NONASSOCIATED LIQUIDS WITH STRONG INTERMOLECULAR INTERACTIONS

Effect of temperature	Relaxation time - lifetime plot	Self-diffusion coefficient	Two components of potential *U*
Fig. 2.7	Fig. 2.8, 2.9	Fig. 2.10 a, b	Fig. 2.10 c, d

ASSOCIATED HYDROGEN-BONDED SYSTEMS

Homogeneous systems		H_2O molecules bounded by a protein molecule		Aqueous solutions	
Liquid water	Ice I	Humidified macro-molecule, skin	Hemoglobin in aqueous solution	Electro-lytes	Solution of DMSO
Fig. 2.11, 2.12	Fig. 2.13	Fig. 2.14 - 2.17	Fig. 2.18	Fig. 2.19 2.21	Fig. 2.22, 2.23

Homogeneous systems		H_2O molecules bounded by a protein molecule		Aqueous solutions	
Liquid water	Ice I	Humidified macro-molecule, skin	Hemoglobin in aqueous solution	Electro-lytes	Solution of DMSO
Fig. 2.11, 2.12	Fig. 2.13	Fig. 2.14 - 2.17	Fig. 2.18	Fig. 2.19 2.21	Fig. 2.22, 2.23

Fig. 2.1 The dispersion of dielectric permittivity / absorption in liquid and gas states: scheme of plots, presented in Chap. 2.

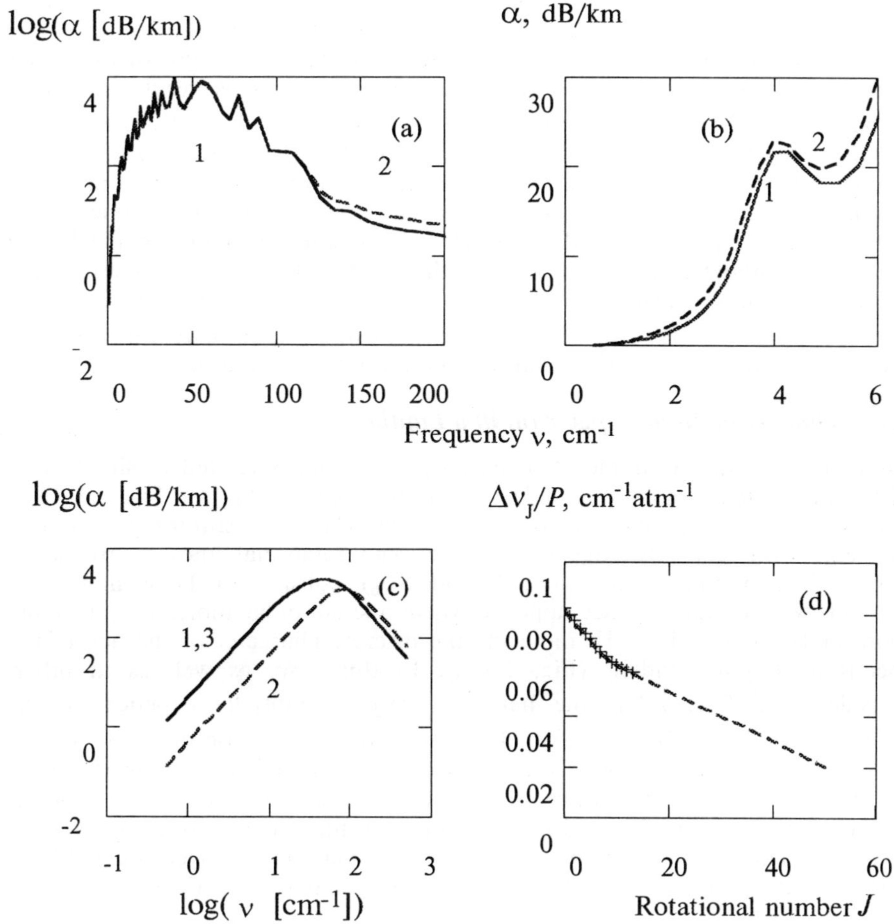

Fig. 2.2 **(a, b, c):** The theoretical rotational absorption spectrum of CO-N_2 gas mixture. Pressures of the buffer N_2 gas are 10 (a, b) and 1000 (c) atm, T = 20°C. 1 - The quantum extended diffusion model with Gross induced distribution F_G, 2 - The set of Lorentz lines, 3 - The classical rotational extended diffusion model. (d): $\Delta\nu$ vs. J dependence used in this calculation; dots - experimental values of $\Delta\nu_J$. The calculation procedure is described in Sec. 16.2.2.

In these calculations the concentration N of CO molecules was taken equal to the N value for pure CO at $P = 1$ atm. Hence, the alteration of the pressure in case of Fig. 2.1 is achieved by alteration of pressure of the buffer gas (e.g. of N_2). At $P = 10$ atm both collision models result in almost coincident $\alpha(\nu)$ spectra, separate rotational lines being distinctly observed at this P value (Fig. 2.2 a, b). At higher pressure $P = 1000$ atm (Fig. 2.2 c) the discrete rotational spectrum disappears because of widening of rotational lines. Simultaneously the difference arises between these calculation schemes (e. g. for F_G and F_B), in other words, - between the J diffusion model (curve 1) and the set of Lorentz lines (curve 2). For the former (J-diffusion model), which seems to be more adequate, the absorption peak α is higher than for the Lorentz lines and is shifted to lower frequencies. The classical approach practically coincides with the quantum one (cf. curves 1 and 3).

2.1.2 Quasi-Rotational Spectrum in a Liquid

Another interesting example of quantum phenomena is related to simple gas-like liquids, such as solutions of HF, HCl and DCl in SF_6 (see Fig. 2.3). The lightest polar component of these solutions (HF) constitutes a diatomic molecule with relatively large distance between rotational lines on frequency scale. The nonpolar component (liquid SF_6) consists of large molecules, between which enough space appears, where a small polar molecule can rotate more or less freely. For HF molecule the discrete character of the absorption spectrum is quite evident (Fig. 2.3 a). In this case, as well as in other, considered in Sec. 2.1.1, the halfwidth $\Delta\nu_J$ substantially depends on the quantum number J. Namely, $\Delta\nu_J$ decreases with the increase of J (like in CO gas). Since the broadening of rotational lines and their overlapping is greater for *smaller numbers J,* first lines are not observed experimentally. For instance, in HCl-SF_6 solution (Fig. 2.3 b) the "vanishing lines" correspond to $J = 0$, 1, 2, and 3. This part of the rotational spectrum behaves "in a classic fashion". For the heavier DCl molecule the quantum specificity practically disappears (see Fig. 2.3 c).

2.1.3 The Effect of Molecular Rotation on the Infrared Vibration Spectrum

Rotational motion of polar molecules can be observed also in the infrared (IR) region as a *fine structure of the vibration line* of polar gases. In Fig. 2.4a the P- and R-branches of such spectrum are shown for a gaseous CO at low pressure. Here upper and lower dashed lines are envelopes for maximal and minimal absorption values. The solid line presents the quantum rotational band that arises with account of broadening of individual vibration-rotational lines until

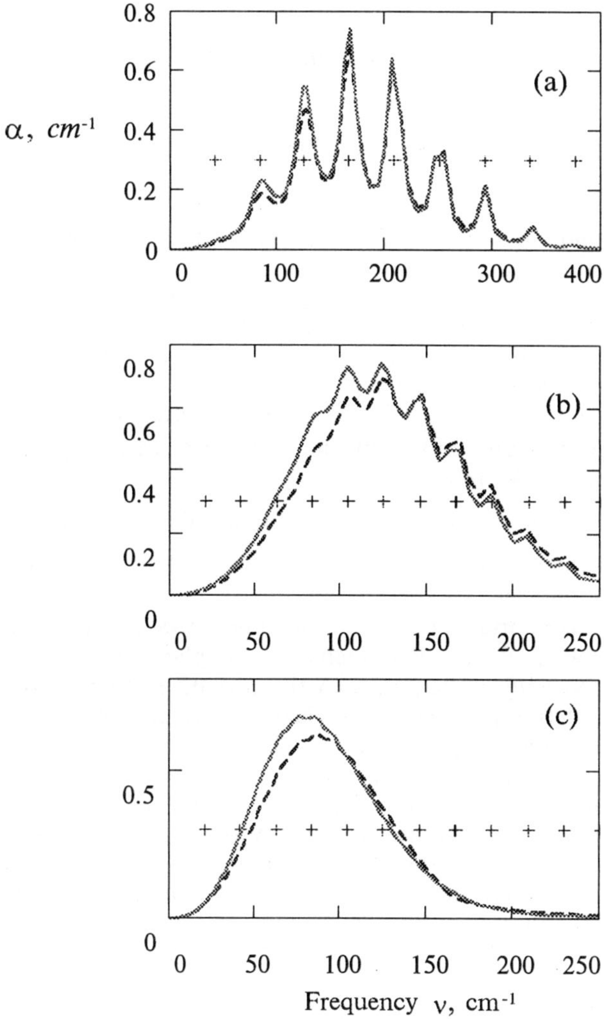

Fig. 2.3 Absorption spectra of liquid mixtures HF-SF_6 (a), HCl-SF_6 (b) and DCl-SF_6 (c), calculated for the Gross collision model (solid curves) and for the set of Lorentz lines (dashed curves). Crosses denote frequencies of rotational transitions. T = 273 K. The description of the calculation procedure is given in Sec. 16.4.

they fuse into a smooth contour. Both branches correspond to transitions with $\Delta v = \pm 1$, v being the vibration quantum number, $v = 0, 1, 2, \ldots$. For the P-branch the rotational transitions obey the rule $\Delta J = -1$. Thus, the upper and lower vibration states correspond to rotational numbers J and $J - 1$, respectively. For the R-branch these numbers are J and $J + 1$, so that $\Delta J = 1$. For the envelope of the IR spectrum the absorption maximum is greater for the R- branch. No absorption line appears between these two branches, since the $J = 0 \leftarrow 0$ rotational transition is forbidden. The maximum of the P-branch corresponds to the $6 \leftarrow 7$, transition, while the maximum of the R-branch corresponds to the $7 \leftarrow 6$ transition.

It should be noted that comparing with original publications, that are cited in Part V of the book, some Figures (e.g. 2.4 - 2.6, 2.19 - 2.21) are simplified, since in this chapter we are concerned only with a *qualitative* consideration of dielectric behavior of polar fluids.

Fig. 2.4 b illustrates the evolution of the vibration-rotational spectrum in a gaseous CO-Ar mixture, caused by the increase of density ρ_{Ar}. This value is measured in amagat. We remark that

$$1 \text{ amagat} = L_0 = 2.69 \bullet 10^{19} \text{ molecules in } 1 \text{ cm}^3, \qquad (2.2)$$

that is equal to the concentration of an ideal gas at normal conditions (L_0 is the Loschmidt number).

When ρ_{Ar} rises, individual rotational lines widen, and some absorption peak appears between P- and R-branches on the place of the above mentioned gap. This peak *formally* corresponds to the forbidden $\Delta J = 0$ transition. The physical nature of this peak becomes clear, if one considers the high-density and the low-temperature spectrum, for which the phase transition to liquid CO-Ar occurs. In a liquid the form of vibration line substantially differs from that, observed in gas mixture (cf. Fig. 2.4 c and 2.4 b), since in liquid the prohibition of the vibration transition with $\Delta v = 1$ and $\Delta J = 0$ is removed. The new peak corresponds to the transformation of relatively free rotation, occurred in a gas state, into librational motion *in a liquid.* The quantum theory, elaborated for free rotors, is invalid for the latter. Consequently, the quantum rule $\Delta J = \pm 1$ does not govern dielectric behavior in a liquid.

Another (more formal) evidence of such alteration of the type of molecular rotation is the following. If we take ν to be a small frequency shift relative to the center of the vibration transition, then one can find that in a gas state the ratio $\chi''(\nu) / \nu$ increases with the rise of ν, while in a liquid this ratio decreases with ν.

The same phenomena are characteristic for dielectric spectra, ν being the frequency of EM radiation in the latter case.

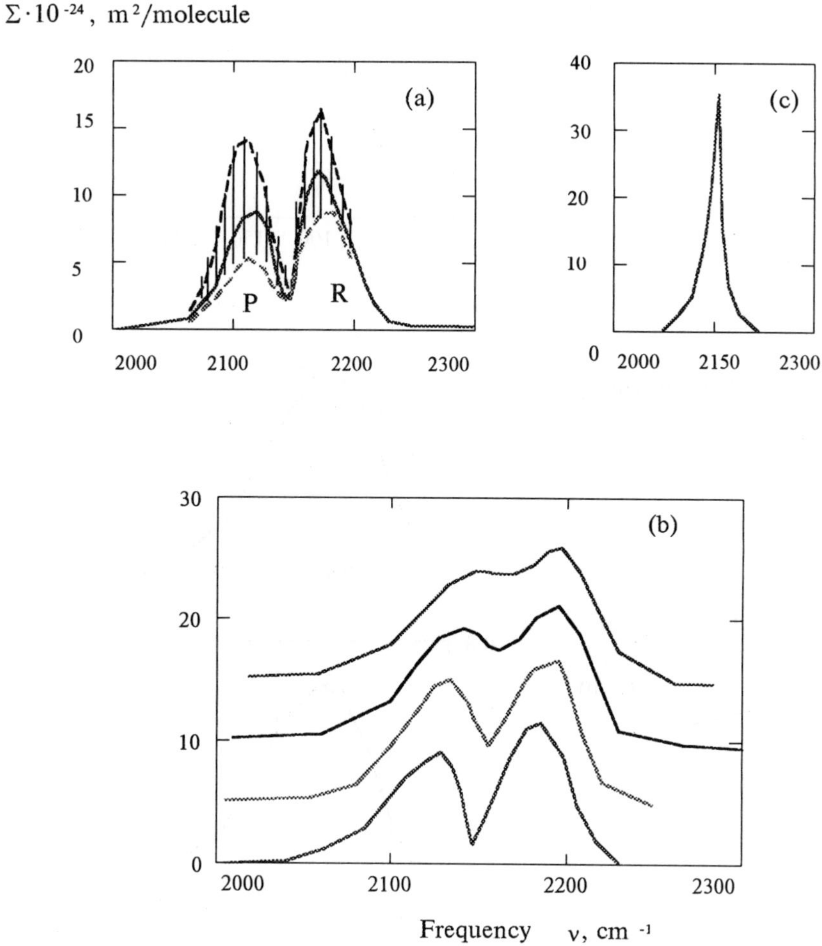

Fig. 2.4 Vibration-rotational fundamental band spectrum for gas phase of CO. $T = 298$ K. Σ is the absorption cross section. From (Berens and Wilson 1981). Explanation is given in the text.

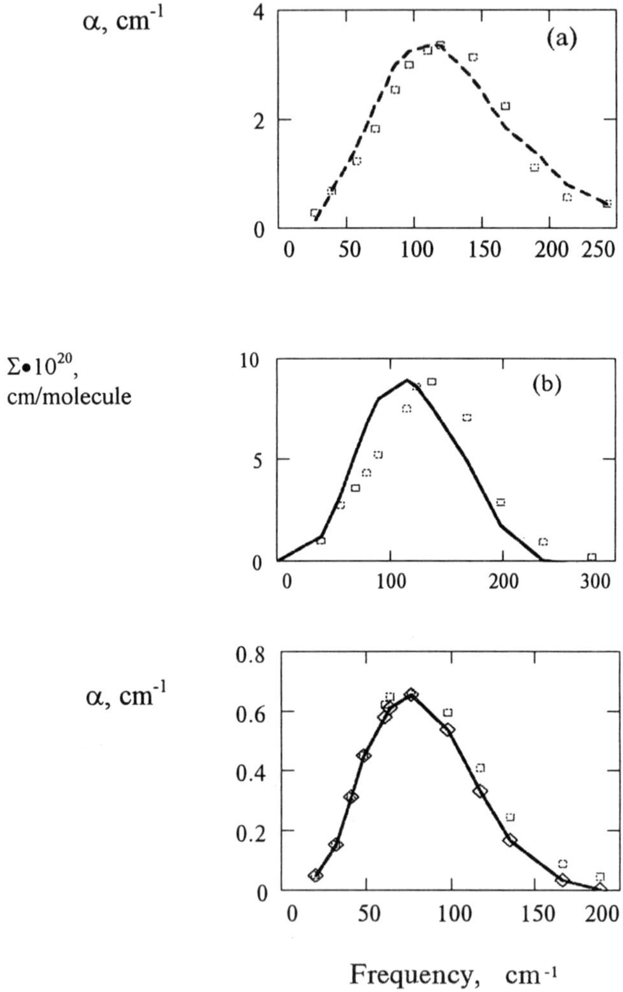

Fig. 2.5 Absorption vs. ν . calculated by Kalmykov and McConnell (1993). For solutions HCl in cyclohexane (a), HCl in CCL_4 (b), and DCl in SF_6 (c). Boxes - the extended rotational diffusion model with account of finitness of a collision time τ_c. Curves - the model of free rotation.

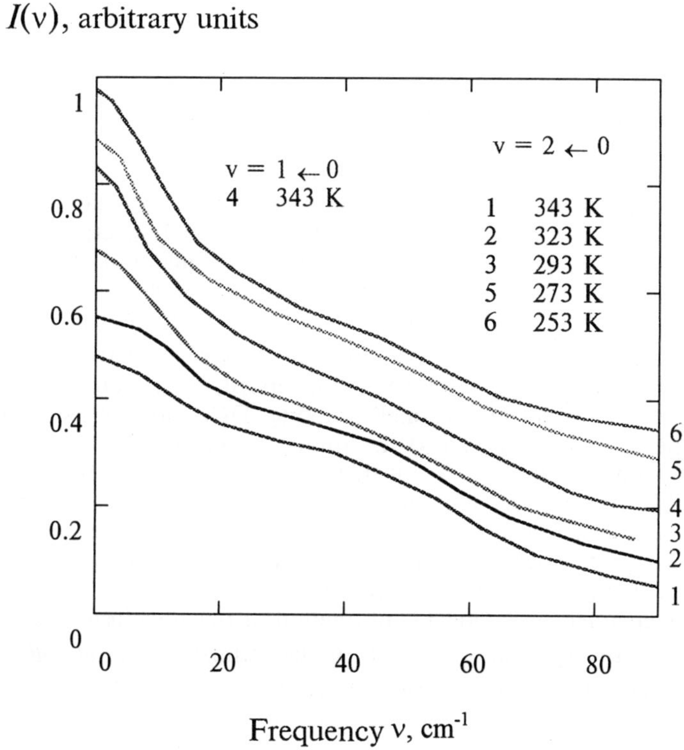

Fig. 2.6 Vibration band in CO - CCl_4 solution. Curves 1 to 6 - from bottom to top. From (Filippov, Vaculin and Orlova 1986)

In Fig. 2.4 b the evolution of the spectrum of CO-Ar mixture is pictured for $T = 298$ K, when density ρ_{Ar} rises: $\rho_{Ar} = 14$, 143, 364, and 510 amagat (from bottom to top). Concentration of CO corresponds to 1 atom CO per 17 atoms of Ar at $\rho_{Ar} = 14$ amagat and to 1 atom of CO per 55 atoms of Ar at greater ρ_{Ar} values. For Fig. 2.4 c the spectrum of CO-Ar solution is calculated for $T = 97$ K, $\rho_{Ar} = 750$ amagat. Concentration of CO corresponds to 1 CO atom per 55 Ar atoms.

2.2 Diluted Solutions

2.2.1 Absorption Spectra

We have touched briefly on this kind of liquids when discussing quantum phenomena, which characterize rotational (Fig. 2.3) and vibration-rotational (Fig. 2.4) spectra. Now we shall give few illustrations of a purely classic behavior of a polar molecule in nonpolar solvent (Fig. 2.5 - 2.6).

In Fig. 2.5 c we reproduce again the rotational spectrum of DCl-SF_6 solution. In this drawing the fine structure due to quantum effects is ignored (cf. Fig. 2.5 c and Fig. 2.3 c, where this structure is slightly seen). The shift of the observed solid curve relative to points, drawn for the model of free rotating molecules in the limit $\tau \to \infty$, is small. This fact is indicative of weak *spatial (in view of stereometrey) restrictions,* influenced by a large nonpolar SF_6 molecules of a solvent on a small rotating DCl molecule. For the HCl-cyclohexane solution (Fig. 2.5 a) the similar frequency shift (between dashed curve and points) is larger; it is still substantially greater for HCl-CCl_4 solution (Fig. 2.5 b). Indeed, the aforementioned shift is about 7, 12 and 26% in Figures c, a and b, respectively. It seems, that in the last case strong collisions are most frequent, since an "empty" cavity, where a polar molecule is placed, probably has the smallest mean dimension.

An independent theoretical confirmation of this conclusion can be drawn, if one regards the τ_c/τ ratio, where τ_c is the estimated *time of collision.* Fitted lifetime τ and collision time τ_c (in ps) are: 0.115 and 0.0357 (a); 0.0689 and 0.0443 (b); 0.224 and 0.0395 (c), respectively. Thus, we find that $\tau_c/\tau = 0.2$, 0.3 and 0.6 for Fig. 2.5 c, a, and b, respectively.

We note that in a gas $\tau_c/\tau \ll 1$, while in a pure liquid, where the aforementioned shift is much greater, τ becomes *less* than τ_c. So, the greater the shift of the absorption curve $\alpha(\nu)$ relative that for a free rotor, the closer the $\alpha(\nu)$ dependence to that, corresponding to a typical pure liquid. In other words, the greater this shift, the more frequent are collisions of a polar molecule with its neighbors, since the "collision time" τ_c approaches (or even exceeds) the mean time τ between collisions.

On the other side, the picture of molecular motion in the aforementioned solutions substantially differs from a typical picture of Brownian motion. In contrast to the Debye diffusion model (see Sec. 1.4), the collision of a polar molecule with its neighbors are rather rare in our case of dilute solutions, while a large "Brownian particle" collides extremely frequently with its smaller neighbors.

2.2.2 Vibration-Rotational Spectra

We shall consider these spectra for the CO-CCL_4 solution at various temperatures. In Fig. 2.6 the symmetrized envelop $I(\nu)$ is presented, which is obtained from experimental data $I^{\exp}(\nu)$ by the following procedure:

$$I(\nu) = (1/2)[I^{\exp}(\nu) + I^{\exp}(-\nu)]. \tag{2.3}$$

Here $I(\nu)$ is the intensity of the infra -red spectrum and $\nu = \omega/(2\pi c)$ is the frequency shift relative to the center of the $1 \leftarrow 0$ or $2 \leftarrow 0$ vibration transition. Just like in the case of CO - Ar system, the central Q-branch and the shoulder appear (cf. Fig. 2.6 with Fig. 2.4 b). The two parts of a vibration-rotational line are observed especially distinctly at low temperature. In this case strong collisions are more rare, so that the solution behaves more like a gas.

Theoretical calculations of these spectra agree with the experiment (Filippov, Vaoulin, and Orlova 1986). The field model, viz. the double well cosine squared potential $U(\vartheta)$, was employed, in which $U \propto \cos^2\vartheta$. It was shown that the constant field (CF) potential, for which $U(\vartheta) \propto \cos\vartheta$, does not fit in this case.

This result might be expected, since the mean dipole moment of a dielectric sample should be zero, that is not valid for the CF model. The estimations yield several times the value of $k_B T$ for the potential welldepth U_0. This interesting result shows that intermolecular interactions are rather strong even at small concentration of polar component of a solution. Still greater U_0 value probably can be met in *pure liquids* where intermolecular interactions should be stronger.

2.3 Individual Nonassociated Liquids

We shall consider briefly dielectric spectra of such liquids. The results of these and other calculations are presented in Chap. 14.

2.3.1 The Evolution of Dielectric Spectra due to the Temperature Alteration

Typical picture of loss $\varepsilon''(\nu)$ and of absorption coefficient $\alpha(\nu)$ plots are shown in Fig. 2.7 for liquid CH_3F, that is characterized by rather high dipole moment μ. Molecular rotation is strongly hindered by nearest neighbors of a molecule because of strong intermolecular interactions. Correspondingly, dielectric spectra exhibit the following features.

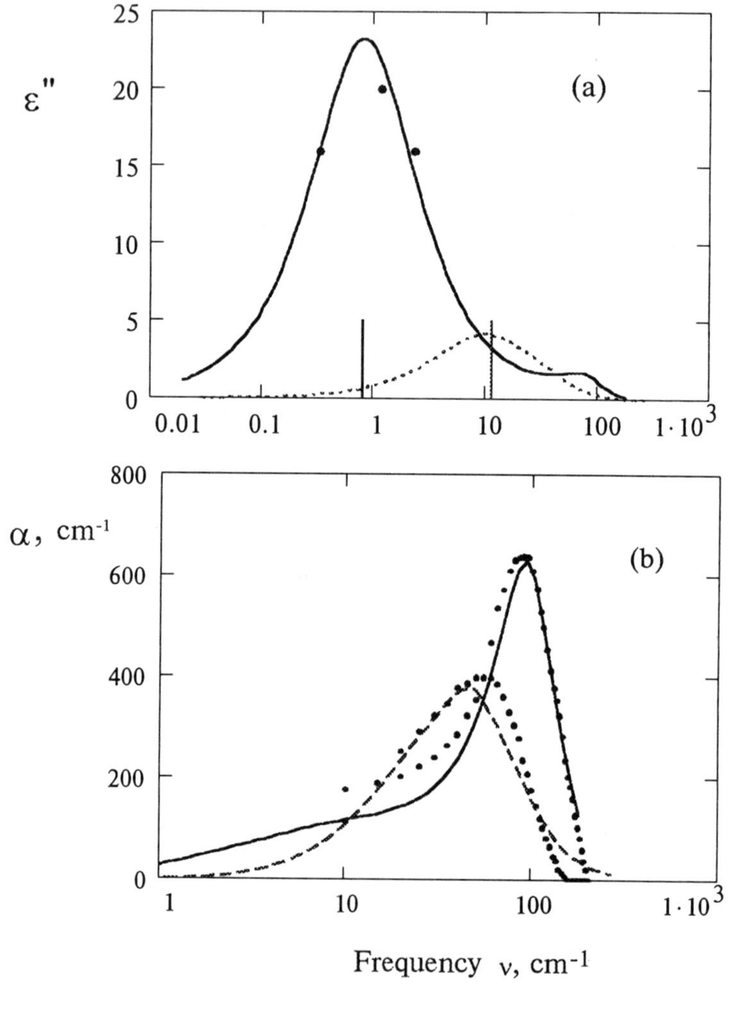

Fig. 2.7 Dielectric loss (a) and absorption coefficient (b) vs. ν, calculated for liquid fluoromethane CH_3F at T = 133 K (solid lines) and 293 K (dashed lines). The model parameters are fitted for the best agreement with the observed data. The experimental (Gerschell, Dimicoli, Jaffre, and Riou 1976) data are marked by dots and the observed positions of the Debye loss - by vertical lines. The calculation procedure is described in Sec. 14.1.

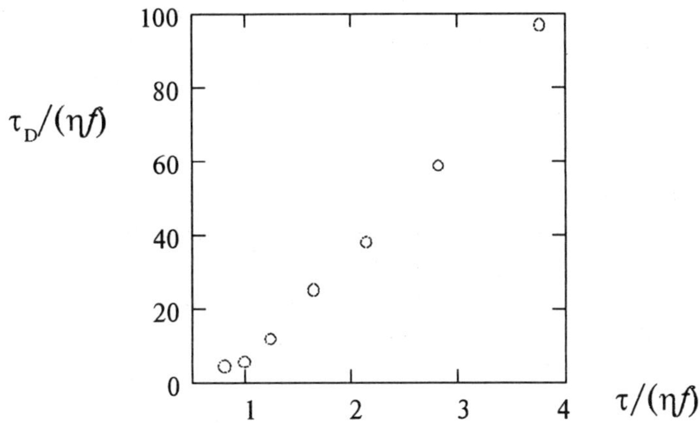

Fig. 2.8 The τ_D vs. τ plot for liquid CH_3F. Points relate to T = 133, 153, 173, 193, 233, 270, and 293 K (from top to bottom). The parameters of the confined rotator model (lifetime τ and libration angle β) are fitted for these values of T; $f = 2\beta/\pi$. The experimental relaxation time τ_D was taken from the aforecited work of Gerschel et al. (see the capture to Fig. 2.7).

a. A broad absorption band (of about 100 cm^{-1} on the level 1/2) with a maximum absorption α_L, ranging from 200 to 700 cm^{-1}, the value α_L being proportional to the concentration N.

b. The maximal absorption α_L and the corresponding frequency ν_L decrease with increasing of temperature. The last feature is a definite characteristic of a liquid state, since in a gas ν_L increases, if temperature T rises. This fact indicates, that a molecule in a liquid is permanently influenced by torque caused by neighboring molecules. The libration amplitude of a dipole increases with temperature, since intermolecular interactions weaken with T. Correspondingly, libration frequency decreases. On the contrary, in a gas the mean frequency of molecular rotation increases with the rise of T.

c. At frequencies of about ν_L the refractive index $n(\nu)$ is at minimum, this being an indicative of a resonance character of absorption. On the other side, the loss curve $\varepsilon''(\nu)$ exhibits a feature (shoulder), which is more pronounced at low temperature (Fig. 2.7a, solid curve).

d. The HF slope of the $\alpha(\nu)$ curve is rather steep; the observed minimum of this curve corresponds to the HF edge of the orientation-relaxation region. At greater ν, viz. in the IR and optical regions, the $\alpha(\nu)$ and $\varepsilon(\nu)$ dispersions are provided by the deformation of molecules and / or electron sheaths.

e. In the low frequency range (usually at microwaves, for ν about 1 - 5 cm^{-1}), the loss $\varepsilon''(\nu)$ goes through the Debye maximum ε''_D. If T varies from the triple point T_{tr} to the critical temperature T_{cr},, e.g. in the range, in which liquid state exists, then the maximal loss ε''_D shifts substantially to higher frequencies and becomes several times lower.

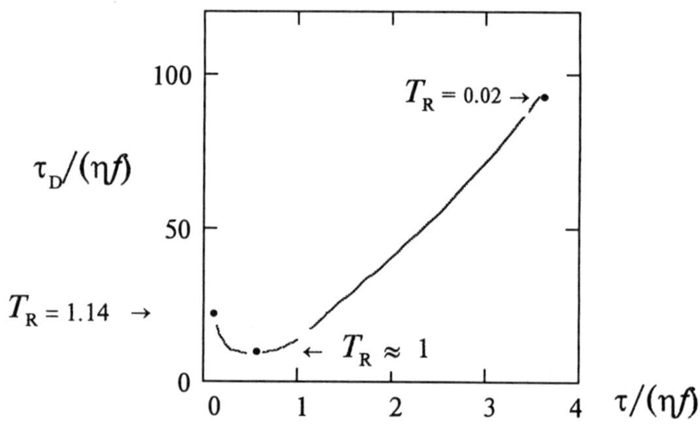

Fig. 2.9 The same function, as in Fig. 2.8, but predicted in a wide range of normalized temperatures $T_R = (T - T_{tr})/(T_{cr} - T_{tr}$. The right and left end points of the curve correspond to $T_R = 0.02$ and 1.14, respectively, while its minimum - to $T_R = 0.94$, that approximately coincides with T_{cr}. The detail of calculations see in Sec. 14.1.

2.3.2 *Temperature Dependence of the Relaxation Time and of the Lifetime*

The relaxation time τ_D is related to the measured frequency ν_D of the loss maximum ε''_D, since $\tau_D = (2\pi c\nu_D)^{-1}$. It seems that the fitted parameter τ of the confined rotator model means the lifetime of a near surroundings of a given molecule, to which the prescribed potential well $U(\vartheta)$ corresponds.

It is reasonable to suggest that during the relaxation time τ_D at least few changes of temporary-equilibrium states occur. Then the condition $\tau_D/\tau >> 1$ should hold. *The Debye relaxation time thus should be much longer than the lifetime of librations.* When temperature increases, both times (τ_D and τ) decrease, as well as the τ_D/τ ratio. The dependence of τ_D on τ is shown in Fig. 2.8, 2.9. As it follows from the theory (Fig. 2.9), this dependence has two branches, the left and the right, arising from the minimum. The calculated relaxation time τ_D increases for the left branch, when the lifetime τ decreases. Since for such dependence there is no physical ground, the minimum probably corresponds to the liquid - gas transition. Our calculations confirm this idea (see Chap. 14).

2.3.3 The Self-Diffusion Coefficient and Translational Motion

In this book we concentrate our attention on molecular phenomena related to *reorientation.* In liquids the latter are accompanied by translational motion. Since the self-diffusion coefficient D is the measure of translations, it may be supposed that D is somehow related to librations.

The self-diffusion rises with the increase of T (see Fig. 2.10 a, b). It is possible to calculate the $D(T)$ dependence, if one assumes that in a liquid some potential barrier U_v exists, which *hinders translational motion.* Only fast particles, overcoming this barrier, participate in translations. One may consider, that the latter are determined by an isotropic component of the intermolecular potential, while the anisotropic component $U(\vartheta)$, depending on the angular shift ϑ of a dipole, determines molecular rotation. Theoretical estimations show the following features:

a. $U_v < U_0$; spatial restrictions in liquids are *less* for translations, than for molecular rotation, see Fig. 2.9 d, where u is the ratio of U_0 to $k_B T$ and V is the similar ratio for the isotropic component of the intermolecular potential. The opposite inequality is characteristic for molecular crystals, for which the self-diffusion coefficient is extremely low, but librations do not alter much compared with liquid state. As an example, one can compare the D values in liquid water and ice. Indeed, these differ on several orders of magnitude, while the FIR absorption spectra are similar in both substances, see further Sec. 2.4.3.

b. The ratio U_v / U_0, e.g. the ratio V/u, is approximately a linear function of temperature T and is weakly dependent on the type of a liquid (Fig. 2.10 c).

c. When temperature increases, both potential barriers decrease, see Fig. 2.10 c for the $V(T)$ dependence.

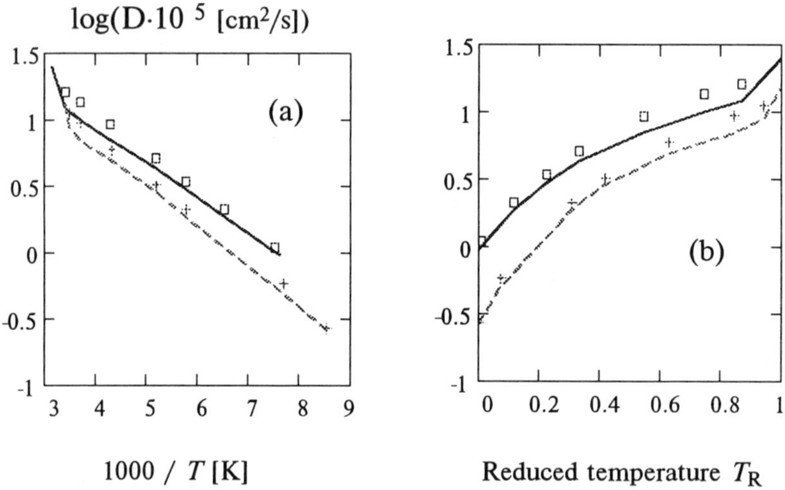

Fig. 2.10 The temperature dependencies of the self-diffusion coefficient (a, b), of the isotropic part V of the reduced potential (c) and of the V/u ratio (d), where u is the reduced ϑ-dependent part of the potential. Solid and dashed curves are calculated values for liquid CH_3F and CHF_3, respectively, points - experimental data of D. The calculation scheme is described in Sec. 14.8.

2.4 H-Bonded Network

2.4.1 Anomalies of Liquid Water

Dielectric spectra of liquid water must in some way reflect the existence of water anomalies. To their number, as is known, belong (Zatsepina 1974, Belaya and Levadnii 1987, Gaiduk 1991):

a) Water H_2O is a liquid, though similar (heavier) substances H_2S and H_2Se are gases. For this "exception" life on Earth is obliged.

b) The density of a solid phase (ice Ih), equal to 0.92 $g{\cdot}cm^{-3}$, is lower than that of water. Because of this property exist fauna in freezing basins. On lowering temperature below 4°C the water density ρ decreases, not increases as in other liquids on cooling. On decreasing of T from 4 to 0°C the density decreases from 1 only to 0.9998 g·cm^{-3}.

c) Very large anomalies include temperatures of boiling and melting and interval $\Delta T = 100$°C between them.

d) Large anomalies include the static dielectric permittivity of water: ε_S = 80; ε_S of ice is still greater (95).

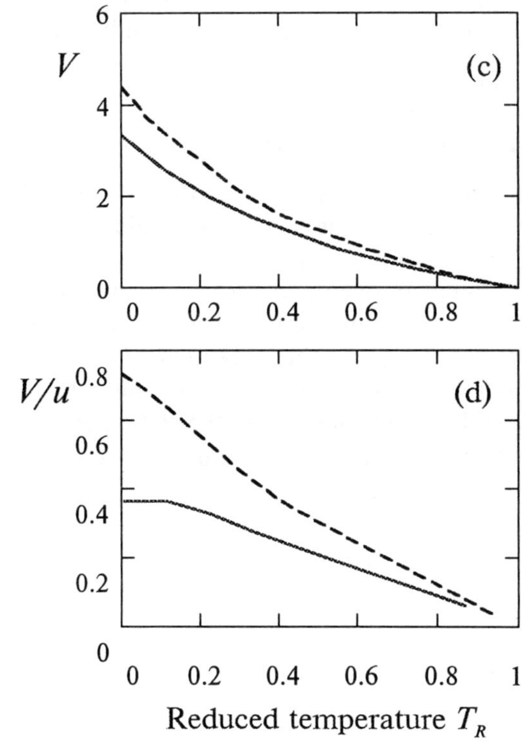

Fig. 2.10 **c, d:** Caption to this Figure is given on the previous page.

These anomalies are due to directed hydrogen bonds (H-bonds), to the cooperation of H-bonds and, probably, to the tunneling of a proton along the network of hydrogen bonds. Cooperation means that, if H_2O molecule has one

H-bond, then the second one is formed more easily. The spins of electrons, forming H-bonds, are directed contrariwise. In each H_2O molecule there stays one more pair of uncompensated electrons of the O atom. Compensating their spins, these electrons create the hydrogen bond with the H atom of another molecules: Thus, each oxygen atom can form two covalent and two hydrogen bonds. The ability of a water molecule both to give back and to join an electron, allows it to behave simultaneously like a donor and an acceptor. Thanks to this property molecules of liquid water can form a conglomerate (cluster), and in ice they can form a crystal. In the latter four neighboring H_2O molecules are situated in apexes of a triangular pyramid, and the fifth is in its center. On uniting, the pyramids form a spatial hexagonal structure of ice Ih, in which each H_2O molecule is bounded with other neighboring water molecules by four bonds.

2.4.2 Dielectric Spectra of Liquid Water

The wideband low frequency (e.g. at microwaves) spectra $\varepsilon'(\nu)$, $\varepsilon''(\nu))$, $\alpha(\nu)$ are shown in Fig. 2.11 a, b, c and the corresponding FIR spectra - in Fig. 2.12 a, b, c. The Cole-Cole plots $\varepsilon''(\varepsilon')$ are pictured respectively in Fig. 2.11 d and 2.12 d for two aforementioned parts of wideband spectra. In this example we take $T = 1$ and 27°C (solid and dashed curves, respectively).

When temperature T rises, the maximum loss factor ε'', which is displayed in the microwave region, decreases and shifts to higher frequencies; the relaxation time τ_D becomes shorter and the static permittivity ε_s decreases. The dispersion $\varepsilon'(\nu)$ of the real part of permittivity occurs mainly in the low frequency region. When ν increases to $\approx$ 10 cm^{-1}, ε' falls from about 80 to the limiting (in terms of Debye theory) value ε_∞ ($\varepsilon_\infty \approx 5$). One conditionally may regard that in this region the Debye plateau is placed (see Fig. 2.11 c). When ν approaches the submillimeter region, the $\alpha(\nu)$ absorption dependence leaves this plateau.

In liquid water there exists two absorption maxima, which are observed in the FIR region. On the contrary, in "simple" (nonassociated) liquids usually one $\alpha(\nu)$ maximum is observed, cf. Fig. 2.12 c with Fig. 2.7 b. In water the α maxima correspond to the so-called "translational" and librational bands, having peak frequencies $\nu_R \approx 200$ cm^{-1} and $\nu_L \approx 700$ cm^{-1}, respectively. Further we term these bands as R- and L-bands. The quantities, related to them, we supply by the same subscripts, R and L. The widths $\Delta\nu_R$ and $\Delta\nu_L$ are about peak frequencies ν_R and ν_L.

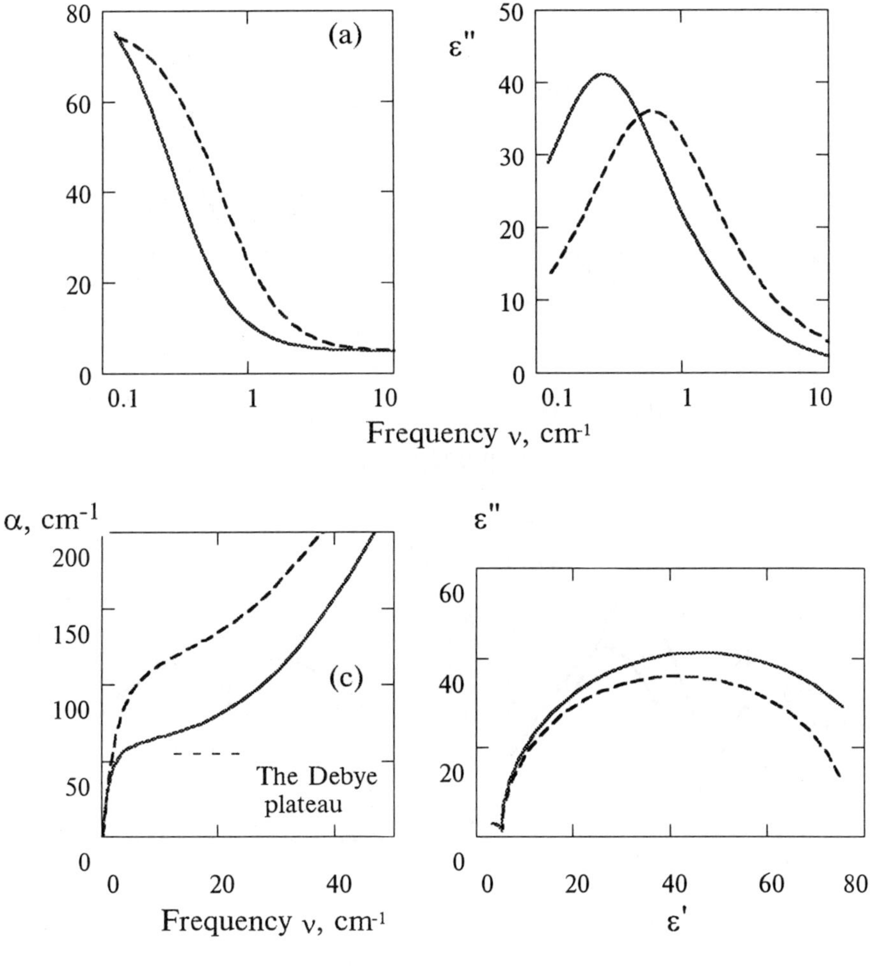

Fig. 2.11 The low frequency dielectric spectra (in the Debye region) of liquid water (a, b, c) and the corresponding Cole-Cole plot (d). Solid lines for $T = 1^{o}C$, dashed lines for $27^{o}C$. Detail see in Sec. 15.1-15.2.

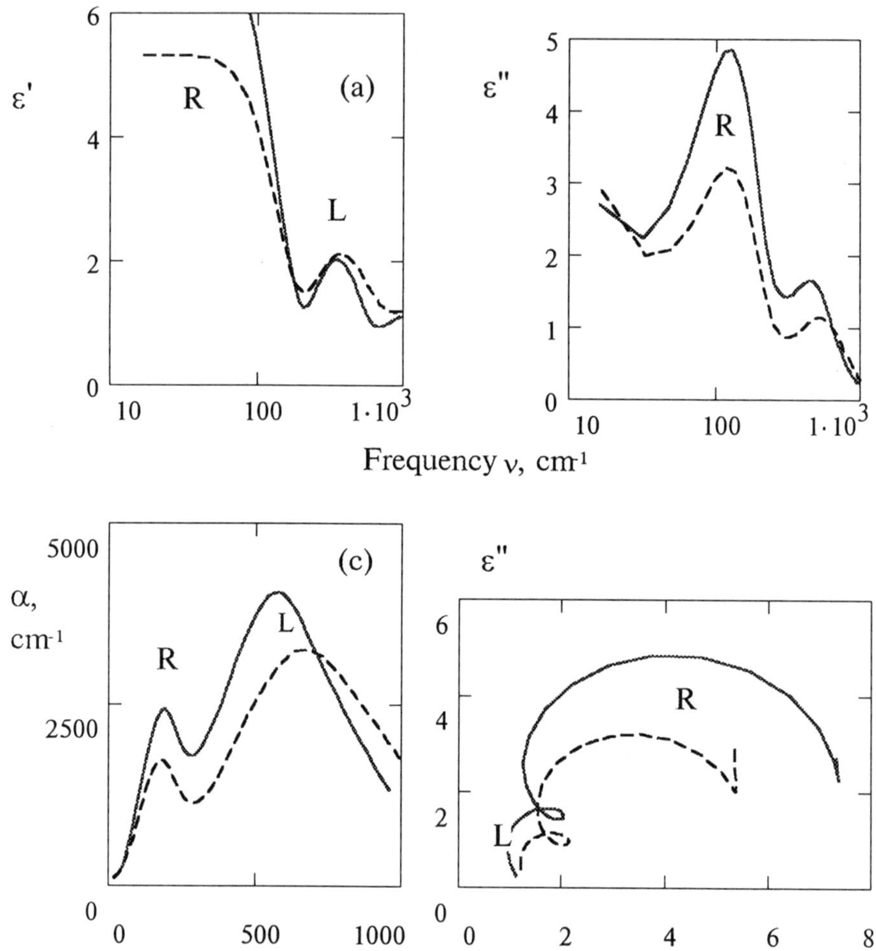

Fig. 2.12 The same, as in Fig. 2.11, but for the FIR region. R and L denote spectral peculiarities devoted to two subensembles of dipoles (viz. to "rotators' and "librators").

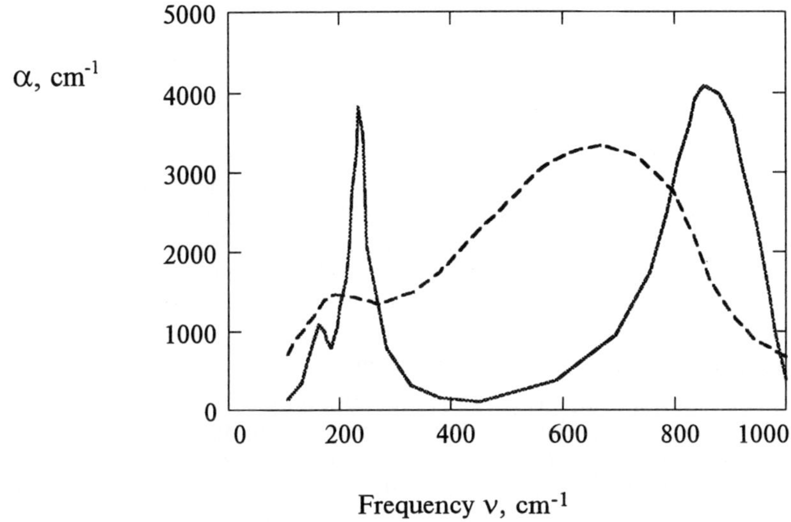

Fig. 2.13 Absorption coefficient vs. frequency for ice at 100 K (solid line) and for liquid water at 300 K (dashed line). This comparison of experimental data is taken from (Gaiduk, Gaiduk, Novskova, Tseitlin and McConnell 1995).

Liquid water: quasi-resonance dielectric spectra

R-band	L-band
The properties of a "hot" band are revealed. Molecules, contributing to this band, behave like molecules in a gas state.	Molecules, contributing to this band, behave like in nonassociated liquids.
With increasing of T in the interval [1, 30°C]:	
a. The peak absorption α_R rises several times.	**a**. The peak absorption α_L decreases about 10%.
b. The peak frequency ν_R remains practically constant ($\nu_R \approx 200$ cm^{-1}).	**b.** The peak frequency ν_L decreases about 10% ($\nu_L \approx 700$ cm^{-1}).

The temperature dependencies related to these bands are quite opposite (see the scheme above). The non-monotone curves $\varepsilon'(\nu)$ and $\varepsilon''(\nu)$, where each curve has two (R and L) dispersion regions, correspond to the two-humped FIR absorption $\alpha(\nu)$ dependence (see Fig. 2.12 a, b, c). Correspondingly, the Cole-Cole plot (Fig. 2.12 d) exhibits two "curls" that are additional to the main Debye semicircle (the latter is pictured in Fig. 2.11d).

2.4.3 Dielectric Spectra of Ice Ih

Several modifications (from I to VIII) of ice crystals are known. At the lowest temperature / pressure a monocrystalline ice Ih is formed. Observations are very often made with the polycrystals of ice I. The two resonance absorption bands of ice I fall into the FIR range (Fig. 2.13, solid line). These bands are similar to those, observed in water (Fig. 2.13, dashed line) and therefore correspond to R- and L-bands in water, but are appreciably narrower. Besides, the R-band in ice I is split. However, the Debye relaxation in ice I drastically differs from that in water, since the frequency ν_D of the loss peak falls into the kHz region. Consequently, the Debye relaxation time τ is 7 - 8 orders of magnitude longer than that in liquid water. This fact accompanies the corresponding decrease of the self-diffusion coefficient *D*. The resemblance of FIR spectra in water and ice I, as well as the close values of the static permittivity in these substances ($\varepsilon_S^{ice} > \varepsilon_S^{water}$), compels us to suppose that, as in water, the dielectric spectra in ice I are mainly caused by reorientation of polar H_2O molecules in terms of molecular rotation / translation.

2.4.4 Molecular Model of Reorientation in Liquid Water

The natural way to explain an "unusual" (compared with nonassociated liquids) absorption spectra of water consists in an attempt to regard water as a *bimodal system*. One may suppose, that molecules of the main L-ensemble, which are responsible for the L-band, librate under influence of a relatively deep potential well *U*, say, such as is employed in the confined rotator (CR) model. A smaller proportion *r* of molecules ($r \leq 10\%$) with greater rotational and probably translational mobility is responsible for the R-band. The contribution of R-molecules to the dielectric susceptibility $\chi(\omega)$ may be found, say, from the extended diffusion (ED) model.

Recently the author (Gaiduk 1994), (Gaiduk and Gaiduk 1997) suggested an idea that *both subensembles* reveal the specific property of the H-bond, so that the concentration *r* of more mobile R molecules is a function of the $U(\vartheta)$ potential profile. The latter was supposed to be a "hat-like". We shall return to these questions in Chap. 13 and 15, where we shall try to show that L- and R-subensembles are respectively the low-energy librators and high-energy hindered rotators. These subensembles are oriented by a strong conservative field, produced by a hat-like potential well. One can regard the hat potential as a generalized $U(\vartheta)$ representation of the intermolecular potential, applicable to many aqueous systems, as well as to nonassociated liquids. The distinction of the latter from water is probably obliged to the lower moment of inertia I in water. This, in turn, leads to larger frequency separation of the non-resonance Debye spectrum from FIR quasi-resonance spectrum. Besides, due to H-bond network librations in water are more pronounced, than in "simple liquids", such as CH_3F.

The theoretical bandwidth of the L-band, calculated for liquid water, *exceeds* the observed one. This result one may interpret as follows. Librations in water are actually *not* free (as it is assumed in the CR model), since they are influenced by some intermolecular electric field $E_0(\vartheta)$. The latter is obviously stronger in ice I, than in liquid water. That is why the CR model is quite inapplicable to ice.

2.5 Water Bounded by a Macromolecule

2.5.1 Critical Humidity of a Bound Water

In biological structures there appears at least one more state of water, viz. *bound water*, to which an additional Debye relaxation region corresponds. The additional loss peak is positioned in the range of meter waves. Correspondingly, the relaxation time τ_D is longer by two orders of magnitude, than in the case of liquid water. At microwaves variations of bound water state reflect the alteration of temperature and / or of the structure of biological molecules.

Due to existence of H-bonds, directed in space, a macromolecule, embedded in aqueous medium, acquires necessary conformation and ability to fulfill its specific biological functions. When a well dried protein molecule is moistened, its "animation" takes place at the critical humidity h_{cr} (usually h is measured in g of water per g of protein).

Uneven "turning on" of enzyme activity of protein was discovered by Khurgin (1976). Rebuilding of protein structure at moistening, close to the critical point, is accompanied by formation of bound water layer on the surface of a protein. This process *resembles the phase transition.* The hydrate layer constitutes, figuratively speaking, a "lace" at critical moistening h_{cr}, when molecules of bounded water acquire some *collective properties.* H_2O molecules form bridges between polar groups on the protein's surface. A protein becomes labilious at this point. As was fixed by Raleigh dispersion of γ-quanta, the average-quadratic oscillation amplitude of non-hydrogen atoms sharply increases at h_{cr}, and mechanical properties of a protein also change.

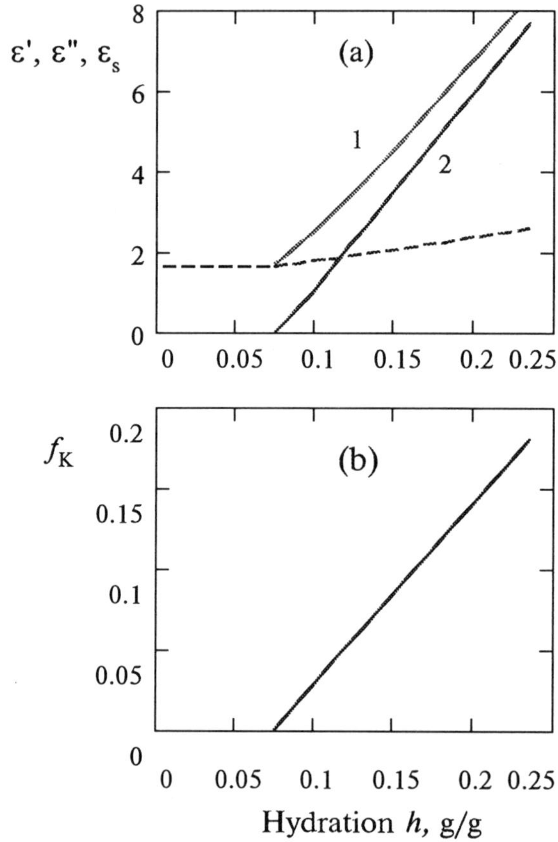

Fig. 2.14 The weakly humidified protein.
(a): Real (curve 1) and imaginary (curve 2) parts of the complex permittivity vs. hydration h, calculated for frequency $f = 0.14$ MHz, corresponding to the loss maximum. Dashed curve - calculated static permittivity.
(b): Dependence on h of the factor f_K, proportional to the Kirkwood correlation factor g. The zero-value part of the curve $f_K(h)$ and of curve 2 in Fig. a corresponds to humidity for which all H_2O molecules exist in the state of structural water. The calculation scheme is described in Sec. 15.9.2.

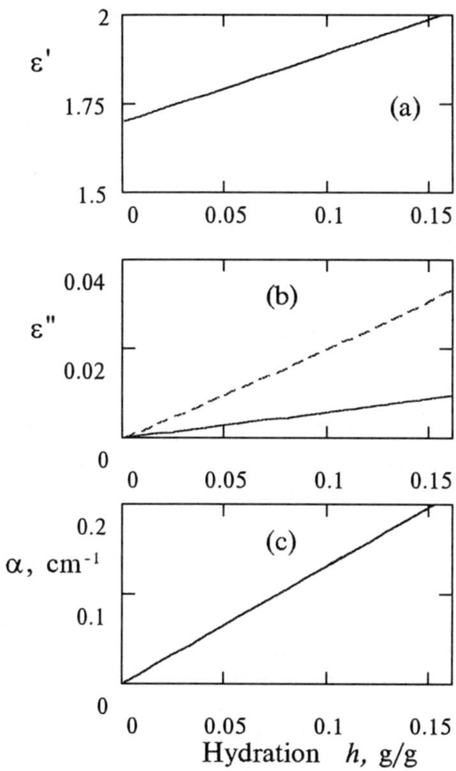

Fig. 2.15 Weakly humidified protein. Real (a), imaginary (b) parts of the complex permittivity and absorption (c) vs. hydration h, calculated for $\lambda = 2$ mm (solid curves) and $\lambda = 7$ mm (dashed curve).

At critical humidity h_{cr} the hydrate layer is rather thin, while at $h > h_{cr}$ "bulk water" appears (its properties resemble those of liquid water). At large h the thickness of the hydrate layer may exceed several times that at h_{cr}. Correspondingly, since rotational / translational mobility of H_2O molecules rises at $h > h_{cr}$, the Kirkwood correlation factor g rapidly increases with the humidity h, and real / imaginary parts of the permittivity ε^* become strongly dependent on h. These considerations are illustrated for two proteins by Fig. 2.14 - 2.15.

2.5.2 Dielectric Properties of Humidified Collagen as a Model of a Skin

After these introductory remarks let us examine some results of calculations. We consider dielectric spectra of strongly humidified collagen (Fig. 2.16). Now the three-fractional model was employed which is based on the hybrid model and permits to estimate (on a semi-microscopic level) absorption of electromagnetic radiation by skin of man and animals. Indeed, collagen constitutes the main building material of skin and connective tissues. For $\lambda > 1$ mm we see in Fig. 2.16a two loss ε'' peaks and correspondingly - two dispersion regions of $\varepsilon^*(\omega)$.

One region is approximately the same, as in liquid water, cf. Fig. 2.16a and 2.11b. The other (lower-frequency) Debye relaxation range is conditioned by water, bounded by a protein surface. Maximal loss comes here on the wavelength λ falling into meter range. Therefore, the relaxation time $\tau_D = \lambda/(2\pi c)$ of bound water is about one nanosecond. This τ_D value is two orders of magnitude longer than in the first region (and than in liquid water), where τ_D is about 10 ps. This difference testifies that protein bounds water molecules stronger, than one H_2O molecule bounds another one by means of H-bond. The contribution of different fractions to loss ε'' is illustrated by Fig. 2.17a. Because of this phenomenon the theory predicts (Fig. 2.17 b) noticeably lesser absorption of strongly humidified collagen (and of skin), than that, calculated for the same concentration of H_2O molecules, if these are supposed to behave like in a liquid water.

The reflection coefficient R of skin substantially decreases with the increase of frequency ν or with the decrease of wavelength λ (see Fig. 2.17 a and 2.17 b, respectively). Both dependencies, $\alpha(\nu)$ and $R(\nu)$, are smooth. The absorbing layer is considered to be homogeneous in the aforementioned simple model. However, one may regard, that skin comprises several layers, which possess different humidity and dielectric permittivity. Experiments show that the observed $\alpha(\nu)$ dependence is also smooth. Consequently, because of strong humidity no interference effects are usually revealed in skin (such effects principally might be related to the presence of several layers).

2.5.3 Hemoglobin Molecule in an Aqueous Solution

Let us turn to another system, viz. to aqueous solution of hemoglobin (Hb). This protein has a smaller molecular weight, than collagen, and dissolves in water. In Fig. 2.18 we present the results of calculations, made for the three-fractional model. Two fractions (these employ the hybrid model) describe dielectric behavior of water and one fraction regards reorientation of a protein as a whole. For the latter fraction the extended diffusion model is used. We see again two regions, γ and δ, related to the relaxation of bulk water (γ) and to bound water (δ). What is interesting, that the second dispersion region is in Hb much less intensive than in collagen. In Hb a new spectral feature also appears, viz. at *very low frequencies* (near 1 MHz) a third relaxation region β arises, conditioned by the Brownian rotation of a protein molecule, performed together with the hydrate layer.

The loss peak ε'' due to bound water, that appears in the δ-region, may be described, if to suppose that in the hydrate layer approximately 3000 H_2O molecules come on one Hb molecule. To such concentration of bound water the hydration h about 0.8 g of water per 1 g of solution corresponds.

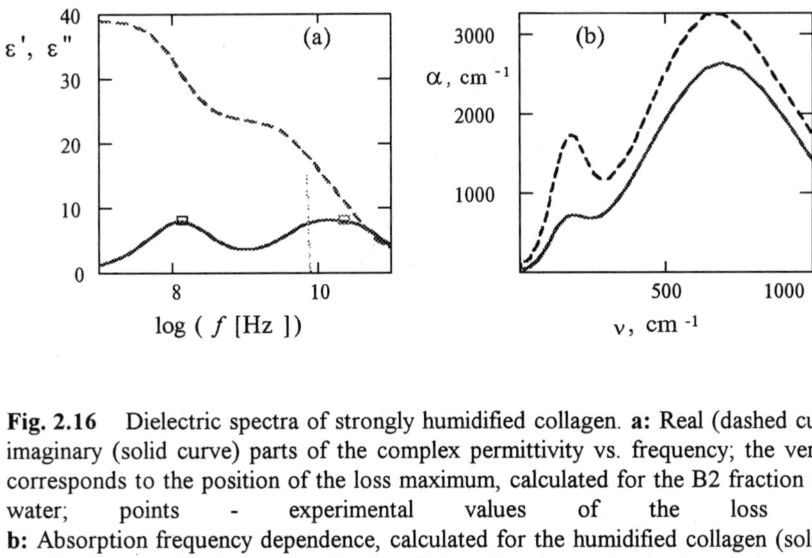

Fig. 2.16 Dielectric spectra of strongly humidified collagen. **a:** Real (dashed curve) and imaginary (solid curve) parts of the complex permittivity vs. frequency; the vertical line corresponds to the position of the loss maximum, calculated for the B2 fraction of bound water; points - experimental values of the loss maxima. **b:** Absorption frequency dependence, calculated for the humidified collagen (solid curve) and liquid water (dashed curve). $T = 300\ K$. Detailed description is given in Sec. 15.9.1.

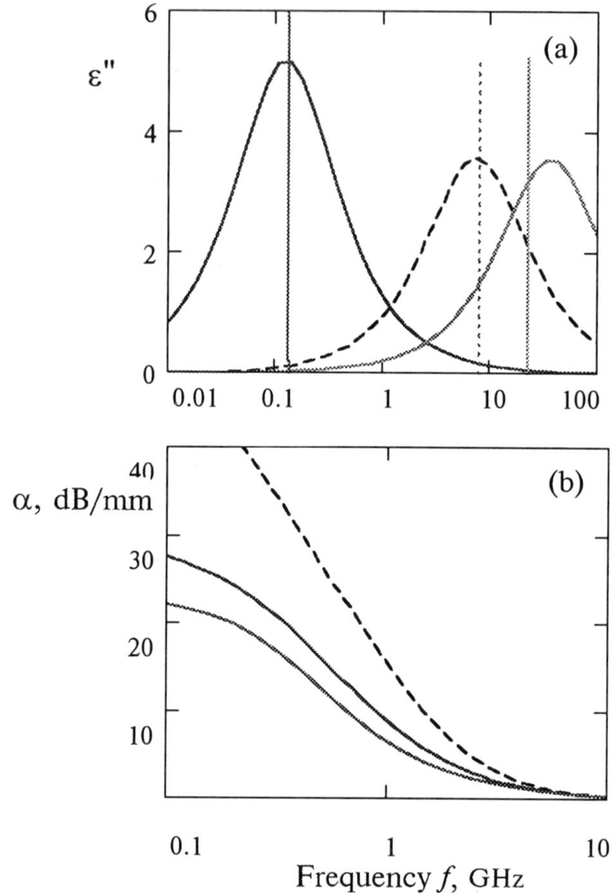

Fig. 2.17 Strongly humidified collagen (see Sec. 15.9.1). **(a):** Calculated contribution to dielectric loss due to fractions 1, 2 of bound water and due to the bulk water fraction (from left to right). Vertical lines indicate frequencies of loss maxima, presented for these fractions in Table 15.12. **(b):** The estimated microwave absorption vs. frequency due to humidified collagen, to the model of skin and to liquid water (from bottom to top). T = 300 K. **(c, d):** Reflection coefficient at microwaves, estimated for the model of skin (solid curves) and for liquid water (dashed curves) vs. frequency (c) or wavelength (d).

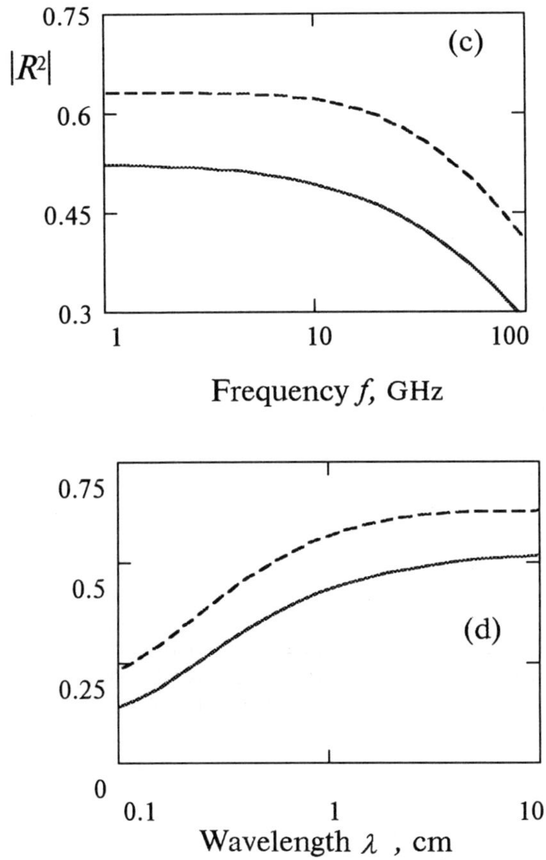

Fig. 2.17 c, d: Caption to this Figure see on the previous page.

The latter quantity is several times greater, than the h value, estimated with use of some other methods (these yield information only about the *first* hydrate layer). Thus, interpretation of dielectric spectra, observed in Hb-water solution, testifies considerable quantity of water molecules, modified by a protein.

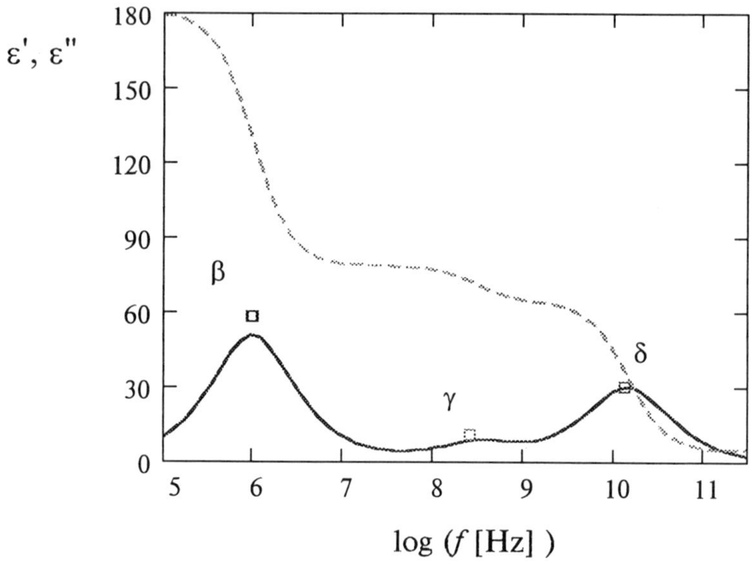

Fig. 2.18 Real (dashed curve) and imaginary (solid curve) parts of the complex permittivity vs. frequency, calculated for 20% aqueous Hb solution. Points - experimental loss from (Grant, Sheppard and South 1978). $T = 300$ K. Details of t hese calculations are given in Sec. 15.8.3.

Application of the above microscopic model yields better results, than *formal* description of dielectric spectra, in which three Debye terms are employed with different relaxation times. In particular, this is true for the δ–region, stipulated by reorientation of bound water.

2.6 Aqueous Solutions

2.6.1 FIR Spectra of Water, Modified by Ions

It was observed experimentally, that FIR absorption spectra of aqueous electrolyte solutions resemble spectra of liquid water. This fact becomes evident from Fig. 2.19 - 2.21, related to aqueous solutions of NaCL, NaI and KI. These spectra were calculated similarly to those for liquid water. The hybrid model was employed in these calculations.

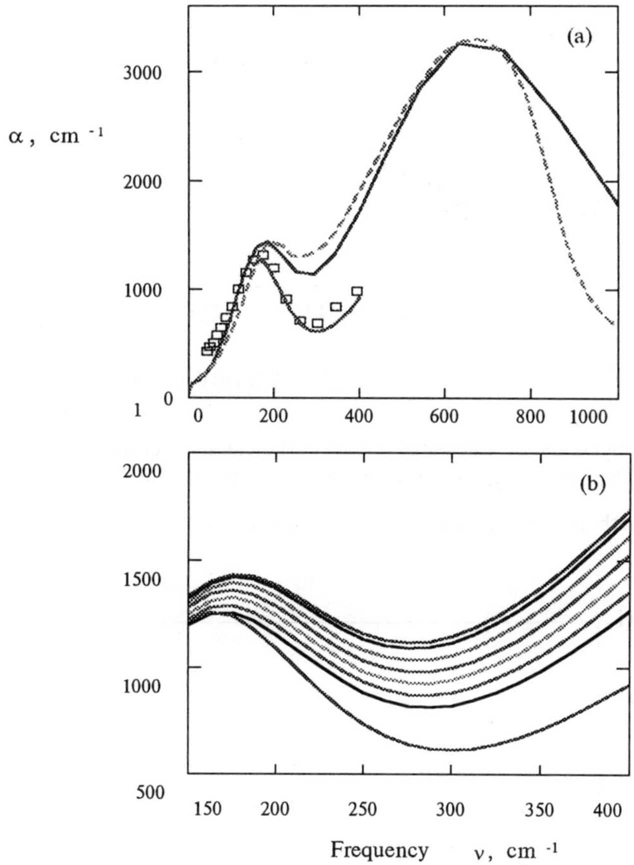

Fig. 2.19 Absorption spectra of liquid water and NaCl-H_2O solutions. **a:** experimental data for liquid water at T = 27°C from (Downing and Williams 1975) (dashed curves) and for 5 M solution from (Bennouna, Cachet, Lestrade, and Birch 1981) at T = 25°C (dots) Solid curves - calculated spectra of liquid water (curve 1) and of 5 M solution (curve 8). The calculation procedure is described in Sec. 15.4.2.
b: Calculated $\alpha(\nu)$ curves. The concentration C = 0, 0.5, 1, 1.5, 2, 2.5, 3, and 5 mole/l (from top to bottom), T = 30°C.

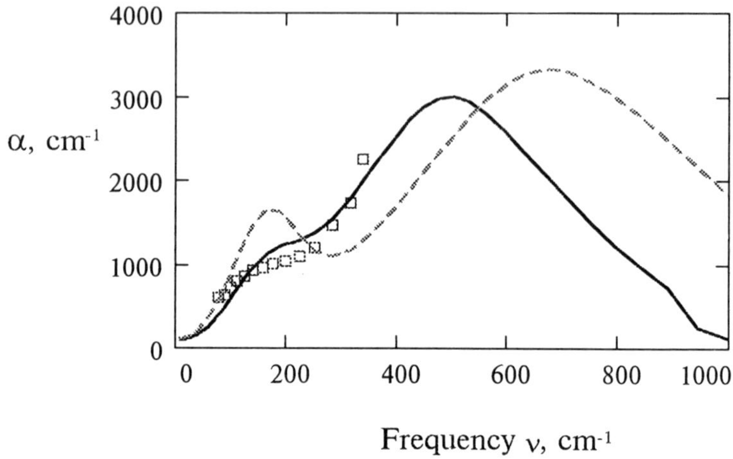

Fig. 2.20 Comparison of calculated (solid curve) and experimental (dots) absorption spectra for 8M NaI-H_2O solution; dashed curve - results of calculation for liquid water. $T = 20^oC$. The method of calculation is described in Sec. 15.5.

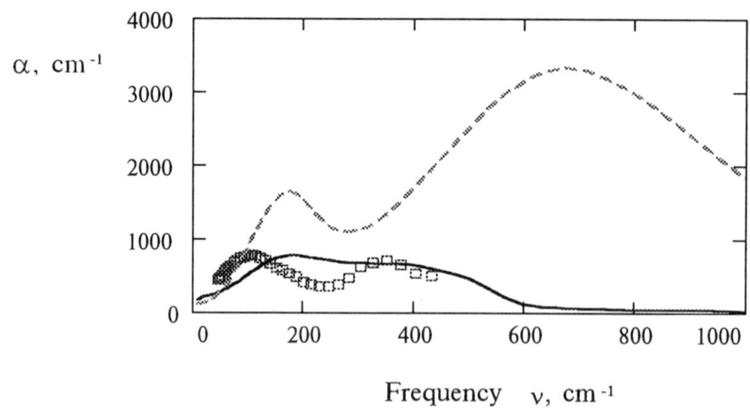

Fig. 2.21 Comparison of calculated (solid line) and experimental (dots) absorption curves for 6M KI-H_2O solution; dashed curve - the result of calculation for liquid water. $T = 20°C$. Detail of this calculation see in Sec. 15.5.

Additionally, the contribution of ions into the low frequency loss $\varepsilon''(\nu)$ was taken into account on the basis of a phenomenological approach. The latter can be justified by the following arguments. Typical spectral dependencies, observed in the FIR range, may be related mainly to the alteration of rotational / librational motion of H_2O molecules within the ionic hydration sheath. Because of their inertia ions do not contribute much to the $\alpha(\nu)$ and $\varepsilon(\nu)$ dependencies in the range 100 - 400 cm^{-1}. A model of molecular rotation within the ionic hydration sheaths can be suggested on the basis of the same approach that was applied above to liquid water (see Sec. 2.4.2).

It will be shown in Chap. 15, that the fitting parameters of the semi-microscopic model are placed in a definite order. This order, being an indicative of the so called positive and negative hydration, is the same, as was previously found from the study of the self-diffusion and other phenomena in electrolyte solutions. Thus, an independent confirmation is found of the idea, that translational and rotational motions are closely related in liquids.

2.6.2 Aqueous Solution of Dimethyl Sulfoxide (DMSO): the Concentration Dependence

The last example of our short review concerns the solution of two polar liquids, viz. aqueous solution of DMSO. The DMSO molecule $H_3C\text{-}S\text{-}CH_3$ is relatively rigid. Being strongly polar ($\mu \approx 4$ D), it replaces H_2O molecules in H-bond network. Liquid DMSO is soluble in water in any proportion; it is an organic (aproton) solvent, in which O or S atoms can form H-bonds with protons or water molecules. Thus, DMSO-H_2O complexes appear in the solution, and dielectric properties of the solution may considerably differ from those of liquid DMSO or H_2O. Fig. 2.22 illustrates experimental dependence of the relaxation time τ_D on the volume-part concentration C_V of DMSO. At small C_V practically all DMSO molecules are bound by water molecules. The relaxation time τ_D of the solution is *substantially longer* in some C_V -range than that in *any* of two pure liquids (water or DMSO).

The low-frequency relaxation spectrum of the solution by no means reduces to the superposition of similar spectra of the solution's components. If C_V is near 1, the relaxation time decreases with the increase of C_V, since almost all H_2O molecules are probably bounded by DMSO. The excess concentration of DMSO molecules corresponds to the "bulk" DMSO fraction whose dielectric properties may differ from those, corresponding to pure liquid DMSO.

Theoretical explanation of these phenomena is given in Sec. 15.4 for the simplified two-fractional semi-microscopic model. Each fraction is described with use of the hybrid model. In one fraction dielectric response of water molecules is found, the free parameters (lifetime and potential welldepth) of the model being modified comparing with those, fitted for liquid water. Another fraction corresponds to DMSO molecules, with τ and U_0, differing from values, fitted for pure solute. The model under consideration describes the recorded $\tau_D(C)$ dependence, shown in Fig. 2.22. Another interesting phenomenon is related to the appearance in the range of millimeter - submillimeter wavelengths of the so called "*absorption deficit*", see Fig. 2.23.

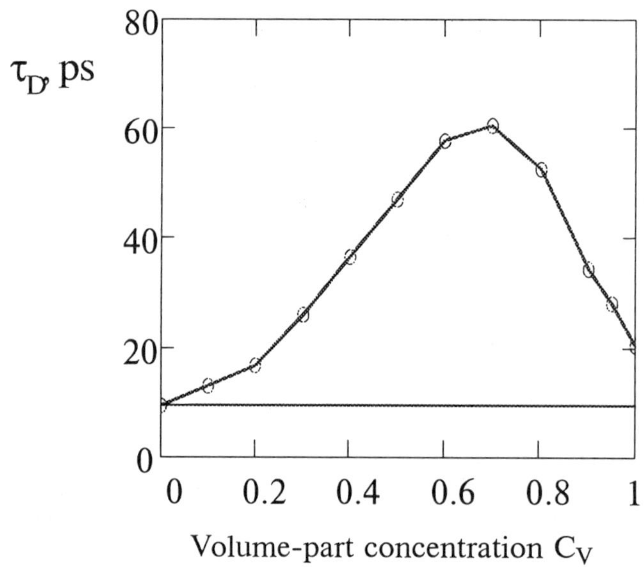

Fig. 2.22 The observed effect of concentration on the relaxation time in DMSO-H_2O solution. From (Shakhparonov and Galiyarova 1980). The horizontal line corresponds to the τ_D value in liquid water. T = 20°C.

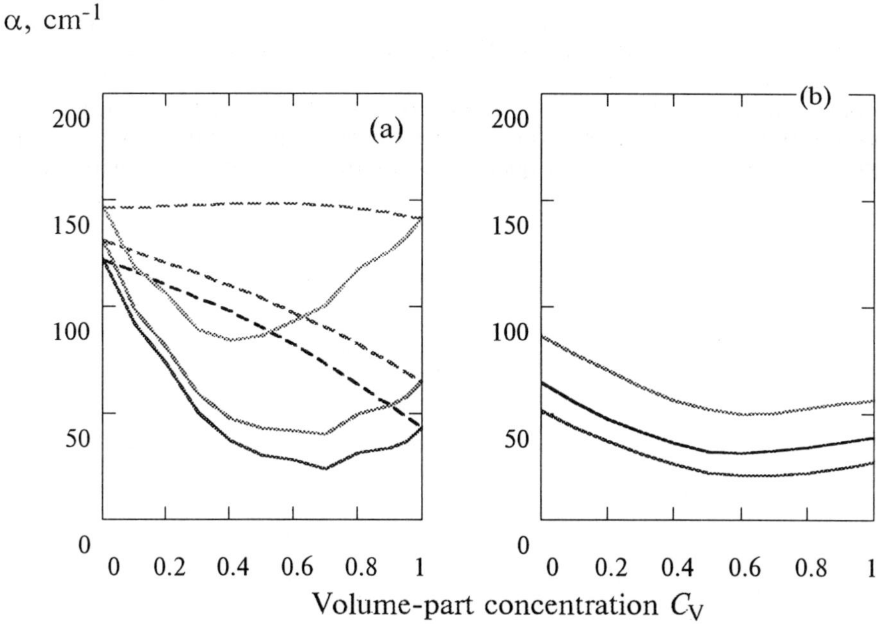

Fig. 2.23 Calculated (a) and observed (b) concentration dependencies of DMSO-water solutions; radiation wavelengths 0.41, 0.69, and 1.1 mm (from top to bottom). Solid / dashed lines in Fig. a - for parameters of the hybrid model, depending or not depending on concentration of DMSO. Details are discussed in Sec. 15.4.

This result confirms the idea, that the *ability of water to absorb microwave radiation decreases, when hydration appears.* This phenomenon is stipulated by the *decrease of rotational mobility of water molecules, bounded by DMSO molecules.* We shall return to this system in the end of Chap. 15.

* * *

In this chapter we have briefly considered dielectric spectra, characteristic for polar fluids. In the next Part II we start a *systematic study* of linear response dielectric theory. In Parts III and IV we consider some molecular models of orientational relaxation. In Part V we apply these models to some of the above examples and to many others. We shall return to the calculation procedures / interpretation of dielectric spectra in some classes of polar fluids. We shall try to give a quantitative or at least qualitative description of observed phenomena.

PART TWO

THE DYNAMIC METHOD

CHAPTER 3
BASIC EQUATIONS AND THEOREMS

3.1 Maxwell's Equations and Wave Equation for a Plane Electromagnetic Wave

Maxwell's equations for electromagnetic field, related to charges in *vacuo* and waves, propagating in a non-magnetic medium, are

$$\mathrm{curl}\boldsymbol{E} = -\frac{1}{c}\frac{\partial \boldsymbol{B}}{\partial t}, \quad \mathrm{curl}\boldsymbol{B} = \frac{1}{c}\left(\frac{\partial \boldsymbol{E}}{\partial t} + 4\pi\boldsymbol{J}\right), \tag{3.1a}$$

$$\mathrm{div}\ \boldsymbol{E} = 4\pi\rho. \tag{3.1b}$$

These equations do not explicitly contain the characteristics of the medium, wherein the electromagnetic wave propagates. Later the charge and current densities ρ and $\boldsymbol{J}$ will be expressed in terms of the dipole moment $\mu(t)$ of a polar particle. The time dependence of the vector $\mu(t)$ denotes reorientation of a particle, the value of this vector is supposed to be independent on t : $\mu = |\mu(t)| \equiv \mathrm{const}$. So, we assume that dipoles are *rigid.* From Eq. 3.1a we obtain the wave equation for the electric field vector $\boldsymbol{E}(t)$:

$$\mathrm{curlcurl}\ \boldsymbol{E} = (\mathrm{grad\ div} - \Delta)\ \boldsymbol{E} = -\frac{1}{c^2}\frac{\partial}{\partial t}\left(\frac{\partial \boldsymbol{E}}{\partial t} + 4\pi\boldsymbol{J}\right). \tag{3.2}$$

For the study of the properties of a polar medium it is convenient to express the field $\boldsymbol{E}$ and current density $\boldsymbol{J}$ in the form of a homogeneous plane wave:

$$\boldsymbol{E}(t) = \mathrm{Re}\left(\hat{\boldsymbol{E}}\exp(i\Phi)\right) = (1/2)\left(\hat{\boldsymbol{E}}\exp(i\Phi) + \hat{\boldsymbol{E}}^*\exp(-i\Phi^*)\right), \quad (3.3a)$$

$$\boldsymbol{J}(t) = \mathrm{Re}\left(\hat{\boldsymbol{J}}\exp(i\Phi)\right) = (1/2)\left(\hat{\boldsymbol{J}}\exp(i\Phi) + \hat{\boldsymbol{J}}^*\exp(-i\Phi^*)\right), \quad (3.3b)$$

where the complex phase is

$$\Phi = \omega t - \boldsymbol{kr}, \quad (3.4)$$

$i = \sqrt{-1}$, $\hat{\boldsymbol{E}}$ and $\hat{\boldsymbol{J}}$ are complex amplitudes, which are independent of $\boldsymbol{r}$. The wave vector $\boldsymbol{k}$ may be represented in the form: $\boldsymbol{k} = (k' - ik'')\boldsymbol{s}$, where $\boldsymbol{s}$ is a unit vector, $s^2 = 1$. The *propagation constant* $k = k' - ik''$, *which is the very object of our study*, is supposed to be complex; * denotes the complex conjugate, thus $k^* = k' + ik''$. For thermodynamic equilibrium $k'' > 0$ (absorbing medium).

Higher harmonics of current density $\boldsymbol{J}(t)$ may be omitted here, because they are important only in the case of a nonlinear medium. So, we assume, that in the medium field $\boldsymbol{E}(t)$ remains monochromatic. Therefore, if $\boldsymbol{J}(t)$ is expanded in a series of harmonics, only the first one is retained.

For a plane wave operator $(\partial/\partial t)\boldsymbol{E}$ in (3.2) can be replaced by $i\omega$, graddiv$\boldsymbol{E}$ - by - $\boldsymbol{k}(\boldsymbol{kE})$, and $\Delta\boldsymbol{E}$ - by $-k^2\boldsymbol{E}$. Substituting Eq. 3.3 into Eq. 3.2, we obtain equation, relating amplitudes $\hat{\boldsymbol{E}}$ and $\hat{\boldsymbol{J}}$:

$$\left(k^2 - \frac{\omega^2}{c^2}\right)\hat{\boldsymbol{E}} - \boldsymbol{k}(\boldsymbol{k}\hat{\boldsymbol{E}}) = -\frac{4\pi i\omega}{c^2}\hat{\boldsymbol{J}}. \quad (3.5a)$$

The Poisson equation (3.1b) takes the form

$$-i\boldsymbol{k}\hat{\boldsymbol{E}} = 4\pi\hat{\rho}. \quad (3.5b)$$

For *transverse electromagnetic waves*, considered in this book, we have $\boldsymbol{k}\perp\hat{\boldsymbol{E}}$, hence $\boldsymbol{k}\hat{\boldsymbol{E}} = 0$. Then the wave equation (3.5a) takes the form

$$\left(k^2 - \frac{\omega^2}{c^2}\right)\hat{\boldsymbol{E}} = -\frac{4\pi i\omega}{c^2}\hat{\boldsymbol{J}}\,. \tag{3.6}$$

Correspondingly, for the complex conjugates we have

$$\left(k^{*2} - \frac{\omega^2}{c^2}\right)\hat{\boldsymbol{E}}^* = \frac{4\pi i\omega}{c^2}\hat{\boldsymbol{J}}^*\,. \tag{3.6a}$$

In Chap. 4 we shall describe the self-consistent approach, in which the r.h.s. of Eq. 3.6a is expressed through the propagation constant k^*. So, we can derive from (3.6a) the *dispersion equation* for $k^*(\omega)$, which describes dielectric properties of the medium. Note, since we use k^* instead of k, positive imaginary part k'', proportional to the absorption coefficient, is $k'' = \mathrm{Im}\,\{k^*\}$.

3.2 The t_0 Theorem

We assume, that in Maxwell's equations the current density $J(t)$ varies periodically in time. This density arises from charges or dipoles moving, respectively, with translational or angular velocities $\mathbf{v}(t)$ or $\Omega(t)$. These velocities can be found as the solution of equations of motion of a particle, affected by radiation field $E(t)$ and conservative field, produced by other particles. Let $Q(t)$ be *complex power*, consumed by one particle at instant t_0; Q corresponds to dynamic variables $\mathbf{v}(t)$ or $\Omega(t)$, pertaining to this particle.

The *dynamic quantities* $\mathbf{v}(t)$, $\Omega(t)$, $Q(t)$ usually are *nonperiodic in time.* In these circumstances it is convenient to employ the t_0 theorem in order to find the mean (over period $2\pi/\omega$) value $\langle\overline{Q}\rangle$ of complex power Q, *which is also averaged over ensemble of N particles, containing in a unit volume*:

$$N\langle\overline{Q}\rangle = \frac{\omega}{2\pi}\int_0^{2\pi/\omega} dt \int Q\,dN\,. \tag{3.7}$$

We assume that instantaneous power $Q(t)$, absorbed by a particle, is an *arbitrary* function of t; Q also depends on phase coordinates Γ, on the lifetime t_v, and on the initial instant t_0. For brevity we write $Q = Q(t, t_0)$. A dipole or a charge, corresponding to fixed t_0, is termed the t_0 *particle*.

Let us choose particles, starting their interaction with periodic electric field $E(t)$, at instants

$$t_0^{(k)} = \left\{ \dots , t_0, t_0 - \frac{2\pi}{\omega}, t_0 - \frac{4\pi}{\omega}, \dots \right\}, \quad k = 1, 2, 3 \dots \tag{3.7a}$$

shifted by the field period $2\pi/\omega$. *An ensemble (3.7a) of t_0 particles is termed the t_0 sequence* (see Fig. 3.1). Since each such particle moves along *the same trajectory*, governed by the same law of motion, power $Q(t, t_0)$ varies in time with the same shift. Consequently,

$$Q\left(t - \frac{2\pi k}{\omega}, t_0 - \frac{2\pi k}{\omega} \right) \equiv Q(t, t_0) \tag{3.8}$$

THE t_0 THEOREM

---

*The complex power $N\langle \overline{Q} \rangle$, defined by Eq. 3.7, that is expended in a unit volume containing N particles by periodic in time field **E(t)**, can be calculated from the following integral*

$$N\langle \overline{Q} \rangle = \frac{\omega}{2\pi} \int_\Gamma \delta N(t_0) \int_{t_0}^{t_0 + t_v} Q(t, t_0)\, dt , \tag{3.9}$$

$$0 \le t_0 \le 2\pi/\omega .$$

Here $\int_\Gamma \delta N(t_0)$ denotes integration over unit volume over all t_0 particles, appearing during *one period.* We also write the normalizing condition

$$\int_\Gamma t_v \, \delta N(t_0) = \frac{2\pi N}{\omega}, \tag{3.10}$$

that follows from Eq. 3.9, if we put $Q \equiv 1$.

REMARK. Unlike $\delta N(t_0)$ in Eq. 3.9, 3.10, dN in Eq. 3.7 denotes *formal* integration over all particles, containing in a unit volume. As it will be shown further, introducing statistical distributions into $\delta N(t_0)$, we can *practically* take integrals (3.9) and (3.10).

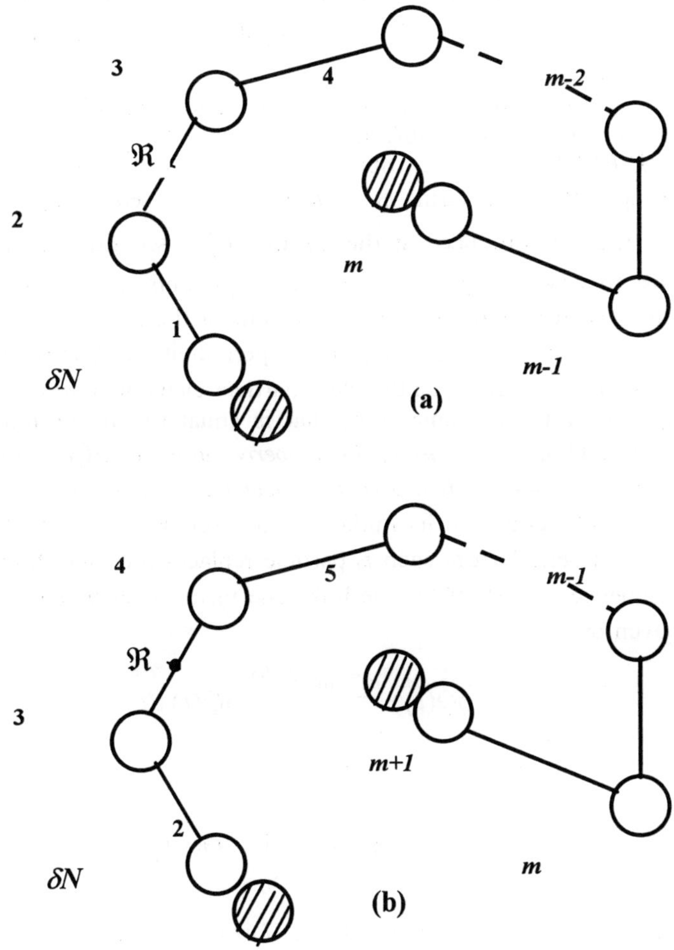

Fig. 3.1 Diagram, pertaining to the proof of the t_0 theorem: the positions of the individual particles of t_0 sequence at instants t_0 (a) and $t_0 + 2\pi/\omega$. (b). Shaded circles denote particles that produce strong instant collisions.

The significance of the t_0 theorem is that it allows one to calculate Fourier amplitudes $\langle \overline{Q_n} \rangle$ in commonly encountered case, when the dynamic variable *Q(t)* is nonperiodic.

Having formulated the theorem, we can, for generality, set Q to be *any additive dynamic variable, satisfying the property* (3.8).

THE PROOF. Let us distribute N particles, containing in a unit volume, into groups, that we term t_0 *sequences.* Each sequence contains $m(t_0)$ particles, that are produced at the instants $t_0^{(k)}$, see Eq. (3.7a), after strong collisions. In view of Fig. 3.1 this sequence presents as if a *chain* of particles, shifted by a field period $2\pi/\omega$. Each element of the chain comprises the same number $\delta N(t_0)$ of particles. Owing to the periodicity of $\boldsymbol{E}(t)$ the trajectories of elements coincide. Consequently, each particle-element passes any point $\mathfrak{R}$ of the trajectory with the same shift, that is equal to one or more periods of alternating field $\boldsymbol{E}(t)$. *Because of this property the power $\delta Q(t)$, absorbed by all elements of the sequence, is a periodic function of t*, since after a time interval $2\pi/\omega$ the chain returns to its initial position (cf. Fig. 3.1 a with Fig. 3.1 b); after the field period $2\pi/\omega$ each t_0 particle replaces the preceding one and the whole sequence as if shifts by one link. The mean (over period $2\pi/\omega$) value of δQ is given by

$$\overline{\delta Q(t_0)} = \frac{\omega}{2\pi} \int_0^{2\pi/\omega} \delta Q(t)\, dt$$

$$= \frac{\omega}{2\pi}\, \delta N(t_0) \sum_{k=1}^{m(t_0)} \int_{t_0}^{t_0+\frac{2\pi}{\omega}} Q(t, t_0^{(k)})\, dt \,. \tag{3.11}$$

Let index $k = 1$ corresponds to the *last* particle, produced at the instant $t_0^{(1)}$ with a new value of its momentum. At the end $t = t_0 + \dfrac{2\pi}{\omega}$ of the integration interval in (3.11) this particle will take up the position, which at the instant $t_0^{(1)}$ was occupied by the *second* (preceding) particle, produced at the instant $t = t_0 - \frac{2\pi}{\omega}$. Therefore, the contribution to (3.11) of the first two terms of the sum over k can be found, if the first term of the sum is integrated over a double-time interval, namely, for $t \in [t_0\ ,\ t_0 + 4\pi/\omega]$, and so on. Thus, instead of Eq. 3.11 we get:

$$\overline{\delta Q(t_0)} = \frac{\omega}{2\pi}\ \delta N(t_0) \int_{t_0}^{t_0+\frac{2\pi}{\omega}} Q(t,t_0)\, dt\ . \tag{3.12}$$

The resultant averaged power will be found, if we integrate this expression over t_0 in the interval $[0,\ 2\pi/\omega])$ and over all points of the phase volume Γ. So we get Eq. 3.9.

THE GENERALIZATION OF THE THEOREM. Let the ensemble average of Q is represented in the form of the Fourier series:

$$\langle Q \rangle = \sum_{n=0}^{\infty} \langle Q_n \rangle \exp(in\omega t).$$

Fourier amplitudes $\langle Q_n \rangle$ can be found in an way, analogous to that, given before:

$$N\langle Q_n \rangle = \frac{\omega}{2\pi} \int_\Gamma \delta N(t_0) \int_{t_0}^{t_0+t_v} Q(t,t_0) \exp(-in\omega t)\, dt\,, \tag{3.13a}$$

$$0 < t_0 < 2\pi/\omega. \tag{3.13b}$$

For $n = 0$ this expression reduces to (3.9): $\langle Q_0 \rangle = \langle \overline{Q} \rangle$.

For $n \neq 0$ one may regard amplitudes $\langle Q_n \rangle$ to describe a non-linear response of a medium to the time-alternating field ***E(t)***.

PHYSICAL RESTRICTIONS. The t_0 theorem is applicable under following assumptions:

(i) radiation field is time-periodic;

(ii) the selected ensemble consists of *numerous* particles. Namely, strong collisions are so frequent, that one may construct the t_0 sequence for *any* t_0 particle.

These assumptions practically disregard field and charge *fluctuations*, which obviously are most important in case of rarefied media, on tails of distributions and / or for very high frequencies.

3.3 The Average Perturbation (AP) Theorem

Let us consider an ensemble of particles *at equilibrium*. This ensemble is characterized by a *single-particle steady state distribution* W over energies H of particles. We shall use normalized quantities: time $\varphi = t/\eta$, energy $h = H/(k_B T)$, generalized (canonical) coordinates s_i and momenta p_i, i being the i th degree of freedom of a particle; $\eta = \sqrt{\dfrac{I}{2k_B T}}$, I being the moment of inertia of a dipole. *Interactions between particles are taken into account only implicitly* through averaged (over time) conservative potential. We remark, that cross-correlation (i.e. correlated motion of particles at equilibrium) will be roughly taken into account only on the final stage of our theory (see Chap. 6), viz. by introducing the Kirkwood correlation factor g (that we consider to be known from the experiment).

First, we regard motion of a single particle; then we find an ensemble average of the quantity δq, where δq is the alteration of the directional cosine q, induced by radiation $\boldsymbol{E(t)}$.

Let

$$s_i \equiv s_i^0(\varphi) \text{ and } p_i \equiv p_i^0(\varphi) \tag{3.14}$$

be the law of motion of a particle in a *conservative* system. This law is determined by the Hamilton equations

$$\dot{p}_i^{\,0} = -\frac{\partial h}{\partial s_i}, \ \dot{s}_i^{\,0} = \frac{\partial h}{\partial p_i}. \tag{3.15}$$

Now we add an external AC force to the interparticle interaction. This force is characterized by the potential $\tilde{V} = \dfrac{\tilde{U}(\varphi)}{k_B T}$, *explicitly* depending on time φ. The following Hamiltonian of a particle corresponds to this non-conservative system:

$$\tilde{h}\ (\varphi, s_i, p_i) \equiv h(s_i, p_i) + \tilde{V}(\varphi, s_i, p_i). \tag{3.16}$$

Instead of (3.15) equations of motion are now given by

$$\dot{p}_i^{\,0} = -\frac{\partial(h+\tilde{v})}{\partial s_i}, \quad \dot{p}_i^{\,0} = \frac{\partial(h+\tilde{v})}{\partial p_i}. \tag{3.17}$$

The external (depending *explicitly* on time (φ)) potential $\tilde{v}$ can be expressed through variables s_i, p_i:

$$\tilde{v} = \tilde{v}\left[\varphi,\, s_i(\varphi),\, p_i(\varphi)\right]. \tag{3.18}$$

In the AC theorem we consider function $q(\varphi) \equiv \mu_E/\mu$. It is convenient to represent the time dependence $q(\varphi)$ as the sum

$$q(\varphi) = q[s_i^{\,0}(\varphi), p_i^{\,0}(\varphi),] + \delta q(\varphi), \tag{3.19}$$

where the first term corresponds to the steady state and the increment is obliged to the external potential $\tilde{v}(\varphi)$.

THE AP THEOREM

Let the linear-response condition

$$|\tilde{v}| << 1 \tag{3.20}$$

be satisfied and let the increment δq be caused by external potential $\tilde{v}(\varphi)$. Then the ensemble average $\langle \delta q \rangle$, defined as

$$\langle \delta q \rangle \equiv \int_\Gamma \delta q W(h)\, d\Gamma, \tag{3.21}$$

is determined by the unperturbed quantities (i.e. at equilibrium):

$$\langle \delta q \rangle = \int_\Gamma q[s_i^{\,0}(\varphi), p_i^{\,0}(\varphi)] \frac{dW}{dh}\, d\Gamma \int_0^\varphi \left(\frac{d}{d\varphi'} - \frac{\partial}{\partial \varphi'}\right) \tilde{v}^{0}(\varphi')\, d\varphi', \tag{3.22}$$

where in distinction from (3.18)

$$\tilde{v}^{0}(\varphi) = v\left[\varphi, s_i^{\,0}(\varphi), p_i^{\,0}(\varphi)\right]. \tag{3.22a}$$

THE PROOF OF THE AP THEOREM. Let us represent, by analogy with (3.19), the time dependence of the generalized coordinates and momenta in the form

$$s_i(\varphi) = s_i^{\,0}(\varphi) + \delta s_i(\varphi), \; p_i(\varphi) = p_i^{\,0}(\varphi) + \delta p_i(\varphi). \tag{3.23}$$

In view of the smallness of $|\tilde{\nabla}|$, Eq. 3.20, we can represent the increment δq as the sum

$$\delta q \approx \sum_i \left(\frac{\partial q}{\partial s_i} \partial s_i + \frac{\partial q}{\partial p_i} \partial p_i \right), \tag{3.24}$$

where derivatives refer to the steady state. Comparing (3.15) and (3.17) with regard of (3.23), we write equations of motion for the increments δs_i and δp_i:

$$\delta \dot{p}_i = -\frac{\partial \tilde{\nabla}}{\partial s_i}, \; \delta \dot{s}_i = \frac{\partial \tilde{\nabla}}{\partial p_i}. \tag{3.25}$$

From these equations and from (3.24) we get:

$$\frac{d}{d\varphi}\delta q \equiv \delta \dot{q} \approx \sum_i \left(\frac{\partial q}{\partial s_i} \delta \dot{s}_i + \frac{\partial q}{\partial p_i} \delta \dot{p}_i \right) = -\sum_i \{q, \tilde{\nabla}\}_{p_i, s_i}, \tag{3.25a}$$

where $\{\bullet\}$ denote the Poisson brackets:

$$\{q, \tilde{\nabla}\}_{p_i, s_i} \equiv \left(\frac{\partial q}{\partial p_i} \frac{\partial \tilde{\nabla}}{\partial s_i} - \frac{\partial q}{\partial s_i} \frac{\partial \tilde{\nabla}}{\partial p_i} \right). \tag{3.25b}$$

Integrating Eq. 3.25a over time, we have

$$\delta q = -\int_0^{\varphi} \sum_i \{q, \tilde{\nabla}(\varphi')\}_{p_i, s_i} d\varphi'. \tag{3.26}$$

Hence, the average (3.22) takes the form

$$\langle \delta q \rangle = -\int_{\Gamma} W(h) d\Gamma \int_0^{\varphi} \sum_i \{q, \tilde{\nabla}(\varphi')\}_{p_i, s_i} d\varphi'. \tag{3.27}$$

If we set initial values $s_i(0)$, $p_i(0)$ as phase variables, then the phase-volume element transforms to

$$d\Gamma = \prod_i ds_i(0)\, dp_i(0)\,. \tag{3.28}$$

According to the Liouville theorem (Landau and Lifshits 1973, p.184) we may equate the integral (3.27) over Γ, with $d\Gamma$ given by (3.28), to the integral over Γ with the element

$$d\Gamma = \prod_i ds_i(\varphi)\, dp_i(\varphi)\,, \tag{3.29}$$

corresponding to a particle, moving in the conservative system, describing by Eq. 3.15. The Poisson bracket (3.25a) in Eq. 3.26 may be transformed as follows:

$$\Big\{q(\varphi),\, \nabla(\varphi')\Big\}_{p_i, s_i} = \frac{\partial}{\partial p_i}\left[q(\varphi)\frac{\partial \nabla(\varphi')}{\partial s_i}\right] - q(\varphi)\frac{\partial^2 \nabla(\varphi')}{\partial s_i\, \partial p_i}$$

$$- \frac{\partial q(\varphi)}{\partial s_i}\frac{\partial \nabla(\varphi')}{\partial p_i} = \frac{\partial}{\partial p_i}\left[q(\varphi)\frac{\partial \nabla(\varphi')}{\partial s_i}\right] - \frac{\partial}{\partial s_i}\left[q(\varphi)\frac{\partial \nabla(\varphi')}{\partial p_i}\right]. \tag{3.30}$$

We substitute this into the phase integral in (3.27), integrate the first term over p_i and the second term over s_i, replace the energy h in the distribution $W(h)$ by the Hamiltonian representation $h(s_i,\ p_i)$. It may be shown that all ensembles, considered further, are "good" in the following sense. Any partial derivative of dynamic variables vanish at the boundaries of the intervals $[s_i]$, $[p_i]$ of integration. Consequently,

$$\int_{[p_i]} \frac{\partial}{\partial p_i} q(\varphi)\frac{\partial \nabla(\varphi')}{\partial s_i} W\big[h(s_i, p_i)\big] dp_i = -\int_{[p_i]} q(\varphi)\frac{\partial \nabla(\varphi')}{\partial s_i}\frac{dW}{dh}\frac{\delta h}{\partial p_i} dp_i\,,$$

$$-\int_{[s_i]} \frac{\partial}{\partial s_i} q(\varphi)\frac{\partial \nabla(\varphi')}{\partial p_i} W\big[h(s_i, p_i)\big] ds_i = \int_{[s_i]} q(\varphi)\frac{\partial \nabla(\varphi')}{\partial p_i}\frac{dW}{dh}\frac{\delta h}{\partial s_i} dp_i\,. \tag{3.31}$$

Taking account of (3.15), we may express the integrands in terms of s_i and p_i. Hence, following (3.30) -(3.31), we write the phase integral in the form

$$\int_\Gamma W(h)\,\{q(\varphi),\tilde{v}(\varphi')\}_{p_i,s_i}\,d\Gamma$$

$$= -\int_\Gamma q(\varphi)\left(\frac{\partial\,\tilde{v}(\varphi')}{\partial s_i}\dot{s}_i + \frac{\partial\,\tilde{v}(\varphi')}{\partial s_i}\dot{p}_i\right)\frac{dW}{dh}\,d\Gamma\,. \qquad (3.32)$$

Summing the integrands in (3.32) over all i degrees of freedom and using the representation (3.18) for the external potential, we get:

$$\int_\Gamma W(h)\sum_i\{q(\varphi),\tilde{v}(\varphi')\}_{p_i,s_i}\,d\Gamma$$

$$= -\int_\Gamma\sum_i q(\varphi)\left(\frac{\partial\tilde{v}(\varphi')}{\partial s_i}\dot{s}_i + \frac{\partial\tilde{v}(\varphi')}{\partial s_i}\dot{p}_i\right)\frac{dW}{dh}\,d\Gamma$$

$$= -\int_\Gamma q(\varphi)\left(\frac{d}{d\varphi} - \frac{\partial}{\partial\varphi}\right)\tilde{v}(\varphi)\frac{dW}{dh}\,d\Gamma\,.$$

Comparing this result with Eq. 3.26, we get Eq. 3.22. Hence, the AP theorem is proved.

REMARK. In Eq. 3.22 the integral over φ' actually presents the variation $(-\delta h)$ of the energy, caused by the external force. Indeed,

$$\delta h = \int_0^\varphi \left(\frac{d}{d\varphi'} - \frac{\partial}{\partial\varphi'}\right)\tilde{v}(\varphi')\,d\varphi'\,. \qquad (3.33)$$

In view of representation (3.18) the operator $\left(\frac{d}{d\varphi} - \frac{\partial}{\partial\varphi}\right)$ means, that we *exclude the contribution due to explicit dependence* $\tilde{v}$ *on* φ in order to obtain the average $\langle\delta q\rangle$. Thus, we regard *only* the change of q, caused by *perturbations of coordinates and momenta* of particles. This becomes clear, if we write the explicit expression for $\tilde{v}$:

$$\tilde{v} = -\frac{\mu E(t)}{k_B T} = -\frac{\mu E_m q(\varphi)}{k_B T}\sin(x\varphi + \gamma), \qquad (3.34)$$

so that

$$\left(\frac{d}{d\varphi} - \frac{\partial}{\partial\varphi}\right) \nabla(\varphi') = -\frac{\mu E_m \dot{q}(\varphi')}{k_B T} \sin(x\varphi' + \gamma).$$

Consequently,

$$\delta h = \frac{\mu E_m}{k_B T} \int_0^{\varphi} \dot{q}(\varphi') \sin(x\varphi' + \gamma)\, d\varphi' . \tag{3.35}$$

The energy of a dipole changes in alternating electric field along with the change of its orientation. In accord with the AP theorem

$$\langle \delta q \rangle = -\int_\Gamma q(\varphi) \frac{dW}{dh} \delta h(\varphi) d\Gamma \equiv -\int_\Gamma q(\varphi)\ \delta W(\varphi) d\Gamma. \tag{3.36}$$

It is seen by comparing this expression with (3.21), that the AP theorem actually means *the transfer of the disturbance* from a dynamic quantity $q(\varphi)$ to the statistical distribution $W(\Gamma)$:

$$\langle \delta q \rangle = \int_\Gamma \delta q\ W(h) d\Gamma = -\int_\Gamma q(\varphi)\ \delta W(\varphi) d\Gamma . \tag{3.37}$$

We remark finally that a more visual formulation of the AP theorem is the following:

$$\left\{ \begin{matrix} \text{the average} \\ \text{perturbation} \\ \langle \delta f(\bullet) \rangle \end{matrix} \right\} = \frac{1}{k_B T} \left\{ \begin{matrix} \text{non-perturbed} \\ \text{function } f^0(t) \end{matrix} \right\} \left\{ \begin{matrix} \text{work, made by} \\ \text{external field } \boldsymbol{E}(t) \\ \text{during time} \\ \text{interval } [0,\ t] \end{matrix} \right\} \tag{3.37a}$$

3.4 Rotation of a Polar Molecule in an Axisymmetric Potential

We shall write down classical equations of motion for linear and symmetric-top molecules and find for few cases their closed-form solutions. Although the final result often can be obtained by simpler derivations comparing with those, given in this section, we deliberately trade this simplicity for the sake to remind the Reader basic relations concerning the problem under consideration.

3.4.1 Linear Molecule: Equations of Motion

Let $T_{\rm kin}$ be kinetic and U potential energy of a molecule in a conservative system, and let Z be the symmetry axis of the potential U. Taking into account only *rotational* degrees of freedom, we employ the spherical coordinate system θ, ψ (see Fig. 3.2). We assume that the potential function does not depend on ψ, i.e. $U = U(\theta)$. The Hamiltonian H and Lagrangian $\mathbb{L}$ of a particle are given by

$$H = T_{\rm kin} + U = \frac{I}{2}\left(\frac{d\theta}{dt}\right)^2 + \frac{I}{2}\left(\frac{d\psi}{dt}\right)^2 \sin^2\theta + U(\theta), \tag{3.38}$$

$$\mathbb{L}\left(\theta, \frac{d\theta}{dt}, \frac{d\psi}{dt}\right) = T_{kin} - U(\theta)$$

$$= \frac{I}{2}\left(\frac{d\theta}{dt}\right)^2 + \frac{I}{2}\left(\frac{d\psi}{dt}\right)^2 \sin^2\theta - U(\theta) \ . \tag{3.39}$$

The angular momenta P_θ and P_ψ correspond to angles θ and ψ:

$$P_\theta = \frac{\partial \boldsymbol{L}}{\partial\left(\frac{d\theta}{dt}\right)} = I\frac{d\theta}{dt}, \ P_\psi = \frac{\partial \boldsymbol{L}}{\partial\left(\frac{d\psi}{dt}\right)} = I\frac{d\psi}{dt}\sin^2\theta . \tag{3.40}$$

In view of Eq. 3.38, 3.39 Lagrange equations take form:

$$\frac{dP_\theta}{dt} = \frac{\partial \boldsymbol{L}}{d\theta}, \ \text{i.e.} \ I\frac{d^2\theta}{dt^2} = I\left(\frac{d\psi}{dt}\right)^2 \sin\theta\cos\theta - \frac{\partial U}{\partial\theta}, \tag{3.41}$$

$$\frac{dP_\psi}{dt} = \frac{\partial \boldsymbol{L}}{d\psi}, \ \text{i.e.} \ I\frac{d}{dt}\left(\frac{d\psi}{dt}\sin^2\theta\right) = 0 . \tag{3.42}$$

It follows from the last equation that the angular momentum P_ψ is time-independent. This property we denote as const(t):

$$P_\psi = I\frac{d\psi}{dt}\sin^2\theta = \text{const}(t). \tag{3.43}$$

Again we use dimensionless variables:

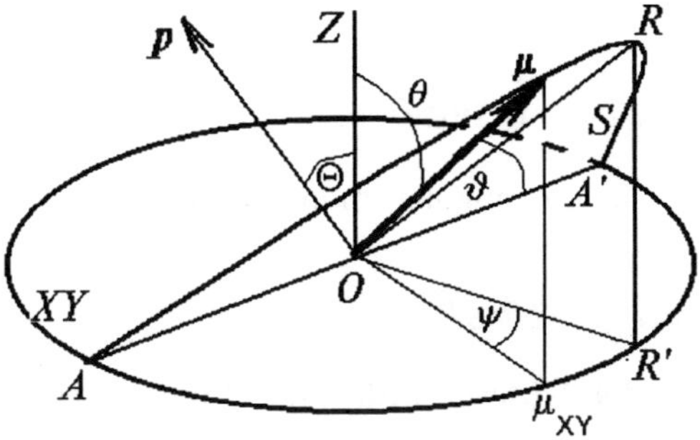

Fig. 3.2 Geometric scheme, pertaining to a linear molecule.

$$h = \frac{H}{k_B T}, \ \Lambda = \frac{\boldsymbol{L}}{k_B T}, \varphi = \frac{t}{\eta}, \dot{\theta} \equiv \frac{d\theta}{d\varphi}, \ \dot{\psi} \equiv \frac{d\psi}{d\varphi}, u = \frac{U}{k_B T}, \tag{3.44}$$

in which quantities, having dimensions of energy or time, are divided respectively by $k_B T$ or $\eta = \sqrt{I/(2k_B T)}$, I being the moment of inertia of a molecule and dots over θ and ψ denote $d/d\varphi$. Thus,

$$h(\theta,\dot{\theta}, \dot{\psi}) = \dot{\theta}^2 + \dot{\psi}^2 \sin^2\theta + u(\theta) \ ; \tag{3.45a}$$

$$\Lambda(\theta,\dot{\theta}, \dot{\psi}) = \dot{\theta}^2 + \dot{\psi}^2 \sin^2\theta - u(\theta) \ . \tag{3.45b}$$

We also introduce dimensionless angular momenta

$$p_\theta(\theta, \dot{\theta}, \dot{\psi}) = \frac{\partial \Lambda}{\partial \dot{\theta}} = 2\dot{\theta}, \ p_\psi(\theta,\dot{\theta}, \dot{\psi}) = \frac{\partial \Lambda}{\partial \dot{\psi}} = 2 \dot{\psi} \sin^2\theta \ . \tag{3.46}$$

The last quantity does not vary with time. To simplify expressions, we introduce l *to designate* half of the normalized angular momentum p_ψ:

$$l = p_\psi / 2 = \dot{\psi} \sin^2\theta \quad . \tag{3.47}$$

It follows from Eq. 3.46, that l is the angular momentum P_ψ, divided by $\sqrt{2Ik_BT}$

$$l \equiv P_\psi \big/ \sqrt{2Ik_BT} \ . \tag{3.47a}$$

We rewrite Lagrange equations (3.41)-(3.42) with use of our dimensionless quantities:

$$\dot{p}_\theta = \frac{\partial\Lambda}{\partial\dot{\theta}}, \text{ i.e. } \frac{d^2\theta}{d\varphi^2} = -\frac{du}{d\theta} + 2\dot{\psi}^2 \sin\theta\cos\theta \quad , \tag{3.48}$$

$$\dot{p}_\psi = \frac{\partial\Lambda}{\partial\dot{\psi}}, \text{ i.e. } l = \dot{\psi} \sin^2\theta = \text{const}(\varphi) \quad . \tag{3.49}$$

3.4.2 Linear Molecule: the Law of Free Rotation

We put $u \equiv 0$ in Eq. 3.48, multiply both sides of this equation by $\sin\theta$ and express ψ through l from (3.49):

$$\frac{d^2\theta}{d\varphi^2}\sin\theta = \frac{l^2\cos\theta}{\sin^2\theta} \quad . \tag{3.50}$$

Let us take

$$q_{||} = \cos\theta \tag{3.51}$$

as a new variable. By differentiation we easily prove the identity of (3.50) and the following equation:

$$\frac{d^2 q_{||}}{d\varphi^2} + \left(\dot{\theta}^2 + \frac{l^2}{\sin^2\theta}\right) q_{||} = 0 \quad .$$

Since u do not depend on θ (we take $u \equiv 0$), then, in view of Eq. 3.45a, the term in (•) presents the Hamiltonian h. Consequently, since in a conservative system h is constant in time, we get the following equation for $q_{\|}$:

$$\frac{d^2 q_{\|}}{d\varphi^2} + h q_{\|} = 0 \quad , \tag{3.52}$$

where

$$h = \dot{\theta}^2 + \frac{l^2}{\sin^2\theta} = \dot{\theta}_0^2 + \frac{l^2}{\sin^2\theta_0} \quad , \tag{3.53}$$

θ_0 refer to some "initial" instant $\varphi = 0$. The solution of Eq. 3.52 is harmonic in time:

$$q_{\|}(\varphi) = q_{\|0} \cos p\varphi + \frac{\dot{q}_{\|0}}{p} \sin p\varphi \, , \tag{3.54a}$$

$$\dot{q}_{\|}(\varphi) = -q_{\|0}\, p \sin p\varphi + \dot{q}_{\|0} \cos p\varphi \, , \tag{3.54b}$$

where $q_{\|0}$ and $\dot{q}_{\|0}$ are values of $q_{\|}$ and $\dot{q}_{\|}$ at $\varphi = 0$, while the normalized angular frequency

$$p = \sqrt{h}\, . \tag{3.55}$$

The law of motion (3.54) corresponds to rotation of a dipole in a plane $\boldsymbol{S}$. The normal $\boldsymbol{N}$ to this plane makes the angle Θ with the axis Z (see Fig. 3.2). Let $\vartheta(\varphi)$ be the angular deviation of a rotor on the plane $\dot{\boldsymbol{S}}$. Projecting the dipole moment μ first on the line AA′ of intersection of planes $\boldsymbol{S}$ and XY and then onto the Z axis, we have:

$$\cos\theta = \cos\vartheta \sin\Theta, \tag{3.56}$$

where

$$\vartheta = p(\varphi + \varphi_0); \; \cos\vartheta = \cos p(\varphi + \varphi_0). \tag{3.57}$$

Equations 3.54 are compatible with (3.56)-(3.57), if

$$\sin\Theta = \sqrt{\cos^2\theta_0 + \left(\dot{\theta}_0^{\,2}\Big/p^2\right)\sin^2\theta_0}\ .$$

Taking account of Eq. 3.53 and 3.55, we make sure that

$$\sin\Theta = \sqrt{1-l^2\big/p^2}\ \text{, hence, } \cos\Theta = l\,/\,p, \tag{3.58}$$

as it should be.

We remark, that $\dot{\psi}$ is the projection of the angular velocity of a dipole onto XY plane. For an ensemble under consideration (spatial rotators) the phase-volume element $d\Gamma$ and the ranges of phase variables are given respectively by

$$d\Gamma = dh\ d\varphi_0\ dl\ d\psi_0; \tag{3.59}$$

$$[0 \le h < \infty],\ [0 \le \varphi_0 \le 2\pi\,/\sqrt{h}\,],\ [0 \le l \le \sqrt{h}],\ [0 \le \psi_0 \le 2\pi]. \tag{3.60}$$

Now we consider the case of axial symmetry. Taking into account Eq. 3.58, we can omit integration over ψ_0 and may replace integration over l by that over Θ in the interval $[0, \pi/2]$:

$$dl = -\ p \sin\Theta\ d\Theta. \tag{3.61}$$

We can omit the minus sign as any constant multiplier. Then

$$d\Gamma = dh\ d\varphi_0\ \sqrt{h}\ \sin\Theta\ d\Theta, \tag{3.62}$$

$$h \in [0, \infty],\ \varphi_0 \in [0, 2\pi/\sqrt{h}],\ \Theta \in [0, \pi/2]. \tag{3.63}$$

The normalizing coefficient C of the Boltzmann (Maxwellian in this case) steady state distribution $W = C\exp(-h)$ can be found from the equation

$$\frac{1}{C} = \int_0^\infty \exp(-h)\,dh \int_0^{2\pi/\sqrt{h}} d\varphi_0 \int_0^{\pi/2} \sqrt{h}\sin\Theta\ d\Theta = 2\pi;\ W = \frac{\exp(-h)}{2\pi}\ . \tag{3.64}$$

Let alternating field be directed along the Z axis. Using (3.55)-(3.57), we find the averaged square of direction cosine:

$$\left\langle q_{\parallel}^2 \right\rangle = \frac{1}{2\pi}\int_0^\infty \exp(-h)\,dh \int_0^{\frac{2\pi}{\sqrt{h}}} \cos^2\!\left(\sqrt{h}\ \varphi\right)d\varphi \int_0^{\pi/2} \sqrt{h}\sin^3\Theta\ d\Theta = \frac{1}{3}. \tag{3.65}$$

The same result is obtained, if amplitude $\boldsymbol{E}_{\mathrm{m}}$ is normal to the Z axis. We suppose that $\boldsymbol{E}_{\mathrm{m}}$ is placed on XY plane and is directed along Y axis. Then

$$q_{\perp}(\varphi) = \rho(\varphi) \cos\left[\psi(\varphi) + \psi_0\right], \tag{3.66a}$$

where

$$\rho(\varphi) = \sqrt{1 - q_{\parallel}^2(\varphi)} \tag{3.66b}$$

and ψ_0 is the angle between Y and the projection OR′ onto the XY plane of the line OR, that is normal to the trace AA′ of the plane $\boldsymbol{S}$. Then

$$\left\langle q_{\perp}^2 \right\rangle = \left(1 - \left\langle q_{\parallel}^2 \right\rangle\right)\left\langle \cos^2(\psi + \psi_0)\right\rangle = \frac{1}{2}\left(1 - \left\langle q_{\parallel}^2 \right\rangle\right) = \frac{1}{3}. \tag{3.66c}$$

Indeed, since the ensemble under consideration is isotropic, *any* in-space direction can be chosen as "the symmetry axis" Z. Hence, the mean value of μ_Z^2 should be the same for any Z. We shall prove further that the spectral function $L(z)$ correspondingly does not depend on the choice of Z.

The mean normalized kinetic energy is given by

$$\langle h \rangle = \frac{1}{2\pi} \int_0^{\infty} h \exp(-h)\, dh \int_0^{2\pi} d\left(\sqrt{h}\, \varphi_0\right) \int_0^{\pi/2} \sin\Theta\, d\Theta = 1 \quad . \tag{3.67}$$

The corresponding non-normalized quantity is then equal to

$$\langle H \rangle = k_B T, \tag{3.68}$$

as it should be for ensemble with two degrees of freedom.

3.4.3 Linear Molecule: Rotation in a Plane

This particular case is widely discussed in this book. The time-dependent angle in a plane will be denoted as ϑ:

$$\left\{\begin{matrix}\text{Libration in space} \\ \text{is characrezied by} \\ \theta(\varphi)\end{matrix}\right\} \to \left\{\begin{matrix}\text{Libration in plane} \\ \text{is characrezied by} \\ \vartheta(\varphi)\end{matrix}\right\} \tag{3.69}$$

Since in this case we set zero angle ψ and angular velocity $\dot{\psi}$, the normalized Hamiltonian is

$$h = \dot{\vartheta}^2 + u(\theta). \tag{3.70}$$

Since $h = \text{const}(\varphi)$, we can find the solution of this equation in quadratures. For a constant u, e.g. for $u \equiv 0$, the law of free rotation in a plane is given by

$$\dot{\vartheta} = \sqrt{h} = \text{const}(\varphi);\ \vartheta = \sqrt{h}\left(\varphi + \varphi_0\right). \tag{3.71}$$

This law of motion is employed in the *planar extended rotational diffusion model.* The phase-volume element is represented as the product of two quantities (cf. with Eq. 3.62):

$$d\Gamma = dh\ d\varphi_0\ ;\ h \in [0, \infty],\ \varphi_0 \in [0, 2\pi/\sqrt{h}]. \tag{3.72}$$

The normalizing coefficient C of the Maxwellian distribution $W = C\exp(-h)$ can be found from the relation

$$\frac{1}{C} = \int_0^\infty \exp(-h)\,dh \int_0^{2\pi/\sqrt{h}} d\varphi_0 = 4\pi \int_0^\infty \exp(-t^2)\,dt = 2\pi^{3/2};$$

$$W = \frac{\exp(-h)}{2\pi^{3/2}}. \tag{3.73}$$

The mean energy of a plane rotator is *half* of that, found above for a linear molecule:

$$\langle h \rangle = \frac{1}{2\pi^{3/2}} \int_0^\infty h\exp(-h)\,dh \int_0^{2\pi/\sqrt{h}} d\varphi_0 = \frac{2}{\sqrt{\pi}} \int_0^\infty t^2 \exp(-t^2)\,dt = \frac{1}{2};$$

$$\langle H \rangle = \frac{k_B T}{2}. \tag{3.74}$$

This distinction, being obliged to the decrease of the number of degrees of freedom, is reflected also in the form of dielectric spectra (see. Sec. 8.1.3).

3.4.4 *Equations of Motion of a Symmetric-Top Molecule*

Now we consider an ensemble with three degrees of freedom. The geometry of the problem is illustrated by Fig. 3.3 a. Here axes X, Y, Z are fixed in space, while the rotating axes X', Y', Z' are fixed to a molecular top.

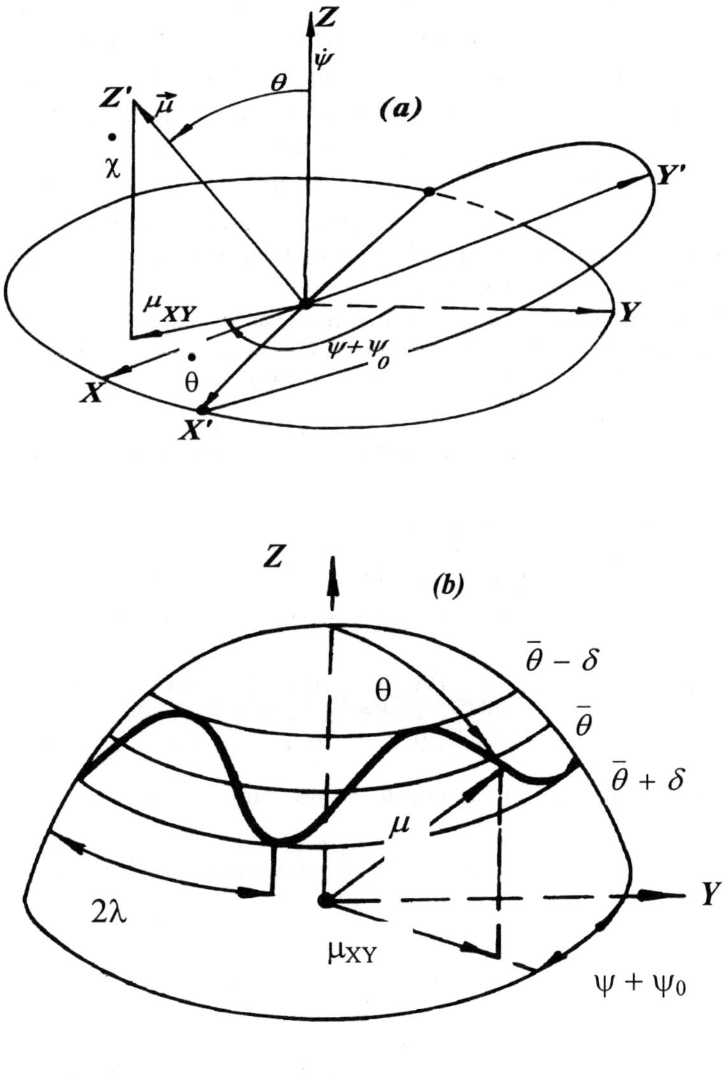

Fig. 3.3 Symmetric top in an axisymmetric potential: coordinate system (a) and hodograph of a dipole moment vector (b).

Let Z' be the symmetry axis of the top, directed along its dipole moment μ. We denote I_c the moment of inertia I_Z around this axis. Two other moments of inertia are equal one to another: $I_{X'} = I_{Y''} = I$. The three Euler angles are used to describe the law of motion of a top (Fig. 3.3). The angles θ and ψ characterize reorientation of a top. The angle χ (it should not be confused with the susceptibility χ) describes rotation of a top around its symmetry axis Z'. The angle ψ describes rotation of a top around the fixed symmetry axis Z of a potential $U = U(\theta)$. Dielectric response is mainly caused by alteration of angles θ and ψ.

Using the same normalized quantities, as in Eq. 3.9 -3.9, we write down standard expressions for the normalized Hamiltonian h and Lagrangian Λ of a top, see e.g. (Landau and Lifshits 1973, p.141):

$$\begin{Bmatrix} h \\ \Lambda \end{Bmatrix} = \{\dot{\theta}^2 + \dot{\psi}^2 \sin^2\theta + \frac{I}{I_c}(\dot{\chi} + \dot{\psi}\cos\theta)^2 \pm u\}.. \tag{3.75}$$

According to Lagrange equations

$$\frac{dp_\theta}{d\varphi} = \frac{\partial\Lambda}{\partial\theta}, \frac{dp_\psi}{d\varphi} = \frac{\partial\Lambda}{\partial\psi}, \frac{dp_\chi}{d\varphi} = \frac{\partial\Lambda}{\partial\chi}, \tag{3.76}$$

$$p_\theta = \frac{\partial\Lambda}{\partial\dot{\theta}}, p_\psi = \frac{\partial\Lambda}{\partial\dot{\psi}}, p_\chi = \frac{\partial\Lambda}{\partial\dot{\chi}}. \tag{3.77}$$

Since $\Lambda = \mathrm{const}(\psi, \chi)$, p_ψ and p_χ do not vary with time φ:

$$p_\psi = 2\dot{\psi}\sin^2\theta + 2\frac{I}{I_c}\left(\dot{\chi} + \dot{\psi}\cos\theta\right)\cos\theta \equiv 2l, \tag{3.78}$$

$$p_\chi = 2\frac{I}{I_c}\left(\dot{\chi} + \dot{\psi}\cos\theta\right) \equiv 2m. \tag{3.79}$$

The projection of the vector μ/μ on the symmetry axis Z is denoted $q_{||}$. Let ρ be the projection $(\mu/\mu)_{XY}$:

$$q_{||}(\varphi) \equiv \cos\theta(\varphi), \tag{3.80}$$

$$\rho^2 = \sin^2\theta = 1 - q_{||}^2. \tag{3.81}$$

The angle between vector μ and axis Y is denoted $\psi + \psi_0$, where ψ_0 is the aforementioned angle at the instant $\varphi = 0$ and ψ is its increment during time φ. The projection of μ/μ on the Y axis we denote $q_\perp(\varphi)$. In accordance with Fig. 3.3 a

$$q_\perp(\varphi) = \rho(\varphi) \cos [\psi(\varphi). + \psi_0]. \tag{3.82}$$

Let projection μ_{XY} turn through 2λ during the period Φ of function $q_\|(\varphi)$. Since λ is generally incommensurable with the angle 2π, $q_\perp(\varphi)$ is nonperiodic (see Fig. 3.3 b). (In Sec. 5.6.1 we represent $\psi(\varphi)$ as the sum of a periodic and secular terms). We have:

$$\dot{\psi} \sin^2 \theta + m \cos\theta = l , \tag{3.83}$$

$$\dot{\psi} = \frac{l - m\cos\theta}{\sin^2 \theta} = \frac{l - m q_\|(\varphi)}{1 - q_\|^2(\varphi)} \tag{3.84}$$

The law of motion $\psi(\varphi)$is thus determined by the integral

$$\psi(\varphi) = \int_0^\varphi \frac{l - m q_\|(\varphi')}{1 - q_\|^2(\varphi')} d\varphi' . \tag{3.85}$$

Now we shall derive equation for $\theta(\varphi)$, describing precession of a top. Taking account of Eq. 3.78, 3.79, we have from the first equations in (3.45)-(3.46):

$$\frac{d^2\theta}{d\varphi^2} \sin\theta = \frac{\cos\theta}{\sin^2\theta} (l - m\cos\theta)^2 - m(l - m\cos\theta) - \frac{u'(\theta)}{2} \sin\theta \quad .(3.86)$$

We can calculate $\theta(\varphi)$ from (3.86), then $\psi(\varphi)$ from (3.85). If need, we may also find $\chi(\varphi)$ from (3.79). However, it follows from (3.80), (3.82), that $q_\|$, $q_\perp$ and consequently dielectric response (viz., the spectral function) does not depend on χ.

Thus, one can find in quadratures the solution of Euler equations for a top, rotating in the potential $u(\theta)$. In Sec.3.4.6 we shall consider a particular case, in which this solution can be expressed in an explicit form.

3.4.5 *Some Symmetry Properties of the Orientational Autocorrelation Function*

The quantity, which is further of interest to us, is the averaged product $q(\varphi)q(0)$, $q(\varphi)$ being the projection of unit vector μ/μ onto $\boldsymbol{E}_{\mathrm{m}}$, i. e. onto the direction of alternating field. More precisely, the spectral function *L(z)*, which is introduced in Chap. 5, is linearly related to the spectrum (= to the Fourier transform) of the dipolar autocorrelation function $\langle q(\varphi)q(0)\rangle$. We shall prove that

---

One can content oneself with the choose of positive or zero angular momenta, viz.

$$l \geq 0 \text{ and } m \geq 0. \tag{3.87}$$

We use Eq. 3.75 for h, Eq. 3.78 - 3.79 for m and l.

For the *longitudinal orientation* of the external field, for which $q_{||} \equiv \cos\theta$, we express the Hamiltonian h in terms of quantities $q_{||}$, l and m:

$$h = \frac{q_{||}^2 + \left(l - mq_{||}\right)^2}{1 - q_{||}^2} + m^2 (b+1) + u(q_{||}), \tag{3.88}$$

where we introduce the form-factor

$$b = (I/I_c) - 1. \tag{3.88a}$$

One can easily see, that the Hamiltonian h and functions $q_{||}(\varphi)$, $q_{||}(0)$ do not alter, if integrals of motion l and m *both* change their sign. On the contrary, the sign of $q_{||}(\varphi)$ and $q_{||}(0)$ changes, if only one of the momenta (l or m) alters its sign:

$$q_{||}(\varphi, l, m) \equiv q_{||}(\varphi, -l, -m); \tag{3.89a}$$

$$q_{||}(\varphi)\,\big|_{l/m>0} = -q_{||}(\varphi)\,\big|_{l/m<0}. \tag{3.89b}$$

Thus, the integral of the product $q_{||}(\varphi)q_{||}(0)$ over l and m remains the same for any sign of the ratio l/m; the statement (3.88) is valid.

The same conclusion is true for the transverse orientation of alternating field. According to Eq. 3.82

$$q_\perp(\varphi)q_\perp(0) = \rho(\varphi)\,\rho(0)\,\cos\left[\psi(\varphi) + \psi_0\right]\cos\psi_0 . \tag{3.90}$$

Averaging over ψ_0 , we find, that

$$\left\langle q_\perp(\varphi)\, q_\perp(0)\right\rangle_{\psi_0} = \frac{1}{2}\,\rho(\varphi)\rho(0)\cos\,\psi(\varphi) \tag{3.91}$$

is an even function of ψ.

On the other hand, taking into account integral (3.85), we find that

$$\psi(\varphi,\ -l,\ -m) \equiv -\ \psi(\varphi,\ l,\ m). \tag{3.92}$$

Thus, the ensemble averaging of product (3.91) yields for negative l and m the same result, as for positive l, m.

If one changes the sign of any of the phase variables, e.g., the sign of m, then, in view of property (3.89b), the integrand in (3.85) remains the same,

$$\psi(l,-m) = \frac{l + mq_{\|}(l,-m)}{1 - q_{\|}^2(l,-m)} = \frac{l - mq_{\|}(l,m)}{1 - q_{\|}^2(l,m)} = \psi(l,m). \tag{3.93}$$

It follows from Eq. 3.91, that integration of (3.90) yields the same result, as for l and m, having the same sign. The rule (3.87) is proved.

This rule will be applied to the symmetric top molecule with the purpose to determine its spectral function in terms of Fourier amplitudes.

THE SQUARE OF THE TOTAL ANGULAR MOMENTUM k^2. According to (3.75) the first summand of Eq. 3.88 is the normalized squares of angular momenta relative to X′ and Y′ axes, which present also kinetic energy components $h_{X'}$ and $h_{Y'}$, corresponding to rotation around these axes:

$$h_{X'} + h_{Y'} = \dot{\theta}^2 + \dot{\psi}^2\,\sin^2\theta = p_{X'}^2 + p_{Y'}^2 . \tag{3.94a}$$

We remark that the physical (with dimension of energy) quantity is given by

$$H_{X'} + H_{Y'} = k_B T\left(h_{X'} + h_{Y'}\right) = \frac{I}{2}\left[\left(\frac{d\theta}{dt}\right)^2 + \left(\frac{d\psi}{dt}\right)^2 \sin^2\theta\right]. \tag{3.94b}$$

However, the coincidence, similar to that, given by Eq. 3.94 a, is *not valid* for the rotation around the symmetry axis Z'.

The kinetic energy of rotation around the symmetry axis Z' of a top is equal to square of normalized angular momentum relative to Z, multiplied by I/I_c. (3.95)

Indeed, the second term in (3.88) is, according to definition (3.88a), given by

$$h_{Z'} = \frac{H_{Z'}}{k_B T} = m^2 (b + 1) \equiv m^2 \frac{I}{I_c} . \tag{3.96}$$

Taking account of Eq. 3.75, 3.88, 3.88 a, we find that the square k^2 of the *total angular momentum* is related as follows to the Hamiltonian h:

$$k^2 = \dot{\theta}^2 + \left(\frac{l - m\cos\theta}{\sin\theta}\right)^2 + m^2 = h - bm^2 - u(\theta) . \tag{3.97}$$

Unlike h, this quantity, k^2, is a function of time because $\theta = \theta(\varphi)$. But for homogeneous medium, e.g. for $u \equiv 0$, the total angular momentum (and its square) becomes constant:

$$k = h - bm^2 = \text{const}(\varphi). \tag{3.98}$$

In the next section we shall show that for a free rotating top

$$p = k = \sqrt{(h - bm^2)} \tag{3.99}$$

is the fundamental angular frequency of precession. The Reader should not mix the quantity $k(\varphi)$, relating to the top, with the propagation constant $k(\omega)$ of a polar fluid.

3.4.6 The Law of Free Rotation of a Symmetric-Top Molecule

We put in (3.75) $u \equiv 0$ and account for that the constant (in time) value of the energy is the Hamiltonian:

$$h = \dot{\theta}^2 + \dot{\psi}^2 \sin^2\theta + \frac{I}{I_c}(\dot{\chi} + \dot{\psi}\cos\theta)^2 . \tag{3.100}$$

We express $\dot{\theta}^2$ from this equation and rewrite Eq. 3.86 in terms of variable $q_{\|}(\varphi) = \cos\theta(\varphi)$:

$$\frac{d^2}{d\varphi^2} q_{\|} + p^2 q_{\|} = ml, \tag{3.101}$$

where we define the constant p^2 as

$$p^2 \equiv h + m^2 (1 - I / I_c). \tag{3.102}$$

If we introduce the formfactor b of a molecule, defined in (3.88a), then we confirm Eq. 3.99:

The angular frequency p of a top is equal to its total angular momentum k. (3.103)

We remark that the energy h is given by

$$h = p^2 + bm^2. \tag{3.104}$$

For $m = 0$ we have a formal transition to the case of a *linear molecule*, since Eq. 3.101 coincides with (3.52) and Eq. 3.99 - with (3.55). Additionally, the top rotates around the symmetry axis Z'. We shall see that dielectric response of a linear molecule can be found, if transition is made to *infinitely prolate* top, viz. if the formfactor $b \to \infty$. In this case the moment of inertia I_c of a top around its symmetry axis is much lesser than two other moments of inertia (these are equal to I). For $b \to \infty$ rotation of a top around Z' is so fast, that dielectric spectra of a molecular ensemble does not reflect this rotation.

Two other important particular cases are: *spherical top* ($b = 0$) and extremely oblate top (*disk*), for which the b-value of is minimal ($b = b_{\min} = -1/2$).

The following general solution satisfies Eq. 3.101:

$$q_{\|}(\varphi) = a \cos [\, p(\varphi + \varphi_0)\,] + ml/p^2, \tag{3.105}$$

where a and φ_0 are two arbitrary constants; the projection of a vector μ on Z may be represented as the sum of a constant and a harmonic (in time) function.

Since we have already introduced three arbitrary constants (generalized momenta) h, l and m, we cannot take a as an independent arbitrary constant. Let us show that

$$a = \sqrt{\left(p^2 - m^2\right)\left(p^2 - l^2\right)} \Big/ p^2 \tag{3.106}$$

and

$$\dot{q}_{\|}^2 + \left(l - m q_{\|}\right)^2 = \left(p^2 - m^2\right)\left(1 - q_{\|}\right)^2 . \tag{3.107}$$

Indeed, we may rewrite Eq. 3.100 in the form

$$h \sin^2\theta = \dot{\theta}^2 \sin^2\theta + \dot{\psi}^2 \sin^4\theta + \frac{I}{I_c} m^2 \sin^2\theta, \tag{3.108}$$

that is,

$$h\left(1 - q_{\|}\right)^2 = \dot{q}_{\|}^2 + \left(l - m q_{\|}\right)^2 + \frac{I}{I_c} m^2 \left(1 - q_{\|}^2\right) .$$

We have got Eq. 3.107. Substitution of Eq. 3.105 into (3.107) shows that the last equation is satisfied, if a is related to constants p, l, and m in accordance with Eq. 3.106.

If we put $m = 0$, then $p = \sqrt{h}$; in this case the solution (3.105) coincides with the law of motion 3.56, 3.58 of a linear molecule:

$$q_{\|}(\varphi) = \cos\theta(\varphi) = \sin\Theta \cos p(\varphi + \varphi_0). \tag{3.109}$$

If the potential u is constant, the change of the Z-axis direction, being related to the alteration of the form of the solution (3.105), should not influence the essential physical properties of molecular ensemble under consideration. Thus,

a) If Z axis is normal to the angular momentum vector, then $l = 0$ and

$$q_{\|}(\varphi) = \sqrt{1 - \frac{m^2}{p^2}} \cos(\varphi + \varphi_0).$$

b) If Z axis coincides with the fixed (in space) constant angular momentum of a freely rotating top, then $|l|$ has maximal possible value ($|l| = p$); the angle θ does not alter being equal to the precession angle $\theta_p = \mathrm{acos}(m/p)$ of the top; $q = m/p$; $a = 0$.

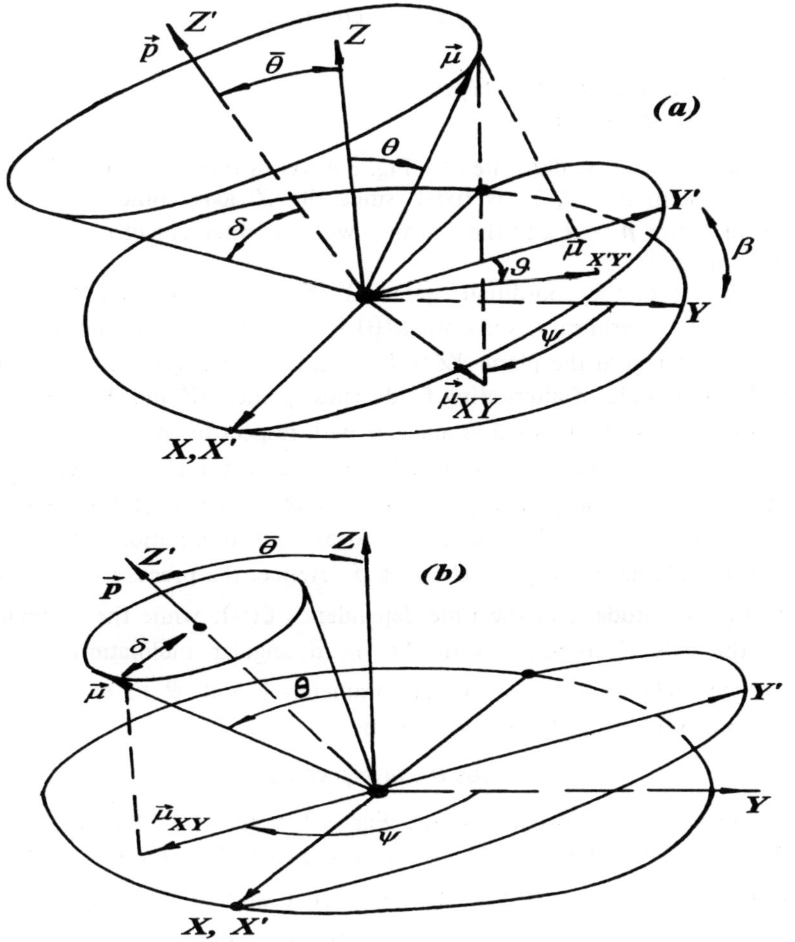

Fig. 3.4 Geometric scheme pertaining to the free rotation of a molecular top. The precession angle δ is greater (a) or less (b) than the angle θ between axes Z and Z'.

(c) The precession also vanishes, if rotation of a top around its symmetry axis occurs with the maximal angular velocity:

$$|k| = p,\ q \equiv l/p = \text{const}(\varphi).$$

3.4.7 Motion of a Symmetric Top: a More Detailed Picture

Free rotation of a top is illustrated by Fig. 3.4. As distinct from Fig. 3.3a, the coordinate system X', Y', Z' is fixed, since the Z' axis coincides with the direction of vector $\boldsymbol{p} = \boldsymbol{k}$ and the X' axis - with the intercept of plane XY and precession plane $X'Y'$.

The choice of the coordinate system X, Y, Z and, consequently, of the pair X', Y' axes, is arbitrary, since for $u(\theta) = 0$ the medium is isotropic. For convenience we regard the plane YZ to be common for angular momentum $\boldsymbol{p}$ and for $\boldsymbol{E}_m$ amplitude of alternating field. Thus, planes YZ and $Y'Z'$ coincide, while Z and Z' axes are inclined at angle $\bar{\theta}$ with respect to each other.

The cone, describing precession of μ around vector $\boldsymbol{p}$, is shown in Fig. 3.4a, 2δ being the opening of the cone. It is evident, that projection $\mu_{X'Y'}$ on precession plane is directed along Y' at two extreme orientations of vector μ relative to the Z axis, namely, at $\theta = \bar{\theta} \pm \delta$. Hence, the precession angle δ is equal to the amplitude a of the time dependence $\theta(\varphi)$, while the inclination of $\boldsymbol{p}$ to the axis Z coincides with the mean angular inclination $\theta = \bar{\theta}$. Consequently, relations of geometrical parameters δ and $\bar{\theta}$ to m and l, the angular momentum projections, are given by

$$l = p\cos\bar{\theta}\ ,\ m = p\cos\delta. \tag{3.110}$$

One may consider two cases (see Fig. 3.4 a and 3.4 b) of precession, depending on relation between $\bar{\theta}$ and δ. In the first (Fig. 3.4 a) $\delta > \bar{\theta}$ and projections m and l of vector $\boldsymbol{p}$ *respectively* on μ and on Z are such, that $m < l$. On the contrary, $\delta < \bar{\theta}$ in Fig. 3.4 b and correspondingly $m > l$.

We now compare the free rotation at $u = 0$ with the motion in the potential $u(\theta)$ (the latter case is illustrated by Fig. 3.3 b). For this comparison it is more convenient to employ Fig. 3.4 b, where angles $\bar{\theta}$ and δ are correlated in the same way, as in Fig. 3.3 b. Under influence of the potential u the free-top precession transforms into *nutation* around the median $\theta = \bar{\theta}$. We may imagine this nutation to be the precession of vector μ around an instant direction of vector $\boldsymbol{k}$. The latter, in turn, executes precession with $\theta = \bar{\theta}$ around axis Z. In contrast, vector $\boldsymbol{k}$ is fixed for a free rotating top.

THE STEADY STATE DISTRIBUTION. Substituting (3.110), (3.106) into (3.105), we write the law of motion $q_{\|}(\varphi)$ and expression (3.104) for the energy h:

$$q_{\|}(\varphi) = \cos\bar{\theta}\,\cos\delta + \sin\bar{\theta}\,\sin\delta\,\cos p(\varphi+\varphi_0), \tag{3.111a}$$

$$\beta\gamma + \cos p(\varphi+\varphi_0)\cdot\sqrt{\left(1-\beta^2\right)\left(1-\gamma^2\right)}, \tag{3.111b}$$

where

$$\gamma = \cos\,\bar{\theta}\,,\ \beta = \cos\,\delta; \tag{3.111c}$$

$$h = p^2\,(1 + b\,\cos^2\delta) = p^2\,(1 + b\,\beta^2\,). \tag{3.112}$$

Since for rotating top $\vartheta = p(\varphi + \varphi_0)$, the increment $d\varphi = d\vartheta/p$. We regard also that the projections $|l|$ and $|m|$ vary from zero to p and that quantities, that are of interest to us, are *even functions* of l and m (see Sec. 3.4.5). Then we can chose, as new phase variables,

$$p,\ \vartheta,\ \beta,\ \gamma\,; \tag{3.113}$$

and may set the following ranges of these variables:

$$h \in ([0, \infty],\ \vartheta \in ([0, 2\pi],\ \beta \in ([0, 1],\ \gamma \in ([0, 1]. \tag{3.114}$$

Hence, we can take the phase-volume element $d\Gamma$ in the form

$$d\Gamma = dh\,d\vartheta\,dl\,dm = |D|\,dp\,d\vartheta\,d\beta\,d\gamma. \tag{3.115}$$

Ttransition to new variables is given by the Jacobian

$$D = \frac{\partial(h,\vartheta,l,m)}{\partial(p,\vartheta,\beta,\gamma)} = \begin{vmatrix} 2p(1+b\beta^2) & -\vartheta p^2 & \gamma & \beta \\ 0 & p^{-1} & 0 & 0 \\ 0 & 0 & p & 0 \\ 2p^2 b\beta^2 & 0 & 0 & p \end{vmatrix} = 2p^2(1 + b\beta^2) - 2p^2 b\beta^2 = 2p^2.$$

Thus,

$$d\Gamma = 2p^2 dp\, d\vartheta\, d\beta\, d\gamma. \tag{3.116}$$

At first, let us calculate the normalizing coefficient C of the steady state distribution function

$$W(\Gamma) = C \exp(-h) = C \exp(-p^2 - \beta m^2) = C \exp\left[-p^2(1 + b\beta^2)\right]: \tag{3.117}$$

$$\frac{1}{C} = 2\int_0^\infty p^2 \exp(-p^2)dp \int_0^1 \exp(-p^2 b\beta^2)d\beta \int_0^{2\pi} \frac{d\vartheta}{p} \int_0^1 d\gamma. \tag{3.118}$$

Integral over β yields

$$\int_0^1 \exp(-p^2 b\beta^2)dp = \sqrt{\frac{\pi}{b}}\,\frac{\operatorname{erf}(p\sqrt{b})}{2p}. \tag{3.119}$$

Noting that

$$2\int_0^\infty p\, \exp(-p^2)\operatorname{erf}(p\sqrt{b})\, dp = \sqrt{\frac{b}{b+1}}, \tag{3.120}$$

we obtain

$$C = \sqrt{\frac{1+b}{\pi^3}}, \text{ so that } W = \sqrt{\frac{1+b}{\pi^3}} \exp\{-p^2(1 + b\,\beta^2)\}. \tag{3.121}$$

We remark that the integral over ϑ of the value of $q_\parallel^2$ can be easily found with use of (3.111a):

$$\int_0^{2\pi} q_\parallel^2\, d\vartheta = \pi\left[2\gamma^2\beta^2 + (1-\beta^2)(1-\beta^2)\right].$$

Averaging of this equation over γ yields $(2/3)\pi$.

Hence, with regard of (3.121) we have for ||orientation:

$$\left\langle q_\parallel^2 \right\rangle = \frac{2}{3}\sqrt{\frac{1+b}{\pi}}\; 2\int_0^\infty p^2 \exp(-p^2)\, dp \int_0^1 \exp(-p^2 b\beta^2)d\beta\;. \tag{3.122}$$

Taking account of (3.119), (3.120), we find:

$$\left\langle q_{\parallel}^{2} \right\rangle = 1/3, \tag{3.122a}$$

as it should be for an isotropic medium. The same value is obtained for $<q_{\perp}^{2}>$:

$$\left\langle q_{\perp}^{2} \right\rangle = (1/2)\,(1 - \left\langle q_{\parallel}^{2} \right\rangle) = 1/3. \tag{3.122b}$$

The results of Sec. 3.4 will be applied in Chap. 7.

3.5 Libration of a Polar Molecule in a Parabolic Potential Well

In this molecular ensemble it is possible to obtain analytic (in terms of elementary functions) expressions for the law of motion. As a result, simple description of dielectric behavior can be obtained for some particular field models. In these we employ the following potentials

$$U(\theta) = C_n U_0(1 - \cos^n\theta) \qquad \text{for spatial motion,} \tag{3.123a}$$

$$U(\vartheta) = U_0(1 - \cos^n\vartheta) \qquad \text{for rotation in a plane,} \tag{3.123b}$$

where usually $n = 1, 2$; θ or ϑ is the angle between the dipole moment vector μ and symmetry axis Z of the potential, U_0 is the potential welldepth. We shall study in detail the constant field (CF) and the cosine squared (CS) models, in which

$$U(\vartheta) = (1/2)\, U_0(1 - \cos\vartheta) \qquad \text{for the CF model} \tag{3.123c}$$

$$U(\vartheta) = U_0(1 - \cos^2\vartheta) \qquad \text{for the CS model} \tag{3.123d}$$

$$U_0 = \max U(\vartheta) \text{ or } U_0 = \max U(\theta) \tag{3.123e}$$

If the well is deep, i.e. if

$$U_0 >> k_B T, \tag{3.124}$$

then almost all dipoles librate *near the bottom* of the well. For such dipoles one can substantially simplify description of motion by introducing the *parabolic potential* (PP) as a limiting case of Eq. 3.123 c, d:

$$U(\theta) = \text{const}_1 - \theta^2\bullet\text{const}_2 \qquad \text{for spatial motion,} \tag{3.125a}$$

$$U(\vartheta) = \text{const}_1 - \vartheta^2\bullet\text{const}_2 \qquad \text{for rotation in a plane,} \tag{3.125b}$$

3.5.1 Harmonic Librations in a Plane

For the PP the torque (restoring force acting on a dipole) is proportional to the angular shift ϑ or θ from the equilibrium position. The PP corresponds to the *elastic bond (EB) approximation* in which it is assumed that a dipole performs *harmonic librations.* Let Ω_m be the angular frequency of librations and p_m - be corresponding normalized frequency:

$$p_m = \eta\Omega_m \text{ , where } \eta = \sqrt{\frac{I}{2k_B T}} . \qquad (3.126a)$$

Further we shall relate p_m (and Ω_m) to other parameters of prescribed molecular models. We shall see in Part IV, that for a more realistic potentials (3.123 c, d)

$$p_m \textit{ is the maximal librational frequency } . \qquad (3.126b)$$

Thus,

$$\lim_{\vartheta \to 0} \frac{U(\vartheta)}{k_B T} = \frac{U_0 \vartheta^2}{k_B T} \begin{Bmatrix} 1/4 \\ 1 \end{Bmatrix} = p_m^2 \vartheta^2 \text{ for the } \begin{Bmatrix} \text{CF} \\ \text{CS} \end{Bmatrix} \text{ models,} \qquad (3.126d)$$

so that

$$p_m = \begin{Bmatrix} 1/4 \\ 1 \end{Bmatrix} \sqrt{\frac{U_0}{k_B T}} \text{ for the } \begin{Bmatrix} \text{CF} \\ \text{CS} \end{Bmatrix} \text{ models,} \qquad (3.126e)$$

Proceeding from Eq. 3.45 to normalized Hamiltonian, we change notation for angular shift ($\theta \to \vartheta$) and omit the term, proportional to $\dot{\psi}^2$, since the tangential velocity, corresponding to rotation around the symmetry axis Z, vanishes for rotation in a plane. Thus, instead of Eq. 3.45 we have

$$h = \dot{\vartheta}^2 + p_m^2 \vartheta^2 , \qquad (3.127)$$

where

$$\cdot \equiv d/d\varphi, \ \varphi \equiv t/\eta. \qquad (3.128)$$

The harmonic (in time φ) law of motion corresponds to Eq. 3.127:

$$\vartheta = \frac{\sqrt{h}}{p_m} \sin [p_m(\varphi + \varphi_0)], \tag{3.129}$$

h and φ_0 being two arbitrary constants. The libration amplitude β of a dipole depends on its energy h:

$$\beta(h) = \sqrt{h}/p_m\,. \tag{3.130a}$$

We shall suppose that the corresponding mean value is small:

$$\bar{\beta} = \langle \beta(h) \rangle = \frac{\langle \sqrt{h} \rangle}{p_m} << 1. \tag{3.130b}$$

For simplicity we shall also assume that:

(i) the energy h of a dipole changes in infinite interval, $h \in [0, \infty]$;

(ii) the angular shift is small,

$$\vartheta << 1. \tag{3.131}$$

Then we can obtain from Eq. 3.129 the time dependencies of longitudinal and transversal direction cosines:

$$q_{\|} = \cos\vartheta = \cos\{\beta \sin p_m(\varphi + \varphi_0)\} \approx 1 - (\beta^2/2) \sin^2 p_m(\varphi + \varphi_0)$$

$$\approx 1 - (\beta^2/4) + (\beta^2/4) \cos 2p_m(\varphi + \varphi_0). \tag{3.132a}$$

$$q_{\perp} = \sin\vartheta \approx \beta \sin p_m(\varphi + \varphi_0). \tag{3.132b}$$

Both quantities, $q_{\perp}(\varphi)$ and $q_{\|}(\varphi)$, are periodic functions of time φ; their periods are respectively

$$\Phi \equiv \Phi_{\|} = \pi/p_m\,, \quad \Phi_{\perp} = 2\pi / p_m. \tag{3.133}$$

Let us find now the normalizing constant C of the distribution function $W = C\exp(-h)$. We have:

$$\frac{1}{C} = \int_0^{\infty} \exp(-h)\,dh \int_0^{\Phi} d\varphi_0 = \frac{\pi}{p_m}, \tag{3.134}$$

so that

$$W = (p_m/\pi)\exp(-h); \tag{3.135}$$

$$\langle q_{\parallel} \rangle = \int_0^{\infty} \left(1 - \frac{\beta^2}{4}\right) \exp(-h)dh = 1 - \frac{\overline{\beta^2}}{4}, \tag{3.135a}$$

$$\langle q_{\parallel}^2 \rangle = \int_0^{\infty} \left(1 - \frac{\beta^2}{2}\right) \exp(-h)dh = 1 - \frac{\overline{\beta^2}}{2} + \frac{3}{32}\overline{\beta^4}, \tag{3.135b}$$

$$\langle q_{\perp}^2 \rangle = \frac{p_m}{\pi} \int_0^{\infty} \exp(-h)dh \int_0^{\pi/p_m} \beta^2 \sin^2 p_m(\varphi + \varphi_0)\, d\varphi_0 \approx \frac{\overline{\beta^2}}{2}, \tag{3.136}$$

where

$$\overline{\beta^2} = \frac{1}{p_m^2} \int_0^{\infty} h \exp(-h)dh = \frac{1}{p_m^2}; \; \overline{\beta^4} = \frac{2}{p_m^4} \int_0^{\infty} h^2 \exp(-h)dh = \frac{2}{p_m^4}. \tag{3.137}$$

Consequently,

$$\langle q_{\parallel} \rangle = 1 - \frac{1}{4p_m^2}; \; \langle q_{\parallel}^2 \rangle = 1 - \frac{1}{2p_m^2} + \frac{3}{16p_m^4}; \; \langle q_{\perp}^2 \rangle = \frac{1}{2p_m^2}. \tag{3.138}$$

For the single potential well (i.e. for the CF model) we also need to find the dispersion of the longitudinal direction cosine. From Eq. 3.138 we have:

$$\Delta = \left\langle \left(q_{\parallel}(\varphi) - \langle q_{\parallel} \rangle \right)^2 \right\rangle = \langle q_{\parallel}^2 \rangle - \langle q_{\parallel} \rangle^2 \tag{3.138a}$$

$$1 - \frac{1}{2p_m^2} + \frac{3}{16p_m^4} - \left(1 - \frac{1}{4p_m^2}\right)^2 = \frac{1}{8p_m^4}. \tag{3.138b}$$

In the other (double well potential) model, where

$$\langle q_{\parallel} \rangle = 0, \tag{3.138c}$$

$$\Delta = \langle q_{\parallel}{}^2 \rangle \approx 1 - 1/(2p_m{}^2). \tag{3.138d}$$

We remark that since the mean energy of a dipole is

$$\langle h \rangle = \int_0^{\infty} h \exp(-h) dh = 1 \quad , \tag{3.138e}$$

the condition (3.130b) of the applicability of the EB approximation reduces to

$$p_{\mathrm{m}} >> 1. \tag{3.139}$$

We shall see later that the criterion (3.130b) and consequently the non-equality (3.139) do not take into account such property as *nonlinear dependence of energy h on libration frequency*. Because of this the EB approximation is not good for *rare collisions* even if condition (3.139) holds.

The mean libration amplitude β is actually small, if Eq. 3.139 is valid. Indeed,

$$\bar{\beta} = \frac{\langle h \rangle}{p_m} = \frac{1}{p_m} \int_0^{\infty} \sqrt{h} \exp(-h) dh = \frac{\sqrt{\pi}}{2p_m}. \tag{3.140}$$

Hence, $\bar{\beta} << 1$, if $p_{\mathrm{m}} >> 1$.

3.5.2 Harmonic Librations in Space

We rewrite Eq. 3.45 for the PP:

$$h = \dot{\theta}^2 + \frac{l^2}{\sin^2 \theta} + p_m^2 \theta^2, \tag{3.141}$$

where the projection l of the angular momentum on the symmetry axis Z is nonzero, unlike for motion in a plane. Assuming again, that the angular shift θ is small,

$$\theta << 1, \tag{3.142}$$

we may rewrite Eq. 3.141 in the form

$$\theta^2 \dot{\theta}^2 = h\theta^2 - l^2 - p_m^2 \theta^4 \equiv \frac{h^2}{4p_m^2} - l^2$$

$$= \frac{h^2}{4p_m^2} - l^2 - \left(p_m \theta^2 - \frac{h}{2p_m} \right)^2. \tag{3.143}$$

Integrating, we have

$$\varphi + \varphi_0 = \frac{1}{2}\int_0^{\theta^2} \left[\frac{h^2}{4p_m^2} - l^2 - \left(p_m\theta^2 - \frac{h}{2p_m}\right)^2\right]^{-1/2} d\theta^2$$

$$= -\frac{1}{2p_m}\operatorname{acos}\left(\frac{p_m\theta^2 - \dfrac{h}{2p_m}}{\sqrt{\dfrac{h^2}{4p_m{}^2} - l^2}}\right).$$

Thus,

$$\theta^2 = \frac{1}{2p_m{}^2}\left\{h + \sqrt{h^2 - 4p_m{}^2 l^2}\,\cos 2p_m(\varphi + \varphi_0)\right\}. \qquad (3.144)$$

For small θ we find, that $q_{\|}(\varphi) = \cos\theta \approx 1 - \theta^2/2$ reduces to

$$q_{\|}(\varphi) \approx 1 - \frac{1}{4p_m{}^2}\left\{h + \sqrt{h^2 - 4p_m{}^2 l^2}\,\cos 2p_m(\varphi + \varphi_0)\right\} \qquad (3.145)$$

Without lack of generality we may put for simplicity $\varphi_0 = 0$ in Eq. 3.144, 3.145. This means that we chose zero instant φ, when θ is equal to the median value $\bar{\theta} = h/(2p^2)$. This assumption does not conflict with employed physical picture of librations.

In accordance with Eq. 3.82

$$q_{\perp}(\varphi) = \rho(\varphi)\cos[\psi(\varphi) + \psi_0] = \sin\theta(\varphi)\cos[\psi(\varphi) + \psi_0]$$

$$\approx \theta(\varphi)\cos[\psi(\varphi) + \psi_0], \qquad (3.145a)$$

where $\theta(\varphi)$ is determined by Eq. 3.144. From Eq. 3.49 we find the angular shift of the projection μ_{XY} of a dipole moment vector on XY plane (see Fig. 3.2):

$$\psi(\varphi) = l\int_0^{\varphi}\frac{d\varphi'}{\sin^2\theta} \approx l\int_0^{\varphi}\frac{d\varphi'}{[\theta(\varphi')]^2}$$

$$\approx 2p_m{}^2 l \int_0^{\varphi} \frac{d\varphi'}{h + \sqrt{h^2 - 4p_m{}^2 l^2} \cos 2p_m\varphi'} .$$

For $a^2 > b^2$ we have:

$$\int \frac{dx}{a + b\cos x} = \frac{2}{\sqrt{a^2 + b^2}} \operatorname{a\,tan}\left(\frac{\sqrt{a^2 - b^2} \tan \frac{x}{2}}{a + b} \right);$$

$$\psi(\varphi) = \operatorname{atan}\left(\frac{2p_m l \tan p_m\varphi}{h + \sqrt{h^2 - 4p_m{}^2 l^2}} \right) . \quad (3.146)$$

During the period $\varphi = \Phi = \pi/p_{\mathrm{m}}$ of the $q_{\|}(\varphi)$ function the angular shift 2λ becomes equal to $\psi(\Phi) = \pi$. Hence,

$$\lambda = \pi/2. \quad (3.146a)$$

We shall see later, that this λ value corresponds to transition *from spatial to planar motion* (for the latter $l = 0$). Thus, the angular momentum is for small shift θ almost perpendicular to the symmetry axis Z, so that l is close to zero. We may expect that dielectric behavior of planar and spatial ensembles should also be close.

Let us find the normalizing constant C of the distribution function:

$$\frac{1}{C} = \int_0^{\infty} \exp(-h) dh \int_0^{l_{\max}} dl \int_0^{\pi/p_m} d\varphi_0 = \frac{\pi}{p_m} \int_0^{\infty} \exp(-h) l_{\max}(h)\, dh . \quad (3.147)$$

We see from Eq. 3.144 that at

$$l = l_{\max}(h) = h/(2p_m) \quad (3.148)$$

the angular shift θ becomes minimal (attains zero value).

Substituting $l_{\max}(h)$ into (3.147), we have

$$\frac{1}{C} = \frac{\pi}{2p_m{}^2} \int_0^{\infty} h \exp(-h) dh = \frac{\pi}{2p_m{}^2};\ W = \frac{2p_m{}^2}{\pi} \exp(-h). \quad (3.149)$$

At last, we find the averaged values:

$$\left\langle q_{\|}\right\rangle \approx 1-\frac{\langle h\rangle}{4p_m^2} \qquad \text{(for SWP);}$$

$$\left\langle q_{\|}^2\right\rangle \equiv \left\langle \cos^2\theta\right\rangle \approx 1-\frac{\langle h\rangle}{2p_m^2}+\frac{\left\langle 3h-4p_m^2 l^2\right\rangle}{32p_m^4}\,. \tag{3.150}$$

In accordance with Eq. 3.82

$$q_{\perp}(\varphi) \equiv \rho(\varphi)\cos[\psi(\varphi)+\psi_0],$$

so that

$$\left\langle q_{\perp}^2\right\rangle \equiv \left\langle \rho^2(\varphi)\cos^2[\psi(\varphi)+\psi_0]\right\rangle = \frac{1}{2}\left\langle \rho^2(\varphi)\right\rangle \tag{3.151}$$

$$\approx \frac{1}{2}\left\langle \sin^2\theta\right\rangle \approx \frac{1}{2}\left\langle \theta^2\right\rangle \approx \frac{\langle h\rangle}{4p_m^2}\,, \tag{3.152}$$

where

$$\langle h\rangle = \frac{2p_m^2}{\pi}\int_0^\infty \frac{h^2}{2p_m}\exp(-h)dh\int_0^{\pi/p_m} d\varphi_0 = \int_0^\infty h^2\exp(-h)\,dh = 2, \tag{3.153}$$

$$\left\langle l^2\right\rangle = \frac{2p_m^2}{\pi}\int_0^\infty \frac{\pi}{p_m}\exp(-h)dh\int_0^{h/(2p_m)} l^2\,dl$$

$$= 2p_m\int_0^\infty \frac{h^3}{24p_m^3}\exp(-h)\,dh = \frac{1}{2p_m^2} \tag{3.154}$$

$$\left\langle h^2\right\rangle = \int_0^\infty h^3\exp(-h)\,dh = 6,\ \left\langle 3h^2-4p^2l^2\right\rangle = 16\,. \tag{3.155}$$

Hence,

$$\left\langle q_{\parallel}\right\rangle = 1 - \frac{1}{2p_m^2} \text{ (for SWP);} \tag{3.156}$$

$$\left\langle q_{\parallel}^2\right\rangle = 1 - \frac{1}{p_m^2} + \frac{1}{2p_m^4}; \tag{3.157}$$

$$\left\langle q_{\perp}^2\right\rangle = \frac{1}{2p_m^2}. \tag{3.158}$$

From Eq.3.144 we find approximate value of the libration amplitude:

$$\beta = \theta_{\max} = \frac{\sqrt{h + \sqrt{h^2 - 4p_m{}^2 l^2}}}{\sqrt{2}p_m}$$

The mean value is

$$\bar{\beta} = \frac{1}{\sqrt{2}p_m} \frac{2p_m^2}{\pi} \frac{\pi}{p_m} \int_0^\infty \exp(-h)\, dh \int_0^{h/(2p_m)} \sqrt{h + \sqrt{h^2 - 4p_m{}^2 l^2}}\, dl$$

Introducing $t = \sqrt{h^2 - 4p_m^2 l^2}$ as a new variable, we transform the integrand: $[\bullet]^{1/2}\, dl = -\frac{t\, dt}{2p_m\sqrt{h-t}}$. Then we find:

$$\bar{\beta} = \frac{1}{\sqrt{2}\, p_m} \int_0^\infty \exp(-h)\, dh \int_0^{h/(2p_m)} \frac{t\, dt}{\sqrt{h-t}}$$

$$= \frac{4/3}{\sqrt{2}\, p_m} \int_0^\infty h^{3/2} \exp(-h)\, dh = \frac{\sqrt{\pi/2}}{p_m}. \tag{3.159}$$

Comparing with Eq. 3.140, we conclude that

$$\bar{\beta}_{\text{spatial}} / \bar{\beta}_{\text{planar}} \approx \sqrt{2}. \tag{3.160}$$

In spatial ensemble the mean librational amplitude of dipoles, calculated for the elastic-bond approximation, is $\sqrt{2}$ *times larger, than in planar ensemble.* (3.161)

The other averages also alter for the plane → space transition.

It should be noted that, if condition (3.142) holds, then all the above obtained results are approximately applicable also for symmetric top molecules, since the effect on spectra of a deep conservative potential $U(\theta)$ exceeds that of the form of a molecule.

For the SWP the dispersion Δ is found from Eq. 3.156 - 3.157:

$$\Delta = 1 - \frac{1}{p_m^2} + \frac{1}{2p_m^4} - \left(1 - \frac{1}{2p_m^2}\right)^2 = \frac{1}{4p_m^4} \qquad \text{for SWP} \tag{3.162}$$

For the double well potential we have:

$$\Delta = \left\langle q_{\parallel}^2 \right\rangle \approx 1 - \frac{1}{p_m^2} \qquad \text{for DWP.} \tag{3.163}$$

The results of Sec. 3.5 will be used in Chap. 10.

CHAPTER 4
RELATION OF SUSCEPTIBILITY TO COMPLEX POWER

4.1 The Dispersion Equation

Let us obtain the frequency dependence of complex propagation constant $k^*(\omega)$, which describes dielectric properties of a polar medium. After scalar multiplication of Eq. 3.6a by $\hat{\boldsymbol{E}}$ we have

$$\left(k^{*2} - \frac{\omega^2}{c^2}\right) \hat{\boldsymbol{E}}\hat{\boldsymbol{E}}^* = 4\pi i \frac{\omega}{c^2} \hat{J}_E^* \hat{E}. \tag{4.1}$$

Hereinafter, the subscript E denotes the projection on $\boldsymbol{E}_m$, the amplitude of alternating field $\boldsymbol{E}(t)$. From representation (3.3) it follows that

$$\overline{J_E(t)\hat{E}\exp(i\Phi^*)} \equiv \frac{\omega}{2\pi}\int_0^{2\pi/\omega} J_E(t)\hat{E}\exp(i\Phi^*)dt = \frac{1}{2}\hat{J}_E^*\hat{E}. \tag{4.2}$$

Substitution of Eq. 4.2 into Eq. 4.1 yields

$$\left(k^{*2} - \frac{\omega^2}{c^2}\right) \hat{\boldsymbol{E}}\hat{\boldsymbol{E}}^* = 8\pi i \frac{\omega}{c^2} \overline{J_E(t)\hat{E}\exp(i\Phi^*)} \tag{4.3}$$

4.1.1 The Dispersion Equation in Terms of Discrete Charges

We regard the point particles to be the source of AC field. The current density $\boldsymbol{J}$ may be defined as the sum of δ-functions:

$$J(t, r) = \sum_m \mathbf{v}_m q_m \delta(\boldsymbol{r} - \boldsymbol{r}_m),$$

where $\boldsymbol{r}_m(t)$ and $\mathbf{v}_m(t)$ are coordinate and velocity of the m th charge. Then we integrate Eq. 4.3 over a *physically small volume* V. Denoting this integral over V as $\langle \bullet \rangle_V$, one has

$$k^{*2} - \frac{\omega^2}{c^2} = 8\pi i \frac{\omega}{c^2} \left\langle \hat{E} \hat{E}^* \right\rangle_V^{-1} \sum_m \overline{Q_m(t)}, \tag{4.4}$$

$$Q_m(t) \equiv q_m \mathbf{v}_m \hat{E} \exp(i\Phi^*), \tag{4.5}$$

where $Q_m(t)$ is the *complex power*, expended by external field $\boldsymbol{E}(t)$ and related to the m th particle. The summation over m is performed inside the volume V.

We can *formally* set $V = 1$ and drop the symbol V, then N means *concentration,* i.e. the number density. $Q_m(t)$ is a nonperiodic function of time, defined for an *arbitrary* law of motion of a charged (or polar) particle. In contrast, it is supposed that the sum in (4.4) varies in time *periodically*, since it describes the response of the medium to the *harmonic* radiation field $\boldsymbol{E}(t)$, see Eq. 3.3.

4.1.2 *The Dispersion Equation in Terms of Integral $\int Q\, dt$*

In this section we shall derive formula for the time average

$$\sum_m \overline{Q_m(t)}. \tag{4.6}$$

It is convenient to express (4.6) in terms of the time-dependent dynamic quantity $Q(t)$, *related to one particle of the medium.* The power Q is expended by AC external field, when a single charged (or polar) particle moves (or rotates) during its lifetime between strong collisions. The variable $Q(t)$ is known, if a solution is found of the equations of motion of a particle under action of *total, i.e. AC external + conservative field.* This section is devoted to the transformation of Eq. 4.6 to a form, that allows one to find a *formal solution* of the problem.

We take a time interval $[t_0, t_0 + t_v]$ between two strong collisions, where the random quantity t_v is the lifetime (the mean lifetime $\tau \equiv \langle t_v \rangle$ will be introduced later). We apply the t_0 theorem in order to find the mean value

(4.6) over period $2\pi/\omega$ of electromagnetic field. The theorem was proved in Chap. 3, Sec. 3.2. According to this theorem, the averaged (over period $2\pi/\omega$) value $\overline{U}$ of any additive quantity $U(t)$, defined as

$$\overline{U} = \frac{\omega}{2\pi} \int_0^{2\pi/\omega} U(t)\, dt, \qquad \text{where} \qquad U(t) = \sum_{m=1}^{N} u_m(t), \tag{4.7}$$

can be transformed to the form

$$\overline{U} = \frac{\omega}{2\pi} \int_\Gamma \delta N(t_0) \int_{t_0}^{t_0+t_v} u(t)\ dt, \quad t_0 \in [0, 2\pi/\omega]. \tag{4.8}$$

where $\delta N(t_0)$ is the number of particles, corresponding to one link of the t_0 sequence (see Fig. 3.2). The instant t_0 fixes an ensemble of t_0 particles. The integration over t_0 is carried for one period of the AC field. The dynamic variable $u(t)$ is represented here as a *continuous* function of time and phase variables.

If we set $u(t) = 1$, then we obtain the normalization condition:

$$\int t_v\, \delta N(t_0) = \frac{2\pi}{\omega} N\,. \tag{4.9}$$

Next, we set $u(t) \equiv Q(t)$. Then (4.8) is replaced by

$$\sum_m \overline{Q_m(t)} \equiv \frac{\omega}{2\pi} \int_\Gamma \delta N(t_0) \int_{t_0}^{t_0+t_v} Q(t)\, dt\,. \tag{4.9a}$$

From Eq. 4.4 we find that the propagation constant k is related to the complex power, averaged over ensemble of particles:

$$k^{*2} - \frac{\omega^2}{c^2} = 4i\frac{\omega^2}{c^2}\left\langle \hat{E}\hat{E}^* \right\rangle^{-1} \int_\Gamma \delta N(t_0) \int_{t_0}^{t_0+t_v} Q(t)\, dt\,. \tag{4.10}$$

For ensemble of *free charges*, in accordance with (4.5),

$$Q(t) \equiv q\mathbf{v}(t)\hat{\mathbf{E}}\exp(i\Phi^*) = q v_E(t)\hat{E}\exp(i\Phi^*)\,. \tag{4.11}$$

For a *polar medium* we transform Eq. 4.15, noting that positive and negative charges of each particle are bound one to another. Thus, we may rewrite Eq. 2.9 as

$$Q_m(t) \equiv q_m\left(\mathbf{v}_m - \mathbf{v}_{-m}\right)\hat{\boldsymbol{E}}\exp(i\Phi^*),$$

where $\mathbf{v}_m$ and $\mathbf{v}_{-m}$ are displacement velocities, relating to $+q$ and $-q$ charges of the m th dipole ($q > 0$). Omitting subscript m, *we regard Q as a continuous function of time t and of phase variables.* By analogy with (4.11) we obtain

$$Q(t) \equiv \exp(i\Phi^*)\,\hat{\boldsymbol{E}}\,\frac{d\boldsymbol{\mu}(t)}{dt} = \hat{E}\,\exp(i\Phi^*)\frac{d\mu_E(t)}{dt} \tag{4.12a}$$

$$\boldsymbol{M} \bullet \Omega\,(t), \tag{4.12b}$$

where

$$\mu(t) = q[\boldsymbol{r}_+(t) - \boldsymbol{r}_-(t)], \tag{4.13a}$$

$$\boldsymbol{M} = \mu(t) \times \hat{\boldsymbol{E}}\exp(i\Phi), \tag{4.13b}$$

$\mu(t)$ is the dipole moment of a single particle, $\Omega(t)$ is the angular velocity of $\mu(t)$ vector and $\boldsymbol{M}$ is the complex torque, that acts on a molecule. To derive Eq. 4.12 we write

$$\hat{\boldsymbol{E}}q\left(\frac{d\boldsymbol{r}_+(t)}{dt} - \frac{d\boldsymbol{r}_-(t)}{dt}\right) \equiv \hat{\boldsymbol{E}}\,\frac{d\boldsymbol{\mu}(t)}{dt}$$

and use two vector identities:

$$\hat{\boldsymbol{E}}\,\frac{d\boldsymbol{\mu}(t)}{dt} \equiv \hat{\boldsymbol{E}}\bullet\left(\Omega(t) \times \mu(t)\right) \equiv \left(\mu(t) \times \hat{\boldsymbol{E}}\right)\bullet\Omega(t). \tag{4.14}$$

It is seen from Eq. 4.12 that Q is actually the *complex power.*

4.2 The Steady State and Induced Distributions

In Eq. 4.10 we have an integral over phase coordinates Γ, initial instants t_0 and lifetimes t_v. Accordingly, we present δN in (4.10) as the product

$$\delta N(t_0) = N\,\tau^{-1}\,W(\Gamma)\,\delta\Gamma\,F(t_0,\Gamma)\,dt_0\,\Psi_v(t_v)\,dt_v. \tag{4.15}$$

Here

$$W = C\exp[-H/(k_B T)] \tag{4.16}$$

is the normalized steady state distribution over energies H of particles, such that

$$\int_\Gamma W(\Gamma)\,d\Gamma = 1; \tag{4.17}$$

$\Psi_v(t_v)$ is the normalized *exponential distribution over lifetimes*:

$$\Psi_v(t_v) = \tau^{-1}\exp(-t_v/\tau), \tag{4.18}$$

where τ is the mean lifetime

$$\tau = \int_{0}^{\infty} t_v \Psi_v(t_v)\,dt_v \tag{4.19}$$

further τ is termed simply *lifetime*. $F(t_0, \Gamma)$ is the distribution, induced by radiation field $\boldsymbol{E}(t)$. For simplification we write $F(t_0, \Gamma) \equiv F(t_0)$. This distribution F accounts for the non-conservative addition $[-\boldsymbol{\mu}\bullet\boldsymbol{E}(t_0)]$ to the energy H, where

---

$$H \equiv \textit{energy of a particle in a conservative system.} \tag{4.20}$$

Then the effect of $\boldsymbol{E}(t)$ field on the distribution function F may be described by the exponent (more rigorous this statement will be proved in Chap. 6)

$$\exp\left[-\frac{H - \boldsymbol{\mu}\bullet\boldsymbol{E}(t_0)}{k_B T}\right] \propto WF. \tag{4.21}$$

Thus the product WF appears in Eq. 4.15.

Referring to Eq. 4.15 - 4.18, one can easily see that the normalization condition (4.9) holds, N/τ in (4.15) being the normalization constant. Substituting (4.11), (4.15) and (4.18) into (4.10), we may relate the propagation constant k to the dynamic variable $d\mu_E/dt$, averaged over the phase volume Γ, initial instants t_0 and lifetimes t_v. *We do not take into account the spatial dispersion.* Indeed, the phase shift of the wave on the *length of a dipole* is so small in the orientational-relaxation region, that

$$\exp(i\Phi^*) \approx \exp(i\omega t), \tag{4.21a}$$

since

$$\left|k^*\left[\boldsymbol{r}(t_v + t_0) - \boldsymbol{r}(t_0)\right]\right| << 1.$$

In this approximation the r.h.s. of Eq. 4.10 does not depend *explicitly* on k, and we derive the following expression (viz. dispersion equation) for the propagation constant k :

$$k^{*2} - \frac{\omega^2}{c^2} = \frac{4i\omega^2 N}{c^2 \hat{E}\hat{E}^*} \int_\Gamma \frac{W(\Gamma)d\Gamma}{\tau} \int_0^{\frac{2\pi}{\omega}} F(t_0)dt_0 \int_0^\infty e^{-\frac{t_v}{\tau}} \frac{dt_v}{\tau} \int_{t_0}^{t_0+t_v} \hat{E}e^{i\omega t} \frac{d\mu_E(t)}{dt} dt. \tag{4.22}$$

4.3 The Effective Susceptibility

Thus, neglecting the spatial dispersion, we represent alternating field at any point of the medium in a form

$$\boldsymbol{E}(t) = \boldsymbol{E}_m \sin \omega t. \tag{4.23}$$

In accord with Eq. 3.3 we have

$$\hat{\boldsymbol{E}} = -i\boldsymbol{E}_m. \tag{4.24}$$

We shall study the dispersion $k(\omega)$ in the frequency range of orientational polarization. For this purpose a microscopic model should be invoked in order to calculate the integral in the r.h.s. of Eq. 4.22. In this integral we do not account for in detail the effect of molecular or electron-shell deformations. The latter are responsible for a dispersion only in the optical range. The corresponding (optical) part of the r.h.s. in (4.22) can be roughly accounted for, if we replace it by the term

$$(\omega^2 / c^2)(n_\infty^2 - 1), \tag{4.25}$$

where n_∞ is the optical refractive index, viz. n_∞ is found on the edge of the dispersion range under consideration; we assume that n_∞ does depend on ω.

The propagation constant $k(\omega)$ may be expressed in terms of the *effective dielectric susceptibility* $\chi(\omega)$, stipulated by reorientation of dipoles:

$$k^{*2} = (\omega^2 / c^2)(4\pi\chi^* + n_\infty^2). \qquad (4.26)$$

Consequently,

$$k^{*2} - (\omega^2 / c^2) = \{\text{integral in Eq. 4.22} + (\omega^2 / c^2)(n_\infty^2 - 1)\}. \qquad (4.27)$$

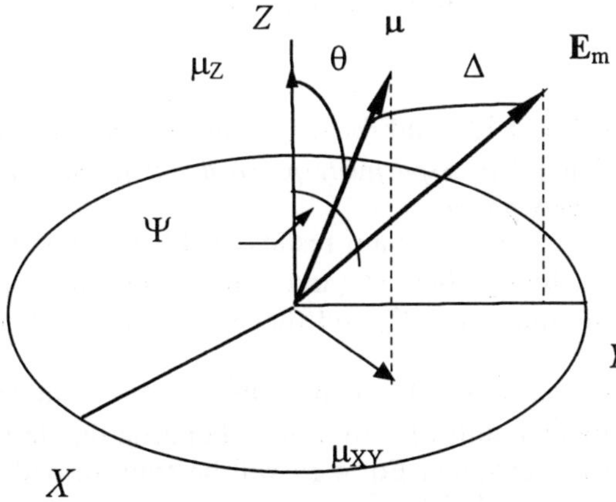

Fig. 4.1 Geometry pertaining to the case of an anisotropic polar medium.

Shifting the time variable t in such a way that $t \to t + t_0$, introducing the «initial» phase of radiation field

$$\gamma = \omega t_0 \qquad (4.28)$$

and using Eq. 4.22 and 4.27, we derive finally the following expression for χ^*:

$$\chi^*(\omega) = \frac{2N}{E_m}\int_0^{2\pi} F(\gamma) e^{i\gamma} \frac{d\gamma}{2\pi} \int_\Gamma \frac{W(\Gamma)d\Gamma}{\tau} \int_0^\infty e^{-\frac{t_v}{\tau}} \frac{dt_v}{\tau} \int_0^{t_v} e^{i\omega t} \frac{d\mu_E(t)}{dt} dt. \qquad (4.29)$$

Here motion of any dipole is considered to be statistically independent on motion of other particles (this is *one-particle approximation*).

We regard *linear response*, in which the result (4.29) is evidently independent on the field amplitude E_m , since for $\mu E_m << k_B T$ the integral is proportional to E_m.

From Eq. 4.26 and 4.29 we can obtain the real (k') and imaginary (k'') components of the propagation constant k. Consequently, the absorption coefficient α and the phase velocity v_{ph} of plane wave, propagating in a polar medium, are given respectively by

$$\alpha(\omega) = 2 \operatorname{Im}(k^*) = 2k'', \tag{4.30a}$$

$$v_{ph} = \omega / k'. \tag{4.30b}$$

Taking account of Eq. 4.12a and 4.29, we find that *imaginary / real part of the effective susceptibility* χ^* *is respectively proportional to the real / imaginary part of the averaged complex power.*

Eq. 4.26 and 4.29 are the main results of this section. They are applicable to an *anisotropic* medium, when χ^* (and consequently k) depend on the angle Ψ between the field amplitude $\boldsymbol{E}_m$ and the symmetry axis Z of the conservative potential $U(\theta)$.

The geometry pertaining to such a disposition is illustrated by Fig. 4.1. Here $\theta(t)$ is the angular shift of a rotating / librating dipole relative Z.

In two following chapters Eq. 4.29 will be transformed to a form, that is more convenient for calculations.

4.4 Susceptibility / Permittivity Relation for the Case of Isotropy

Let us consider a small part of an isotropic medium. In this part the symmetry axis Z of the potential $U(\theta)$ has a definite direction but directions Z of *various parts* are inclined at various angles Ψ to the field direction $\boldsymbol{E}_m$. However, the wave vector $\boldsymbol{k}$ remains the same for any Z. It seems that the polar medium is isotropic, if a local order does not appear on a spatial scale $\delta r \approx \lambda/n$, where λ/n is the radiation wavelength *in the medium* and n is the refractive index.

We may introduce two *scalar* complex quantities for a polar isotropic medium, viz. the permittivity $\varepsilon^* = \varepsilon' + i\,\varepsilon''$ and susceptibility $\chi^* = \chi' + i\chi''$. The Maxwell's equations can be written for such a medium in terms of the permittivity ε:

$$\text{curl}\ \hat{\boldsymbol{E}} = -\frac{i\omega}{c}\hat{\boldsymbol{B}}, \quad \text{curl}\ \hat{\boldsymbol{B}} = i\omega\,\varepsilon(\omega)\,\hat{\boldsymbol{E}}, \tag{4.31}$$

The effect of microscopic charges / dipoles on field is now accounted for by the quantity $\varepsilon(\omega)$. The field equations (4.31) and (3.1) become equivalent, if the equation

$$(i\,\omega\,/\,4\pi)\,(\varepsilon - 1)\,\hat{\boldsymbol{E}} = \hat{\boldsymbol{J}} \tag{4.32a}$$

holds. Instead of the permittivity ε we introduce the complex dielectric susceptibility

$$\chi = \chi' - i\chi'' = (\varepsilon - 1)\,/(4\pi). \tag{4.32b}$$

In terms of complex amplitudes the field equations reduce to

$$i\,\omega\,\chi(\omega)\,\hat{E} = \hat{J}\ , \tag{4.33}$$

while in terms of time-dependent quantities - to

$$\boldsymbol{J}(t) = \text{Re}\left\{\chi(\omega)\,\frac{\partial \boldsymbol{E}(t)}{\partial t}\right\}. \tag{4.34}$$

In view of Eq. 4.33 the complex amplitudes of the current density and electric field are proportional one to another; Eq. 4.33 presents the *differential-form Ohm's law*. The $\boldsymbol{E}(t)$ and $\boldsymbol{J}(t)$ vectors are collinear in an isotropic medium (this statement becomes invalid for an anisotropic medium), but differ in phase, if the loss factor $\chi'' \neq 0$. It follows from (4.32b), that

$$\varepsilon(\omega) = 1 + 4\pi\chi(\omega) \tag{4.35}$$

or equivalently

$$\varepsilon^*(\omega) = 1 + 4\pi\chi^*(\omega). \tag{4.36}$$

In order to accounts for molecular and electron deformations, viz. the contribution to permittivity of the process of atomic / electron polarization, we should replace (4.36) by

$$\varepsilon^*(\omega) = n_\infty^2 + 4\pi\chi^*(\omega). \tag{4.37}$$

The susceptibility $\chi(\omega)$ of an isotropic medium is related to the transverse and longitudinal susceptibilities, which may be calculated from Eq. 4.29.

Up to this point we neglected the distinction between the averaged (macroscopic) radiation field $\boldsymbol{E}(t)$ and the alternating field

$$\boldsymbol{E}_c(t) = \mathrm{Re}\,[\,\hat{E}_c \exp(i\omega t)\,], \tag{4.38}$$

acting on a polar molecule in a dielectric. For an approximate account of this difference we introduce the *internal field correction*. Namely, we find the complex permittivity ε^* in terms of susceptibility χ^* from the quadratic equation

$$\frac{\varepsilon^* - n_\infty^2}{4\pi}\,\frac{2\varepsilon^* + n_\infty^2}{3\varepsilon^*} = \chi^*(\omega). \tag{4.39}$$

This equation relates the "macroscopic" characteristic $\varepsilon^*(\omega)$ of the medium to the "microscopic" quantity $\chi^*(\omega)$.

In this book we shall calculate the susceptibility $\chi^*(\omega)$ for a number of *microscopic models.* After this we can obtain from Eq. 4.39 the dispersion $\varepsilon^*(\omega)$. The comparison of Eq. 4.37 with Eq. 4.39 shows, that the neglect of the internal field correction is equivalent to the following transformation of Eq. 4.39:

$$\frac{2\varepsilon^* + n_\infty^2}{3\varepsilon^*} \to 1. \tag{4.39a}$$

Following Frohlich (1958, p.163) or Hippel (1954, p.64), one can derive Eq. 4.39 by considering the *static* problem. The latter concerns a dipole, placed into the spherical hole, which is embedded into a homogeneous dielectric medium. This solution is approximately valid also for the *low frequency* (Debye) region. In this region one can find some distinction between the dispersion $\varepsilon^*(\nu)$, described (for the same χ^*) by Eq. 4.37 or 4.39. For $\nu \gg (2\pi c\tau_D)^{-1}$ this distinction disappears.

Examples show, that *spectral dependencies $\varepsilon^*(\omega)$, found from* (4.37) *or* (4.39), *are similar and both yield reasonable dielectric spectra.* If we fix the molecular model, then the main difference concerns the *values of fitted parameters of this model.* Therefore,

---

For a qualitative study of a microscopic molecular model it is reasonable to employ simple Eq. 4.37, for which the loss factors ε'' and χ'' are proportional one to another. (4.40)

We remark, that important macroscopic characteristics, related to the complex permittivity, that are determined experimentally, are:

a) The static permittivity $\varepsilon_S \equiv \varepsilon^*(0)$.

b) The loss maximum $\varepsilon''(\omega)$, that we denote ε''_D. This maximum is observed at the "Debye" frequency $\omega_D = \tau_D^{-1}$; hence, the reduced frequency $\nu_D = \omega_D / (2\pi c)$ is related to the Debye relaxation time as $\nu_D = (2\pi c \tau_D)^{-1}$.

Solving quadratic equation (4.39), we can find ε^* in terms of χ^* :

$$\varepsilon^*(\nu) = \frac{1}{4}\left[12\chi^* + n_\infty^2 + \sqrt{\left(12\chi^* + n_\infty^2\right)^2 + 8n_\infty^4}\right]. \tag{4.41}$$

The propagation constant k^* is related to permittivity ε^*:

$$k^{*2} = (\omega/c)^2\,\varepsilon^* = (\omega/c)^2(n + i\kappa), \tag{4.42}$$

where $n + i\kappa$ is the complex refractive index [the Reader should not mix the Greek $\kappa \equiv \text{Im}\{n^*\}$ with the Roman k (propagation constant)!].

In view of Eq. 4.30a we have:

$$n = \text{Re}\,\{\sqrt{\varepsilon^*}\}; \tag{4.43}$$

$$\alpha = 2k'' = \frac{\omega}{c}\frac{\varepsilon''}{n}. \tag{4.44}$$

Hence, taking account of Eq. 4.42, we can calculate the refractive index n and the absorption coefficient α from the following relations

$$n = \frac{1}{\sqrt{2}}\left\{\sqrt{\varepsilon'^2 + \varepsilon''^2} + \varepsilon'\right\}^{1/2}, \tag{4.45}$$

$$\alpha = \frac{2\pi\nu\varepsilon''}{n} = 4\pi\nu\,\text{Im}\{\sqrt{\varepsilon^*}\} \tag{4.46}$$

$$= \frac{4\pi\nu}{\sqrt{2}}\sqrt{\sqrt{\varepsilon'^2 + \varepsilon''^2} - \varepsilon'}\,. \tag{4.47}$$

CHAPTER 5

THE SPECTRAL FUNCTION

5.1 Representation in Terms of Perturbed Motion of a Dipole in Radiation Field

We now introduce the direction cosine q between vectors $\mu(t)$ and E_m, the angle Δ between them varies in time (Fig.5.1a):

$$q(t) \equiv \cos \Delta(t) = \mu(t)/|\mu| = \mu(t)/\mu. \tag{5.1}$$

We also introduce the following dimensionless quantities:

current time	$\varphi = t/\eta$
AC field frequency	$x = \omega\eta$
the mean collision frequency	$y = \eta/\tau$
complex frequency	$z = x + iy$
radiation field amplitude	$v = \mu E_m/(4k_B T)$
concentration of molecules	$G = \mu^2 N/(3k_B T)$
Hamiltonian of a particle in conservative intermolecular field	$h = H/(k_B T)$ (5.2)

The just introduced quantities q and v should not be confused with the point charge q and its velocity v (see Chap. 3 and 4). We denote by dot derivatives with respect to φ (e.g. $\dot{q} \equiv dq/d\varphi$) and rewrite Eq. 4.29 in terms of quantities (5.1) and (5.2):

$$\chi^* = \frac{3G}{2x\mathrm{v}} \int_{\Gamma} W(h)\, y^2 \, d\Gamma \int_0^{2\pi} F(\gamma,\Gamma)\exp(i\gamma)$$

$$\times \frac{d\gamma}{2\pi} \int_0^{\infty} \exp(-y\varphi)\, d\varphi \int_0^{\varphi\bullet} q(\varphi')\exp(i\varphi')\, d\varphi'. \tag{5.3}$$

After integration by parts over φ this expression reduces to

$$\chi^* = \frac{3G}{2x\mathrm{v}} \left\langle yW(\Gamma)F(\gamma,\Gamma)\exp(i\gamma) \int_0^{\infty} q(\varphi)\exp(iz\varphi)\, d\varphi \right\rangle_{\Gamma,\gamma}; \tag{5.4}$$

subscripts Γ and γ denote corresponding averages.

We write:

$$q(\varphi) = q^0(\varphi) + \delta q(\varphi). \tag{5.5}$$

The first term $q^0(\varphi)$ is obliged to rotation of a dipole in a *conservative field*, the second term $\delta q(\varphi)$ - to rotation, *perturbed by radiation*. Further it will be shown that the deviation of the distribution F from the homogeneous one (from $F = 1$) is proportional to a small AC amplitude v, viz. $F - 1 \propto \mathrm{v}$. The same is true for the field-induced perturbation $\delta q(\varphi)$. So, we may put $F = 1$ in the product $F\,\delta q$:

$$\chi^* = \frac{3G}{2x\mathrm{v}} \left\langle yW(\Gamma)F(\gamma,\Gamma)\exp(i\gamma) \int_0^{\infty} q(\varphi)\exp(iz\varphi)\, d\varphi \right.$$

$$\left. + \, yW(\Gamma)\exp(i\gamma) \int_0^{\infty} \delta q(\varphi)\exp(iz\varphi)\, d\varphi \right\rangle_{\Gamma,\gamma}. \tag{5.6}$$

In order to simplify notations we omitted in the first line the symbol "0"; the same simplification is also used later:

$$q^0(\varphi) \rightarrow q(\varphi). \tag{5.6a}$$

We integrate by parts over φ the second term in (5.6). Since $\delta q(\varphi = 0) = 0$, we obtain

$$\chi^* = \frac{3G}{2x\mathrm{v}} \left\langle yW(\Gamma)F(\gamma,\Gamma)\exp(i\gamma) \int_0^{\infty} q(\varphi)\exp(iz\varphi)\, d\varphi \right\rangle_{\Gamma,\gamma} + \frac{z}{x}\,\chi_B^*; \tag{5.7}$$

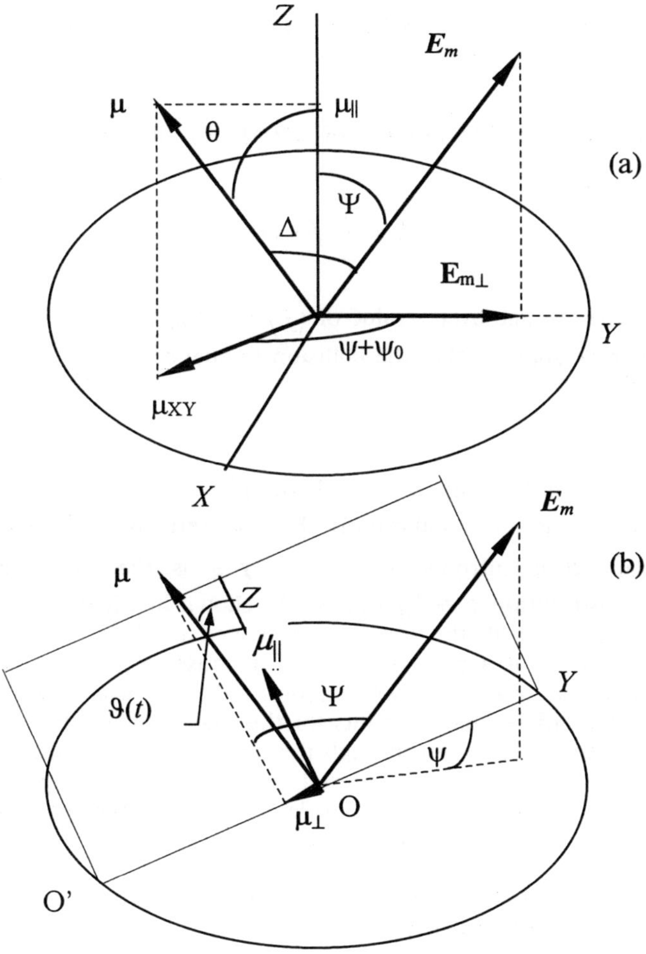

Fig. 5.1 **(a):** Geometric scheme pertaining to rotation of a linear molecule in space. **(b):** Quasi-spatial ensemble, pertaining to the description of an isotropic medium; *YZ* is a plane, in which a dipole **μ** moves.

$$\chi_B^* = 3G\left\langle q^2 \right\rangle L(z), \tag{5.8}$$

where the spectral function (SF) *L(z)* is defined as

$$L(z) = -\frac{1}{2\nu\left\langle q^2 \right\rangle}\left\langle yW(\Gamma)\exp(i\gamma)\int_0^\infty q(\varphi)\exp(iz\varphi)\,d\varphi\right\rangle_{\Gamma,\gamma}, \tag{5.9}$$

$\left\langle q^2 \right\rangle$ being the ensemble average of $q^2(\varphi)$. In the sequel we denote such averaging simply as $\langle \bullet \rangle$:

$$\langle f \rangle \equiv \int_\Gamma f(\Gamma)\,W(\Gamma)d\Gamma. \tag{5.10}$$

We denote the average over γ value of $\langle f \rangle$ as $\langle f \rangle_\gamma$.

Further we suppose that the collision frequency y does not depend on the phase variables Γ:

$$\langle yf \rangle_\gamma \equiv y\langle f \rangle_\gamma \,. \tag{5.11}$$

Unlike its counterpart in Eq. 5.4, $q(\varphi)$ is an *unperturbed* function (i.e. it is calculated at thermal equilibrium). The last term in (5.7) is caused by δq; the subscript B is introduced, because χ_B^* is the susceptibility for the Boltzmann distribution $F = F_B$ (this will be shown in Chap. 6).

Thus, the spectral function *L(z)* is related to dielectric response due to undamped motion of a polar particle in a conservative potential well, such, that the polarization ***P***(*t*) follows radiation field ***E(t)*** ***without a phase lag.*** It is the first term in (5.7), that accounts for this lag. As it will be shown in Chap. 6, this term can be expressed through the same SF *L(z)*. Consequently, the SF *L(z)* can serve as a basic spectral dependence, appropriate for semi-phenomenological models under consideration. The larger part of this book is devoted to calculation of spectral functions.

5.2 Representation in Terms of Undamped Motion for Thermal Equilibrium

The induced part of the dimensionless Hamiltonian, i.e. the alternating potential $\tilde{V}(\varphi)$), is defined as follows:

$$\tilde{V}(\varphi) = -\frac{\mu(t)\bullet E(t)}{k_B T} = -4\mathrm{v}\left[q(\varphi) + \delta q(\varphi)\right]w(\varphi); \tag{5.12}$$

$$w(\varphi) = \sin(x\varphi + \gamma) \tag{5.13}$$

is the time dependence of alternating field. In accord with the average perturbation (AP) theorem (see Chap. 3) the average value

$$\langle \delta q(\varphi)\rangle \equiv \int_{\Gamma} \delta q(\varphi) W(h)\, d\Gamma, \tag{5.14}$$

is related to the work, made by alternating field:

$$\langle \delta q(\varphi)\rangle = -4v \int_{\Gamma} q(\varphi)\frac{dW}{dh} d\Gamma \int_0^{\varphi} w(\varphi')\dot{q}(\varphi')\, d\varphi'. \tag{5.15}$$

We can rewrite Eq. 5.9 in terms of the average $\langle \delta q(\varphi)\rangle$:

$$L(z) = -\frac{1}{2\mathrm{v}\langle q^2\rangle}\int_0^{\infty} y\, d\varphi \int_0^{2\pi} \langle \delta q(\varphi)\rangle \exp(iz\varphi + i\gamma)\, \frac{d\gamma}{2\pi}. \tag{5.16}$$

Substitution (5.15) with account of (5.13) and averaging over γ yields:

$$\langle \exp(i\gamma)\sin(x\varphi+\gamma)\rangle_{\gamma} = -\frac{i}{2}\exp(-ix\varphi'). \tag{5.17}$$

$$L(z) = -\frac{1}{\langle q^2\rangle}\int_{\Gamma}\frac{dW}{dh}\, d\Gamma \int_0^{\infty} y\, q(\varphi)\, e^{iz\varphi}\, d\varphi \int_0^{\varphi} \dot{q}(\varphi')\, e^{-ix\varphi'}\, d\varphi'. \tag{5.18}$$

Hence, dielectric dispersion appears only in an ensemble, characterized by a *non-homogeneous distribution W(h)*, since $L(z) = 0$, if $W = \mathrm{const}(h)$. Further we use the steady state (Boltzmann) distribution, for which

$$W(h) = C\exp(-h). \tag{5.19}$$

In this case $dW(h)/dh = -W$; hence, Eq. 5.18 transforms to

$$L(z) = \frac{\langle \Im(z)\rangle}{\langle q^2\rangle}, \tag{5.20}$$

$$\Im(z) = y\int_0^\infty q(\varphi)\exp(iz\varphi)\,d\varphi\int_0^\varphi q(\varphi')\exp(-ix\varphi')\,d\varphi'. \qquad (5.21)$$

5.2.1 Transition to the Spectrum of the Autocorrelation Function

In order to simplify Eq. 5.21 we chose as one (say, j th) pair of phase variables s_i, p_i the total energy h and integral of motion φ_0, additive to φ, such that $q = q(\varphi + \varphi_0)]$. Then

$$d\Gamma = dh\,d\varphi_0 \prod_{i\neq j} ds_i\,dp_i\,. \qquad (5.22)$$

We consider microscopic models with a periodic time dependence, corresponding to finite motion. Let Φ be the period of such a motion, then

$$q(\varphi + \varphi_0\,) = q(\varphi + \varphi_0 + \Phi). \qquad (5.23)$$

On integrating over ensemble, we take Φ as the interval of integration over φ_0: $\varphi_0 \propto [0,\ \Phi]$. It follows from (5.23) that

$$\int_0^\Phi\left[\dot{q}(\varphi)q(\varphi') + q(\varphi)\dot{q}(\varphi')\right]d\varphi_0 = \int_0^\Phi \frac{d}{d\varphi}\,q(\varphi)q(\varphi')\,d\varphi_0\,. \qquad (5.24)$$

In accord with (5.23) the integral (5.24) vanishes:

$$\int_0^\Phi\left[\dot{q}(\varphi)q(\varphi') + q(\varphi)\dot{q}(\varphi')\right]d\varphi_0 = 0\,. \qquad (5.24a)$$

In particular, putting $\varphi = \varphi'$, we have

$$\int_0^\Phi \dot{q}(\varphi)q(\varphi)\,d\varphi_0 = 0\,. \qquad (5.25)$$

We shall use Eq. 5.24 and 5.25 for the transformation of Eq. 5.20. We integrate (5.21) by parts, first over φ and then over φ':

$$\Im(z) = \frac{iy}{z}\int_0^\infty d\varphi\,\dot{q}(\varphi)e^{ix\varphi}\exp(-y\varphi)\left[\dot{q}(\varphi)q(\varphi)e^{-ix\varphi} + \int_0^\varphi \dot{q}(\varphi')\exp(-ix\varphi')\,d\varphi'\right]$$

$$= \frac{iy}{z}\int_0^\infty \Big[\, 2\dot{q}(\varphi)q(\varphi) - q(0)\,\dot{q}(\varphi)\exp(ix\varphi) + ix\,\dot{q}(\varphi)\exp(ix\varphi)$$

$$\times \int_0^\varphi q(\varphi')\exp(-ix\varphi')\,d\varphi' \,\Big]\exp(-y\varphi)\,d\varphi\ .$$

Noting (5.25), we can omit here the first term. Then

$$\Im(z) = \frac{iy}{z}\int_0^\infty \exp(iz\varphi)\left[\, ix\,\dot{q}(\varphi)\int_0^\varphi q(\varphi')\,\exp(-ix\varphi')\,d\varphi' - \dot{q}(\varphi)\,q(0)\,\right]d\varphi\,. \quad (5.26)$$

Furthermore, in accord with (5.24a)

$$\left\langle \dot{q}(\varphi')q(\varphi)\right\rangle \equiv -\left\langle \dot{q}(\varphi)q(\varphi')\right\rangle. \quad (5.27)$$

Later we perform averaging in Eq. 5.20. Therefore, in view of (5.27) we can rewrite (5.21) in the form

$$\Im(z) = -y\int_0^\infty \dot{q}(\varphi)\exp(iz\varphi)\,d\varphi\,\int_0^\varphi q(\varphi')\,\exp(-ix\varphi')\,d\varphi'\ . \quad (5.28)$$

Comparing this with Eq. 5.26, we find that

$$\Im(z) = \Im(z)\,\frac{x}{z} - \frac{iy}{z}\,q(0)\,\int_0^\infty \dot{q}(\varphi)\exp(iz\varphi)\,d\varphi\ . \quad (5.29)$$

Integrating this by parts and using the identity

$$q(0) = -iz\,q(0)\,\int_0^\infty \exp(iz\varphi)\,d\varphi\,, \quad (5.30)$$

we find finally that

$$\Im(z) = iz\,q(0)\,\int_0^\infty \big[q(\varphi) - q(0)\big]\exp(iz\varphi)\,d\varphi\,. \quad (5.31)$$

Since φ and φ_0 are additive in the law of motion, we may write

$$\dot{q}_0 \equiv \left(\frac{dq}{d\varphi}\right)_{\varphi=\varphi_0} = \frac{dq_0}{d\varphi_0}, \tag{5.31a}$$

where we denote $q(0)$ as q_0. Substituting (5.29) and (5.31) into (5.8) and (5.20), we have as a result three equivalent representations for the SF $L(z)$ (5.8) and for the susceptibility χ_B^* , viz. for

$$\chi_B^* / G = 3\left\langle q^2 \right\rangle L(z) \tag{5.32}$$

(all three will be later employed):

$$\chi_B^* = -3G\left\langle q(0) \int_0^{\infty} \dot{q}(\varphi) \exp(iz\varphi)\, d\varphi \right\rangle, \tag{5.33}$$

$$\chi_B^* = 3G\left\langle \dot{q}(0) \int_0^{\infty} q(\varphi) \exp(iz\varphi)\, d\varphi \right\rangle, \tag{5.34}$$

$$\chi_B^* = 3Giz \left\langle q(0) \int_0^{\infty} \left[\, q(\varphi) - q(0)\, \right] \exp(iz\varphi)\, d\varphi \right\rangle \tag{5.35}$$

The equivalence of (5.33) and (5.34) follows from Eq. 5.24.

Thus, we have expressed the spectral function (SF) $L(z)$ in terms of undamped law of motion $q(\varphi)$ in the potential $U(\theta)$. This means that in contrast to Eq. 5.3 the complex susceptibility is expressed in terms of *unperturbed motion at equilibrium.* Therefore, using (5.33)-(5.35) for calculation eliminates the task of solving equation of motion in presence of external field $\boldsymbol{E}(t)$.

The complex susceptibility is linearly related to the spectrum of the dipolar autocorrelation function (ACF)

$$\left\langle q(0) q(\varphi) \right\rangle . \tag{5.36}$$

Indeed, let us introduce the normalized ACF $C^0(t)$ *for undamped motion* and its spectrum $\tilde{C}^0(\omega)$

$$C^0(t) = \frac{\langle q_0\, q(t)\rangle}{\langle q_0^2 \rangle} = \frac{\langle \mu_E(0)\mu_E(t)\rangle}{\langle \mu_E^0(0) \rangle}, \tag{5.37}$$

$$\tilde{C}^0(\omega) = \int_0^\infty C^0(t)\exp(i\omega t)\,dt. \tag{5.38}$$

We notice that

$$\tilde{C}^0(\omega + i/\tau) = \int_0^\infty C^0(t)\exp[i(\omega + i/\tau)t]dt. \tag{5.39}$$

Then we can rewrite Eq. 5.35 in the form

$$\chi_B^* = 3G\langle q^2\rangle\left[1 + i(\omega + i/\tau)\tilde{C}^0(\omega + i/\tau)\right]. \tag{5.40}$$

Our statement (5.36) is proved. Comparing (5.40) with (5.32) we see that

$$L(z) = 1 + i\,(\omega + i/\tau)\;\tilde{C}^0(\omega + i/\tau)\;. \tag{5.41}$$

The superscript "0" in formulas (5.37) - (5.41) for the ACF is equivalent to the subscript "B" in formula (5.32) for the susceptibility and hence has the same meaning. Namely, the ACF (5.37) is determined by the collision-free reorientation of a dipole, performed during lifetime τ in the conservative potential $U(\theta)$. Consequently, the ACF spectrum $\tilde{C}^0(\omega + i/\tau)$ and the spectral function $L(z)$ characterize collisionless component of reorientation at equilibrium.

In accordance with the representation (5.35) of the SF we may also relate χ^* (and the SF L) to the spectrum $\tilde{D}^0(\omega)$ of the normalized dipole relaxation function $D^0(t)$:

$$\chi_B^* = 3G\left\{\langle q^2\rangle - \langle q\rangle^2\right\}\left[1 + i(\omega + i/\tau)\tilde{D}^0(\omega + i/\tau)\right], \tag{5.42}$$

where

$$D^0(t) = \frac{\left\langle q_0\, q(t) - [q(0)]^2 \right\rangle}{\left\langle [q(0)]^2 \right\rangle - \left\langle q(0) \right\rangle^2} \;, \tag{5.43a}$$

$$\tilde{D}^0(\omega) = \int_0^\infty D^0(t) \exp i\omega t \, dt . \tag{5.43b}$$

The $D^0(t)$ function corresponds, just as $C^0(t)$ does, to the regular (collisionless) motion of a particle under influence of a potential $U(\theta)$.

It is evident from Eq. 5.8, 5.42 and 5.43 that $L(z)$ and χ_B^* is nonzero only for *moving* dipoles / charges; the same is also true for the susceptibility χ^* at an *arbitrary* induced distribution F, see the next chapter. This conclusion may be drawn just, when one finds the average value of $q(0)[q(t) - q(0)]$. The same conclusion follows from our starting point, Eq. 4.10 and 4.11. Indeed, *for dipoles / charges at rest the complex power Q vanish.*

THE AUTOCORRELATION FUNCTION IN THE GENERAL CASE. In the limit $\tau \to \infty$ Eq. 5.40 takes form

$$\chi_B^* = \chi_S\left[1 + i\omega\, \tilde{C}^0(\omega)\right] , \tag{5.44}$$

where

$$\chi_S = \chi_B^*(\omega = 0) = 3G\left\langle q^2 \right\rangle \tag{5.45}$$

is the static susceptibility. Although Eq. 5.44 was obtained for a particular case of a *long-lived rotational motion*, its range of application is quite general. Actually, the susceptibility χ_B^* in Eq. 5.44 we may substitute by χ^*. The first (χ_B^*) is determined for the Boltzmann induced distribution F_B, the second (χ^*) - for an *arbitrary induced distribution F.* Correspondingly, we may substitute the specific ACF spectrum $\tilde{C}^0(\omega)$, determined for "undamped" rotation, by $\tilde{C}(\omega)$*:*

$$\chi^*(\omega) = \chi_S\left[1 + i\omega\, \tilde{C}(\omega)\right] . \tag{5.46}$$

Consequently, this relation is valid also, when strong "collisions" are taken into account. Indeed, the notions "instantaneous collisions" and "average lifetime τ of rotational motion" are the specific terms, appropriate to the dynamic method (we have introduced the quantity τ and the term "strong collisions" in order to avoid a detailed stochastic description of reorientation process).

Thus Eq. 5.46 formulates the principle of a linear-reaction theory:

The complex susceptibility $\chi^*(\omega)$ is linearly related to the spectrum $\tilde{C}(\omega)$ of the dipolar autocorrelation function $C(t)$. (5.47)

However, the question arises, *how to calculate* this spectrum $\tilde{C}(\omega)$ for an appropriate molecular model. *The dynamic method actually gives us such a possibility.* Indeed, we can find *first* $\chi^*(\omega)$ and only then $\tilde{C}(\omega)$ and $C(t)$. This scheme will be realized at the end of Chap. 6. We picture it as a 5-step procedure:

Step 1	**Step 2**	**Step 3**
Consideration of motion of a dipole in a conservative potential $U(\theta)$; determination of the law of motion $q(\varphi)$	Determination of the spectrum $\chi_B^*(\omega)$ from Eq. 5.33. We take $F = F_B$ at this stage.	Determination of the spectrum $\chi^*(\omega)$ for some collision model (at $F \neq F_B$.

Step 4	**Step 5**
Determination of the ACF spectrum from the equation $\tilde{C}(\omega) = \frac{1}{i\omega\chi_s}\left[\chi^*(\omega) - \chi_s\right]$	Calculation of the ACF $C(t)$ using the inverse Fourier transform.

5.3 Relation to the Direction of Alternating External Field

5.3.1 Planar and Spatial Ensembles

Microscopic motion, accounting for intermolecular interactions, is determined by axially symmetric potential $U(\theta)$. It is supposed that the torque, which acts on a dipole due to other molecules, depends on the angular shift θ of the dipole relative to some physically distinguished direction Z (viz. on the symmetry axis). If this axis is fixed relative to laboratory coordinate system, the dipolar medium is actually anisotropic, like a liquid crystal of a certain class. We shall introduce the angle Ψ between Z and the amplitude $\boldsymbol{E}_{\mathrm{m}}$ of external alternating field $\boldsymbol{E}(t)$, see Fig. 5.1.

In an *anisotropic medium* the angle Ψ is fixed, so the SF L depends on Ψ: $L = L(z, \Psi)$. To find this dependence, one must express $q(\varphi)$ in Eq. 5.33 - 5.35 through the direction cosines $q_{\|}$ and $q_{\perp}$. These are the projections of a dipole moment μ respectively on the symmetry axis Z and on axis Y. The latter coincides with the $\boldsymbol{E}_{\mathrm{m}}$ projection on the XY plane (see Fig. 5.1). Applying the rule of scalar multiplication, we write

$$\mu \bullet \boldsymbol{E}_{\mathrm{m}} = \mu E_m q = \mu E_m \,(q_{\|} \cos \Psi + q_{\perp} . \sin \Psi).$$

Thus,

$$q = q_{\|} \cos \Psi + q_{\perp} . \sin \Psi. \tag{5.48}$$

It will be shown later that

$$\langle q_{\|} \, q_{\perp} \rangle = 0. \tag{5.49}$$

Thus, substituting (5.48) into (5.35), we derive formulas for the SF and susceptibility χ^* (for the case of Boltzmann distribution F_{B}):

$$\overset{*}{\chi}_B(z, \Psi) = 3G\langle q^2 \rangle L(z,\Psi) = \overset{*}{\chi_{\|}}(z) \cos^2 \Psi + \overset{*}{\chi_{\perp}}(z) \sin^2 \Psi \quad ,(5.50)$$

where

$$\overset{*}{\chi}_{\| B}(z) = 3iGz \left\langle q_{\|0} \int_0^{\infty} \left(q_{\|} - q_{\|0}\right) \exp(iz\varphi)\, d\varphi \right\rangle \tag{5.51a}$$

$$\overset{*}{\chi}_{\perp B}(z) = 3iGz \left\langle q_{\perp 0} \int_0^{\infty} \left(q_{\perp} - q_{\perp 0}\right) \exp(iz\varphi)\, d\varphi \right\rangle \tag{5.51b}$$

The last two quantities, $\overset{*}{\chi_{\|}}$ and $\overset{*}{\chi_{\perp}}$, may be termed the *longitudinal and transverse effective susceptibilities*. Indeed, $\chi^* = \overset{*}{\chi_{\|}}$ at $\Psi = 0$ (for $\boldsymbol{E}_{\mathrm{m}}$ along Z) and $\chi^* = \overset{*}{\chi_{\perp}}$ at $\Psi = \pi/2$ (for $\boldsymbol{E}_{\mathrm{m}}$ normal to Z).

We shall now introduce, by analogy with Eq. 5.50, two specific spectral functions $L_{\|}(z)$ and $L_{\perp}(z)$, which fully characterize dielectric response of an anisotropic medium:

$$\overset{*}{\chi}_{\|B}(z) = 3G\left\langle q_{\|}^{2}\right\rangle L_{\|}(z);\ \overset{*}{\chi}_{\perp B}(z) = 3G\left\langle q_{\perp}^{2}\right\rangle L_{\perp}(z). \qquad (5.52)$$

Later these SFs will be obtained for a number of molecular models.

Let the angle Ψ be fixed; we relate the SF $L(z,\Psi)$ to $L_{\|}(z)$ and $L_{\perp}(z)$. In accord with Eq. 5.48 the mean values $\left\langle q_{\|}^{2}\right\rangle$ and $\left\langle q_{\perp}^{2}\right\rangle$ are related to $\left\langle q^{2}\right\rangle$ in the same manner, as χ^* does to $\overset{*}{\chi}_{\|}$ and $\overset{*}{\chi}_{\perp}$:

$$\left\langle q^{2}\right\rangle = \left\langle q_{\|}^{2}\right\rangle \cos^{2}\Psi + \left\langle q_{\perp}^{2}\right\rangle \sin^{2}\Psi . \qquad (5.53)$$

Taking this into account, we find from Eq. 5.50 - 5.52:

$$K(z,\ \Psi) = 3\left\langle q^{2}\right\rangle L(z,\Psi) \qquad (5.54a)$$

$$= 3\left[\left\langle q_{\|}^{2}\right\rangle L_{\|}(z)\cos^{2}\Psi + \left\langle q_{\perp}^{2}\right\rangle L_{\perp}(z)\sin^{2}\Psi\right] = \frac{\overset{*}{\chi}_{B}}{G}. \qquad (5.54b)$$

We shall term K also the "spectral function"; in most cases the theoretical results are preferable to express through this quantity.

5.3.2 The Quasi-Spatial Ensemble

We consider here an *artificial* ensemble of dipoles, that combines properties of planar and spatial ensembles. Namely, we correct the dielectric-response theory, elaborated for relatively simple *planar rotation / libration,* with account of *statistics, valid for in-space rotors.* This approach yields the better description of *experimental* spectra, than it is possible with help of a pure planar ensemble.

Namely, we consider a *continuum* of rotation planes *YZ*. Each plane has an arbitrary orientation relative to the external field vector E_{m} (see Fig. 5.1b) and to its own symmetry axis *Z*. Relative to the latter a dipole performs a regular librational / rotational motion, governed by the prescribed potential function $U(\theta)$. This axis may alter due to strong collision, interrupting motion of a particle in an appropriate potential well.

Such an ensemble is termed "quasi-spatial". As seen from Fig. 5.1 b, the orientation of any *YZ* plane relative to E_{m} vector is determined by angles Ψ (between E_{m} and *Z*) and ψ (between *Y* axis and the E_{m} projection on fixed plane *XY*). In view of Fig. 5.1 b we may replace Eq. 5.48 by

$$q = \mu_E / \mu = q_{\parallel} \cos\Psi + q_{\perp} \sin\Psi \cos\psi \ . \tag{5.54c}$$

From physical reasoning we may suppose, that dielectric properties of the ensemble under consideration should not depend on angle ψ. Averaging over ψ, we obtain instead of (5.53) and (5.54a):

$$K(z,\Psi) = K_{\parallel}(z)\cos^2\Psi + \frac{1}{2} K_{\perp}(z)\sin^2\Psi, \tag{5.54d}$$

$$q^2(\Psi) = q_{\parallel}^2 \cos^2\Psi + \frac{1}{2} q_{\perp}^2 \sin^2\Psi. \tag{5.54e}$$

Here multiplier 1/2 appears due to averaging over ψ.

In this book the quasi-spatial ensemble is actually employed for consideration of dielectric spectra in isotropic media, see the next section, where an additional averaging is performed.

5.4 The Spectral Functions in the Case of an Isotropic Medium

We recall the Reader that the Z axis characterizes in this important particular case a rather *small local-symmetry zone* (e.g. the *near order* region). We suppose that a dielectric sample consists of parts, *analogous to domains.* Each such «domain» is characterized by its local potential $U(\theta)$, whose symmetry axis Z is oriented at an arbitrary angle Ψ relative to the field amplitude vector $\boldsymbol{E}_{\mathrm{m}}$ (see Fig. 5.1). Within each domain particles move under action of its potential $U(\theta)$. The so constructed isotropic medium presents a sort of a *locally anisotropic (*in other words, of a *micro-heterogeneous)* medium, since we suppose, that there is no *regular* (in terms of some conservative potential) interaction between particles, belonging to different domains. If linear dimension of each domain is much lesser than the wavelength in the medium, *the wave will "see" the latter as a homogeneous one,* which is characterized by some averaged electrodynamic parameters. Using these physical arguments, we may suggest, that dielectric properties of such isotropic medium will fall *between* those for the longitudinal and transverse orientations of an alternating field.

We suggest that the axes Z of different parts are *homogeneously distributed in space.* For such a medium the susceptibility χ^* is found by averaging of Eq. 5.50 or 5.54d over angles Ψ. *This averaging we denote by dash over the* *corresponding quantity*. According to this reasoning we establish the following *hierarchy* for averaging.

First. We find the average of an arbitrary quantity $f(\Psi, \Gamma)$ over the phase space Γ of *anyone* domain, which is denoted as

$$\langle f(\Psi)\rangle \equiv \int_{\Gamma} f(\Psi,\Gamma)\, d\Psi . \tag{5.55}$$

Second. We find the average of $\langle f(\Psi)\rangle$ for *various* domains, which is denoted as $\overline{\langle f\rangle}$:

$$\overline{\langle f\rangle} \equiv \frac{1}{2\pi}\int_0^{2\pi} \langle f(\Psi)\rangle\, d\Psi \qquad \text{(for planar ensemble)}, \tag{5.56a}$$

$$\overline{\langle f\rangle} \equiv \frac{1}{2}\int_0^{\pi} \langle f(\Psi)\rangle \sin\Psi\, d\Psi \quad \text{(for spatial and quasi-spatial ensembles)}. \tag{5.56b}$$

Below we consider rotation *in space*, *in a plane* and in an intermediate case, called *quasi-spatial ensemble*, which we shall widely use in order to simplify calculations of the spectral functions.

5.4.1 Reorientation in Space

Since the subensembles are statistically independent, each of them is characterized by a number of its own phase variables Γ. We may consider the angle Ψ as an *additional phase variable*. In accord with (5.56b) we have

$$\chi_B^*(z) \equiv \frac{\int_0^{\pi} \chi_B^*(z,\Psi)\sin\Psi\, d\Psi}{\int_0^{\pi} \sin\Psi\, d\Psi}. \tag{5.57}$$

We note that

$$\left\langle \cos^2\Psi\right\rangle = \int_0^{\pi/2} \cos^2\Psi \sin\Psi\, d\Psi = \frac{1}{3} , \tag{5.58a}$$

$$\left\langle \sin^2 \Psi \right\rangle = \int_0^{\pi/2} \sin^3 \Psi \, d\Psi = \frac{2}{3}, \tag{5.58b}$$

$$\overline{\left\langle q^2 \right\rangle} = \overline{\left\langle \cos^2 \Delta \right\rangle} = \frac{1}{3}. \tag{5.59}$$

It follows from (5.50) that

$$\overset{*}{\chi}_B(z) = \frac{1}{3} \overset{*}{\chi}_{B\|}(z) + \frac{2}{3} \overset{*}{\chi}_{B\perp}(z) . \tag{5.60}$$

$$L(z) = \left\langle q_\|^2 \right\rangle L_\|(z) + 2 \left\langle q_\perp^2 \right\rangle L_\perp(z). \tag{5.61}$$

Eq. 5.52, 5.60 yield the following relation between the spectral functions of isotropic and two specific molecular ensembles.

To calculate the SF L(z) of an isotropic ensemble we should first obtain two specific anisotropic SFs $L_\|(z)$ and $L_\perp(z)$. (5.62)

5.4.2 Reorientation in a Plane

If a macroscopically isotropic ensemble has a single rotational degree of freedom, with a local symmetry axis and vector $\boldsymbol{E}_m$, lying on one and the same plane (see Fig. 5.2), the Ψ-average is found without $\sin\Psi$ in the integrand of Eq. 5.56. Since

$$\int_0^{\pi/2} \sin^2 \Psi \, d\Psi = \int_0^{\pi/2} \cos^2 \Psi \, d\Psi = \frac{1}{2}; \left\langle q^2 \right\rangle = \frac{1}{2}, \tag{5.63}$$

we have

$$\overset{*}{\chi}_B(z) = \frac{1}{2} \overset{*}{\chi}_{B\|}(z) + \frac{1}{2} \overset{*}{\chi}_{B\perp}(z) . \tag{5.64}$$

On account of (5.62) and of general relation (5.8), viz.

$$\overset{*}{\chi}_B(z) = 3\left\langle q^2 \right\rangle GL(z), \tag{5.65}$$

we find that

$$L(z) = \frac{\langle q_{\|}^2 \rangle L_{\|}(z) + \langle q_{\perp}^2 \rangle L_{\perp}(z)}{2\langle q^2 \rangle} = \langle q_{\|}^2 \rangle L_{\|}(z) + \langle q_{\perp}^2 \rangle L_{\perp}(z) \cdot \quad (5.66)$$

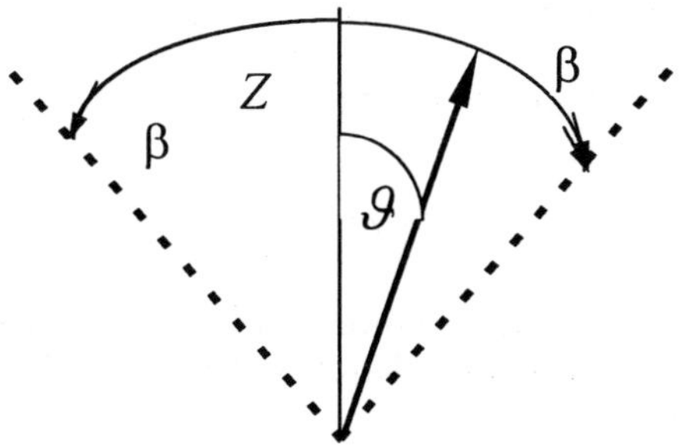

Fig. 5.2 Geometric scheme for libration / rotation of a dipole in a plane

5.4.3 The Quasi-Spatial Ensemble

We take into attention Eq. 5.54 d and remark that the direction cosines $q_{\|}$ and $q_{\perp}$, corresponding to planar motion, obey the relation

$$q_{\|}^2 + q_{\perp}^2 = 1. \quad (5.67)$$

Since Ψ is the spatial angle, Eq. 5.58 holds for averages; for the angle ψ in a plane we have: $\langle \cos^2 \psi \rangle = 1/2$. It then follows from (5.54 c), that

$$\overline{\langle q^2 \rangle} = \frac{1}{3}\langle q_{\|}^2 \rangle + \frac{2}{3}\frac{1}{2}\langle q_{\perp}^2 \rangle = \frac{1}{3} . \quad (5.68)$$

In view of Eq. 5.54 b, d we find:

$$\overset{*}{\chi}_B(z) = \frac{\int_0^{\pi} \sin\Psi\, d\Psi \int_0^{2\pi} \overset{*}{\chi}_B(z,\Psi,\psi)\, d\psi}{\int_0^{\pi} \sin\Psi\, d\Psi \int_0^{2\pi} d\psi}, \tag{5.69}$$

$$\overset{*}{\chi}_B(z,\Psi,\psi) = \overset{*}{\chi}_{B\|}(z)\cos^2\Psi + \overset{*}{\chi}_{B\perp}(z)\sin^2\Psi\cos^2\psi\,. \tag{5.70}$$

After integration over Ψ and ψ we obtain in view of Eq. 5.52:

$$= \frac{\overset{*}{\chi}_{B\|} + \overset{*}{\chi}_{B\perp}}{3} = G\left[\left\langle q_\|^2 \right\rangle L_\|(z) + \left\langle q_\perp^2 \right\rangle L_\perp(z)\right], \tag{5.71}$$

where the quantities in the r.h.s. correspond to a planar ensemble. Comparing this expression with the general relation (5.8), we find the SF of an isotropic quasi-spatial ensemble:

$$L(z) = \left\langle q_\|^2 \right\rangle L_\|(z) + \left\langle q_\perp^2 \right\rangle L_\perp(z) \tag{5.72}$$

or, in other terms,

$$3\,\overline{\left\langle q^2 \right\rangle} L(z) = K_\|(z) + K_\perp(z). \tag{5.73}$$

We shall see further, that for $\overset{*}{\chi}_B$, given by Eq. 5.72, the integrated absorption Π is only half of that, found for dipoles, rotating in space. This distinction arises because the planar motion of dipoles is now considered. Indeed, in the quasi-spatial ensemble we regard the only degree of freedom; we do not introduce momenta, canonically conjugated with angles Ψ and ψ (note: these quantities are *not* true phase variables).

One may go a step toward removing this drawback by doubling the *L(z)* spectral function and consequently the susceptibility:

For better comparison with experiment we may set

$$L_{\text{corr}}(z) = 2L(z), \tag{5.74a}$$

$$\chi_{\text{Bcorr}}(z) = 2G\,L(z). \tag{5.74b}$$

---

where $L(z)$ is found from Eq. 5.72 or 5.73.

5.5 Fourier Series for the Spectral Function: Planar motion

5.5.1 General Case of Periodic Libration / Rotation

We start with an important class of rigid rotors, that posses one degree of freedom. Let us suppose that a dipole μ rotates on a fixed plane XZ under action of an uniaxially symmetric potential $U(\vartheta)$, ϑ being an instantaneous angle between μ and the symmetry axis Z, see Fig. 5.2. Let Φ be the period of the function

$$q_{||}(\varphi) = \cos\vartheta(\varphi), \text{ hence } q_{||}(\varphi + \Phi) \equiv q_{||}(\varphi). \tag{5.75}$$

We take the total energy h of a dipole and its initial phase φ_0 as a pair of phase variables. We may account for the contribution of *all particles* to χ^* by choosing φ_0 in the interval $[0, \Phi]$. This choice of variables $\{h, \varphi_0\}$ is convenient since it is the energy h, that usually determines the character of motion of a particle.

At sufficiently small h a dipole librates in a certain range of angles ϑ, namely

$$|\vartheta(\varphi)| << \beta, \tag{5.76}$$

β being *libration amplitude* (see Fig. 5.2).

At larger energies h a dipole performs *hindered rotation*, during the course of which the direction of its angular velocity remains constant, the dependence $\vartheta(\varphi)$ being a monotone function.

Thus, we distinguish two dipolar subensembles, *librators and hindered rotators* for rotation in a plane, the corresponding symbols be $^{\vee}$ and $^{\circ}$.

For instance, we employ function $\overset{\vee}{q}(\varphi)$ for $h \in [h_{\min}, h_{\text{th}}]$ and function $\overset{\circ}{q}(\varphi)$ for $h \in [h_{\text{th}}, \infty]$, h_{th} being the *threshold energy,* dividing the two classes of particles. We shall see that the angular velocity of a librator / rotator tends to zero at $h = h_{\text{th}}$.

We consider further only those potentials $U(\vartheta)$, that cause a *periodic (in time) change of μ_Z, e.g. of the projection of μ on the symmetry axis.* It is evident (see Fig. 5.3) that in the librator-subensemble

---

The period of the function $\mu_Z(\varphi) = \mu \check{q}_{||}(\varphi)$ *is equal to the half of the periods of functions* $\check{\vartheta}(\varphi)$ *and* $\check{q}_{\perp}(\varphi) = \sin[\check{\vartheta}(\varphi)]$. (5.77)

So, the libration frequency $\check{p}$ is related to the period $\check{\Phi}$ as

$$\check{p} = 2\pi / 2\check{\Phi} = \pi / \check{\Phi}; \check{q}(\varphi + \check{\Phi}) \equiv \check{q}(\varphi) . \tag{5.78}$$

Using this expression and the schematic, shown in Fig.5.3, we conclude that

---

The fundamental frequency in trigonometric series for $\check{q}_{\perp}(\varphi)$ *function is* $\check{p}$, *while in series for* $\check{q}_{||}(\varphi)$ *it is equal to* $2\check{p}$. (5.79)

Series for $\check{q}_{\perp}(\varphi)$ *and* $\check{q}_{||}(\varphi)$ *may be represented respectively by odd and even harmonics of* $\check{p}$ *:* (5.80)

$$\check{q}_{||}(\varphi) = \sum_{n=0}^{\infty} a_{2n} \cos[2n \check{p}(\varphi + \varphi_0)] , \tag{5.81a}$$

$$\check{q}_{\perp}(\varphi) = \sum_{n=0}^{\infty} b_{2n+1} \sin[(2n+1) \check{p}(\varphi + \varphi_0)] . \tag{5.81b}$$

In Fig. 5.3b for the comparison with $\check{q}_{\perp}(\varphi)$ function we also draw $\sin \check{p}\varphi$ and $\sin 2\check{p}\varphi$, see dash-dotted and dashed curves respectively to the right and to the left of the ordinate axis. It is evident that $\sin 2\check{p}\varphi$ and $\check{q}_{\perp}(\varphi)$ have the same signs in the interval $[-\check{\Phi}/2, 0]$ and opposite signs -

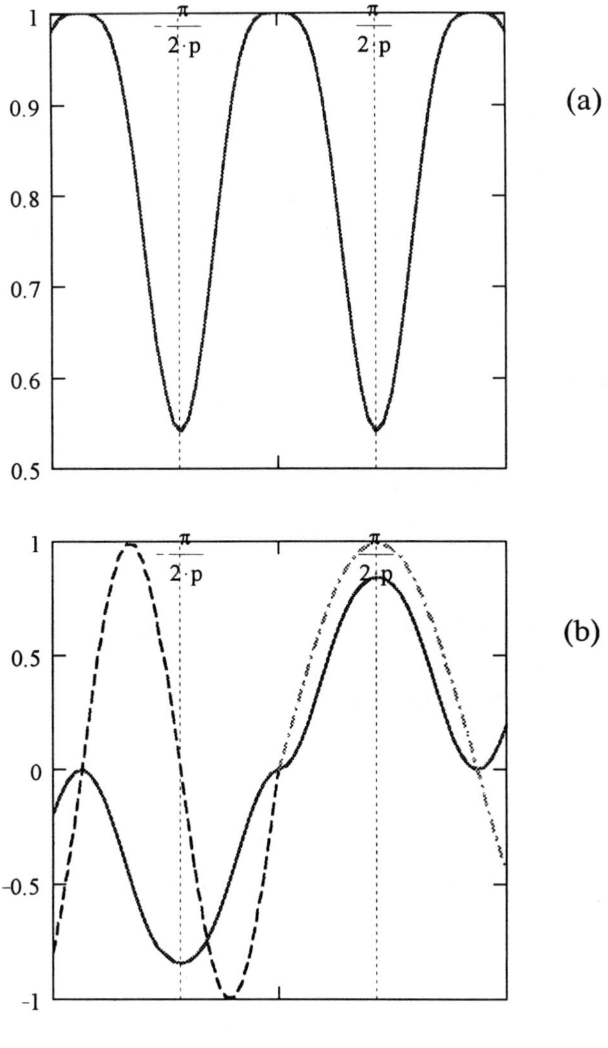

Fig. 5.3 Graph, pertaining to the derivation of the law of motion of librators in terms of Fourier amplitudes.

in $[-\check{\Phi}, -\check{\Phi}/2]$. Thus, Fourier amplitude b_2, that is expressed by the integral over φ of the product $\check{q}_{\|}(\varphi)\sin 2\check{p}\varphi$, vanishes. On the contrary, the product $\check{q}_{\perp}(\varphi)\sin\check{p}\varphi$ does not change its sign in both intervals of integration over φ, hence $b_1 \neq 0$.

Similar conclusions are valid for $n > 1$, if one considers $\sin 2n\check{p}\varphi$ and $\sin(2n+1)\check{p}\varphi$.

It follows from Eq. 5.81 that Fourier amplitudes are determined by the laws of motion $\check{q}_{\|}(\varphi)$ and $\check{q}_{\perp}(\varphi)$ as integrals

$$\begin{Bmatrix}\check{a}_n\\ \check{b}_n\end{Bmatrix} = \left[1 \pm (-1)^n\right]\frac{1}{\check{\Phi}}\int_0^{\check{\Phi}}\begin{Bmatrix}\check{q}_{\|}\cos\check{p}n\varphi\\ \check{q}_{\perp}\sin\check{p}n\varphi\end{Bmatrix}d\varphi .$$

Without loss of generality we can choose the instant φ_0 such, that $\check{q}_{\|}(\varphi)$ and $\check{q}_{\perp}(\varphi)$ become symmetrical and antisymmetrical functions of φ, determined in the interval $[-\check{\Phi}/2, \check{\Phi}/2]$:

$$\check{q}_{\|}(-\varphi) \equiv \check{q}_{\|}(\varphi),\ \check{q}_{\perp}(-\varphi) \equiv -\check{q}_{\perp}(\varphi).$$

Therefore, we may reduce the interval of integration:

$$\begin{Bmatrix}\check{a}_n\\ \check{b}_n\end{Bmatrix} = \left[1 \pm (-1)^n\right]\frac{2}{\check{\Phi}}\int_0^{\check{\Phi}/2}\begin{Bmatrix}\check{q}_{\|}\cos\check{p}n\varphi\\ \check{q}_{\perp}\sin\check{p}n\varphi\end{Bmatrix}d\varphi . \qquad \begin{Bmatrix}(5.82a)\\ (5.82b)\end{Bmatrix}$$

We shall apply these formulas in Chap.8 in the case of the planar confined rotator model.

The average over the subensemble has the form

$$\langle u\rangle \equiv \int_{h_{\min}}^{h_{\text{th}}} W\,dh \int_0^{\check{\Phi}} \check{u}\,d\varphi_0 . \tag{5.83}$$

Integrating over φ_0, we have:

$$\int_0^{\check{\Phi}} \left\{ \begin{matrix} q_{\|0}\left[q_{\|}(\varphi) - q_{\|0}\right] \\ q_{\perp 0}\left[q_{\perp}(\varphi) - q_{\perp 0}\right] \end{matrix} \right\} d\varphi_0 = \int_0^{\pi/\check{p}} \sum_{m,n=0}^{\infty} \left\{ \begin{matrix} \check{a}_{2m}\check{a}_{2n} \cos 2m\check{p}\varphi_0 \\ \check{b}_{2m+1}\check{b}_{2n+1} \sin(2m+1)\check{p}\varphi_0 \end{matrix} \right.$$

$$\times \left[\begin{matrix} \cos 2n\check{p}(\varphi + \varphi_0) - \cos 2n\check{p}\varphi_0 \\ \sin(2n+1)\check{p}(\varphi + \varphi_0) - \sin(2n+1)\check{p}\varphi_0 \end{matrix} \right] \Bigg\} d\varphi_0$$

$$= \frac{\pi}{2\check{p}} \sum_{n=0}^{\infty} \begin{pmatrix} \check{a}_{2n}^2 \\ \check{b}_{2n+1}^2 \end{pmatrix} \left\{ \begin{matrix} \cos 2n\check{p}\varphi - 1 \\ \cos(2n+1)\check{p}\varphi - 1 \end{matrix} \right\} \tag{5.84}$$

We represent the spectral function $L(z)$ as the sum

$$L(z) = \check{L}(z) + \mathring{L}(z) . \tag{5.85}$$

In accordance with Eq. 5.35 we must integrate Eq. 5.84 over φ. Using the standard integral

$$\int_0^{\infty} \exp(-px)\sin^2 bx\,dx = \frac{2b^2}{p}\,\frac{1}{p^2 + 4b^2}, \tag{5.85a}$$

we find the contributions $\check{L}_{\|}(z)$, $\check{L}_{\perp}(z)$ due to *librational motion*:

$$\left\langle q_{\|}^{2}\right\rangle \check{L}_{\|}(z) = \frac{\pi}{2}\int_{h_{\min}}^{h_{\text{th}}} dh\, W \check{p} \sum_{n=1}^{\infty} \frac{(2n)^2 \check{a}_{2n}^{2}}{\left(2n\check{p}\right)^2 - z^2}, \tag{5.86}$$

$$\left\langle q_{\perp}^{2}\right\rangle \check{L}_{\perp}(z) = \frac{\pi}{2}\int_{h_{\min}}^{h_{\text{th}}} dh\, W \check{p} \sum_{n=1}^{\infty} \frac{(2n-1)^2 \check{b}_{2n-1}^{2}}{\left((2n-1)^2\check{p}^2\right)^2 - z^2}. \tag{5.87}$$

We can rewrite both expressions (5.86) and (5.87) in the following generalized form:

$$\left\{\begin{matrix}\left\langle q_{\|}^{2}\right\rangle \check{L}_{\|}(z) \\ \left\langle q_{\perp}^{2}\right\rangle \check{L}_{\perp}(z)\end{matrix}\right\} = \frac{\pi}{4}\int_{h_{\min}}^{h_{\text{th}}} dh\, W \check{p} \sum_{n=1}^{\infty} \frac{n^2\left[1 \pm (-1)^n\right]}{\check{p}^2 n^2 - z^2} \tag{5.88}$$

We shall now determine the contributions $\overset{\circ}{L}_{\|}(z)$, $\overset{\circ}{L}_{\perp}(z)$ of *hindered rotors*. The periodic (in time) angular velocity of a planar rotor with a given energy h does not change its sign. In contrast, each of the functions $\overset{\circ}{q}_{\|}(\varphi)$, $\overset{\circ}{q}_{\perp}(\varphi)$ alters its sign periodically. Both functions have the same period $\overset{\circ}{\Phi}$ which determines the fundamental frequency $\overset{\circ}{p} = 2\pi / \overset{\circ}{\Phi}$.

Thus, we may write
Fourier series as

$$\left\{\begin{matrix}\overset{\circ}{q}_{\|}(\varphi) \\ \overset{\circ}{q}_{\perp}(\varphi)\end{matrix}\right\} = \sum_{n=0}^{\infty}\left\{\begin{matrix}\overset{\circ}{a}_n \cos pn(\varphi + \varphi_0) \\ \overset{\circ}{b}_n \sin pn(\varphi + \varphi_0)\end{matrix}\right\}. \tag{5.89}$$

We have by analogy with (5.86) - (5.88):

$$\begin{Bmatrix} \left\langle q_{\parallel}^2 \right\rangle \overset{\circ}{L}_{\parallel}(z) \\ \left\langle q_{\perp}^2 \right\rangle \overset{\circ}{L}_{\perp}(z) \end{Bmatrix} = \pi \int_{h_{\text{th}}}^{\infty} dh\, W \overset{\circ}{p} \sum_{n=1}^{\infty} \begin{Bmatrix} \overset{\circ}{a}{}_n^2 \\ \overset{\circ}{b}{}_n^2 \end{Bmatrix} \frac{n^2}{\overset{\circ}{p}{}^2 n^2 - z^2} \tag{5.90}$$

We find averages $\left\langle q_{\parallel}^2 \right\rangle$ and $\left\langle q_{\perp}^2 \right\rangle$ from (5.81) and (5.89) after integration over φ_0 and h:

$$\left\langle q_{\parallel}^2 \right\rangle = \pi \int_{h_{\min}}^{h_{\text{th}}} \frac{W}{\overset{\vee}{p}} \left\{ \overset{\vee}{a}{}_0^2 + \frac{1}{2} \sum_{n=1}^{\infty} \overset{\vee}{a}{}_{2n}^2 \right\} dh + 2\pi \int_{h_{\text{th}}}^{\infty} \frac{W}{\overset{\circ}{p}} \left\{ \overset{\circ}{a}{}_0^2 + \frac{1}{2} \sum_{n=1}^{\infty} \overset{\circ}{a}{}_{2n}^2 \right\} dh, \tag{5.91}$$

$$\left\langle q_{\perp}^2 \right\rangle = \frac{\pi}{2} \int_{h_{\min}}^{h_{\text{th}}} \frac{W}{\overset{\vee}{p}} \sum_{n=1}^{\infty} \overset{\vee}{b}{}_{2n-1}^2 \, dh + \pi \int_{h_{\text{th}}}^{\infty} \frac{W}{\overset{\circ}{p}} \sum_{n=1}^{\infty} \overset{\circ}{b}{}_n^2 \, dh. \tag{5.92}$$

At last, taking account of Eq. 5.66, we obtain from Eq. 5.88 and 5.90 the planar spectral function for an isotropic medium:

$$L(z) = \frac{\pi}{4} \int_{h_{\min}}^{h_{\text{th}}} dh\, W \overset{\vee}{p} \sum_{n=1}^{\infty} \frac{n^2}{\overset{\vee}{p}{}^2 n^2 - z^2} \left[\overset{\vee}{a}{}_n^2 + \overset{\vee}{b}{}_n^2 + (-1)^n \left(\overset{\vee}{a}{}_n^2 - \overset{\vee}{b}{}_n^2 \right) \right]$$

$$+ \pi \int_{h_{\text{th}}}^{\infty} dh\, W \overset{\circ}{p} \sum_{n=1}^{\infty} \frac{n^2}{\overset{\circ}{p}{}^2 n^2 - z^2} \left(\overset{\circ}{a}{}_n^2 + \overset{\circ}{b}{}_n^2 \right). \tag{5.93}$$

In the far infrared region the dispersion of a polar medium is obliged to *resonance denominators* in the integrands in (5.88), (5.90) and (5.93). The frequency dependence $L''(z) = \text{Im}\,[L''(z)]$ of the loss spectral function usually describes a *quasi-resonant absorption spectrum in the FIR region.*

For librators the resonant denominators are $4n^2 \overset{\vee}{p}{}^2 - z^2$ for longitudinal and $(2n-1)^2 \overset{\vee}{p}{}^2 - z^2$ for transverse components of alternating external field.

If $\boldsymbol{E}_m \perp Z$, then the main mechanism of interaction is related to the fundamental frequency $\check{p}$ of librations.

For hindered rotators the main effect of the field-medium interaction is related to the fundamental harmonic in the resonant denominator $n^2 \mathring{p}^2 - z^2$. This mechanism will be illustrated immediately for the ensemble of polar molecules, freely rotating in the interval τ between strong collisions.

5.5.2 Free Rotation between Strong Collisions

As a simple example we shall determine the spectral function $L(z) = L_F(z)$ in the absence of the potential $U(\vartheta)$. This is the extended rotational diffusion (ED) model. Since $U = 0$, for our normalized quantities the angular momentum squared coincides with the constant kinetic energy (see Sec.3.4.3):

$$\dot{\vartheta}^2 = h, \text{ hence } \vartheta(\varphi) = \sqrt{h}\,(\varphi + \varphi_0) \tag{5.94}$$

The librational subensemble yields no contribution to $L(z)$, thus

$$h_{\text{th}} = h_{\text{min}} = 0, \check{L} = 0, L_F(z) = \mathring{L}(z), \begin{Bmatrix} q_{\|}(\varphi) \\ q_{\perp}(\varphi) \end{Bmatrix} = \begin{Bmatrix} \cos\left[\sqrt{h}\,(\varphi + \varphi_0)\right] \\ \sin\left[\sqrt{h}\,(\varphi + \varphi_0)\right] \end{Bmatrix}. \tag{5.95}$$

Noting, that the frequency $\mathring{p}$ of periodic motion coincides with the constant angular momentum

$$\mathring{p} = p = \sqrt{h}\,, \tag{5.96}$$

we find the distribution $W(h) = C\exp(-h)$ and the average $\langle q^2 \rangle$:

$$\frac{1}{C} = \int_0^\infty \exp(-h)\,\mathring{\Phi}\,dh = 4\pi \int_0^\infty \exp(-x^2)\,dx = 2\pi^{3/2}; \tag{5.97a}$$

$$\left\langle q_{\|}^2 \right\rangle = \left\langle q_{\perp}^2 \right\rangle = \frac{\pi}{2\pi\sqrt{\pi}} \int_0^\infty \frac{\exp(-h)\,dh}{\sqrt{h}} = \frac{1}{2}. \tag{5.97b}$$

Substituting (5.96), (5.97) into (5.90) and taking into account, in accord with (5.95), that $a_1 = b_1 = 1$; $a_n = b_n = 0$ at $n > 1$, we have

$$L_{\|}(z) = L_{\perp}(z) = L(z) = \frac{2}{\sqrt{\pi}} \int_0^{\infty} \exp(-p^2) \frac{p^2\, dp}{p^2 - z^2} \equiv L_F(z). \quad (5.98)$$

The r.h.s. can be expressed through the *probability integral of complex argument*, see (Abramowitz and Stegun 1964):

$$L_F(z) = 1 + i\sqrt{\pi}\, z w(z); \quad (5.99)$$

$$w(z) = \frac{i}{\pi} \int_{-\infty}^{\infty} \frac{\exp(-t^2)\, dt}{z - t} \quad (5.100)$$

$$= \exp(-z^2)\left(1 + \frac{2i}{\sqrt{\pi}}\right) \int_0^{\infty} \exp(t^2)\, dt = \exp(-z^2)\big(1 + \operatorname{erf}(iz)\big). \quad (5.101)$$

In view of Eq. 5.64 we obtain the susceptibility for the Boltzmann distribution F_B :

$$\chi_B^* = \frac{3}{2} G L_F(z) = \frac{\mu^2 N}{2k_B T}\left[1 + i\sqrt{\pi}\, z w(z)\right]. \quad (5.102)$$

For the quasi-spatial ensemble of rotors we replace $(3/2)G$ by G:

$$\chi_B^* = G L_F(z) = \frac{\mu^2 N}{3k_B T}\left[1 + i\sqrt{\pi}\, z w(z)\right]. \quad (5.103)$$

The spectral properties of the ED model will be discussed in Chap. 8.

5.6 Fourier Series for the Spectral Function: Motion in Space

We shall consider dipoles, having the form of a symmetric-top molecule and performing rotational motion in the presence of an axisymmetric potential $u(\theta)$. Every particle is characterized by three degrees of freedom. We use the coordinate system, shown in Fig. 3.3a.

Initially we write trigonometric series for the periodic function $q_{\|} = \cos \theta(\varphi)$:

$$q_{\|}(\varphi) = \sum_{n=0}^{\infty} a_n \cos\left[pn(\varphi + \varphi_0)\right]; \; p \equiv \frac{2\pi}{\Phi}, \tag{5.104}$$

Φ being the period, $q_{\|}(\varphi) \equiv q_{\|}(\varphi + \Phi)$, and p being the fundamental frequency.

The function $q_{\perp}(\varphi)$ depends, unlike $q_{\|}(\varphi)$, on two angular coordinates:

$$q_{\perp}(\varphi) \equiv \rho(\varphi) \cos(\psi + \psi_0), \tag{5.104a}$$

where

$$\rho^2(\varphi) = 1 - q_{\|}^2(\varphi). \tag{5.104b}$$

The axis Y in Fig. 3.3a coincides with direction of the field amplitude $\boldsymbol{E}_{\mathrm{m}}$, taken for $\boldsymbol{E}(t) \perp Z$. In the XY plane we introduce the angle $\psi + \psi_0$, where ψ_0 is the angle between the projection $\mu_{XY}(\varphi = 0)$ and the axis Y; ψ is the angular shift of this projection relative to its initial position. Thus, in accord with (5.104a) $q_{\perp}(\varphi)$ is the normalized projection μ_Y/μ. This function is nonperiodic in time φ, since the angular shift 2λ of this projection during time $\varphi = \Phi = 2\pi/p$ in general is non-commensurable with π (see Fig. 3.3b).

If we employ *trigonometric series* for the laws of motion $q_{\|}(\varphi)$ and $q_{\perp}(\varphi)$, we can, as it will be shown, perform integration over initial phase φ_0, angle ψ_0 and time φ. The integral over initial coordinate χ_0 may be omitted, since the integrand in (5.35) does not depend on χ_0. As a result, the spectral function is expressed through the triple integral over h (energy), l [angular momentum around the symmetry axis Z of a potential $U(\theta)$] and m (angular momentum around the symmetry axis Z' of a top).

For the particular case of a *linear molecule* the integral over m should be omitted, so the spectral function is calculated from the *double integral* with variables h and l.

Using similar Fourier representation for a *planar rotator*, one should take an *ordinary integral over energy* h.

Consequently, the possibility to elaborate theory for any concrete molecular model actually reduces to the possibility of obtaining simple expressions for functions $a_n(h, l, m)$ and $b_n(h, l, m)$, i.e. for Fourier amplitudes a_n and b_n of reorienting polar molecules.

We shall show now, that for the *longitudinal* orientation of alternating field the general expression for $L_{||}(z)$ is similar to that, found in Sec. 5.5 for planar rotators. For the transverse SF $L_{\perp}(Z)$ the situation is a little more complex, but the final result is *formally* rather simple.

5.6.1 Longitudinal Orientation of External Field

The projection $q_{||}(\varphi)$ of a unit vector μ_E/μ on the symmetry axis Z changes periodically in time with the period Φ. We represent $q_{||}(\varphi)$ as the series (5.104). Then

$$\int_0^{\Phi} q_{||}(\varphi) q_{||}(0)\, d\varphi = \frac{\pi}{p} \sum_{n=0}^{\infty} a_n^2 \cos pn\varphi \,. \tag{5.105}$$

It follows from these equations, that the average (found for one period) value

$$\overline{q_{||}(\varphi)} = a_0; \int_0^{\Phi} q_{||}^2(0)\, d\varphi_0 = \frac{\pi}{p} \left(2a_0^2 + \sum_{n=1}^{\infty} a_n^2 \right) .$$

Integrating over other phase variables, we have

$$\left\langle q_{||}^2 \right\rangle = \pi \int \frac{W}{p} \left(2a_0^2 + \sum_{n=1}^{\infty} a_n^2 \right) dh\, dl\, dm \,. \tag{5.106}$$

With account of (5.105) and (5.35) we obtain

$$\left\langle q_{||}^2 \right\rangle L_{||}(z) = iz \int_0^{\infty} \left\langle q_{||}(\varphi) q_{||}(0) - q_{||}^2(0) \right\rangle \exp(iz\varphi) d\varphi$$

$$= iz\pi \int dh\, dl\, dm \frac{W}{p} \sum_{n=0}^{\infty} a_n^2 \int_0^{\infty} \left(\cos pn\varphi - 1 \right) \exp(iz\varphi)\, d\varphi \,. \tag{5.107a}$$

Using (5.85), we find, that here the integral over φ is

$$\int_0^{\infty} \sin^2 \frac{pn\varphi}{2} \exp(iz\varphi)\, d\varphi = \frac{1}{2iz} \frac{p^2 n^2}{p^2 n^2 - z^2} \,. \tag{5.107b}$$

Thus,

$$\left\langle q_{\|}^{2}\right\rangle L_{\|}(z) = \pi \int W\, p S_{\|}\, dh\, dl\, dm; \qquad S_{\|} \equiv \sum_{n=1}^{\infty} \frac{n^2 a_n^2}{p^2 n^2 - z^2} \quad . \tag{5.108}$$

5.6.2 Transverse Orientation of External Field

We use here Euler equations of motion, given in Sec. 3.4.4. We have shown there, that the function $q_{\perp}(\varphi)$ is usually nonperiodic. Yet, we can select the periodic component $\tilde{\psi}(\varphi)$ from $\psi(\varphi)$ and thus can expand $\psi(\varphi)$ in trigonometric series.

Let 2λ be the angular shift of the vector $(\mu/\mu)_{XY}$ on XY plane during time interval Φ. Then the mean angular velocity is given by

$$\overline{\dot{\psi}} \equiv \frac{1}{\Phi} \int_0^{\Phi} \dot{\psi}(\varphi) d\varphi = \frac{2\lambda}{\Phi} = \frac{\lambda p}{\pi} \tag{5.109}$$

Let us consider the difference

$$\frac{d}{d\varphi}\tilde{\psi}(\varphi) = \dot{\psi} - \overline{\dot{\psi}} .$$

By definition, this difference is a periodic function of time, that has no constant component. Consequently, the integral of this function over φ is also a periodic function with the same period Φ:

$$\tilde{\psi}(\varphi) = \int^{\varphi} \left[\dot{\psi}(\varphi') - \frac{\lambda p}{\pi} \right] d\varphi' \equiv \tilde{\psi}(\varphi + \Phi) \ , \tag{5.109a}$$

so that

$$\psi(\varphi) = \int_0^{\varphi} \left[\dot{\psi}(\varphi') - \frac{\lambda p}{\pi} + \frac{\lambda p}{\pi} \right] d\varphi' \equiv \frac{p\lambda}{\pi}\varphi + \tilde{\psi}(\varphi) - \tilde{\psi}(0) \ ; \tag{5.110}$$

$\psi(\varphi)$ can be represented as the sum of a periodic function and of a secular term, proportional to φ.

We now turn our attention to Eq. 3.82. Using Eq. 5.110, we write down the average over ψ_0 of the product $q_{\perp}(\varphi)q_{\perp}(0)$:

$$\frac{1}{2\pi}\int_0^{2\pi} q_\perp(\varphi)q_\perp(0)\,d\psi_0$$

$$= \frac{\rho\rho_0}{2}\cos\left[\tilde{\psi}(\varphi) - \tilde{\psi}(0) + \frac{p\lambda\varphi}{\pi}\right] \tag{5.111}$$

$$\frac{\rho\rho_0}{2}\left\{\cos\left[\tilde{\psi}(\varphi) - \tilde{\psi}(0)\right]\cos\frac{p\lambda\varphi}{\pi} - \sin\left[\tilde{\psi}(\varphi) - \tilde{\psi}(0)\right]\sin\frac{p\lambda\varphi}{\pi}\right\}. \tag{5.112}$$

We write Fourier series of

$$\rho\begin{Bmatrix}\cos\tilde{\psi}(\varphi)\\ \sin\tilde{\psi}(\varphi)\end{Bmatrix} = \begin{Bmatrix}c_0/2\\ 0\end{Bmatrix} + \sum_{n=1}^{\infty}\begin{Bmatrix}c_n\cos pn(\varphi+\varphi_0)\\ s_n\sin pn(\varphi+\varphi_0)\end{Bmatrix} \tag{5.113}$$

It is evident that for $\varphi = 0$

$$\rho_0\begin{Bmatrix}\cos\tilde{\psi}(0)\\ \sin\tilde{\psi}(0)\end{Bmatrix} = \begin{Bmatrix}c_0/2\\ 0\end{Bmatrix} + \sum_{n=1}^{\infty}\begin{Bmatrix}c_n\cos pn\varphi_0\\ s_n\sin pn\varphi_0\end{Bmatrix} \tag{5.114}$$

We integrate over φ_0 the product of two last expressions:

$$\int_0^{\Phi}\rho\rho_0\cos\left[\tilde{\psi}(\varphi) - \tilde{\psi}(0)\right]d\varphi_0$$

$$= \int_0^{\Phi}\left\{\left[\frac{c_0}{2} + \sum_{n=1}^{\infty} c_n\cos pn(\varphi+\varphi_0)\right]\left[\frac{c_0}{2} + \sum_{n=1}^{\infty} c_n\cos pn\varphi_0\right]\right.$$

$$\left. + \sum_{n=1}^{\infty}\sum_{m=1}^{\infty} s_n s_m\sin pn(\varphi+\varphi_0)\sin pn\varphi_0\right\}d\varphi_0$$

$$= \frac{\Phi}{2}\left[\frac{c_0^2}{2} + \sum_{n=1}^{\infty}\left(c_n^2 + s_n^2\right)\cos pn\varphi_0\right]; \tag{5.115}$$

$$\int_0^{\Phi}\rho\rho_0\sin\left[\tilde{\psi}(\varphi) - \tilde{\psi}(0)\right]d\varphi_0$$

$$= \quad = \int_0^{\Phi}\left\{ \sum_{n=1}^{\infty} s_n \sin pn(\varphi+\varphi_0)\left[\frac{c_0}{2} + \sum_{n=1}^{\infty} c_n \cos pn\varphi_0\right]\right.$$

$$\left. - \sum_{n=1}^{\infty} s_n \sin pn\varphi_0\left[\frac{c_0}{2} + \sum_{n=1}^{\infty} c_n \cos pn(\varphi + \varphi_0)\right] \right\} d\varphi_0$$

$$= \Phi \sum_{n=1}^{\infty} c_n\, s_n \sin pn\varphi \ . \qquad (5.116)$$

Substituting (5.115) and (5.116) into (5.112), we have

$$\int_0^{\Phi} \langle q_\perp(\varphi) q_\perp(0)\rangle_{\psi_0}\, d\varphi_0 = \frac{\Phi}{4}\left\{ \frac{c_0^2}{2}\cos\frac{p\lambda\varphi}{\pi} + \sum_{n=1}^{\infty}\left[\left(c_n^2 + s_n^2\right)\cos pn\varphi\, \cos\frac{p\lambda\varphi}{\pi}\right.\right.$$

$$\left.\left. -2c_n\, s_n \sin pn\varphi\, \sin\frac{p\lambda\varphi}{\pi} \right]\right\} = \frac{\Phi}{8}\left\{ c_0^2 \cos\frac{p\lambda\varphi}{\pi}\right.$$

$$\left. + \sum_{n=1}^{\infty}\left[\left(c_n + s_n\right)^2 \cos\left(n+\frac{\lambda}{\pi}\right) p\varphi + \left(c_n - s_n\right)^2 \cos\left(n-\frac{\lambda}{\pi}\right) p\varphi\right] \right\} .$$

This result can be rewritten in a more compact form with account of the symmetry properties

$$c_{-n} = c_n;\ s_{-n} = -s_n,$$

that follow from the definitions of c_n and s_n, that correspond to Eq. 5.114. Then we have:

$$\int_0^{\Phi} \langle q_\perp(\varphi) q_\perp(0)\rangle_{\psi_0}\, d\varphi_0 = \frac{\pi}{p} \sum_{n=-\infty}^{\infty} b_n^2 \cos\left(n - \frac{\lambda}{\pi}\right) p\varphi, \qquad (5.117)$$

where we set

$$b_n = \left(c_n - s_n\right)/2 \qquad (5.118)$$

It is convenient to calculate Fourier amplitudes b_n directly by employing functions $\rho(\varphi)$ and $\psi(\varphi)$. In accord with the definition (5.113)

$$C_n - S_n = \frac{2}{\Phi}\int_0^{\Phi} \rho\left[\cos\tilde{\psi}(\varphi)\cos pn(\varphi+\varphi_0) - \sin\tilde{\psi}(\varphi)\sin pn(\varphi+\varphi_0)\right] d\varphi_0$$

$$= \frac{2}{\Phi}\int_0^{\Phi} \rho(\varphi)\cos\left[\tilde{\psi}(\varphi) + pn(\varphi+\varphi_0)\right] d\varphi_0 \equiv 2b_n \ . \qquad (5.119)$$

The Fourier amplitude b_n of the transverse SF is equal to the mean (over the period Φ) value of the function

$$\rho(\varphi)\cos\left[\tilde{\psi}(\varphi) + pn(\varphi+\varphi_0)\right] \ . \qquad (5.120)$$

It follows from (5.117), that

$$\int_0^{\Phi} \left\langle q_\perp^2(0)\right\rangle_{\psi_0} d\varphi_0 = \frac{\pi}{p}\sum_{n=-\infty}^{\infty} b_n^2 \qquad (5.120a)$$

Combining this expression with (5.117) and taking integral over φ, we get by analogy with (5.107a):

$$\left\langle q_\perp^2\right\rangle L_\perp(z) = iz\int_0^{\infty}\left\langle q_\perp(\varphi)q_\perp(0) - q_\perp^2(0)\right\rangle \exp(iz\varphi)\,d\varphi$$

$$= iz\pi\int\frac{W}{p}dh\,dl\,dm\,d\varphi\ \exp(iz\varphi)\sum_{n=0}^{\infty} b_n^2\left[\cos\left(n-\frac{\lambda}{\pi}\right)p\varphi - 1\right].$$

Finally we have:

$$\left\langle q_\perp^2\right\rangle L_\perp(z) = \pi\int dh\,dl\,dm\ p\,W\,S_\perp \ ; \qquad (5.121a)$$

$$S_\perp = \sum_{n=-\infty}^{\infty}\frac{\left(n-\frac{\lambda}{\pi}\right)^2 b_n^2}{\left(n-\frac{\lambda}{\pi}\right)^2 p^2 - z^2} \ ; \qquad (5.121b)$$

$$\left\langle q_\perp^2 \right\rangle = \pi \int dh\, dl\, dm\, \frac{W}{p} \sum_{n=-\infty}^{\infty} b_n^2 \,. \tag{5.122}$$

5.6.3 Frequencies of Resonant Interaction.

Introducing $d\Gamma' \equiv dh\ dl\ dm$, we find in accordance with Eq. 5.106, 5.108, 5.121, 5.122, that the longitudinal and transverse SFs are expressed respectively through Fourier coefficients a_n and b_n :

$$\left\langle q_\parallel^2 \right\rangle L_\parallel(z) = \pi \int_{\Gamma'} d\Gamma' \, Wp \sum_{n=1}^{\infty} \frac{n^2 a_n^2}{p^2 n^2 - z^2} \quad ; \tag{5.123a}$$

$$\left\langle q_\perp^2 \right\rangle L_\perp(z) = \pi \int_{\Gamma'} d\Gamma' \, pW \sum_{n=-\infty}^{\infty} \frac{\left(n - \frac{\lambda}{\pi}\right)^2 b_n^2}{\left(n - \frac{\lambda}{\pi}\right)^2 p^2 - z^2} ; \tag{5.123b}$$

$$\left\langle q_\parallel^2 \right\rangle = \pi \int_{\Gamma'} d\Gamma' \frac{W}{p} \left(2a_0^2 + \sum_{n=1}^{\infty} a_n^2 \right) , \tag{5.124a}$$

$$\left\langle q_\perp^2 \right\rangle = \pi \int_{\Gamma'} d\Gamma' \frac{W}{p} \sum_{n=-\infty}^{\infty} b_n^2 \tag{5.124b}$$

The fundamental frequencies of resonant interaction correspond to $n = 1$ in denominators of Eq. 5.123. These frequencies are

$x \approx p$ and $x \approx p\left(1 - \frac{\lambda}{\pi}\right)$ for $\|$- and $\perp$-orientations, respectively, $p =$

$2\pi/\Phi$ being the fundamental frequency of the periodic function $q_\|(\varphi)$. Thus, the unnormalized resonant frequencies are

$$\omega_n^{\text{res}} = \frac{p}{\eta} \begin{pmatrix} n \\ n - \frac{\lambda}{\pi} \end{pmatrix} \text{ for } \begin{pmatrix} L_\parallel(z) \\ L_\perp(z) \end{pmatrix}, \quad n = 1,2,3, \ldots \,. \tag{5.125}$$

For the transverse orientation of alternating field the fundamental frequency is shifted with respect to p, the shift being the mean precession frequency $p\lambda/\pi$. Eq. 5.125 corresponds to the *resonance condition of synchronous interaction* of librating / rotating dipole with one of two circularly polarized waves. We remark, that their sum represents linearly polarized field $E(t)$.

5.6.4 The Static SF and the Susceptibility at Rare Collisions

Here we consider the long lifetime limit, $\tau \to \infty$ (that is, $y = 0$). For the zero-frequency case ($x = 0$) we put $z = 0$ in (5.123) and compare the result with (5.124). With account of (5.32) we have:

$$L_{\perp}(0) = \pi \int_{\Gamma'} d\Gamma' \frac{W}{p\langle q_{\perp}^2\rangle} \sum_{n=-\infty}^{\infty} b_n^2 = 1, \tag{5.126a}$$

$$\chi_{\perp B}(0) = 3G\langle q_{\perp}^2\rangle L_{\perp}(0) = 3G\langle q_{\perp}^2\rangle. \tag{5.126b}$$

In the case of rare collisions the static transverse susceptibility is equal to $3G$ times the mean square of the corresponding direction cosine

$$q_{\perp} = \mu_{E\perp} / \mu \tag{5.127}$$

On the other side, noting that $2\pi/p = \Phi$ and $d\Gamma' d\Phi = d\Gamma$, we have for the *longitudinal orientation:*

$$\langle q_{\parallel}^2\rangle L_{\parallel}(0) = \pi \int_{\Gamma'} d\Gamma' \frac{W}{p} \sum_{n=1}^{\infty} a_n^2 = \int_{\Gamma} d\Gamma W \sum_{n=1}^{\infty} a_n^2 = \left\langle \sum_{n=1}^{\infty} a_n^2 \right\rangle \tag{5.128}$$

We recall, that in accord with the definition (5.104)

$$a_0 = \overline{q_{\parallel}(\varphi)} \equiv \overline{q_{\parallel}}\,.$$

Thus,

$$\overline{\left[q_{\|}(\varphi)\right]^2} \equiv \frac{1}{\Phi}\int_0^{\Phi}\left[\sum_{n=0}^{\infty} a_n \cos pn(\varphi+\varphi_0)\right]^2 d\varphi$$

$$= \left(\overline{q_{\|}}\right)^2 + \frac{1}{2}\sum_{n=1}^{\infty} a_n^2\,; \tag{5.129}$$

$$\overline{\left[q_{\|}(\varphi) - \overline{q_{\|}}\right]^2} \equiv \frac{1}{\Phi}\int_0^{\Phi}\left[\sum_{n=1}^{\infty} a_n \cos pn(\varphi+\varphi_0) - \overline{q_{\|}}\right]^2 d\varphi$$

$$= \frac{1}{2}\sum_{n=1}^{\infty} a_n^2 = \left\langle \tilde{q}_{\|}^2 \right\rangle \tag{5.130}$$

This quantity characterizes the *localization* of a particle on its trajectory. Hence,

$$\frac{\chi_{\| B}(0)}{3G} = \left\langle q_{\|}^2 \right\rangle L_{\|}(0) = \left\langle q_{\|}^2 - \overline{q_{\|}}^2 \right\rangle \equiv \left\langle \tilde{q}_{\|}^2 \right\rangle; \tag{5.131}$$

$$L_{\|}(0) = 1 - \frac{\left\langle \overline{q_{\|}}^2 \right\rangle}{\left\langle q_{\|}^2 \right\rangle} \equiv \frac{\left\langle \tilde{q}_{\|}^2 \right\rangle}{\left\langle q_{\|}^2 \right\rangle} \tag{5.132}$$

In the case of rare collisions the static longitudinal susceptibility is equal to 3G times the mean value of $\tilde{q}_{\|}^2$. (5.133)

In view of (5.127) the same conclusion is valid for the transverse static susceptibility $\chi_{\perp B}(0)$. Indeed, as seen from Eq. 5.82b, 5.89 for rotation / libration in a plane, the constant component of the corresponding direction cosine is equal to zero:

$$\overline{q_{\perp}(\varphi)} = 0\,. \tag{5.134}$$

This expression holds also for rotation in space. Thus, with account of Eq. 5.126b

$$\frac{\chi_{\perp B}(0)}{3G} = \left\langle q_\perp^2 \right\rangle L_\perp(0) = \left\langle q_\perp^2 - \overline{q_\perp}^2 \right\rangle \equiv \left\langle \tilde{q}_\perp^2 \right\rangle \tag{5.135}$$

We see again that for particles at rest, when $q(\varphi) \equiv \overline{q}$, the static susceptibility is equal to zero.

5.6.5 About an Ensemble Averaging

The normalizing coefficient C of the distribution $W = C \exp(-h)$ is given by the expression

$$\frac{1}{C} = \int_\Gamma \exp(-h)d\Gamma = \int_{\Gamma'} \exp(-h)d\Gamma' \int_0^{2\pi/p} d\varphi_0 = 2\pi \int_{\Gamma'} \frac{\exp(-h)}{p} d\Gamma', \tag{5.136a}$$

where

$$d\Gamma = dh\, dl\, dm \frac{d\psi_0}{2\pi}\, d\varphi_0 = d\Gamma' \frac{d\psi_0}{2\pi}\, d\varphi_0. \tag{5.136b}$$

The Euler angle χ is not included into the phase volume, since in view of Eq. 3.80, 3.82 functions $q_{||}(\varphi)$ and $q_\perp(\varphi)$ do not depend on this coordinate. Taking into account the symmetry relations, involving phase variables l and m (see Sec. 3.4.6), *one can take only positive l and m in integrals over Γ'.*

We remark that for some models instead of h it is useful to introduce

$$g = h - bm^2 \tag{5.137}$$

as a phase variable. An example was given in Sec. 3.4.5, where $u \equiv 0$ and $p = \sqrt{g}$ coincides with the total angular momentum of a free rotating symmetric-top molecule.

5.6.6 Transition to Ensembles, Having Reduced Number of Degrees of Freedom

ENSEMBLE OF LINEAR MOLECULES. Using in (5.137) g as a phase variable instead of h, we can easily find from Eq. 5.123, 5.124 the SF $L(z)$ for spatial ensemble of rigid rotors. Since the moment of inertia I_c relative the symmetry axis is equal to zero, we put for this transition:

$$b = \frac{I}{I_c} - 1 \to \infty \tag{5.138}$$

We remark that the steady state (Boltzmann) distribution is given by

$$W = C \exp\left[-\left(g + bm^2\right)\right] \tag{5.139}$$

For a free rotating top the SF will be found in Chap. 7.

Another approach for ensemble of linear molecules consists in the *reduction of the phase volume*. Indeed, we may omit the dm multiplier in Eq. 5.136b and put $m = 0$ in dependencies $p(m)$, $a_n(m)$, $b_n(m)$ and in the integrands in (5.123) - (5.124). This is valid, because the angular momentum $\boldsymbol{k}$ of a rotor is normal to the plane of rotation, so that the projection of $\boldsymbol{k}$ on the Z' axis is zero (see Fig. 3.3a).

ENSEMBLE OF PLANAR ROTORS. We now perform transition from (5.123), (5.124) to Eq. 5.88, 5.90 - 5.93, derived for the planar rotor. One should cancel dl out the phase-volume element $d\Gamma = dh\ dl$ of a linear molecule and put $l = 0$, since the angular momentum of the planar rotor is normal to the plane, containing symmetry axis. In addition one has to double the right-hand side of Eq. 5.123 for $L_{\perp}(z)$, because

$$\frac{1}{\left\langle \cos^2 \psi_0 \right\rangle} = 2\,.$$

THE SPECTRAL FUNCTIONS FOR LIBRATORS. We start from the *transverse orientation* of alternating field. Let us consider the change of μ_X, the X-projection of the vector μ, due to transition from one extreme orientation to another: from $\vartheta = \beta$ to $\vartheta = -\beta$, see Fig. 5.2 (we remind the Reader that for planar rotation we denote the angle as ϑ instead of θ for motion in space). It may be seen in Fig. 5.2 that the $q_{\perp} = \mu_X/\mu$ value alters from $\sin\beta$ to $-\sin\beta$, when the aforementioned transition occurs. The latter takes time $\varphi = \Phi$; correspondingly the rotation of the μ_{XY}-projection (see Fig. 5.3) is described by

$$\psi(\varphi = \Phi) = 2\lambda = \pi.$$

Thus, one may select the subensemble of *librators* in the doubled expression (5.123b) for $L\ (z)$ by taking $\lambda = \pi/2$ and by putting the fundamental frequency p equal to $2\check{p}$, since the time dependence in (5.81) must be analogous to that in (5.104). After that we transform integrand in (5.123b) by the following manner:

$$2p\sum_{n=-\infty}^{\infty}\frac{\left(n-\frac{1}{2}\right)^2 b_n^2}{\left(n-\frac{1}{2}\right)^2 p^2-z^2}=\check{p}\sum_{n=-\infty}^{\infty}\frac{(2n-1)^2 b_n^2}{(2n-1)^2\check{p}^2-z^2}$$

$$=\check{p}\sum_{n=1}^{\infty}\frac{(2n-1)^2 b_n^2}{(2n-1)^2\check{p}^2-z^2}+\check{p}\sum_{n=1}^{\infty}\frac{(2n+1)^2 b_{-n}^2}{(2n-1)^2\check{p}^2-z^2}$$

$$\check{p}\sum_{n=1}^{\infty}\frac{(2n-1)^2\left(b_n^2+b_{-n+1}^2\right)}{(2n-1)^2\check{p}^2-z^2} \tag{5.140}$$

We take $\Gamma' = [h] = [h_{\min}, \infty]$, integrate over h and finally get Eq. 5.88 for $L_\perp(z)$, where

$$\check{b}_{2n-1}^2 \to 2\left(b_n^2+b_{1-n}^2\right). \tag{5.141}$$

In order to obtain this equality one may use Eq. 5.82b for the time dependence $\check{q}_\perp(\varphi)$ with $\check{b}_{2n+1}$ coefficients in the integrand of the l.h.s. of Eq. 5.117. Noting that $\Phi = 2\pi/p = \pi/\check{p}$, we calculate this integral, double the r.h.s. of Eq. 5.140 (since the averaging over ψ_0 is not performed now) and substitute $\lambda = \pi/2$. Then we have:

$$\frac{\pi}{2\check{p}}\sum_{n=0}^{\infty}\check{b}_{2n+1}^2\cos(2n+1)\check{p}\varphi$$

$$=\frac{\pi}{\check{p}}\left[\sum_{n=0}^{\infty}b_n^2\cos(2n-1)\check{p}\varphi+\sum_{n=0}^{\infty}b_{-n}^2\cos(2n+1)\check{p}\varphi\right]. \tag{5.142}$$

Substituting n by n' - 1 and then omitting the dash, we have

$$\sum_{n=1}^{\infty} b_{2n-1}^2 \cos(2n-1)\check{p}\varphi = 2\sum_{n=1}^{\infty}\left[b_n^2 + b_{-n+1}^2\right]\cos(2n-1)\check{p}\varphi \quad .(5.143)$$

Relation (5.141) is proved.

Now let us turn to the *longitudinal susceptibility*. We shall prove the coincidence of $\check{L}_{\|}(z)$, given by Eq. 5.86 and 5.123a. In the latter the integrand can be transformed to $\frac{\check{p}}{2}\sum_{n=1}^{\infty}(2na_n)^2\left[(2pn)^2 - z^2\right]^{-1}$ We get Eq. 5.86 with account, that $p = 2\check{p}$ and that Fourier amplitudes $\check{a}_{2n}$ in the series (5.81a) correspond to even (with respect to $\check{p}$) harmonics: $a_n \to \check{a}_{2n}$.

THE SPECTRAL FUNCTIONS FOR HINDERED ROTATORS. The period Φ of function $\mu_Z(\varphi)$ is equal to the period of hindered rotation of the vector μ and consequently to that of its projection μ_{XY}. Hence,

$$p = 2\pi/\Phi = \check{p} \text{ and } 2\lambda = \int_0^{2\pi}\psi(\varphi)d\varphi = 2\pi.$$

Thus, doubling of Eq. 5.123 b for the *transverse* $\check{L}_{\perp}(z)$ and substitution there λ by π yields the integrand

$$2\sum_{n=-\infty}^{\infty}\frac{(n-1)^2 b_n^2}{(n-1)^2 p^2 - z^2} = 2\sum_{n=1}^{\infty}\frac{n^2\left(b_{n+1}^2 + b_{-n+1}^2\right)}{n^2p^2 - z^2}. \quad (5.144a)$$

Integrating over $\Gamma' = [h]$ in the interval $h \in [h_{th}, \infty]$, we find, that Eq. 5.123b coincides with (5.87), if we put $p = \mathring{p}$ and

$$\mathring{b}_n^2 = 2\left(b_{n+1}^2 + b_{-n+1}^2\right) \quad (5.144b)$$

To prove validity of this identity one may employ function $\overset{\circ}{q}_{\perp}(\varphi)$ from (5.89) as the integrand of Eq. 5.117, integrate over φ_0 and compare the result with the doubled r.h.s. of Eq. 5.140 for $\lambda = \pi$ and $p = \overset{\circ}{p}$. We have by analogy with Eq. 5.143:

$$\Phi \sum_{n=0}^{\infty} b_n^2 \cos n \overset{\circ}{p}\varphi = 2\frac{\pi}{p} \sum_{n=-\infty}^{\infty} b_n^2 \cos(n-1) p\varphi$$

$$= 2\frac{\pi}{p} \sum_{n=-\infty}^{\infty} \left(\sum_{n=0}^{\infty} b_{n+1}^2 + \sum_{n=1}^{\infty} b_{-n+1}^2 \right) \cos np\varphi \, .$$

In view of (5.144) the term with $n = 0$ yields no contribution to the SF. We have got Eq. 5.144.

It is evident that Eq. 5.90 for the *longitudinal* susceptibility $\overset{\circ}{L}_{\|}(z)$ is identical with Eq. 5.123 a, if one takes $p = \overset{\circ}{p}$, $a_n = \overset{\circ}{a}_n$ and $\Gamma' = [h_{\text{th}}, \infty]$.

We conclude: *transition from a symmetric-top molecule to a planar rotator yields correct expressions for all SFs of librating / rotating molecules.*

5.7 Integrated Absorption

General expressions (5.8) and (5.32) satisfy the Gordon sum rule. For an isotropic medium, consisted of linear or symmetric-top molecules, this rule is given by:

$$\Pi = \int_{-\infty}^{\infty} \omega \chi''(\omega) d\omega = \frac{2\pi}{3} \frac{\mu^2 N}{I} \tag{5.145}$$

Taking $x = \omega\eta$ as a dimensionless frequency, we rewrite this expression in the form:

$$\Xi = \frac{1}{G} \int_{-\infty}^{\infty} x \chi''(x) dx = \pi \, . \tag{5.146}$$

We shall first consider an anisotropic medium and then prove the validity of Eq. 5.146 in the particular case of an isotropic medium. In Sec. 5.7 we mean that $\chi^* \equiv \overset{*}{\chi}_B$, that is, we consider the Boltzmann induced distribution F_B. However, all general relations, obtained for integrated absorption, e.g. (5.146), hold for *any* reasonable induced distribution F.

Starting from Eq. 5.8 and 5.33, we have:

$$\Xi = \frac{1}{G}\int_{-\infty}^{\infty} x\chi''(x)dx = -3\int_{-\infty}^{\infty} x dx \int_{\Gamma} q_0 W d\Gamma \int_0^{\infty} \dot{q} \exp(-y\varphi) \sin x\varphi \, d\varphi \,. \quad (5.147)$$

Using one of the definitions of δ-function, viz.

$$\delta(\varphi) \equiv \lim_{x\to\infty} \frac{\sin x\varphi}{\pi\varphi} \,, \quad (5.148)$$

we write

$$\int_{-\infty}^{\infty} x dx \int_0^{\infty} \dot{q} \exp(-y\varphi) \sin x\varphi \, d\varphi = -\int_0^{\infty} \dot{q} \exp(-y\varphi)\, d\varphi \int_{-\infty}^{\infty} x \frac{d}{d\varphi} \cos x\varphi \, dx$$

$$= -\int_0^{\infty} \left[\dot{q} \exp(-y\varphi) \frac{d}{d\varphi} \frac{\sin x\varphi}{\varphi} \, d\varphi \right]_{x\to-\infty}^{x\to\infty} =$$

$$= -2\int_0^{\infty} \dot{q} \exp(-y\varphi)\, d\varphi \frac{d}{d\varphi} \lim_{x\to\infty} \frac{\sin x\varphi}{\varphi}$$

$$= -2\pi \int_0^{\infty} \dot{q}(\varphi)\, \dot{\delta}(\varphi)\, \exp(-y\varphi)\, d\varphi \,. \quad (5.149)$$

Equation (5.149) may be divided by 2 by taking -∞ in the lower limit and $|\varphi|$ in $\exp(-y\varphi)$ instead of φ, since functions $\dot{\delta}(\varphi)$ and $\dot{q}(\varphi)$ are odd. Indeed, in view of (5.147) we have to calculate $\left\langle \dot{q}(\varphi) q_0 \right\rangle$. Averaging over φ_0 yields an even function, since trigonometric series for $q(\varphi)$ give

$$\left\langle \begin{matrix} \cos pn(\varphi + \varphi_0) \cos pn\varphi_0 \\ \sin pn(\varphi + \varphi_0) \sin pn\varphi_0 \end{matrix} \right\rangle_{\varphi_0} = \frac{1}{2} \cos pn\varphi \,.$$

Consequently, $\left\langle \dot{q}(\varphi)\, q(0) \right\rangle_{\varphi 0}$ is odd relative to φ. The derivative $\frac{d}{d\varphi}\exp(-y\varphi)$ vanishes at $\varphi = 0$, as it is odd function of φ. On account of (5.25) we have after substitution (5.149) into Eq. 5.147:

$$\Xi = -3\pi \int_{\Gamma} q_0\, W d\Gamma = \frac{d}{d\varphi}\left[\dot{q}\exp(-y|\varphi|)\right]_{\varphi=0}$$

$$= -3\pi \left\langle q_0 \left[\frac{d^2 q(\varphi)}{d\varphi^2}\right]_{\varphi=0} \right\rangle = 3\pi \left\langle \dot{q} \right\rangle^2 . \tag{5.150}$$

Below we express $\left\langle \dot{q} \right\rangle^2$ in terms of $\left\langle q^2 \right\rangle$.

5.7.1 Rotation in Space

It follows from (3.80), (3.81) and the relation $q_{\|} = \cos\theta$:

$$\left\langle q_{\perp}^2 \right\rangle = \left\langle \sin^2\theta \cos^2(\psi + \psi_0) \right\rangle = \frac{\left\langle \sin^2\theta \right\rangle}{2} = \frac{1 - \left\langle q_{\|}^2 \right\rangle}{2}. \tag{5.151}$$

Then we find:

$$\left\langle \dot{q}_{\perp}^2 \right\rangle = \left\langle \left[\dot{\theta}\cos\theta \cos(\psi + \psi_0) - \dot{\psi}\sin\theta \sin(\psi + \psi_0)\right]^2 \right\rangle$$

$$= \frac{1}{2}\left\langle \dot{\theta}^2 \cos^2\theta + \dot{\psi}^2 \sin^2\theta \right\rangle \tag{5.152}$$

Now we employ variables $\theta, \dot{\theta}$ instead of h, φ_0 for averaging over ensemble. Substituting l for ψ with account of Eq. 3.83, we have in view of Eq. 3.108:

$$h\left(\theta,\dot{\theta},\dot{\psi},m\right) = \dot{\theta}^2 + \dot{\psi}^2 \sin^2\theta + (b+1)m^2 + u(\theta) \tag{5.153}$$

One may write expression for $\left\langle \psi^2 \right\rangle$ without integrating over other variables.

Substituting again l in view of (3.83) for $\dot{\psi}$ and noting that $\int_{-\infty}^{\infty} x^2 \exp(-x^2)\, dx = \frac{1}{2} \int_{-\infty}^{\infty} \exp(-x^2)\, dx$, we write: $dl = \sin^2\theta\, d\dot{\psi}$, so that

$$\left\langle \dot{\psi}^2 \right\rangle_l = \frac{\int_{-\infty}^{\infty} \dot{\psi}^2 \exp\left(-\dot{\psi}^2 \sin^2\theta\right) dl}{\int_{-\infty}^{\infty} \exp\left(-\dot{\psi}^2 \sin^2\theta\right) dl} = \frac{1}{2\sin^2\theta}. \tag{5.154}$$

By analogy we get

$$\left\langle \dot{\theta}^2 \right\rangle = \frac{\int_{-\infty}^{\infty} \dot{\theta}^2 \exp\left(-\dot{\theta}^2\right) d\dot{\theta}}{\int_{-\infty}^{\infty} \exp\left(-\dot{\theta}\right) d\dot{\theta}} = \frac{1}{2}. \tag{5.155}$$

Substituting (3.154) and (3.155) into Eq. 5.152, we have with account of (5.151):

$$\left\langle \dot{q}_\perp^2 \right\rangle = \frac{1}{2}\frac{1}{2}\left\langle 1 + \cos^2\theta \right\rangle = \frac{1}{2}\left[1 - \frac{1}{2}\left\langle \sin^2\theta \right\rangle\right] = \frac{1}{2}\left[1 - \left\langle q_\perp^2 \right\rangle\right]. \tag{5.156}$$

Similarly

$$\left\langle \dot{q}_{\parallel}^{2} \right\rangle = \left\langle \dot{\theta}^{2} \sin^2\theta \right\rangle = \frac{1}{2}\left\langle \sin^2\theta \right\rangle = \frac{1}{2}\left[1 - \left\langle q_{\parallel}^{2} \right\rangle\right] \qquad (5.157)$$

Hence,

$$\left\langle \dot{q}^{2} \right\rangle = \frac{1}{2}\left[1 - \left\langle q^{2} \right\rangle\right] \qquad (5.158)$$

From Eq. 5.146, 5.150 and 5.158 we obtain the Gordon sum rule for an *anisotropic medium:*

$$\Xi = \frac{1}{G}\int_{-\infty}^{\infty} x\chi''(x)dx = \frac{3\pi}{2}\left(1 - \left\langle q^{2} \right\rangle\right) \qquad (5.159)$$

or, in terms of unnormalized quantities,

$$\Pi = \int_{-\infty}^{\infty} \omega\chi''(\omega)d\omega = \frac{\pi\mu^2 N}{I}\left(1 - \left\langle q^{2} \right\rangle\right) \qquad (5.160)$$

For an *isotropic medium* we have Eq. 5.145. Indeed, in view of (5.53) and (5.151) the mean square of the directional cosine is *one third*:

$$\left\langle q^{2} \right\rangle = \frac{1}{3}, \qquad (5.161)$$

as it should be, since

$$\left\langle q^{2} \right\rangle = \left\langle q_{\parallel}^{2} \right\rangle \left\langle \cos^2\psi \right\rangle + \left\langle q_{\perp}^{2} \right\rangle \left\langle \sin^2\psi \right\rangle$$

$$= \int_{0}^{\pi/2} \left(\left\langle q_{\parallel}^{2} \right\rangle \cos^2\psi + \left\langle q_{\perp}^{2} \right\rangle \sin^2\psi \right) \sin\Psi \, d\Psi$$

$$= \frac{1}{3}\left(\left\langle q_{\parallel}^{2} \right\rangle + 2\left\langle q_{\perp}^{2} \right\rangle \right) = \frac{1}{3}\left(\left\langle q_{\parallel}^{2} \right\rangle + 1 - \left\langle q_{\parallel}^{2} \right\rangle \right) = \frac{1}{3}. \qquad (5.162)$$

Substituting into (5.159) and (5.160), we get Eq. 5.145.

5.7.2 Rotation in a Plane

Taking $l = 0$, we write similar to Eq. 5.151, 5.152 :

$$q_{\|} = \cos\vartheta, \quad q_{\perp} = \sin\vartheta, \quad \left\langle q_{\perp}^2 \right\rangle = 1 - \left\langle q_{\|}^2 \right\rangle, \quad W \propto \exp\left(-\dot{\vartheta}^2 \right), \tag{5.163}$$

$$\left\langle \begin{matrix} \dot{q}_{\|}^2 \\ \dot{q}_{\perp}^2 \end{matrix} \right\rangle = \left\langle \begin{matrix} \dot{\vartheta}^2 \sin^2\vartheta \\ \dot{\vartheta}^2 \cos^2\vartheta \end{matrix} \right\rangle = \left\langle \begin{matrix} \dot{\vartheta}^2 \left(1 - q_{\|}^2\right) \\ \dot{\vartheta}^2 \left(1 - q_{\perp}^2\right) \end{matrix} \right\rangle. \tag{5.164}$$

Substituting θ for ϑ, we find that Eq. 5.155 is also valid in this case. We express the averages of derivatives $\dot{q}_{\|}^2$ and $\dot{q}_{\perp}^2$ through averages of $q_{\|}^2$ and $q_{\perp}^2$ with use of the same relation (5.155). Consequently, Eq. 5.159 and 5.160 are again derived from Eq. 5.150 for an *anisotropic* medium:

$$\int_{-\infty}^{\infty} \omega\chi''(\omega)d\omega = \frac{\pi\mu^2 N}{I}\left(1 - \left\langle q^2 \right\rangle\right); \int_{-\infty}^{\infty} x\chi''(x)dx = \frac{3\pi G}{2}\left(1 - \left\langle q^2 \right\rangle\right). \tag{5.165}$$

However, the values $<q^2>$ differ in planar and spatial ensembles. For an *isotropic medium*

$$\left\langle q^2 \right\rangle = \int_0^{\pi/2} \cos^2\vartheta \, d\vartheta \Bigg/ \int_0^{\pi/2} d\vartheta = \frac{1}{2}; \tag{5.166}$$

consequently integrated absorption constitutes 3/4 of the value (5.145):

$$\Pi = \int_{-\infty}^{\infty} \omega\chi''(\omega)d\omega = \frac{\pi\mu^2 N}{2I}. \tag{5.167}$$

5.7.3 The Quasi-Spatial Ensemble

By definition, this ensemble originates from the planar one. In both ensembles the relation $\left\langle q_{\|}^2 \right\rangle + \left\langle q_{\|}^2 \right\rangle = 1$ holds. In accord with Eq. 5.73, 5.159 we get

$$\Xi = \frac{3\pi}{2}\frac{1}{3}\left(1 - \left\langle q_{\|}^2 \right\rangle + 1 - \left\langle q_{\perp}^2 \right\rangle\right) = \frac{\pi}{2}; \Pi = \frac{\pi\mu^2 N}{3I} \tag{5.168}$$

The integrated absorption is half of the exact values (5.145) and (5.146). In view of this property we had proposed the approximate relation (5.74).

5.8 The Kramers-Kronig Rule

We shall now obtain the integral relation between the real and imaginary components of the susceptibility χ_B^*. Omitting again the subscript B, let us consider the integral

$$\int_{-\infty}^{\infty} \chi^*(\xi)\, \frac{d\xi}{\xi - x}.$$

In view of Eq. 5.34 we have an identity

$$\frac{1}{C} \int_{-\infty}^{\infty} \chi^*(\xi)\, \frac{d\xi}{\xi - x} = 3 \int_{-\infty}^{\infty} \frac{d\xi}{\xi - x} \left\langle \int_0^{\infty} \dot{q}_0 q \exp\left[(i\xi - y)\varphi\right] d\varphi \right\rangle$$

$$3 \int_0^{\infty} \exp(-y\varphi)\, d\varphi \left\langle \dot{q}_0 \int_{-\infty}^{\infty} q \exp(i\xi\varphi)\, \frac{d\xi}{\xi - x} \right\rangle \tag{5.169}$$

Having replaced the variable ξ by $t = \xi - x$ and noting that $\int_0^{\infty} \frac{\sin ax}{x}\, dx = \pi$,

we transform the integral over ξ:

$$\int_{-\infty}^{\infty} \exp(i\xi\varphi)\, \frac{d\xi}{\xi - x} = \int_{-\infty}^{\infty} \exp\left[i(t + x)\varphi\right] \frac{dt}{t}$$

$$i \exp(ix\varphi) \int_{-\infty}^{\infty} \sin t\varphi\, \frac{dt}{t} = i\pi \exp(ix\varphi).$$

Substitution this into (5.169) and taking account of (5.34) yields:

$$\frac{1}{G} \int_{-\infty}^{\infty} \chi^*(\xi)\, \frac{d\xi}{\xi - x} = 3i\pi \left\langle \dot{q}_0 \int_0^{\infty} q \exp(iz\varphi)\, d\varphi \right\rangle = \frac{i\pi}{G} \chi^*(x). \tag{5.170}$$

These Kramers-Kronig relations may be rewritten for each of the components of the susceptibility as

$$\int_{-\infty}^{\infty} \begin{Bmatrix} \chi'(\xi) \\ \chi''(\xi) \end{Bmatrix} \frac{d\xi}{\xi - x} = \pi \begin{Bmatrix} -\chi''(x) \\ \chi'(x) \end{Bmatrix} \tag{5.171}$$

We rewrite these expressions as integrals in the half-infinite interval of ξ's by taking into account that χ' and χ'' are respectively even and odd functions of x:

$$\begin{Bmatrix} -\chi''(x) \\ \chi'(x) \end{Bmatrix} = \frac{2}{\pi} \int_0^{\infty} \begin{Bmatrix} x\chi'(\xi) \\ \xi\chi''(\xi) \end{Bmatrix} \frac{d\xi}{\xi^2 - x^2} \; . \tag{5.171a}$$

We shall see in the next section, that in the absence of collisions, that is for $\tau \to \infty$ (i.e. for $y \to 0$), one can usually obtain simple formulas for the loss dependence $\chi''(x)$. In this case it is convenient to get from (5.171a) the frequency dependence $\chi'(x)$ of the real part of the susceptibility as a limit

$$\chi'(x) = \lim_{y \to 0} \frac{2}{\pi} \int_0^{\infty} \xi \chi''(x') \frac{d\xi}{\xi^2 - z^2} \; ; \; z = x + iy \to x \, . \tag{5.171b}$$

Putting here $x = 0$, we obtain integral representation of the static susceptibility through the loss $\chi''(x)$:

$$\chi_s = \chi'(x{=}0) = \lim_{y \to 0} \frac{2}{\pi} \int_0^{\infty} \frac{\chi''(\xi)}{\xi} d\xi . \tag{5.172}$$

5.9 Landau Damping in Polar Medium

Now we shall consider the phenomenon, that in plasma physics is termed *Landau damping*, see (Ginzburg and Rukhadze 1975, p.104). The essence of the phenomenon is the existence of *collisionless absorption* of electromagnetic wave interacting with particles. The term "collisionless", being rather conditional, denotes that *collisions do appear but they are so rare, that the infinitely long lifetime τ limit may be applied.* In other words, collisions are frequent enough to establish thermodynamic equilibrium, described by the steady state distribution *W(h)*.

For a polar medium we use the analogous definition:

$$\{\text{Landau damping}\} \Rightarrow \lim_{y \to 0} \begin{Bmatrix} \chi''(x) \\ x\chi''(x) \end{Bmatrix} \tag{5.173}$$

The frequency dependence of the real part $\chi'(x)$, which corresponds to (5.173), may be obtained for small but still finite values of the collision frequency y. For this purpose we employ the Kramers-Kronig equation (5.171b). Since absorption $x\chi''(x)$ should not depend on the *sign* of frequency x, we suggest that the loss $\chi''(x)$ is odd function. If $z = x + iy$, then

$$\chi'(x) = \frac{1}{\pi}\int_{-\infty}^{\infty} \chi''(\xi)\frac{d\xi}{\xi - x} = \frac{1}{\pi}\int_{-\infty}^{\infty} \chi''(\xi)\left\{\frac{1}{\xi - x} + \frac{1}{\xi + x}\right\} d\xi$$

$$= \frac{2}{\pi}\int_{0}^{\infty} \frac{\xi\chi''(\xi)d\xi}{\xi^2 - z^2}\ . \tag{5.174}$$

In the limit (5.173) we can obtain in some cases a rather simple analytical expressions for the loss $\chi''(x)$ and then $\chi'(x)$ may be calculated from (5.174).

We restrict our consideration by the case of rotation on a *plane*.

For librators we take the limit $y \to 0$ in the n th terms of series (5.88):

$$\lim_{y\to 0} \mathrm{Im}\frac{n^2\check{p}}{\left(n\check{p}\right)^2 - z^2} = n^2\check{p}\lim_{y\to 0}\frac{2xy}{\left[\left(n\check{p}\right)^2 - x^2 + y^2\right] + (2xy)^2}. \tag{5.175}$$

Using the definition

$$\lim_{\lambda\to 0}\frac{\lambda}{\xi^2 + \lambda^2} \equiv \pi\delta(\xi) \tag{5.176}$$

of δ-function, we transform Eq. 5.175 to

$$\pi\check{p}\delta\left(\check{p}^2 - \frac{x^2}{n^2}\right). \tag{5.177}$$

For ensemble of *hindered rotators* we have from series (5.90):

$$\lim_{y\to 0} \operatorname{Im} \frac{n^2 \overset{\circ}{p}}{\left(n\overset{\circ}{p}\right)^2 - z^2} = \pi \overset{\circ}{p}\, \delta\left(\overset{\circ}{p}{}^2 - \frac{x^2}{n^2}\right). \tag{5.178}$$

We replace the variable of integration h by the fundamental librational / rotational frequency $\overset{\vee}{p}$ or $\overset{\circ}{p}$. Consequently,

$$dh = \frac{dh}{d\overset{\vee}{p}{}^2}\, d\overset{\vee}{p}{}^2 + \frac{dh}{d\overset{\circ}{p}{}^2}\, d\overset{\circ}{p}{}^2 . \tag{5.179}$$

Substituting Eq. 5.177 - 5.179 into Eq. 5.88, 5.90 and using the property of δ-function, we find imaginary parts of the SFs (p and h mean here $\overset{\vee}{p}$ or $\overset{\circ}{p}$; $\overset{\vee}{h}$ or $\overset{\circ}{h}$):

$$\lim_{y\to 0}\left\{\begin{matrix}\langle q_{\parallel}^2\rangle \overset{\vee}{L}{}''_{\parallel}(z)\\ \langle q_{\perp}^2\rangle \overset{\vee}{L}{}''_{\perp}(z)\end{matrix}\right\} = \frac{\pi^2}{4}\sum_{n=1}^{\infty}\left\{\begin{matrix}\overset{\vee}{a}{}_n^2\\ \overset{\vee}{b}{}_n^2\end{matrix}\right\}\left[1 \pm (-1)^n\right] W(h_n)\frac{x_n d\overset{\vee}{h}_n}{d\overset{\vee}{p}{}^2}; \tag{5.180}$$

$$\lim_{y\to 0}\left\{\begin{matrix}\langle q_{\parallel}^2\rangle \overset{\circ}{L}{}''_{\parallel}(z)\\ \langle q_{\perp}^2\rangle \overset{\circ}{L}{}''_{\perp}(z)\end{matrix}\right\} = \pi^2\sum_{n=1}^{\infty}\left\{\begin{matrix}\overset{\circ}{a}{}_n^2\\ \overset{\circ}{b}{}_n^2\end{matrix}\right\}\left[1 \pm (-1)^n\right] W(h_n)\frac{x_n d\overset{\circ}{h}_n}{d\overset{\circ}{p}{}^2}. \tag{5.181}$$

$$x_n = \frac{x}{n},\; h_n = h(p = x_n),\; \frac{dh}{dp^2} = \frac{dh}{dp^2}(p = x_n). \tag{5.182}$$

5.9.1 Relation of the Loss Susceptibility to the Steady State Distributions

Eq. 5.180 and 5.181 become more clear, if we introduce the distributions $\overset{\vee}{f}$ and $\overset{\circ}{f}$ over frequencies $\overset{\vee}{p}$ and $\overset{\circ}{p}$. To define these distributions we pick out the zones of librational and rotational energy with use of the normalizing condition (4.17) for *W(h):*

$$\int_{\Gamma} W\,dh\,d\varphi_0 = \int_{h_{\min}}^{\infty} W\Phi\,dh = \int_{h_{\min}}^{h_{th}} W\,\check{\Phi}\,dh + \int_{h_{th}}^{\infty} W\,\mathring{\Phi}\,dh = 1. \qquad (5.183)$$

We introduce $\check{p}$ and $\mathring{p}$ by the following fashion.

On the interval $[h_{\min}, h_{th}]$:

$$\check{p} = \frac{\pi}{\check{\Phi}};\ \check{p}(h = h_{\min}) = \check{p}_{\max} = p_m;\ \check{p}(h = h_{th}) = 0;\ \frac{dh}{d\check{p}} < 0. \quad (5.184)$$

On the interval $[h_{th}, \infty]$:

$$\mathring{p} = \frac{2\pi}{\mathring{\Phi}};\ \mathring{p}(h \to \infty) = \infty;\ \frac{dh}{d\mathring{p}} > 0. \qquad (5.185)$$

In view of Eq. 5.184 - 5.185, we can rewrite (5.183) as integrals over $\check{p}$ and $\mathring{p}$ (for simplicity we denote both quantities as p):

$$\int_0^{p_m} \check{f}(p)dp + \int_{p_m}^{\infty} \mathring{f}(p)dp = 1, \qquad (5.186)$$

where

$$\check{f}(p) = -\frac{\pi}{p}\frac{d\check{h}}{dp}W\left[\check{h}(p)\right],\ \mathring{f}(p) = \frac{2\pi}{p}\frac{d\mathring{h}}{dp}W\left[\mathring{h}(p)\right] \qquad (5.187)$$

Using Eq. 5.187, we can express the Landau damping (5.180) - (5.181) in terms of the distributions $\check{f}$ and $\mathring{f}$:

$$\left\{\begin{array}{l}\langle q_{\|}^{2}\rangle \overset{\vee}{L}''_{\|}(x) \\ \langle q_{\perp}^{2}\rangle \overset{\vee}{L}''_{\perp}(x)\end{array}\right\} = \frac{\pi}{4} x \sum_{n=1}^{\infty} \left\{\begin{array}{l}\overset{\vee}{a}{}_{2n}^{2}(\overset{\vee}{h}_{2n}) \dfrac{\overset{\vee}{f}(x_{2n})}{2n} \\ \overset{\vee}{b}{}_{2n-1}^{2}(\overset{\vee}{h}_{2n-1}) \dfrac{\overset{\vee}{f}(x_{2n-1})}{2n-1}\end{array}\right\}; \quad (5.188a)$$

$$\left\{\begin{array}{l}\langle q_{\|}^{2}\rangle \overset{\circ}{L}''_{\|}(x) \\ \langle q_{\perp}^{2}\rangle \overset{\circ}{L}''_{\perp}(x)\end{array}\right\} = \frac{\pi}{4} x \sum_{n=1}^{\infty} \left\{\begin{array}{l}\overset{\circ}{a}{}_{n}^{2}(\overset{\circ}{h}_{n}) \\ \overset{\circ}{b}{}_{n}^{2}(\overset{\circ}{h}_{n})\end{array}\right\} \frac{\overset{\circ}{f}(x/n)}{n} \quad (5.188b)$$

Thus, absorption $\alpha(x)$ does not vanish in the limit $\tau \to \infty$; $\alpha(x)$ is proportional to $xL''(x)$; it is determined by Fourier amplitudes a_n, b_n and by frequency distributions $\overset{\vee}{f}(p)$ and $\overset{\circ}{f}(p)$, found for periodic motion of dipoles. Two last functions are in its turn determined by a conservative potential $U(\vartheta)$, which governs the dependence of the period of such motion on the energy h of a particle.

Note, that Landau damping underlies the *resonance interaction* of particles with a wave, since the *frequency of maximum absorption is close to the mean librational / rotational frequency.* We remark that another (*non-resonance*) mechanism is due to *collisions,* which lead to the Debye relaxation (see Chap. 6).

Importance of the asymptotic case $y = 0$ may be illustrated as follows:

a. *The case of homogeneous U potential.* For relatively small y values, say, for $y \in [0, 1]$, the resonance absorption curve is *very wide* (usually it falls into the far IR region) and *weakly depends on y.* Consequently, Eq. 5.188 permit to estimate qualitatively dielectric response also for nonzero y (see the next Sec. 5.9.2).

b. *The case of a deep potential U.* In contrast to the previous case Landau damping is characterized by *narrow* absorption / loss lines, having *minimal linewidth.* Eq. 5.188 may be applied for calculation of the *limiting resonant lines.* The decrease of maximum absorption with the rise of y is very appreciable in this case. The Debye relaxation also strongly depends on the collision frequency y: the frequency ω_D of the maximum loss χ'' decreases with the decrease of y.

5.9.2 *The Loss Susceptibility of Free Planar Rotators*

Let us apply Eq. 5.188 b to the extended diffusion model, in which $\check{L} = \mathbf{0}$, $L = \overset{\circ}{L}$ and [see Sec. 5.5.2, Eq. 5.95]:

$$h(p) = p^2, W = \frac{\exp(-p^2)}{2\pi\sqrt{\pi}}, \overset{\circ}{f}(p) = \frac{2\pi}{p}\frac{1}{2\pi\sqrt{\pi}}\frac{d}{dp^2}\exp(-p^2)$$

$$= \frac{2}{\sqrt{\pi}}\exp(-p^2); \left\langle q_{\|}^2 \right\rangle = \left\langle q_{\perp}^2 \right\rangle = \frac{1}{2} \qquad (5.189)$$

Substituting Eq. 5.189 into (5.188b) and noting that Fourier amplitudes $a_1 = b_1 = 1; a_n = b_n = 0$ (for $n \neq 1$), we get

$$L_{\|}''(x) = L_{\perp}''(x) = L''(x) = \sqrt{\pi}\, x \exp(-x^2) \qquad (5.190)$$

Using Eq. 5.72 - 5.74, we find that for the quasi-spatial ensemble the loss $\chi''(x)$ and absorption $\alpha(x)$ are given by :

$$\chi''(x) = 2GL''(x) = 2G\sqrt{\pi}\, x \exp(-x^2)\,, \qquad (5.191)$$

$$\alpha(x) = 2\frac{\omega}{cn}4\pi\chi''(x) \approx \left(2\sqrt{\pi}\right)^3 \frac{G}{c\eta n_\infty} x^2 \exp(-x^2)\,. \qquad (5.192)$$

Here we assume that $n \approx n_\infty$ and $\varepsilon'' \approx 4\pi\chi''$, while multiplier 2 is introduced in accordance with Eq. 5.74.

The maximal absorption α_{max} and the frequencies x_L and $\omega_L = x_L / \eta$, at which it may be observed, are found from Eq. 5.192:

$$x_L = 1, \omega_L = \sqrt{\frac{2k_B T}{I}}, \alpha_{max} = \left(2\sqrt{\pi}\right)^3 \frac{G}{c\eta e n_\infty}. \qquad (5.193)$$

If $y \to 0$, the same result may be obtained from Eq. 5.99 for $L_F(z)$. It is easily seen, that on the level 1/2 the width $\Delta\omega$ of the absorption band is approximately equal to the peak frequency ω_L. This property is the characteristic feature of the extended diffusion model.

CHAPTER 6
COLLISION MODELS

6.1 The Effective Susceptibility at an Arbitrary Non-equilibrium Induced Distribution F

We proceed from Eq. 5.7. In our approximation of linear response to weak alternating electric field $\boldsymbol{E}(t)$ we may omit all terms with order of magnitude lesser than v, see Eq. 5.2. Since function $q(\mathrm{v})$ in the integrand of (5.7) does not depend on v, the only variable, which may change linearly with v, is the difference $dF = F(\gamma) - 1$, where $F(\gamma)$ is the induced distribution function, associated with the effect of alternating field $\boldsymbol{E}(t)$, and γ is the "initial" phase.

Thus, we regard that $(F - 1) \propto \mathrm{v}$. The simplest way to find the latter relation is to express $\delta F(\gamma)$ through $W\left(h + \tilde{\mathrm{v}}(0)\right)$, where $\tilde{\mathrm{v}}(0)$ is the value of the reduced alternating potential $\tilde{\mathrm{v}}(\varphi)$ found at the instant $\varphi = 0$ of strong collision. Using (5.12) and the normalized steady state distribution $W(h) = C\exp(-h)$, we write:

$$W\left(h + \tilde{\mathrm{v}}(0)\right) = C\ \exp\left(-h + 4\mathrm{v}q_0 \sin\gamma\right)$$

$$\approx C\ \exp\left(-h\right)\left(1 + 4\mathrm{v}q_0 \sin\gamma\right) = W(h)\ F(\gamma),$$

that is,

$$F(\gamma) = \approx \left(C/C_0\right)\left(1 + 4\mathrm{v}q_0 \sin\gamma\right). \tag{6.1a}$$

Since $\exp(-h) \approx W(h)\ /\ C_0$, we may now write an obvious normalizing condition for the function:

$$\int_{\Gamma} W\left(h + \tilde{v}(0)\right) d\Gamma \approx \int_{\Gamma} \exp(-h)\left(1 + 4vq_0 \sin\gamma\right) d\Gamma$$

$$= \frac{C}{C^0} \int_{\Gamma} \exp(-h)\left(1 + 4vq_0 \sin\gamma\right) W d\Gamma = \frac{C}{C^0}\left(1 + 4v\langle q\rangle \sin\gamma\right) = 1.$$

Thus, we get expression for the Boltzmann distribution F_B:

$$\frac{C}{C^0} = \frac{1}{1 + 4v\,\langle q\rangle \sin\gamma},$$

$$F_B = \frac{1 + 4vq_0 \sin\gamma}{1 + 4v\langle q\rangle \sin\gamma} \approx 1 + 4v\left(q_0 - q\right)\sin\gamma\ . \tag{6.1b}$$

It is useful to substitute Eq. 6.1b into the main normalizing condition (4.9) and to prove that it is satisfied. Indeed, we have from (4.15)-(4.18), (5.2), (6.1b):

$$\int t_v\, \delta N(t_0) = \int_0^{\infty} \varphi\, yN\, y \exp(-y\varphi) d\varphi = \int_{\Gamma} W(h) dh \int_0^{2\pi} [1$$

$$+\ 4v\left(q_0 - \langle q\rangle\right)\sin\gamma\Big]\frac{d\gamma}{\omega} = \frac{N}{\omega}\int_0^{2\pi}\left[1 + 4v\left(q_0 - \langle q\rangle\right)\sin\gamma\right] d\gamma = 2\pi\frac{N}{\omega} \quad ,(6.1c)$$

where $\varphi \equiv t_v/\tau$. Eq. 6.1b agrees with (4.9).

The following expression for $F(\gamma)$, generalizing Eq. 6.1b, will be used for an arbitrary F:

$$F(\gamma) = 1 + \left[4v\left(q_0 - \langle q\rangle\right)D_1(\gamma) + \dot{q}_0\, D_2(\gamma)\right]. \tag{6.2}$$

Then, substituting Eq. 6.2 into Eq. 5.7, we find the susceptibility χ^*:

$$\chi^* - \frac{z}{x}\chi_B^*$$

$$= 6G\frac{y}{x}\int_{\Gamma} W\, d\Gamma \int_0^{2\pi}\left[\frac{1}{4v} + (q_0 - q)D_1(\gamma) + \dot{q}_0\, D_2(\gamma)\right]\frac{\exp(i\gamma)}{2\pi}\, d\gamma \int_0^{\infty}\dot{q}(\varphi)\, d\varphi$$

$$= 3iG\frac{y}{x}\left\langle\left\{\int_0^{\infty}(q_0 - \langle q\rangle)\,\hat{A} + \hat{B}\,\dot{q}_0\right\} q(\varphi)\exp(iz\varphi)\, d\varphi\right\rangle; \qquad (6.3)$$

$$\left\{\begin{matrix}\hat{A}\\ \hat{B}\end{matrix}\right\} \equiv \frac{1}{i\pi}\int_0^{2\pi}\left\{\begin{matrix}D_1(\gamma)\\ D_2(\gamma)\end{matrix}\right\}\exp(i\gamma)\, d\gamma\,. \qquad \begin{matrix}(6.4a)\\ (6.4b)\end{matrix}$$

In all our models the quantity $\dot{q}$ changes its sign for particles, differing by the sign of their angular velocity and thus tends to zero, having averaged over the ensemble. Hence, we set:

$$\left\langle \dot{q}(\varphi)\,\langle q\rangle\right\rangle = \left\langle \dot{q}\right\rangle\langle q\rangle = 0. \qquad (6.4c)$$

Then the first term in (6.3) yields:

$$= \hat{A}\left\langle q_0 \int_0^{\infty}\dot{q}(\varphi)\exp(iz\varphi)\, d\varphi\right\rangle = \hat{A}\left\langle q^2\right\rangle L(z), \qquad (6.4d)$$

as it follows from (5.33). The second term in (6.3), which contains $\hat{B}$, may be written as follows:

$$= \hat{B}\left\langle \dot{q}_0 \int_0^{\infty}\dot{q}(\varphi)\exp(iz\varphi)\, d\varphi\right\rangle = -\hat{B}\left\langle \dot{q}_0\left(q_0 + iz\int_0^{\infty} q(\varphi)\exp(iz\varphi)\, d\varphi\right)\right\rangle,$$

$$= -\hat{B}\left(\left\langle \dot{q}_0 q_0\right\rangle + iz\left\langle \dot{q}_0\int_0^{\infty} q(\varphi)\exp(iz\varphi)\, d\varphi\right\rangle\right).$$

We note again that the average value of $q\dot{q} = 0$. Comparing the remaining term with (5.33), we have:

$$\hat{B}\left\langle \dot{q}_0 \int_0^\infty q(\varphi)\exp(iz\varphi)\,d\varphi \right\rangle = -iz\hat{B}\left\langle q^2\right\rangle L(z)\,. \qquad (6.4e)$$

Substituting (5.8), (6.4d), (6.4e) into (6.3), we finally find:

$$\chi^* = \frac{3G\left\langle q^2\right\rangle}{x}\left[z - iy\left(\hat{A} + iz\hat{B}\right)\right]L(z). \qquad (6.5)$$

We see, that the susceptibility χ^* is proportional to the SF $L(z)$ and depends on coefficients $\hat{A}$ and $\hat{B}$ and consequently - on the prescribed collision model.

6.2 The Boltzmann Induced Distribution F_B

The comparison of (6.2) with (6.1b) shows, that this distribution corresponds to $D_1 = \sin\gamma$ and $D_2 = 0$. Hence, we have in accord with (6.4 a, b):

$$\hat{A} = 1 \text{ and } \hat{B}.$$

Substitution into (6.5) yields:

$$\chi^* = \frac{3G\left\langle q^2\right\rangle}{x}(z - iy)L(z) = 3G\left\langle q^2\right\rangle L(z) = \chi_B^*\,. \qquad (6.6)$$

Thus, expression (5.8) for χ^* actually corresponds to $F = F_B$.

6.3 The Debye Induced Distribution F_D

The additional phase shift is characteristic for this collision model. To find F_D we apply the Debye theory of rotational diffusion (Debye 1929). The stochastic (Brownian) motion was considered in the latter in the configuration space as perturbed by the time-varying electric field. The Boltzmann equation was employed in this case. In accord with the Debye theory we may write for $F = F_D$:

$$F_D = 1 + 4\mathrm{v}(q_0 - q)\,\mathrm{Im}\left\{\frac{\exp(i\omega\tau_D)}{1 + i\omega\tau_D}\right\}. \qquad (6.7a)$$

The parameter τ_D (Debye relaxation time) characterizes the phase shift (and consequently- collective effects) in a liquid. The symbol Im is used here, since

$$E(t_0) = E_m \sin \omega t_0 = E_m \sin \gamma = -i E_m \,\mathrm{Im}\left\{\exp i\gamma\right\}. \qquad (6.7b)$$

Comparing Eq. 6.7a and 6.2 with account of Eq. 6.4a, we derive:

$$D_1 = \frac{\sin \gamma - \omega \tau_D \cos \gamma}{1 + (\omega \tau_D)^2},\ D_2 = 0,\ \hat{A} = -\frac{1}{1 - i\omega \tau_D},\ \hat{B} = 0; \qquad (6.7c)$$

$$\chi^* = \left(1 + \frac{\tau_D / \tau}{1 - i\omega\tau_D}\right) \chi_B^* = 3G\left\langle q^2 \right\rangle \left(1 + \frac{\tau_D / \tau}{1 - i\omega\tau_D}\right) L(z). \qquad (6.8)$$

6.4 The Self-Consistent Induced Distributions F_G, F_K, F_T

We start from the relation $\hat{P} = \chi \hat{E}$ between the complex amplitude $\hat{P}$ of the polarization and the susceptibility $\chi = \chi' - i\chi''$. That is, taking (4.24) into account, we have

$$P(t_0) = \mathrm{Re}\left\{\hat{P}\exp(i\gamma)\right\} = \mathrm{Re}\left\{\chi \hat{E}\exp(i\gamma)\right\} = \frac{E_m}{2i}\left\{\chi\exp(i\gamma) - \chi^*\exp(-i\gamma)\right\}. \quad (6.9)$$

We remark that amplitude $\hat{P}$ is just the product $\chi \hat{E}$ (but not $\hat{P} = \chi^* \hat{E}$!), owing to which the polarization

$$P(t) = \mathrm{Re}\left\{\hat{P}\exp(i\omega t)\right\} = \mathrm{Re}\left\{\chi \hat{E}\exp(i\omega t)\right\}$$

of a polar medium lags behind the applied field:

$$E(t) = \mathrm{Re}\left\{\hat{P}\exp(i\omega t)\right\} = \mathrm{Re}\left\{\hat{E}\exp(i\omega t)\right\}.$$

Indeed, in accord with (4.24)

$$E(t) = E_m \sin \omega t\ , \qquad (6.9a)$$

while

$$P(t) = E_m\{\chi' \sin\omega t - \chi''\cos\omega t\} = P_m \sin(\omega t - \gamma_1). \qquad (6.9b)$$

Here

$$P_m = E_m|\chi|, \quad \gamma_1 = \mathrm{a}\sin\frac{\chi''}{|\chi|}, \quad 0 \le \gamma_1 \le \frac{\pi}{2}.$$

6.4.1 The Gross Collision Model

It is assumed that an ensemble under consideration consists of non-interacting molecules, viz. we assume that trajectories of particles are statistically independent one from another. The polarization of a unit volume, containing N polar molecules, is defined as $\langle\mu_E\rangle$ multiplied by the concentration N:

$$P(t_0) = N\langle\mu_E\rangle = N\mu\int_\Gamma q_0 FW\, d\Gamma \qquad (\text{at } F = F_G). \qquad (6.10a)$$

Accounting for definition of v (see Eq. 5.2 and 6.9), we rewrite this expression as follows

$$\frac{3G}{4\mathrm{v}}\int_\Gamma q_0 FW\, d\Gamma = \frac{\chi\exp(i\gamma) - \chi^*\exp(-i\gamma)}{2i}. \qquad (6.10b)$$

Substituting F from Eq. 6.2 for $D_2 = 0$ and multiplying both sides by $\exp(i\gamma)$, we have:

$$3G\exp(i\gamma)\left(\frac{\langle q\rangle}{4\mathrm{v}} + D_1\Delta\right) = \frac{\exp(i\gamma)}{2i}\left[\chi\exp(i\gamma) - \chi^*\exp(-i\gamma)\right], \qquad (6.11)$$

where the dispersion (= average fluctuation) of direction cosine q is denoted as

$$\Delta \equiv \langle q^2\rangle - \langle q\rangle^2. \qquad (6.12a)$$

The physical meaning of this quantity becomes evident after it is reduced to the form

$$\Delta \equiv \langle q^2\rangle - 2\langle q\rangle\langle q\rangle + \langle q\rangle^2 = \left\langle q^2 - 2\langle q\rangle\langle q\rangle + \langle q\rangle^2\right\rangle = \left\langle (q - \langle q\rangle)^2\right\rangle. \qquad (6.12b)$$

Averaging (6.11) over γ and noting (6.4a) and (6.5), we derive relation:

$$\chi^* = 3G\hat{A}\left(\left\langle q^2\right\rangle - \langle q\rangle^2\right);$$

$$\hat{A} \equiv \frac{\chi^*}{3G\Delta} \tag{6.13}$$

Substituting into Eq. 6.5 this value of the complex parameter $\hat{A}$ and $\hat{B} = 0$, we obtain:

$$\chi^* \equiv \frac{3G\left\langle q^2\right\rangle z\,L(z)}{x + iyL(z)\left\langle q^2\right\rangle/\Delta} \,. \tag{6.14}$$

6.4.2 The Correlation-Orientation Distribution F_K

It is suggested, that neighboring dipoles rotate to some extent simultaneously. In other words, cross correlation appears in this more realistic collision model. Thus, it is assumed that the total dipole moment of the medium is $gN\langle\mu_E\rangle$, but not $N\langle\mu_E\rangle$, where g is the *Kirkwood correlation factor.* For $g > 1$ the effective dipole moment of a molecule increases due to this correlation, whereas for $g < 1$ this decreases. So, the r.h.s. of Eq. 6.10a should be multiplied by g:

$$P(t_0) = N\mu g \int_\Gamma q_0 FW\, d\Gamma \qquad (\text{at}\, F_K) \tag{6.15a}$$

Correspondingly, Eq. 6.13 changes as follows:

$$\hat{A} \equiv \frac{\chi^*}{3Gg\Delta} \,. \tag{6.15b}$$

For this $\hat{A}$ and $\hat{B} = 0$ we obtain from Eq. 6.5 instead of (6.14):

$$\chi^* \equiv \frac{3Gg\left\langle q^2\right\rangle z\,L(z)}{gx + iy\,L(z)\left\langle q^2\right\rangle/\Delta} \quad (\text{at } F_K). \tag{6.16}$$

This result corresponds to the so-called (see GT, p.190) *correlation-orientation* induced distribution F_K. The latter improves the Gross collision model. In the absence of cross correlation (for $g = 1$) Eq. 6.16 reduces to (6.14).

Let us consider an *isotropic medium* as "constructed" from small dielectric pieces (as if domains), each being characterized by a local anisotropy.

In this and further examples (unlike those considered in Sec. 6.9) we shall assume that in each such anisotropic part the ensemble average of q is equal to zero,

---

$$\langle q \rangle = \int_{\Gamma} q \, d\Gamma \, \Big|_{\text{local-order part}} = 0. \tag{6.16a}$$

Then the average of q^2 over Γ in (6.12) and over the additional variable Ψ [see Sec. 5.3.2 and Eq. 5.55-5.56, 5.68] becomes equal to

$$\overline{\langle q^2 \rangle} = \overline{\Delta} = \frac{1}{3} \tag{6.16b}$$

For *spatial* or *quasi-spatial* ensembles of dipoles Eq. 6.16 yields:

$$\chi^* \equiv \frac{Gg\, z L(z)}{gx + iyL(z)} \tag{6.17a}$$

$$\equiv \frac{g\, z \overset{*}{\chi}_B}{gx + iy\dfrac{\overset{*}{\chi}_B}{G}} \tag{6.17b}$$

These and other expressions will be generalized in Sec. 6.9 for the case when, in contrast to Eq. 6.16a,

$$\langle q \rangle = \int_{\Gamma} q \, d\Gamma \, \Big|_{\text{local-order part}} \neq 0. \tag{6.17c}$$

6.4.3 The Isothermal Collision Model ($F = F_T$)

We consider finally a somewhat more complex collision model. Taking into account the induced perturbation of the velocity distribution, we assume that $D_2 \neq 0$ in Eq. 6.2. Consequently, we put $\hat{B} \neq 0$ in Eq. 6.4b. This self-consistent distribution F_T corresponds to the so called (GT, p.192) *isothermal collision model*, with expression (6.15b) remaining valid. Then in calculating coefficients D_2 and $\check{B}$ *angular velocities unlike orientations are considered to be uncorrelated.* Therefore, the time derivatives over t_0 are taken from Eq. 6.10 by setting $g = 1$. Noting that

$$\frac{dq_0}{dt} = \frac{dq_0}{\eta d\varphi} = \frac{\dot{q}_0}{\eta},$$

we have from Eq. 6.9b and 6.10a:

$$\frac{dP(t_0)}{dt_0} = \frac{N\mu}{\eta} \int_\Gamma q_0 FW\, d\Gamma = \chi \exp(i\gamma) + \chi^* \exp(-i\gamma) \ . \qquad (6.18)$$

Accounting for Eq. 5.158, we have by analogy with Eq. 6.13:

$$\hat{B} \equiv \frac{x\chi^*}{3iG\left\langle \dot{q}^2 \right\rangle} = \frac{2x\chi^*}{3iG\left(1 - \left\langle q^2 \right\rangle\right)}. \qquad (6.19)$$

Substituting Eq. 6.19 into Eq. 6.5, we obtain

$$\chi^* \equiv \frac{3Gg\left\langle q^2 \right\rangle z\, L(z)}{gx + iy L(z)\left\langle q^2 \right\rangle \left(\frac{1}{\Delta} + \frac{2xzg}{3\left(1 - \left\langle q^2 \right\rangle\right)} \right)} \quad \text{(at } F_T\text{)}. \qquad (6.20)$$

For an *isotropic medium* we put

$$\overline{\left\langle q^2 \right\rangle} = \overline{\Delta} = \frac{1}{3} \ .$$

Hence,

$$\chi^* \equiv \frac{Gg\, z\, L(z)}{gx + iy L(z)\, \left(1 + gxz\right)} \qquad (6.21)$$

$$\equiv \frac{g\, z\, \chi_B^*}{gx + iy\left(1 + gxz\right)\chi_B^*/G} \quad \text{(for } F_T\text{)}. \qquad (6.21a)$$

* * *

All collision models, considered above, satisfy the Gordon sum rule (see Sec. 5.7), whereas the last one (for F_T) has a satisfactory behavior al also for the higher order moment:

$$\Sigma_1 = \int_{-\infty}^{\infty} x^3 \chi''(x) dx < \infty \quad \text{for } F_T, \tag{6.22}$$

while

$$\Sigma_1 = \infty \text{ for } F_B, F_G, F_D, F_K. \tag{6.22a}$$

Because of this distinction the isothermal collision model possess better asymptotic behavior on the high frequency edge of the orientational relaxation region, i.e. in the far IR range.

Some other properties of collision models will be considered further.

6.5 The Spectral Function K(z). The Summary Table for the Susceptibility

For an *anisotropic medium* it is convenient to introduce the SF *K(z)*, which differs *from L(z)* only by a constant coefficient, depending on the molecular model:

$$K(z) = K' + iK'' = 3\langle q^2 \rangle L(z) = \frac{\chi_B}{G} \tag{6.23}$$

$$= 3iz \int_0^{\infty} \langle q_0 [q(\varphi) - q_0] \rangle \exp(iz\varphi)\, d\varphi . \tag{6.24}$$

This definition of the *K(z)* spectral function is based on Eq. 5.32 and 5.35.

It is seen from (6.24), that *K(z)* is proportional to the spectrum of the undamped dipolar relaxation function, caused by molecular rotation in a conservative potential *U*. We introduce also the dispersion σ of the direction cosine q at equilibrium:

$$\sigma = 3\left(\langle q^2 \rangle - \langle q \rangle^2\right) \equiv 3\Delta . \tag{6.25}$$

For basic orientations, denoted by the symbols $\parallel$ and $\perp$, we have

$$K = \left\{K_{\|}, K_{\perp}\right\}, \quad \sigma = \left\{\sigma_{\|}, \sigma_{\perp}\right\}. \tag{6.26}$$

Let us consider spatial and quasi-spatial ensembles. For the *titled propagation of electromagnetic wave* we have in accord with Eq. 5.54a, 5.54d for $\begin{pmatrix} \text{spatial} \\ \text{quasi-spatial} \end{pmatrix}$ ensembles:

$$\left\{K, \sigma, \left\langle q^2 \right\rangle\right\} = \left\{K, \sigma, \left\langle q^2 \right\rangle\right\} \cos^2\Psi + \begin{pmatrix} 1 \\ 1/2 \end{pmatrix} \left\{K, \sigma, \left\langle q^2 \right\rangle\right\} \sin^2\Psi. \tag{6.27}$$

AN ISOTROPIC MEDIUM. In this case both SFs are equal one to another. For instance, for spatial or quasi-spatial ensembles

$$K(z) \equiv L(z), \quad 3\left\langle q^2 \right\rangle = 1. \tag{6.28}$$

Let the subscripts «pl» and «sp» refer to planar and spatial ensembles, respectively. We calculate the *L(z)* spectral function of an isotropic medium using the following expressions:

$$L(z) = \begin{Bmatrix} L_{\text{pl}} \\ 2L_{\text{pl}} \\ L_{\text{sp}} \end{Bmatrix} \text{ for } \begin{Bmatrix} \text{quasi-spatial} \\ \text{quasi-spatial with the doubled SF} \\ \text{spatial} \end{Bmatrix} \text{ ensembles} \qquad \begin{matrix} (6.29) \\ (6.30) \\ (6.31) \end{matrix}$$

$$L_{\text{pl}}(z) = \frac{1}{3}\left(K_{\|} + K_{\perp}\right) \text{ with } \begin{Bmatrix} K_{\|} \text{ and } K_{\perp} \text{ calculated} \\ \text{for a motion in a plane} \end{Bmatrix} \tag{6.32}$$

$$L_{\text{sp}}(z) = \frac{1}{3}\left(K_{\|} + 2K_{\perp}\right) \text{ with } \begin{Bmatrix} K_{\|} \text{ and } K_{\perp} \text{ calculated} \\ \text{for spatial motion} \end{Bmatrix} \tag{6.33}$$

THE TABLE FOR $\chi^*(x)$. In Table 6.1 we present a number of expressions just obtained for various collision models. For an anisotropic medium the formulas are given in terms of the *K(z)* spectral function and of the dispersion σ. For an isotropic medium the formulas are given in terms of the SF *L(z)*. In accord with Eq. 6.32, 6.33 *L* is fully determined by the basic SFs $K_{\|}$ and $K_{\perp}$.

THE TABLE FOR $\chi^*(x)$. In Table 6.1 we present expressions just obtained for various collision models. For an anisotropic medium the formulas are given in terms of the $K(z)$ spectral function / dispersion σ. For an isotropic medium the formulas are given in terms of the SF $L(z)$. In accord with Eq. 6.32, 6.33 L is fully determined by the basic SFs $K_\|$ and $K_\perp$.

6.6 The Low-Frequency Approximation

6.6.1 Parameters of the Debye Relaxation Expressed in Terms of the Molecular-Model Parameters

In the low-frequency region, where Debye relaxation usually takes place, we may put $x = 0$ in the SFs $L(z)$ or $K(z)$; these we designate L_0 or K_0 (which are real):

$$L(z) \to L_0 \equiv L(iy);\ K(z) \to K_0 \equiv K(iy). \tag{6.34}$$

This approximation is justified for typical condensed media; the normalized collision frequency y has usually an order of magnitude 1 or 0.1 and the Debye relaxation takes place in the frequency interval $x \in [0, 5x_D]$, where $x_D = \eta / \tau_D << 1$. In Eq. 6.34 we ignore the frequency dependence of the SF $L(z)$; the latter usually reveals itself at higher frequencies, viz. at $x > 0.1$.

We shall show that, if the SFs are calculated in the approximation (6.34), then the theoretical-susceptibility spectrum $\chi^*(z)$, found for all induced distributions F (except F_B), practically coincides with the spectrum, found from the Debye formula

$$\chi^*(x) = \chi_\infty + \frac{\chi_s - \chi_\infty}{1 - x\tau_D/\eta} \quad \text{(for } x << 1\text{)}. \tag{6.35}$$

The parameters of this formula have the following meaning.

(a). The static susceptibility:

$$\chi_s = \chi^*(x)\big|_{x=0}. \tag{6.36}$$

(b). The susceptibility for "infinite frequency" in terms of the rotational-diffusion Debye theory:

$$\chi_\infty = \chi^*(x)\big|_{x\to\infty} \tag{6.37}$$

(c). The Debye relaxation time:

$$\tau_D = \frac{\eta}{\chi_s - \chi_\infty}\lim_{x\to 0}\frac{\chi''(x)}{x} \tag{6.38}$$

Table 6.1 The effective susceptibility χ^* for spatial and quasi-spatial ensembles (the double-well configuration)

F	Anisotropic ensemble	Isotropic ensemble
F_B	$\chi^*_B = GK$	
F_D	$\left(1 + \frac{\tau_D/\tau}{1 - i\omega\tau_D}\right) GK$	
F_G	$\frac{GzK}{x + iyK/\sigma}$	$\frac{GzL}{x + iyL}$
F_K	$\frac{gGzK}{gx + iyK/\sigma}$	$\frac{gGzL}{gx + iyL}$
F_T	$\frac{gGzK}{gx + iy\left(\frac{1}{\sigma} + \frac{2xzg}{3\left(1 - \langle q^2\rangle\right)}\right)K}$	$\frac{gGzL}{gx + iyL(1 + gxz)}$

We remark that τ_D is related to the maximal dielectric loss:

$$\max_x\{\chi''(x)\} \equiv \chi''_D = \frac{1}{2}(\chi_s - \chi_\infty) \qquad \text{at } x = x_D = \frac{\eta}{\tau_D} \tag{6.39}$$

The Debye parameters (6.36) - (6.39) are usually known from *the experiment.* Our approach, realized in frames of the dynamic method, permits to relate Debye parameters to those, that appear in the relevant molecular model. In order to find the corresponding relationships we apply Eq. 6.35 -

6.38 to formulas, presented in Table 6.1. The results, so obtained, are given in Table 6.2.

For the *isothermal* collision model, where $F = F_T$, expression (6.35) is obtained using the first-order (with respect to small x) approximation; neglecting the term, proportional to x^2, we replace by $igxy$ the product gxz in the denominator.

The Debye relaxation is not described by the susceptibility $\chi_B^*(x)$, since *for the Boltzmann distribution* $F = F_B$ the static susceptibility coincides with χ_∞.

For the induced *Debye distribution* F_B the estimated relaxation time τ_D^{est} coincides with the parameter τ_D of this collision model.

For the *self-confident induced distributions* F_G , F_K and F_T the susceptibility $\chi^*(x)$ may be represented as

$$\chi^*(x) = \chi_\infty + \frac{\chi_s - \chi_\infty}{1 - ix\chi_s/(y\chi_\infty)} \quad \text{for } y \ll 1. \tag{6.40}$$

On comparing this with (6.35) we find that the estimated relaxation time τ_D^{est} is proportional to the lifetime τ and to the ratio of the susceptibilities at zero and at very high frequencies:

$$\tau_D^{\text{est}} = \tau\chi_s / \chi_\infty \ . \tag{6.41}$$

Consequently, the relaxation time should increase, if the low-frequency fall of χ' becomes more pronounced.

6.6.2 Interpretation of Debye Relaxation in Terms of Statistical Averages

It is seen from Table 6.2 that the Debye parameters are expressed through:

(i) the mean square $\langle q^2 \rangle$ of the direction cosine,

(ii) the dispersion $\sigma = 3\Delta = 3\left[\langle q^2 \rangle - \langle q \rangle^2\right]$;

(iii) the "static" SFs K_0 or L_0.

Table 6.2 Relation of the Debye relaxation parameters to the parameters of a molecular model

Parameter	F	Anisotropic medium	Isotropic medium
χ_S	for F_B	GK_0	GL_0
	for F_D	$GK_0\{1+\tau_D/\tau\}$	$GL_0\{1+\tau_D/\tau\}$
	for F_G	σG	G
	for F_K, F_T	$g\sigma G$	gG
χ_∞	for $\begin{Bmatrix} F \neq F_T \\ F = F_T \end{Bmatrix}$	$\begin{Bmatrix} 1 \\ \dfrac{1}{1-\zeta} \end{Bmatrix} GK_0$	$\begin{Bmatrix} GL_0 \\ \dfrac{GL_0}{1-y^2L_0} \end{Bmatrix}$
χ''_D	for F_D	$GK_0\,\tau_D/(2\tau)$	$GL_0\,\tau_D/(2\tau)$
	for F_B	0	
	for F_G	$(G/2)\,(\sigma - K_0)$	$(G/2)\,(1 - L_0)$
	for F_K	$(G/2)\,(g\sigma - K_0)$	$(G/2)\,(g - L_0)$
	for F_T	$\dfrac{G}{2}(g\sigma - \dfrac{K_0}{1-\zeta})$	$\dfrac{G}{2}(g - \dfrac{L_0}{1-y^2L_0})$
τ_D	for F_B	Low-frequency relaxation is not described	
	for F_D	τ_D	
	for $\begin{Bmatrix} F_G \\ F_K \end{Bmatrix}$	$\begin{Bmatrix} 1 \\ g \end{Bmatrix} \dfrac{\tau\sigma}{K_0}$	$\begin{Bmatrix} 1 \\ g \end{Bmatrix} \dfrac{\tau}{L_0}$
	for F_T		$\dfrac{\tau g\left(1-y^2L_0\right)}{L_0}$
Note	$\varsigma = (2/3)\,y^2\,K_0\left[1-\left\langle q^2\right\rangle\right]^{-1}$		

The dispersion Δ means, in view of Eq. 6.12b, the average fluctuation of a dynamic variable q. The latter presents the projection of the normalized dipole moment μ_E/μ on the vector $\boldsymbol{E}_m$, where, in accord with Fig. 3.3a, $\mu_Z \propto q_{\|}$ and $\mu_Y \propto q_{\perp}$, respectively for longitudinal and transverse orientations of alternating field. The fluctuation Δ is a measure of the *thermal* dispersion of dipole moments over their median orientations $\bar{\theta}$. In view of Fig. 3.3b $\bar{\theta}$ corresponds to the trajectory of a given particle. The wider distribution of dipole moments over $\bar{\theta}$, the greater Δ. The dependence of the Debye parameters on K_0 or L_0 becomes clear, if one considers a long lifetime limit; this approximation is often applicable for qualitative estimations.

Denoting $\tilde{q} = q - \bar{q}$, let us put $y << 1$ and take for example the Gross collision model. In accord with Eq. 5.132 and 5.135 the "static" spectral function is estimated for this model as $L_0 \approx \langle \tilde{q}^2 \rangle / \langle q^2 \rangle$. We have:

$$\frac{\tau_D}{\tau} = \frac{\sigma}{K_0} = \frac{\Delta}{\langle \tilde{q}^2 \rangle} \equiv \frac{\left\langle \left(q - \langle q \rangle\right)^2 \right\rangle}{\left\langle \left(q - \bar{q}\right)^2 \right\rangle}; \tag{6.42}$$

$$\frac{2\chi_D''}{3G} = \Delta - \langle q^2 \rangle L_0 = \Delta - \langle \tilde{q}^2 \rangle = \left\langle \left(q - \langle q \rangle\right)^2 \right\rangle - \left\langle \left(q - \bar{q}\right)^2 \right\rangle. \tag{6.43}$$

Here $\bar{q}$ is a constant component of the $q(\varphi)$ dependence, that corresponds to a given trajectory of a particle. The difference $\tilde{q} = q - \bar{q}$ characterizes a *regular* deviation of a particle relative to its median orientation $\bar{\theta}$, while $\langle \tilde{q}^{\,2} \rangle$ is a mean squared average of this difference, so that if $|\tilde{q}|$ is small, it is approximately equal to the nutation amplitude δ of a dipole (see Fig. 3.3b).

The relaxation time is proportional to the ratio of "thermal" to "regular" statistic averages; the maximum loss is proportional to their difference, see Eq. 6.43. The greater thermal dispersion relative to the regular one, the greater relaxation time / Debye loss. (6.44)

We conclude. Various molecular models may differ in respect with their Debye relaxation ranges, e.g. they may be characterized by different relaxation times τ_D and loss peaks χ''_D. For a long lifetime τ (this approximation is often applicable for a qualitative description of orientational polarization) the Debye parameters are expressed by two statistical averages:

(a) by the mean squared fluctuation $\left\langle \left(q - \langle q \rangle\right)^2 \right\rangle$, determined by the thermal dispersion of orientations of particles;

(b) and by the localization $\left\langle \tilde{q}^{\,2} \right\rangle$, which has the sense of the square of the libration amplitude of a single particle, performing periodic motion in a potential well.

ISOTROPIC MEDIUM. If $\langle q \rangle = 0$ for a given potential, then a dipole moment of a whole sample does not appear; the average $\Delta = \left\langle \left(q - \langle q \rangle\right)^2 \right\rangle =$ 1/3. In this case only localization $\left\langle \tilde{q}^{\,2} \right\rangle$ is responsible for the Debye parameters. *The narrower localization of a single particle during its regular motion, the greater are τ_D and χ''_D.*

A. Let the intermolecular potential $U = 0$, see Chap.7 devoted to the extended diffusion (ED) model. The Debye relaxation does not appear in an ensemble of freely rotating linear molecules, since a dipole has no localization of orientations. However, the Debye region arises in an ensemble of freely rotating axisymmetric *molecular tops*; τ_D and χ''_D increase, when a top becomes more oblate (i.e. when the form factor b of a molecule decreases). If $U = 0$, *localization of dipoles occurs only due to precession of a top.*

B. The localization $\left\langle \tilde{q}^{\,2} \right\rangle$ and thus the relaxation time τ_D may considerably increase, if motion of a dipole is affected by a strong intermolecular potential $U(\vartheta)$. For example, one may consider hard surfaces, restricting librations of dipoles. Such surfaces constitute the basis of the confined rotator (CR) and cone-confined rotator (CCR) models, see Chap. 8. Hence, the relaxation time τ_D increases in the CR model, if the libration amplitude decreases.

ANISOTROPIC MEDIUM. If a *macroscopic* anisotropy of a dielectric sample appears due to intermolecular / external conservative electric field $\boldsymbol{E}_0$, a new interesting effects arise, which depend on the orientation of alternating field $\boldsymbol{E}(t)$ relative to the symmetry axis Z of a sample.

C. Let us consider the *double well potential* (DWP), that is employed e.g. in the cosine squared model.

For a *longitudinal orientation* ($\boldsymbol{E}_m \parallel Z$) the induced alternating concentration $\delta N(t)$ of dipoles appears near bottoms of the wells due to $\boldsymbol{E}(t)$ field [for low frequencies the latter may be regarded as a quasi-static field]. If $\mu E_0 >> k_B T$, the induced dipole moment $\delta\mu(t)$, corresponding to the excess concentration $\delta N(t)$, is aligned approximately along $\boldsymbol{E}_0$. The relaxation time $\tau_D^{\parallel}$ is much longer than the time $\tau_D^{\perp}$ for the transverse ($\boldsymbol{E}_m \perp Z$) orientation. This distinction enhances, when the welldepth U_0 increases. The increase of $\tau_D^{\parallel}$ is caused by a *large* angle, comparable with π, through which a dipole turns in a condensed medium (overcoming the friction), when quasi-constant $\boldsymbol{E}(t)$ field is switched off. Owing to this rotation the excess concentration $\delta N(t)$ disappears and the steady state orientational distribution of polar molecules is established.

The aforementioned excess concentration practically does not appear in the case of *transverse* field $\boldsymbol{E}(t)$, since perturbations of dipole's trajectories, induced by alternating field, disappear a short time after radiation is switched off.

D. In the single well *potential* [e.g. in the constant field (CF) model] both relaxation times, $\tau_D^{\parallel}$ and $\tau_D^{\perp}$, have the same order of magnitude. In this case there is no other well, to which a dipole could fall after external field $\boldsymbol{E}(t)$ is switched off. Consequently, the dependence of dielectric behavior on the number of wells, characterizing the prescribed potential $U(\vartheta)$, is caused by the distinction between magnitudes of thermal dispersion of orientations in these models. Indeed, for the double well potential (e.g. for the cosine squared well, where $\langle q \rangle = 0$, the dispersion $\Delta = \langle q^2 \rangle$ is great; a dipole turns through an angle $\approx \pi$, while in the CF model this dispersion is small along with the localization parameter $\left\langle \tilde{q}^{\,2} \right\rangle$.

6.6.3 *Other Parameters of the Low-Frequency Behavior*

For description of dielectric relaxation in molecular ensembles with use of the dynamic method the induced distribution $F(\gamma)$ is employed. The latter yields an *implicit* description of stochastic motion. Considering some molecular model, characterized (for the prescribed potential well U) by the localization $\left\langle \tilde{q}^{\,2} \right\rangle$, we actually suppose that spatial restrictions, obliged to intermolecular interactions, may from time to time be overcome by stochastic torques that arise owing to *strong collisions.* We introduce into the theory few parameters (determined from the experiment) as the price paid for such implicit description of collisions. These parameters are:

$$\left\{\begin{array}{l} \varepsilon_S \text{ (static permittivity),} \\ \tau_D \text{ (relaxation time)} \\ g \text{ (Kirkwood correlation factor)} \\ n_\infty \text{ (optical refraction index)} \end{array}\right\} \tag{6.45}$$

Using Eq. 4.39, we have:

$$\frac{\varepsilon^* - n_\infty^2}{4\pi} \frac{2\varepsilon^* + n_\infty^2}{3\varepsilon^*} = \chi^*(x); \tag{6.46}$$

for any ω (or any $x = \omega\eta$) in the region of orientational relaxation the r.h.s. is taken from Table 6.1.

One should fit the "free" parameters of the employed molecular model in order to reach the coincidence of the *calculated* static permittivity / relaxation time with correspondent experimental values (6.45). Consequently, we may write some additional relations between macroscopic and microscopic quantities. In the low-frequency region ($0 < \omega < 5/\tau_D$) we can simplify calculations, if we use in the r.h.s. of Eq. 6.46 expressions, given in Table 6.2.

We remark, in particular, that the calculated static susceptibility is

$$\frac{\varepsilon_S - n_\infty^2}{4\pi} \frac{2\varepsilon_S + n_\infty^2}{3\varepsilon_S} = \chi^*(0) \equiv \chi_S; \ \{\chi_S \text{ from Table 6.2}\}; \tag{6.47}$$

while at the *infinite frequency* (in the sense of the Debye theory, for which $\varepsilon_S \equiv \varepsilon_\infty$) the calculated susceptibility is given by

$$\frac{\varepsilon_\infty - n_\infty^2}{4\pi} \frac{2\varepsilon_\infty + n_\infty^2}{3\varepsilon_\infty} = \chi^*(\infty) \equiv \chi_\infty . \tag{6.48}$$

We present here few comments concerning Eq. 6.47 and 6.48.

ABOUT THE DEBYE DISTRIBUTION F_B. If $x = 0$, we have from Table 6.2 and Eq. 6.46:

$$\frac{\varepsilon_S - n_\infty^2}{4\pi} \frac{2\varepsilon_S + n_\infty^2}{3\varepsilon_S} = G\left(1 + \frac{\tau_D}{\tau}\right) K_0 \ \left\{\begin{matrix}\text{for anisotropic} \\ \text{medium}\end{matrix}\right\}, \tag{6.49}$$

$$\frac{\varepsilon_S - n_\infty^2}{4\pi} \frac{2\varepsilon_S + n_\infty^2}{3\varepsilon_S} = G\left(1 + \frac{\tau_D}{\tau}\right) L_0 \ \left\{\begin{matrix}\text{for isotropic} \\ \text{medium}\end{matrix}\right\}. \tag{6.50}$$

The free parameters of the model must be fitted in such a way that Eq. 6.49 or 6.50 should hold.

ABOUT THE ORIENTATION-CORRELATION DISTRIBUTION F_K. From Table 6.2 and Eq. 6.47 we have for $x = 0$:

$$\frac{\varepsilon_S - n_\infty^2}{4\pi G} \frac{2\varepsilon_S + n_\infty^2}{3\varepsilon_S} \frac{1}{3\left(\left\langle q^2\right\rangle - \left\langle q\right\rangle^2\right)} = g \ \{\text{for anisotropic medium}\}, \tag{6.51}$$

$$\frac{\varepsilon_S - n_\infty^2}{4\pi G} \frac{2\varepsilon_S + n_\infty^2}{3\varepsilon_S} = g \ \{\text{for isotropic medium}\}. \tag{6.52}$$

Eq. 6.52 is satisfied, since it coincides with the definition of g. Consequently, the true value of the static susceptibility is achieved "automatically" in the case of *isotropic* medium. For *anisotropic* medium Eq. 6.51 actually yields the definition of the Kirkwood correlation factor g.

On the other side, from Table 6.2 we have equations to be satisfied in order to reach the coincidence of the calculated and experimental relaxation times:

$$\tau_D = \tau g\sigma / K_0 \qquad \{\text{for anisotropic medium}\}, \tag{6.53}$$

$$\tau_D = \tau g / L_0 \qquad \{\text{for isotropic medium}\}. \tag{6.54}$$

We shall see later that these equations impose serious restrictions on fitted lifetime τ and / or other parameters of the prescribed molecular model. As a result, many models should be rejected, since Eq. 6.53 or 6.54 do not hold *simultaneously* with the corresponding conditions, imposed on the calculated absorption bandwidth $\Delta\nu$ in the far IR spectral region.

FOR THE ISOTHERMAL COLLISION MODEL ($F = F_T$). Eq. 6.51 and 6.52 for the static susceptibility are also valid for F_T (see Table 6.2), while the conditions for the relaxation time becomes:

$$\tau_D = \frac{\tau g \sigma}{K_0}\left(1 - \frac{2}{3}\frac{y^2 K_0}{1 - \langle q^2 \rangle}\right) \qquad \{\text{for anisotropic medium}\}, \tag{6.55}$$

$$\tau_D = \frac{\tau g}{L_0}\left(1 - y^2 L_0\right) \qquad \{\text{for isotropic medium}\}. \tag{6.56}$$

It will be shown later, that using conditions (6.55) and (6.56) for F_T we should fit slightly other (probably, more realistic) parameters of the molecular model, than for F_K from Eq. 6.53, 6.54.

THE DEBYE PERMITTIVITY ε_∞. The physical sense of this parameter is rather vague, since from the viewpoint of a more rigorous theory the Debye limit $\omega \to \infty$ has no fundamental meaning. Eq. 6.48 may be used for the determination of some molecular parameters (if ε_∞ is known) or for the prediction of ε_∞. For instance, for an isotropic medium we have the following equations:

$$\frac{\varepsilon_\infty - n_\infty^2}{4\pi}\,\frac{2\varepsilon_\infty + n_\infty^2}{3\varepsilon_\infty} = G L_0 \begin{Bmatrix} 1 \\ \dfrac{1}{1 - y^2 L_0} \end{Bmatrix} \text{ for } \begin{Bmatrix} F = F_K \\ F = F_T \end{Bmatrix}. \tag{6.57}$$

THE DEBYE PLATEAU α_∞. If Eq. 6.56, 6.57 are satisfied, then for the Debye limit $\omega \to \infty$ we can calculate the α_∞ value of absorption $\alpha(\omega)$:

$$\alpha_\infty = \lim_{\omega \to \infty} \left[\frac{\omega\, \varepsilon''(\omega)}{c\sqrt{\varepsilon^*(\omega)}} \right]_{L \equiv L_0} = \frac{\varepsilon_S - \varepsilon_\infty}{c\tau_D \sqrt{\varepsilon_\infty}} \tag{6.58}$$

In view of Eq. 6.50, 6.56, 6.57 the Debye plateau is directly related to the parameters of the prescribed molecular model. Equations 6.57 and 6.58 are obtained, if

$$L(z) = L(iy);\; x = \omega\eta \le 5\eta / \tau_D << 1. \tag{6.59a}$$

For a *formal* transition to $\omega \to \infty$ the following limits are valid:

$$\varepsilon' \to \varepsilon_\infty,\; \chi' \to \chi_\infty,\; \alpha \to \alpha_\infty,\; \chi'' \to 0. \tag{6.59b}$$

In terms of the dynamic method the Debye the approximations (6.34) does not hold for $\varepsilon' < \varepsilon_\infty$, since one should consider the SFs L and K to be frequency dependent for sufficiently high ω. In other words, the mechanism of dielectric loss, related to dynamic effects / periodic motion of dipoles, is non-operative for the approximation (6.34) However, the aforementioned mechanism *does exist*, since the spectral function does depends on x, $L = L(x + iy)$, and no distinct boundary actually appears between the Debye and Poley frequency regions. Correspondingly, the experimental observations show, that the Debye plateau is usually not pronounced.

The study of the experimental evidences of the Debye rotational diffusion in terms of analytical models appears to be very instructive. Moreover, the coincidence of the experimental and calculated α_∞ or ε_∞ values of can supply additional criteria of applicability of a prescribed molecular model.

6.7 The Resonance Phenomena (The Poley Absorption Region)

6.7.1 The Gordon Sum Rule for the Loss $\chi''(\omega)$

The complex susceptibility $\chi^*(\omega)$ may be expressed, in accord with Table 6.1, in terms of the SFs $K(z)$, $L(z)$ or, in other words, - in terms of the susceptibility $\chi_B^*(\omega)$, calculated for the Boltzmann induced distribution F_B.

For the *spatial* rotational motion the integrated absorption is given by

$$\Pi = \int_{-\infty}^{\infty} \chi''(\omega)\,\omega\,d\omega = \frac{\pi\mu^2 N}{I}\begin{Bmatrix} 1-\langle q^2\rangle \\ 2/3 \end{Bmatrix}, \text{ for } \begin{Bmatrix} \text{anisotropic} \\ \text{isotropic} \end{Bmatrix} \text{ medium} . \quad (6.60)$$

The proof was given in Sec. 5.7 for the Boltzmann distribution F_B (i.e. for $\chi^*(\omega) = \chi_B^*(\omega)$). For the self-consistent induced distributions F_G, F_K, F_T the result (6.60) is also valid, some examples are given in (Gaiduk and Kalmykov 1987). To understand the last result it is useful to note that the main contribution into the integrated absorption gives the Poley frequency range, in which: (i) $\omega >> \tau_D^{-1}$ and (ii) the loss χ'' practically coincides with χ''_B Roughly speaking, one can make the following transition *in the integrand* of (6.60):

$$\chi''(\omega) \to \chi''(\omega \to \infty) \to \chi_B''(\omega) \quad (6.61)$$

Then, using Table 6.1, one can write

$$\Pi = \int_{-\infty}^{\infty} \chi''(\omega)\,\omega\,d\omega = \int_{-\infty}^{\infty} \chi''(\omega \to \infty)\,\omega\,d\omega = \int_{-\infty}^{\infty} \chi_B''(\omega \to \infty)\,\omega\,d\omega \quad (6.62)$$

and for any F in Table I we get the r.h.s. of Eq. 6.60.

The next important step is that for $\omega \to \infty$ the internal field correction may be neglected, so that the following relation between $\varepsilon''(\omega)$ and $\chi''(\omega)$ is valid (cf. with Eq. 6.46):

$$\varepsilon''(\omega \to \infty) = 4\pi\,\chi''(\omega \to \infty) = 4\pi\chi_B''(\omega \to \infty). \quad (6.63)$$

In view of Eq. 6.62 - 6.63 we may rewrite the Gordon some rule in terms of the loss permittivity ε''. Noting that $\varepsilon''(\omega)$ is an odd function, we have:

$$\Pi = \int_0^{\infty} \varepsilon''(\omega)\,\omega\,d\omega = \frac{2\pi^2\mu^2 N}{I}\begin{Bmatrix} 1-<q>^2 \\ 2/3 \end{Bmatrix} \text{ for } \begin{Bmatrix} \text{anisotropic} \\ \text{isotropic} \end{Bmatrix} \text{ medium} \quad (6.64)$$

The integrated absorption is *infinitely large*, if we employ the Debye formula (6.35) that is derived with the neglect of nonlinear dynamics (which is important just for $\eta\omega = x > 1$). This result presents an additional indication that the Debye formula is inapplicable in the far infrared region.

6.7.2 The Estimate of the Dipole Moment from the Experimental value of the Integrated Absorption

The dipole moment μ of a polar molecule usually differs in a liquid from its value μ_0 of an isolated molecule. We introduce a fitting parameter k_μ, using the following relation:

$$\mu = k_\mu \mu_0 \frac{n_\infty^2 + 2}{3} = k_\mu \mu_{\text{Onz}}, \tag{6.65}$$

$$\mu_{\text{Onz}} = \mu_0 \left(n_\infty^2 + 2\right)/3. \tag{6.66}$$

Here μ_{Onz} is the Onzager formula for μ, which corresponds to the value k_μ = 1 and is often used (Frohlich 1963, p.35, Eq. 6.36) in combination with Eq. 6.47, relating the static permittivity ε_S to the Kirkwood correlation factor g:

$$\frac{\varepsilon_S - n_\infty^2}{4\pi} \frac{2\varepsilon_S + n_\infty^2}{3\varepsilon_S} = Gg = \frac{\mu^2 N g}{3k_B T} \tag{6.67}$$

Let us obtain an estimate of the coefficient k_μ with account of the experimental dependencies $\alpha(\nu)$ and $n(\nu)$, $n = \omega/(2\pi c)$ being the reduced frequency (viz. the wave number). We apply the Gordon sum rule (6.64) for an isotropic ensemble of dipoles.

We introduce two frequency intervals: $\omega \in [0, 2\pi c\nu]$ and $[2\pi c\nu\ , \infty]$, the first - for the relaxation spectrum and the second - for the Poley absorption region. We remark, that in accord with (Gerschel et al. 1976) one may take $\nu_1 \approx$ 10 and $\nu_2 \approx$ 200 cm^{-1} for strongly absorbing nonassociated liquids.

We use Debye formula of the type (6.40) for $0 \le \nu \le \nu_1$:

$$\omega\varepsilon|''(\omega) = \omega\,\text{Im}\left\{\varepsilon_\infty + \frac{\varepsilon_S - \varepsilon_\infty}{1 - i\omega\tau_D}\right\} = \omega^2\tau_D \frac{\varepsilon_S - \varepsilon_\infty}{1 + \omega^2\tau_D{}^2} \tag{6.68}$$

For $\nu_1 \le \nu \le \nu_2$ we have in accord with Eq. 4.44

$$\omega\varepsilon''(\omega) = cn(\nu)\alpha(\nu)\ . \tag{6.69}$$

Combining Eq. 6.64, 6.65, 6.88 and 6.69, we obtain expression for the estimated k_μ:

$$k_\mu = \frac{3}{2\pi\mu_0\left(n_\infty^2+2\right)}\sqrt{\frac{3I}{N}}\sqrt{\frac{\varepsilon_S-\varepsilon_\infty}{\tau_D^2}\left(\frac{\nu_1}{\nu_D}-\operatorname{atan}\frac{\nu_1}{\nu_D}\right)+2\pi c^2\int_{\nu_1}^{\nu_2}\alpha(\nu)n(\nu)d\nu}; \quad (6.70)$$

$$\nu_D = \left(2\pi c\tau_D\right)^{-1}; \quad (6.71)$$

$\alpha(\nu)$, $n(\nu)$ dependencies may be taken from the experiment. If the latter, $n(\nu)$, is not known, we may roughly set n to be constant in the far IR region,

$$n(\nu) \approx n_\infty, \text{ for } \nu_1 \le \nu \le \nu_2, \quad (6.72)$$

or obtain it from some appropriate analytical model. Indeed, it is the $\alpha(\nu)$ [and not $n(\nu)$!] far IR frequency dependence, that actually determines the values of μ and k_μ. It is instructive to compare the experimental factor (6.70) with the fitted k_μ value. Their difference may characterise the drawbacks of the prescribed molecular model (see Chap. 14).

6.7.3 The Calculation Scheme for the ACFs

THE ANALYTICAL THEORY. We return to the scheme, presented at the end of Sec. 5.2.1. In accord with the Kubo linear-response theory the spectrum $\tilde{\Phi}(\omega)$ of the dipole autocorrelation function (ACF) is related to the susceptibility $\chi^*(\omega)$ as follows:

$$\chi^*(\omega) = \chi^*(0)\left[1 + i\omega\,\tilde{\Phi}(\omega)\right],\ \chi^*(0) = gG. \quad (6.73)$$

Instead of $\tilde{C}(\omega)$ in Eq. 5.46 we denote here the ACF spectrum as $\tilde{\Phi}(\omega)$ in order to underline that the cross correlation (mutual rotation of dipoles) is approximately accounted for in (6.73). An appropriate collision model is by applied for this purpose (see Table 6.1).

Here we intend to calculate the ACF $\Phi(t)$ and its second derivative $d^2\Phi(t)/dt^2$ *after* the susceptibility $\chi^*(\omega)$ is determined. Taking the Cubo relation (6.73) as a starting point, we have:

$$\tilde{\Phi}(\omega) = \frac{1}{i\omega}\left[\frac{\chi^*(\omega)}{gG} - 1\right]. \quad (6.74)$$

Further we use the frequency dependence $\chi^*(x)$ instead of $\chi^*(\omega)$. Applying the inverted Fourier transform to this equation, we can find the orientational ACF and its second time derivative:

$$\Phi(t) = \frac{1}{\pi} \int_{-\infty}^{\infty} \exp(-\omega t) \operatorname{Re}\left[\tilde{\Phi}(\omega)\right] d\omega = \frac{1}{\pi G g} \int_{-\infty}^{\infty} \frac{\chi''(x)}{x} \exp\left(-\frac{ixt}{\eta}\right) dx, \quad (6.75)$$

$$\frac{d^2\Phi(t)}{dt^2} = -\frac{1}{\pi G g \eta^2} \int_{-\infty}^{\infty} x\chi''(x) \exp\left(-\frac{ixt}{\eta}\right) dx. \quad (6.76)$$

Using the Kramers-Kronig relation (5.172) and the Gordon sum rule (5.146), we note that at $t = 0$ Eq. 6.75 and 6.76 yield, respectively:

$$\Phi(0) = \frac{\chi_s}{gG} = 1; \quad \left.\frac{d^2\Phi(t)}{dt^2}\right|_{t=0} = -\frac{1}{g\eta^2} \int_{-\infty}^{\infty} x\chi''(x)\, dx = \frac{1}{g\eta^2}. \quad (6.77)$$

Consequently, the normalized time dependence

$$\Phi_{rv}(t) = \frac{d^2\Phi(t)}{dt^2} \Bigg/ \left.\frac{d^2\Phi(t)}{dt^2}\right|_{t=0},$$

that means the angular ACF-acceleration, is given by

$$\frac{d^2\Phi(t)}{dt^2} = \frac{1}{\pi G} \int_{-\infty}^{\infty} x\chi''(x) \exp\left(-\frac{ixt}{\eta}\right) dx. \quad (6.78)$$

Since $\chi''(x)/x$ and $x\chi''(x)$ are even functions of x, we can rewrite Eq. 6.75 and 6.78 as follows:

$$\begin{Bmatrix} \Phi(t) \\ \Phi_{rv}(t) \end{Bmatrix} = \frac{2}{\pi G} \int_{0}^{\infty} \begin{Bmatrix} \dfrac{1}{gx} \\ x \end{Bmatrix} \chi''(x) \cos\left(\frac{xt}{\eta}\right) dx. \quad (6.79a)$$

Thus, both functions, $\Phi(t)$ and $\dfrac{d^2\Phi(t)}{dt^2}$, can be obtained by integration in a wide frequency region (practically - in the region of orientational polarization) of the known loss frequency dependence $\chi''(x)$. Examples will be given in Part V.

CALCULATION ON THE BASIS OF THE EXPERIMENTAL SPECTRUM $\varepsilon^*(\omega)$. Similar calculation scheme can be used, if one employs experimental dependencies $\varepsilon''(\omega)$. and $\varepsilon'(\omega)$ of the imaginary and real parts of the complex permittivity. Taking imaginary part of Eq. 6.46 and noting that $\text{Im}\,\varepsilon = -\varepsilon''$, we obtain from (6.79a):

$$\chi''(x) = \text{Im}\left[\frac{\varepsilon^* - n_\infty^2}{4\pi}\,\frac{\left(2\varepsilon^* + n_\infty^2\right)\varepsilon}{3\varepsilon^*\varepsilon}\right] = \frac{\varepsilon''}{12\pi}\left(2 + \frac{n_\infty^2}{|\varepsilon^2|}\right);$$

$$\begin{Bmatrix}\Phi(t)\\ \Phi_{rv}(t)\end{Bmatrix} = \frac{2}{\pi G}\int_0^\infty \begin{Bmatrix}\dfrac{1}{gx}\\ x\end{Bmatrix}\chi''(x)\cos\left(\frac{xt}{\eta}\right)dx\,, \tag{6.79b}$$

where g is related to ε_s, $n_\infty{}^2$ and G, see Eq. 6.52. This calculation scheme is known as the Fattuso-Mason model of the internal field.

6.7.4 The Quasi- Resonant Absorption

The distinction between molecular models, which will be considered in detail later, may become evident in the frequency region, where the *quasi-resonance absorption band* appears. This band shifts to the infra red region when the localization of orientations of dipoles becomes more pronounced. This shift may be enhanced by the decrease of the libration amplitude β in the confined rotator (CR) model or by the increase of the potential depth U_0 in the constant field (CF) and cosine squared (CS) models (the last two are termed *field models*). In Parts III and IV we shall consider in greater detail phenomena, touched briefly in two following (A and B) subsections.

(A). *LONG LIFETIMES* τ, viz. $y \ll 1$. In this case we obtain the so called *limiting line*, for which the absorption bandwidth $\Delta\nu \equiv \Delta\nu_0$ *is minimal* and may considerably differ in various models. For finite but still large lifetime τ the value $\Delta\nu_0$, the position ν_L of the absorption peak, and the ratio τ_D/τ depends completely on the profile $U(\vartheta)$ of a prescribed potential well. Indeed, the absorption band $\Delta\nu_0$ is narrow in the aforementioned (CS and CF) field models, if $U_0/(k_BT) \gg 1$ (see Chap.11, 12). In contrast, the limiting bandwidth $\Delta\nu_0$ is much wider in the extended diffusion (ED) model, describing free rotation, or in the CR model, describing free librations of a dipole between elastic reflecting walls.

We shall see that for $y << 1$ the bandwidth and absorption peak are determined in the field models mainly by the anharmonicity of dipole's regular rotation (and not by its duration).

(B). *SHORT LIFETIMES* τ, viz. $y > 1$.

In the field models:

(1) the absorption band *widens considerably* for greater y,

(2) the effect of the anharmonicity is not pronounced,

(3) the elastic bond (EB) is applicable in a rough approximation, if $U_0 >> k_B T$. In this case the absorption curve $\alpha(\nu)$ is described by the Lorentz frequency dependence, which is very sensible to the value of the lifetime τ.

In the ED and CR models the effect of the latter on the absorption bandwidth is much weaker. The conservative intermolecular potential U does *not* depend now on the current angular shift ϑ of a dipole. For nonassociated liquids, where y is about 1, the field and CR models become commensurable.

If the angular localization of a single particle is pronounced, the maximal absorption α_L depends on the orientation of alternating field ***E(t)*** relative to the symmetry axis *Z*. For example, for large U_0 the quasi-resonance absorption prevails in both field models for the *transverse* orientation: max $\alpha_\perp >>$ max $\alpha_\|$, since interaction of a dipole with radiation is more effective for the latter, that for the *longitudinal* orientation, since charges of a dipole move almost transverse to the conservative ***E***(t) field lines. On the contrary, for transverse orientation charges move almost along AC field lines that makes such interaction more effective, than for ***E(t)*** || *Z*. We remark that this circumstance is reflected in the Gordon sum rule, Eq. 5.160.

6.8 The Extension of the Theory to Multicomponent Media

6.8.1 Systems Possessing a Discrete Set of Relaxation Times

As shown experimentally (see part V), some rather complex media, e.g. mixtures of polar fluids, exhibit *several relaxation processes*, which may be characterized by a number of relaxation times τ_{Dm} , where m = 1, 2,... .

We can describe dielectric spectra of such media by considering a superposition of complex susceptibilities:

$$\chi^* = \sum_m \chi_m^* \tag{6.80}$$

Here the index m is related to a dipole moment μ_m, concentration N_m, proportion r_m of the m th fraction, spectral function $L_m(z_m)$ and so on. To simplify formulas we designate $\sum_m$ simply as $\sum$.

The dispersion of each component is characterized in Eq. 6.80 by its own relaxation time τ_{Dm}, so that χ_m^* can be calculated by the method, given above (Table 6.1).

Consider for example a collision model characterized by the *Debye induced distribution* ($F = F_D$). We have for an isotropic medium

$$\chi^* = \frac{1}{3k_BT}\sum \mu_m^2 N_m \left(1 + \frac{\tau_{Dm}/\tau}{1 - i\omega\tau_{Dm}}\right) L_m = \sum G_m \left(1 + \frac{\tau_{Dm}/\tau}{1 - i\omega\tau_{Dm}}\right) L_m; \tag{6.81}$$

$$G_m = \frac{\mu_m^2 N_m}{3k_BT}, \quad L_m \equiv L_m(z_m), \; z_m = x + i\, y_m \quad , \tag{6.82a}$$

$$x_m = \eta_m \omega, \; y_m = \eta_m / \tau_m, \; \eta_m = \sqrt{\frac{I_m}{2k_BT}} \quad . \tag{6.82b}$$

6.8.2 Media with a Single Relaxation Time (Media, Homogeneous on Microscopic Scale)

SUSCEPTIBILITY FOR THE DEBYE INDUCED DISTRIBUTION F . In many experiments it is difficult to distinguish several Debye relaxation ranges. Let a liquid (e.g. solution) can be considered as a multicomponent medium, characterized by a *single relaxation time* τ_D. Then we come to a simplified form of Eq. 6.81:

$$\chi^* = \sum G_m \left(1 + \frac{\tau_D/\tau}{1 - i\omega\, \tau_D}\right) L_m . \tag{6.83}$$

For an anisotropic medium this equation is rewritten in Table 6.3, which generalizes formulas, given in Table 6.2.

For the Boltzmann induced distribution F_B expression for $\chi^* = \chi_B^*$ is obtained, if one puts here $\tau_D = 0$.

For the Debye collision model ($F = F_D$) the susceptibility of a solution is represented with account of Eq. 6.83 as the superposition of susceptibilities, calculated for various subensembles. Since the relaxation time τ_D usually depends on concentrations N_m of the solution components, we may suppose that the resulting susceptibility depends non-linearly on concentrations N_m.

Each m th term in (6.83) has its maximum loss on the same frequency $\omega_D = \frac{1}{\tau_D}$. Combining expressions for χ^* in Table 6.3 with Eq. 6.47 at $x = 0$, we may relate the static permittivity ε_s to other parameters:

$$\frac{\varepsilon_s - n_\infty^2}{4\pi} \frac{2\varepsilon_s + n_\infty^2}{3\varepsilon_s} = \sum G_m \left(1 + \frac{\tau_D}{\tau_{mD}}\right) K_m^0;\; K_m^0 \equiv K_m(iy_m). \quad (6.84)$$

These must be fitted in such a way, that Eq. 6.84 should be satisfied.

SUSCEPTIBILITY FOR THE SELF-CONSISTENT DISTRIBUTIONS. Very interesting situation occurs, when one employs the self-consistent induced distributions F, e.g. the correlation-orientation (F_K) or isothermal (F_T) distributions. Now the *single relaxation time* τ *becomes an explicit function of the fitting parameters of the prescribed model.*

Further we assume that the correlation factor g and the complex coefficients $\hat{A}$ and $\hat{B}$ (see Eq. 6.4 a, b), responsible for the phase shift of polarization relative to alternating electric field, are characteristics of the *whole solution.* Hence, we assume that g, $\hat{A}$, $\hat{B}$ do *not* depend on the index m, since m refers to the m th component of the solution. Therefore, we generalize Eq. 6.15b and 6.19 to the sum over all components of the solution:

$$\hat{A} = \frac{\chi^*}{g \sum G_m \sigma_m}, \quad (6.85)$$

$$\hat{B} = 0 \text{ for } F_K;\; \hat{B} = -\frac{2i}{\sum G_m \left(1 - \left\langle q_m^2 \right\rangle\right)} \text{ for } F_T; \quad (6.86)$$

$$\sigma_m = 3\left[\left\langle q_m^2\right\rangle - \left\langle q_m\right\rangle^2\right] = 3\Delta_m. \tag{6.87}$$

Inserting these $\hat{A}$ and $\hat{B}$ into Eq. 6.5 and denoting

$$\hat{\omega}_m = \omega + \tau_m^{-1}, \tag{6.87a}$$

we derive expressions presented in Table 6.3 for the complex susceptibility χ^* of the solution.

For an isotropic medium, regarding spatial or quasi-spatial ensembles we have:

$$\left\langle q_m\right\rangle = 0, \left\langle q_m^2\right\rangle = \frac{1}{3}, K_m = L_m, \sigma_m = 1;$$

$$\chi^* = \frac{g\sum G_m \hat{\omega}_m L_m}{g\omega + i\sum \frac{r_m}{\tau_m} L_m} \quad \text{for } F_K, \tag{6.88}$$

$$\chi^* = \frac{g\sum G_m \hat{\omega}_m L_m}{g\omega + i\sum \frac{r_m}{\tau_m}\left(1 + g x_m z_m\right) L_m} \quad \text{for } F_T. \tag{6.89}$$

Here r_m is the normalized concentration of the m th fraction:

$$r_m = \frac{G_m}{\sum G_m} = \frac{N_m^2 \mu_m^2}{\sum N_m^2 \mu_m^2}. \tag{6.90}$$

Let us now relate parameters of the low-frequency (Debye) region with the SF and statistical averages. Putting $x = 0$, we use again the symbol "0" in SFs L and K. For instance,

$$L_m(z_m) \to L_m^0 \equiv L_m(iy_m),\; K_m(z_m) \to K_m^0 \equiv K_m(iy_m),\; \varsigma_m = y_m^2 L_m^0,$$

Then the formulas of Table 6.3 are reduced to the "Debye form":

Table 6.3 Multicomponent systems (solution): the complex susceptibility χ^*

F	$\chi^*(x)$
F_B	$\sum G_m K_m$
F_D	$\sum G_m K_m \left(1 + \dfrac{\tau_D / \tau_m}{1 - \dfrac{ix\tau_D}{\eta}} \right)$
F_K, F_T	$\dfrac{g \sum z_m G_m K_m}{gx + i \dfrac{\sum \xi_m y_m G_m K_m}{\sum \sigma_m G_m}}$
Notation	$\xi_m = 1$ for F_K $\xi_m = 1 + \dfrac{\frac{2}{3} gxz_m \sum G_m \sigma_m}{\sum G_m \left(1 - \langle q_m^2 \rangle \right)}$ for F_T

$$\chi^*(\omega) = \frac{a\omega + ib}{c\omega + id} = \frac{a}{c} + \frac{\dfrac{b}{d} - \dfrac{a}{c}}{1 - \dfrac{i\omega c}{d}} = \chi_\infty + \frac{\chi_s - \chi_\infty}{1 - i\omega\tau_D}, \tag{6.91}$$

$$\chi_s = \frac{b}{d}, \chi_\infty = \frac{a}{c}, \tau_D = \frac{c}{d}. \tag{6.92}$$

The coefficients a, b, c, d are presented in Table 6.4. Combining Eq. 6.47 with expressions for χ_s from Table 6.4, we can relate the correlation factor g of the solution to the parameters of a prescribed model:

$$g = \frac{\varepsilon_s - n_\infty^2}{4\pi} \frac{\varepsilon_s + 2n_\infty^2}{3\varepsilon_s} \frac{1}{\sum G_m \sigma_m} \left\{ \begin{matrix} \text{for anisotropic} \\ \text{medium} \end{matrix} \right\}, \tag{6.93a}$$

$$g = \frac{\varepsilon_s - n_\infty^2}{4\pi} \frac{\varepsilon_s + 2n_\infty^2}{3\varepsilon_s} \frac{1}{\sum G_m} \quad \{\text{for isotropic medium}\}. \qquad (6.93b)$$

From Eq. 6.92 and Table 6.4 we see, that the relaxation time τ_D of the solution is a complicated function of concentrations N_m and of other parameters of the microscopic model. Namely, in the case of an isotropic medium

$$\tau_D = \frac{g \sum G_m \varsigma_m}{\sum \frac{G_m L_m(iy_m)}{\tau_m}}, \; \varsigma_m = \left\{ \begin{array}{c} 1 \text{ for } F_K \\ 1 - y_m^2 L_m(iy_m) \text{ for } F_T \end{array} \right\}. \qquad (6.93c)$$

6.9 Models of Collisions for an Isotropic Medium (Continuation)

6.9.1 Models with a Single Well Potential

We suggest now that the ensemble average of the direction cosine q is non-zero, $\langle q \rangle \neq 0$. If the potential well is deep, then because of this property dielectric behavior radically changes as compared with the case of double wells, wherein $\langle q \rangle = 0$.

For instance, for the *constant field* (CF) model with the potential profile

$$U(\vartheta) = -\mu E_0 \cos\vartheta, \qquad (6.94)$$

for which $\langle q \rangle \neq 0$, the equilibrium could be established only for the angular position $\vartheta = 0$, corresponding to the minimal potential energy [cf. with the cosine squared model, where the double well potential

$$U(\vartheta) = U_0\left(1 - \cos^2 \vartheta\right) \qquad (6.95)$$

is realized; $\langle q \rangle = 0$ and two opposite minimal-energy equilibrium directions ($\vartheta = 0$ and $\vartheta = \pi$) appear].

To elaborate a simplified collision model for the situation, when $\langle q \rangle = 0$, we replace the Ψ-depending quantities K and σ (see in Table 6.1) by their mean values $\overline{K}$ and $\overline{\sigma}$. Here dash means averaging over Ψ, i.e. over the angle between $\boldsymbol{E}_m$ and the symmetry axis Z. We take the correlation-orientation and isothermal induced distributions, F_K and F_T , for which $\left\langle q^2 \right\rangle$ = 1/3 (see Eq. 6.28).

Table 6.4 Multicomponent systems (solution): the parameters of the Debye relaxation

Parameter	τ_D
χ_s	$g\sum G_m\sigma_m$ for F_K, F_T
χ_s	$\sum G_m K_m^0\left(1+\frac{\tau_D}{\tau_m}\right)$ for F_D
a	$g\sum G_m K_m^0$
b	$g\sum y_m G_m K_m^0$
c	$g\zeta$
d	$\dfrac{g\sum y_m G_m K_m^0}{\sum G_m\sigma_m}$
Notation	$\zeta = 1,\ K_m^0 = K_m(iy_m)$ for F_K $\varsigma = 1 - \dfrac{\frac{2}{3}\sum G_m y_m^2 K_m^0}{\sum G_m\left(1 - \left\langle q_m^2 \right\rangle\right)}$ for F_T

Then we derive following expressions for χ^*:

$$\chi^* = \frac{GzL}{gx + iyL/\sigma} \quad \text{for } F_K, \tag{6.96}$$

$$\chi^* = \frac{GzL}{gx + \dfrac{iyL}{\dfrac{1}{\bar{\sigma}} + gxz}} \quad \text{for } F_T; \tag{6.97}$$

$$\bar{\sigma} \equiv 3\bar{\Delta} = 3\left\{ \overline{\langle q^2 \rangle} - \overline{\langle q \rangle \langle q \rangle} \right\} \tag{6.98}$$

$$= \frac{1}{3}\left[\sigma_{\parallel} + \binom{2}{1}\sigma_{\perp}\right] \text{for} \begin{pmatrix}\text{quasi} - \text{spatial} \\ \text{spatial}\end{pmatrix} \text{ensemble} \tag{6.99}$$

$$L = \frac{1}{3}\left[K_{\parallel} + \binom{2}{1}K_{\perp}\right] \quad \text{for} \begin{pmatrix}\text{quasi} - \text{spatial} \\ \text{spatial}\end{pmatrix} \text{ensemble}, \tag{6.100}$$

The comparison with the last column of Table 6.1 shows, that the only distinction from the case $\langle q \rangle = 0$ is related to the appearance $\dfrac{1}{\bar{\sigma}}$ instead of 1 in the denominator of Eq. 6.97. We shall see later that $\dfrac{1}{\bar{\sigma}} >> 1$ for a strong conservative field. For high frequency x, corresponding to quasi-resonance absorption spectrum in a liquid, the term, proportional to iy in the denominator of Eq. 6.96 and 6.97, may be commensurable with the term gx. This means that

---

If a polar medium is described by a model with a deep single-well potential, such that $\langle q \rangle \neq 0$, *then the Poley loss / absorption spectra may considerably differ from those, described by the spectral function L(z).* (6.101)

We remind the Reader that for the double well potential, where $\langle q \rangle = 0$, it is the SF *L(z)* that for $x > 1$ qualitatively describes dielectric behavior in the far IR region. In contrast, for the single well potential *both* factors are essential: the SF *L(z)* and the statistical parameter $\bar{\sigma}$.

6.9.2 The Averaged Susceptibility

Here we briefly describe another (probably, more fundamental) approach to the "construction" of an *isotropic medium* from the domain-like pieces of a dielectric. We suggest that each peace is characterized by some orientational order and some direction of the symmetry axis of the potential U.

Instead of replacing the Ψ-dependent quantities K and σ by their mean values $\overline{K}$ and $\overline{\sigma}$, as was done so far, *we may find the averaged susceptibility* itself by integration χ^* over angle Ψ:

$$\underbrace{\chi^*(x)}_{\text{for isotropic medium}} = \underbrace{\frac{1}{2}\int_0^{\pi} \chi^*(x,\Psi)\sin\Psi\, d\Psi}_{\text{for anisotropic medium}} \qquad (6.102)$$

The idea of such a transition is based on the following reasoning. In view of general relations (4.10) and (4.26) the effective susceptibility is proportional to *consumed power* of electromagnetic wave. Since this power is an *additive* quantity, produced by multiple of charges / dipoles in the medium under consideration, we may find the average of contributions due to subensembles having different directions of symmetry axes. In Sec.6.5 we have seen, that if $\langle q \rangle = 0$, then the low-frequency susceptibility χ^* exhibits a typical Debye relaxation (the parameters of the latter are given in Table 6.2). In the new calculation scheme (6.102)

The integral of the Debye-like frequency dependence $\chi^*(x, \Psi)$ over Ψ may not lead generally to the typical Debye-type relaxation spectrum, characterized by a single relaxation time τ_D. Thus, for an isotropic medium we principally may obtain a non-Debye dependence of the susceptibility vs. x in the low-frequency region.

We remark in conclusion a non-Debye low-frequency $\chi^*(x)$ dependence could arise in a *mixture of polar fluids, differing by the kind of molecular rotation in each molecular subensemble,* that is determined by appropriate potential wells, chosen for these subensembles.

ADDENDUM I

ABOUT THE EVOLUTION OF THE DYNAMIC METHOD

This method, used in the book, presents a consistent classical linear-reaction description of a polar medium. Basically, the theory is founded on two theorems (i) and (ii).

(i) *The t_0 theorem*

---

*Fourier amplitudes of charge / current density, produced by charged particles of a medium, moving non-periodically in a periodic alternating field **E**(t), may be rigorously represented in terms of the law of motion of these particles.* (I.1)

This idea was suggested by the author (1960a) in the field of physical electronics, see also (Gaiduk, Palatov and Petrov 1971, p.113). It is difficult to write rigorous expressions for the *time dependence* $\rho(t)$ or $\boldsymbol{J}(t)$ of these quantities because one particle may overtake another. We introduce the t_0 *sequence*, viz. an infinite set of discrete charges, having the same trajectory and corresponding to different «initial» instants t_0, $t_0 - 2\pi/\omega$, $t_0 - 4\pi/\omega$, ... , shifted by the period $2\pi/\omega$ of alternating field. The total response of charges is found by summing *Fourier amplitudes* of $\rho(t)$ or $\boldsymbol{J}(t)$, corresponding to all sequences. One can calculate by this fashion any *additive* quantity, e.g. energy, consumed / delivered by particles, that are found in a unit volume. Dielectric response of a polar fluid is generally determined by the *average complex power* consumed by rotating dipoles of the medium.

(ii) The averaged perturbation (AP) theorem

The first proof of this theorem was given in GT, p. 373. The revised proof, made by B. M. Tseitlin and the author, is given in Sec. 3.3. The AP theorem shapes the dynamic method into a modern aspect of the linear-reaction dielectric relaxation theory. Indeed,

---

*Previously, see e.g. (Gaiduk and Kalmykov 1987) dielectric susceptibility was calculated, just like power, consumed by charged particles in physical electronics, viz. in terms of trajectories, perturbed by alternating **E(t)** field.* (I.2)

Unlike this, using the AP theorem one can express the complex susceptibility in terms of the law of motion of dipoles, found at thermal equilibrium. (I.3)

THE FIRST STAGE. One can regard, that two early articles by Gaiduk, Loskutov and Sekistov (1976) and by Gaiduk (1976) constitute the origin of the dynamic method. The clue idea was, that using Maxwell equations one can express dielectric response of a polar medium in terms of the *complex power,* consumed by particles during their interaction with alternating field $\boldsymbol{E}(t)$. On this way a real success was achieved only in (Gaiduk, Kalmykov and Tseitlin 1979), where the *closed-form analytical expression* was derived for the complex permittivity $\varepsilon(\omega)$. The authors of this work have realized that:

(a). The exponential distribution $\Psi(t_v)$ of particles over "lifetimes" t_v should be introduced into the theory of polar fluids, just as it was made in the Van Vleck-Weisskopf theory of resonance absorption in polar gases, see e.g. (Townes and Shawlow 1975; Broun 1956). Therefore, the mean lifetime $\langle t_v \rangle \equiv \tau$ appeared in the dynamic method; the reverse value, τ^{-1}, is proportional to the friction coefficient of the medium.

Averaging of calculated quantities over lifetimes t_v makes the theory mathematically elegant and produces its agreement with experimental observations. (I.4)

Hence, the dynamic method was born "between" two fields, viz. those occupied by physical electronics and resonance absorption theory. The former (physical electronics) has usually to do with non-Maxwellian (non-thermal) distributions over velocities, the trajectories of particles being arbitrary. In the latter (resonance theory) approximately harmonic law of motion of particles is usually considered.

We note in passing that Fourier amplitudes of current density are not employed in the resonance theory. Since the Weisskopf-Van Vleck theory is not rigorous, their resonance line does not satisfy the Gordon sum rule and is characterized by a non-physical *infinite* integrated absorption (see Sec. 1.5.7).

(b). The simplest Boltzmann induced distribution F_B provides rather good description of absorption in the far IR spectral region. This distribution is however not applicable at low frequencies. In the aforecited work by Gaiduk, Kalmykov and Tseitlin (1979) the Debye induced distribution ($F = F_D$, in our terms) was involved. This was the first (in frames of the dynamic method) successful attempt to elaborate an appropriate collision model, applicable also in the low-frequency region. The planar rectangular potential well was suggested in this work, that was capable to describe a wideband spectrum of orientational relaxation, characteristic for individual polar liquids. A similar work was published in English (Gaiduk and Kalmykov 1981) after two years, then a number of other articles followed. The first systematic description of the dynamic method was given by Gaiduk and Kalmykov (1987); we denote this review as GK.

THE SECOND STAGE. This stage, characterized by the advantage (I.3), was accomplished in GT after the AP theorem was applied. In this book we follow the approach elaborated in GT. The results, obtained on the basis of the dynamic method for the well known molecular models (e.g. for the extended rotational diffusion model), of course, coincide with those found with use of other methods, based on linear-reaction theory. For instance, the dynamic method yields for ensemble of free rotors the same results as the kinetic equation method, described e.g. in GK. Due to the property (I.3) the dynamic method is much simpler than that described in GK. We stress below two features of the second stage.

ENERGY H AND CONSTANT t_0 *AS AN UNIVERSAL PAIR OF THE PHASE VARIABLES.* In a conservative system, e.g. for a steady-state ensemble of molecules, H and t_0 may be employed as the phase variables, since they are integrals of motion of a given particle and thus do not depend on time t. Instead of H and t_0 we usually employ h and φ_0 as dimensionless quantities. In the law of motion t_0 is met only in the combination $t + t_0$ (or in combination $\varphi + \varphi_0$ for dimensionless variables).

Hence, h and φ_0 are canonically conjugated coordinate and momentum of a particle (Gaiduk 1960b), see also (Gaiduk, Palatov and Petrov 1971, p. 136). In the dynamic method integration over φ_0 can generally be carried over. Consequently, the spectral function $L(z)$ of the prescribed molecular model is expressed for in-plane motion in the form of integral over energy h. For in-space motion double integral appears - over h and over l, where l is projection of the angular momentum on the symmetry axis of intermolecular potential.

ABOUT MODELS OF COLLISIONS. Following GT, we have found in a rather good approximation (see Chap. 6) that takes into account description of correlated rotation of neighboring dipoles. Basically, we introduce the experimentally determined Kirkwood correlation factor g (see Eq. 6.52) as a measure of cross correlation.

Another approach was previously elaborated by Kalmykov (1988), Kalmykov and Limonova (1989). They assumed that correlation of rotational velocities vanishes in ensemble of dipoles unlike correlation of their orientations (in the book we employ actually the same idea). By means of the memory function technique Kalmykov and Limonova had derived the following expression

$$\chi^*(x) = \frac{gGzL(z)}{gx + (z - gx)L(z)} \tag{I.5}$$

for the complex susceptibility of an isotropic polar medium. This formula may be compared with Eq. 6.17a, derived for the correlation-orientation induced distribution F_K. The $\chi^*(x)$ spectra, calculated from Eq. 6.17a and I.5, practically coincide, at least, - for the case of polar strongly absorbing nonassociated liquids.

PART THREE

MODELS OF FREE ROTATION / LIBRATION

This part consists of three chapters. We shall describe:

- The classical *extended diffusion* (ED) model (Chap.7), in which the potential U is homogeneous (not depending on coordinates); we may put $U =$ 0. In the ED model molecules rotate freely by inertia during the time interval t_v between "strong collisions", the mean value $\langle t_v \rangle = \tau$, termed "lifetime", being the free parameter of the model. In the ED model we take into account the form of a molecule by considering dielectric response of symmetric top molecules.

- The *confined rotator* (CR) models (Chap.8), to which a rectangular potential well corresponds. At first, we consider a *planar ensemble* of molecules, in which a dipole librates between two elastic reflecting planes. The result of this consideration is generalized by taking account of the main feature of a spatial system - its *statistics*. Then we elaborate a rigorous spatial model termed the *cone confined rotator* (CCR) model. In the latter reorientation of dipoles are confined by elastically reflecting surface of a cone. The tangential motion, characterized by the projection l of the angular momentum onto the axis of a cone, accompanies "pure" librations, that is, rotation of a dipole from one surface of a cone to another. For the sake of simplicity we consider only linear molecules, which one may regard as extremely prolate molecular tops.

- At last, we elaborate in Chap.9 the *hybrid confined rotator / extended diffusion* model (simply, HM), in which the welldepth U_0 is supposed to be finite (we recall the Reader, that $U_0 \to \infty$ in the CR or CCR models).

Let H be the kinetic energy of a dipole. The walls of the rectangular potential well are assumed to be as if transparent for the high-energy dipoles with $H > U_0$: these dipoles rotate freely like in the ED model. On the contrary, low-energy dipoles with $H > U_0$ librate between walls like in the CR model. The spectral functions $K(z)$ or $L(z)$ for anisotropic or isotropic media are expressed in the form of single or double integrals, wherein integrands are expressed through elementary functions. In many cases these spectral functions reduce to the standard $E_1(z)$ or $w(z)$ functions of complex argument z.

We remark that for a *quantum* ensemble of rotors, which is termed the *J-diffusion model* and is described in the last 16 th Chapter, the integrals over energy are replaced by sums over rotational quantum numbers .

CHAPTER 7
THE EXTENDED ROTATIONAL DIFFUSION (ED) MODEL OF SYMMETRIC TOP MOLECULES

7.1 Ensemble of Linear Molecules

7.1.1 Geometry and the Law of Motion

We start from ensemble of linear molecules, which rotate freely in space between instants of strong collisions. This ensemble presents a simple particular case of symmetric top molecules. The geometry of the system was shown in Fig. 3.2, where $\boldsymbol{S}$ is the plane, in which an arbitrary chosen dipole rotates; $\boldsymbol{AA'}$ is the trace of plane $\boldsymbol{S}$ on XY plane in the laboratory coordinate system XYZ; the angle Θ between axis Z and normalized angular momentum $\boldsymbol{p} = \eta\boldsymbol{\Omega}$ of a dipole can be chosen arbitrary ($\boldsymbol{p} \perp \boldsymbol{S}$); line $\boldsymbol{R}$ on $\boldsymbol{S}$ is perpendicular to $\boldsymbol{AA'}$; $\vartheta(\varphi)$ is current angle on $\boldsymbol{S}$ between $\boldsymbol{R}$ and the dipole moment $\boldsymbol{\mu}$; $\psi(\varphi)$ is current angle between $\boldsymbol{\mu}_{XY}$ and the projection OR' of line $\boldsymbol{OR}$ onto $\boldsymbol{XY}$ plane.

In Sec. 3.4.1 and 3.4.2 equations of motion were written and their solution were found. We note that the value p of the angular momentum is related to the constant normalized energy h of a dipole:

$$p = |\boldsymbol{p}| = \sqrt{h}\,. \tag{7.1}$$

The projection l of $\boldsymbol{p}$ onto Z and the angle Θ are related as follows:

$$\cos\Theta = l/p\,. \tag{7.2}$$

The law of motion in plane $\boldsymbol{S}$ is given by:

$$\vartheta(\varphi) = (\varphi + \varphi_0)\cdot\sqrt{h}\,, \tag{7.3}$$

where constant φ_0 is the integral of motion, which is canonically conjugated to the energy h of a particle.

To obtain expression for the spectral function $K(z)$ we should take integrals over all directions of the angular momentum $\boldsymbol{p}$ and over all h values (we remark that $h = p^2$). In order to apply the theory, elaborated in Part II, we consider the longitudinal and transverse orientations (marked respectively by symbols $\parallel$ and $\perp$) of alternating field $\boldsymbol{E}(t)$ relative to the Z axis. In our case of isotropic medium the terms «longitudinal» and «transverse» are actually *conditional*, since the Z-axis direction is *not* physically distinguished.

Let the Y axis coincide with the amplitude $\boldsymbol{E}_m$ of alternating field [when $\boldsymbol{E}(t) \perp Z$]. In accord with Eq. 3.62 - 3.66 the phase-volume element $d\Gamma$ and $\parallel$- and $\perp$-direction cosines are, respectively:

$$d\Gamma = dh\, d\varphi_0 \sin\Theta\, d\Theta\, \frac{d\psi_0}{2\pi}; \tag{7.4}$$

$$h \in [0, \infty],\ \varphi_0 \in \left[0, \frac{2\pi}{\sqrt{h}}\right],\ \Theta \in \left[0, \frac{\pi}{2}\right],\ \psi_0 \in [0, 2\pi]; \tag{7.5}$$

$$q_\parallel(\varphi) = \cos\vartheta(\varphi)\sin\Theta \equiv \cos\left[(\varphi + \varphi_0)\sqrt{h}\right]\sin\Theta, \tag{7.6}$$

$$q_\perp = \rho\cos(\psi + \psi_0); \tag{7.7}$$

$$\rho = \sqrt{1 - q_\parallel^2}. \tag{7.8}$$

The steady-state distribution and the mean values of the direction-cosines squares are given by

$$W = \frac{\exp(-h)}{2\pi};\ \left\langle q_\parallel^2 \right\rangle = \left\langle q_\perp^2 \right\rangle = \frac{1}{3}. \tag{7.9}$$

7.1.2 The Spectral Functions

In order to find the *longitudinal* SF $L_\parallel(z)$ we compare representation (5.104) with the law of motion (7.6). Since $q_\parallel(\varphi)$ is harmonic, the only one non-vanishing Fourier amplitude is given by

$$a_1 = \sin\Theta,\ a_n \equiv 0 \text{ for } n \neq 1. \tag{7.10}$$

In integral (5.108), where dm should be omitted, $p = \sqrt{h}$, W is given by (7.9) and $dh\,dl = dh\,\sqrt{h}\,\sin\Theta\,d\Theta$; the sum reduces to the only term. Consequently,

$$L_{\|} = \frac{1}{1/3}\,\frac{1}{2\pi}\int_0^\infty \exp(-h)\,\sqrt{h}\,dh \int_0^{\pi/2} \sqrt{h}\,\sin\Theta\,d\Theta\,\frac{\sin^2\Theta}{p^2 - z^2}$$

$$= \int_0^\infty \frac{\exp(-p^2)\,p^2}{p^2 - z^2}\,dp^2 = 1 + z^2\int_0^\infty \exp(-h)\frac{h - z^2}{p^2 - z^2}\,dh = L_F(z). \quad (7.11)$$

Here the subscript F again refers to *free rotation* [cf. with (5.98) for the planar SF]. The SF (7.11) is easily expressed through the integral exponential function $E_1(u)$ of the complex argument u (see e.g. Abramowitz and Stegun, 1964, Chap.5):

$$L_F(z) = 1 + z^2\exp(-z^2)\,E_1(-z^2); \quad (7.12)$$

$$E_1(u) = \int_u^\infty \frac{\exp(-t)}{t}\,dt \quad (7.13)$$

$$= -\gamma - \ln(u) - \sum_{n=1}^{\infty} (-1)^n \frac{u^n}{n n!}, \quad (7.13a)$$

the Euler constant $\gamma = 0.5772$.

We may expect that the transverse SF $L_\perp(z)$ should coincide with $L_{\|}(z)$. To prove this property, we express through the law of motion (7.3) the angle ψ on XY plane (see Fig. 3.2). The projection of μ on line OR' may be found by two ways:

$$\mu_{OR'} = \mu_{OR}\cos\Theta = \mu\cos\Theta\cos\vartheta \quad \text{and} \quad \mu_{OR'} = \mu_{XY}\cos\psi\,.$$

Thus

$$\rho\cos\psi = \cos\Theta\cos\vartheta\,. \quad (7.14)$$

It follows from (7.14) and (7.6) - (7.8), that

$$\rho^2\sin^2\psi = \rho^2 - \cos^2\Theta\,\cos^2\vartheta$$

$$= 1 - \cos^2 \vartheta \left(\sin^2 \Theta + \cos^2 \Theta\right) = \sin^2 \vartheta .$$

Hence,

$$\rho(\varphi) \sin \psi \equiv \sin \vartheta(\varphi). \tag{7.15}$$

We have from (7.7) - (7.8) with account of (7.14) - (7.15):

$$q_\perp = \cos \psi_0 \cos \Theta \cos \vartheta - \sin \psi_0 \sin \vartheta ;$$

$$q_{\perp 0} = \cos \psi_0 \cos \Theta \cos \vartheta_0 - \sin \psi_0 \sin \vartheta_0 .$$

Averaging of the product over ψ_0 in the interval $[0, 2\pi]$ yields:

$$\left\langle q_\perp q_{\perp 0} \right\rangle_{\psi_0} = \frac{1}{2}\left[\cos^2 \Theta \cos \vartheta_0 \cos \vartheta + \sin \vartheta_0 \sin \vartheta\right]. \tag{7.16}$$

Substitution into the general relation (5.35) gives

$$3\left\langle q_\perp^2 \right\rangle L_\perp(z) = 3iz \left\langle \int_0^\infty \left[q_\perp(\varphi) q_{\perp 0} - q_{\perp 0}{}^2\right] \exp(iz\varphi)\, d\varphi \right\rangle. \tag{7.17}$$

The term $q_{\perp 0}^2$ does not depend on φ; the corresponding part of the integral (7.17) yields 1. With account of (7.16) the r.h.s. of (7.17) reduces to

$$1 + 3iz \frac{1}{2\pi} \int_0^\infty \exp(-h)\, dh \int_0^\infty \exp(iz\varphi)\, d\varphi \int_0^{\pi/2} \sin \Theta\, d\Theta$$

$$\times \int_0^{\frac{2\pi}{\sqrt{h}}} \frac{1}{2} \left[\cos^2 \Theta \cos \vartheta_0 \cos \vartheta + \sin \vartheta \sin \vartheta_0\right] d\varphi_0 . \tag{7.18}$$

We introduce the variable $\vartheta_0 = \sqrt{h}\, \varphi_0$ instead of φ_0. Noting that $\vartheta = \sqrt{h}\, \varphi + \vartheta_0$, we integrate over ϑ_0 and Θ in the limits $[0, 2\pi]$ and $[0, \pi/2]$. Thus, (7.17) reduces to

$$1 + 3iz \frac{1}{2\pi} \int_0^\infty \exp(-h)\, dh \int_0^\infty \exp(iz\varphi)\, d\varphi \int_0^{\pi/2} \sin \Theta\, d\Theta$$

$$\times \int_0^{2\pi} \frac{1}{2}\left[\cos^2\Theta \cos\vartheta_0 \cos(p\varphi + \vartheta_0) + \sin\vartheta_0 \sin(p\varphi + \vartheta_0)\right] d\vartheta_0$$

$$1 + 3iz\frac{1}{2\pi}\int_0^{\infty}\exp(-h)\,dh \int_0^{\infty}\exp(iz\varphi)\cos p\varphi\, d\varphi \frac{1}{2}\frac{1}{2}\left(\frac{1}{3}+1\right)2\pi \, .$$

Taking into account that

$$\int_0^{\infty}\exp(iz\varphi)\cos p\varphi\, d\varphi = -\frac{iz}{p^2 - z^2}, \tag{7.18a}$$

we get finally with account of (7.9) the same result as (7.11), as it should be for an isotropic medium:

$$3\left\langle q_{\perp}^2\right\rangle L_{\perp}(z) = L_{\perp}(z) = 1 + z^2\int_0^{\infty}\frac{\exp(-h)}{h - z^2}\,dh = L_F(z). \tag{7.19}$$

In accord with (6.33) and with Eq. 7.9 for $\left\langle q_{\parallel}^2\right\rangle$ and $\left\langle q_{\perp}^2\right\rangle$ values we find, that the SFs $K(z)$ and $L(z)$ are equal one to another:

$$K_{\parallel}(z) = K_{\perp}(z) = K(z) = L_{\parallel}(z) = L_{\perp}(z) = L(z). \tag{7.20}$$

7.1.3 Some Approximate Expressions for the Complex Functions w(z) and $E_1(z)$

The w(z) function describes dielectric response due to free planar rotors, Eq. 5.99, while $E_1(z)$ - due to free spatial rotors, Eq. 7.12. The Reader can find useful formulas e.g. in (Abramowitz and Stegun 1964). We mark planar and spatial ensembles by superscripts "pl" and "sp".

THE LONG LIFETIME LIMIT $\tau \to \infty$ (or, equivalently, $y = 0$). The real and imaginary parts u and v of the complex function $w = u + i$v are expressed respectively through exponent and Doson integral $D(x)$:

$$w(x + i0) = u_0(x) + i\mathrm{v}_0(x), \tag{7.21}$$

$$u_0(x) = \exp(-x^2), \tag{7.22}$$

$$\mathrm{v}_0(x) = \frac{2}{\sqrt{\pi}} D(x), \tag{7.23}$$

$$D(x) = \exp(-x^2) \int_0^x \exp(-t^2)\, dt. \tag{7.24}$$

Thus,

$$L_F^{\mathrm{pl}}(x) = 1 + x\left(-2D(x) + i\sqrt{\pi}\exp(-x^2)\right) \text{ (for planar SF).} \tag{7.25}$$

Correspondingly,

$$E_1(-x \pm i0) = -\,\mathrm{Ei}(x) \mp i\pi\,, \tag{7.26}$$

$$Ei(x) = \gamma + \ln x + \sum_{n=1}^{\infty} \frac{x^n}{n n!} \quad \text{for } x > 0. \tag{7.27}$$

So, for spatial motion

$$L_F^{\mathrm{sp}}(z) = 1 + x^2 \exp(-x^2)\left(\mathrm{Ei}(x^2) + i\pi\right) \tag{7.28}$$

ASYMPTOTIC FORMULAS FOR FREQUENT COLLISIONS. In accord with (Gaiduk, Gaiduk and Tzeitlin 1994):

$$L_F^{\mathrm{pl}}(z) = 1 + i\sqrt{\pi}\, z\, w(z) \approx -\frac{t-7}{t^2 - 10t + 15}, \tag{7.29}$$

$$t = 2z^2, |z| \geq 4. \tag{7.30}$$

As for L_F^{sp},

$$\exp(Z)\, E_n(Z) = \cfrac{1}{Z + \cfrac{n}{1 + \cfrac{1}{Z + \cfrac{n+1}{1 + \cfrac{2}{Z + \cfrac{n+2}{1 + \cfrac{3}{Z + \cdots}}}}}}}$$

In our case $n = 1$, $Z = -z^2$,

$$L = 1 - Z \exp(Z)\, E_1(Z)\,. \tag{7.31a}$$

ASYMPTOTIC EXPRESSION FOR THE "STATIC" SPATIAL SF (Abramowitz and Stegun 1964), Chap. 5. If $x = 0$, $E_1(-z^2) \equiv E_1(y^2)$,

$$E_1(u) = -\ln(u) + \sum_{i=1}^{6} a_i u^i \qquad \text{for } u \le 1; \tag{7.31b}$$

$a_1 = -0.5772\ 1566$ $\quad a_2 = 0.9999\ 9193$ $\quad a_3 = -0.2499\ 1055$

$a_4 = 0.0551\ 9968$ $\quad a_5 = -0.0097\ 6004$ $\quad a_6 = 0.0010\ 7857$

$$u \exp(u)\, E_1(u) = \frac{u^2 + b_1 u + b_2}{u^2 + b_3 u + b_4} \qquad \text{for } u > 1, \tag{7.31c}$$

$b_1 = 2.334733$, $b_2 = 0.250621$, $b_3 = 3.330657$, $b_4 = 1.681534$.

* * *

We remark that the simplest way to find appropriate spectral dependencies usually consists in computing the SFs $L(z)$ *directly* from integrals (5.98) or (7.11). On the other side, it is convenient to employ the expressions, given above, in order to obtain various analytical formulas and in some special cases (e.g. for the limit $y \to 0$). The results of numerical calculations are presented in Fig. 7.1 - 7.4.

GRAPHS FOR THE REAL AND IMAGINARY PARTS OF THE SPECTRAL FUNCTION. First, we consider the ensemble of *planar rotators.* The real (L') and imaginary (L'') parts of the SF $L(z)$ and the plot $xL''(x)$ are shown in Fig. 7.1 as functions of dimensionless frequency x. The latter plot shows more distinctly, than $L''(x)$, the behavior of the SF on the HF edge of the orientation-polarization frequency region. These calculations are made for $y = 0$, 0.5 and 2 (solid curves, dashed curves, and dadots, respectively). The x-dependencies of the planar SF in Fig. 7.1 are closely related to the dielectric behavior of ensemble of rotors, pictured in Fig. 7.2. The specific properties of *spatial rotors* we shall consider in terms of dielectric spectra (see Fig. 7.4).

7.1.4 Dielectric Spectra

Let us choose the Boltzmann and Gross collision models, viz. we take the induced distributions F_B and F_G. In view of Table 6.1 the complex susceptibility $\chi^* = \chi' + i\chi''$ is related to the SF *L(z)* as follows:

$$\chi^* = G\,L(z) \quad \text{at} \quad F_B, \tag{7.32}$$

$$\chi^* = \frac{Gz\,L(z)}{x+iy\,L(z)} \quad \text{at} \quad F_G, \tag{7.33}$$

where $G = \mu^2 N / (3k_B T)$.

Hence, *for the Boltzmann induced distribution* F_B the frequency dependencies *L'(x)* and *L"(x)* in Fig. 7.1 and 7.3 present also dielectric spectra.

For the Gross collision model the spectra are shown in Fig. 7.2 for planar and in Fig. 7.3 for spatial rotors (in Fig. 7.3 we also display curves for F_B). We remark that the susceptibility may be expressed directly through the complex functions *w(z)* and $E_1(-z^2)$. Noting, that

$$\chi^*/G = \frac{L}{1+(iy/z)(L-1)} \quad \text{at } F_G \tag{7.34}$$

and substituting Eq. 5.99 and (7.12) into (7.34), we have for F_G:

$$\chi^*/G = \frac{1+i\sqrt{\pi}zw(z)}{1-\sqrt{\pi}yw(z)} \quad \text{for planar rotors}, \tag{7.35a}$$

$$\chi^*/G = \frac{1+z^2\exp(-z^2)E_1(-z^2)}{1+iyz\exp(-z^2)E_1(-z^2)} \quad \text{for spatial rotors}. \tag{7.35b}$$

In the long lifetime limit ($y = 0$) *there is no difference between collision models*, since

$$\chi^*(x)\big|_{\text{for } F_B} \equiv \chi^*(x)\big|_{\text{for } F_G} \quad \text{at } y = 0. \tag{7.36}$$

The Cole-Cole diagram $\chi''(\chi')$ is shown in Fig. 7.4 a, b, c for plane and in Fig. 7.4 d, e, f - for spatial rotors, the frequency x being the parameter of the curves.

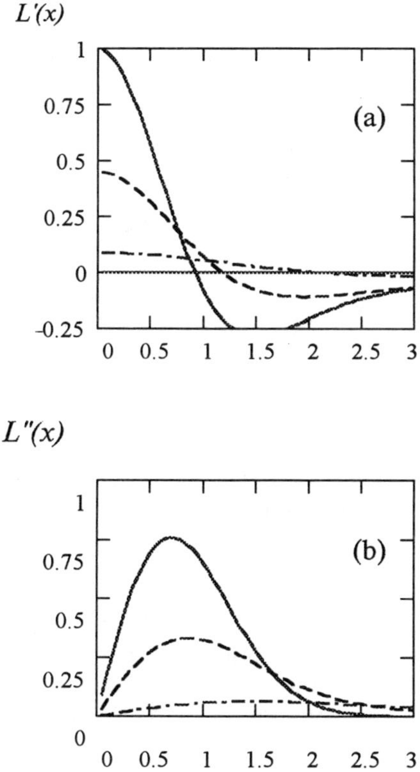

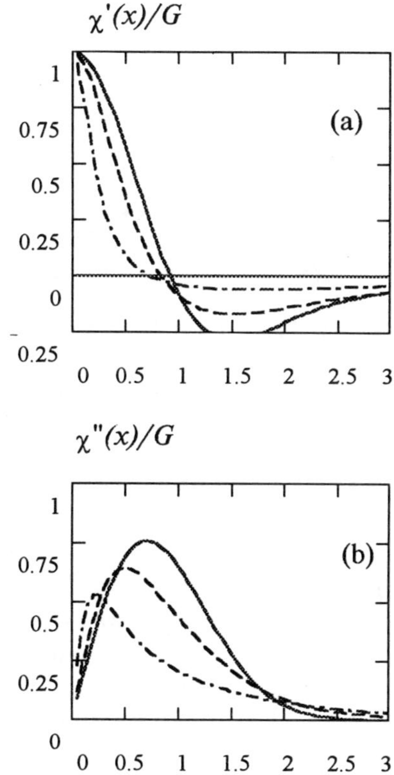

Normalized frequency x

Fig. 7.1 The real (a) and imaginary (b) parts of the planar SF $L(z)$ vs. frequency x; the plot $xL''(z)$ (c). Solid lines for $y = 0$, dashed lines for $y = 0.5$, dadots for $y = 2$.

Fig.7.2 The dielectric spectra of planar rotators, calculated for the Gross collision model. All parameters and designations as in Fig. 7.1.

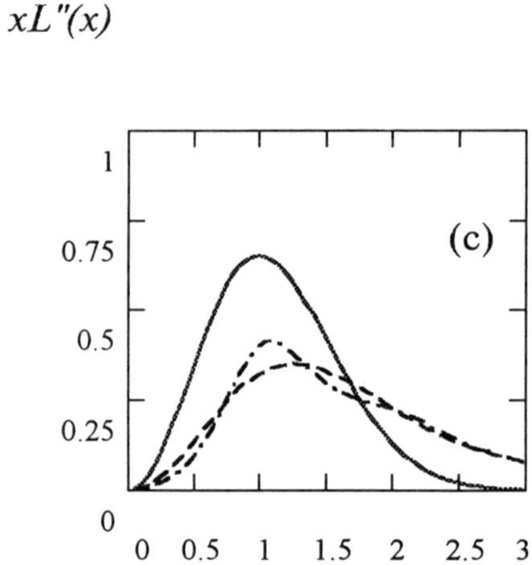

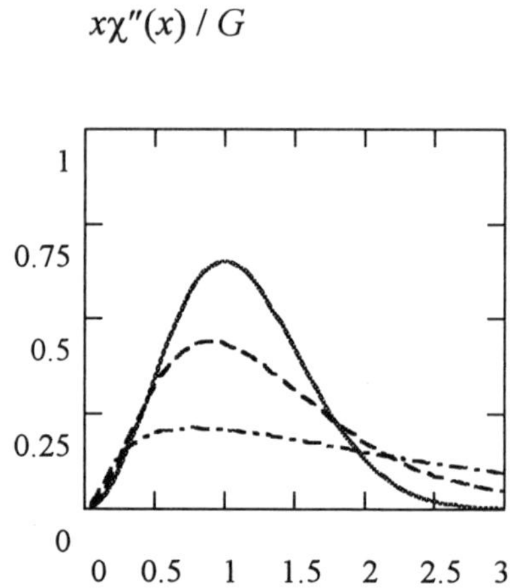

Fig. 7.1 c The caption to this Figure see on the previous page.

Fig. 7.2 c The caption to this Figure see on the previous page.

THE SPECTRA OF PLANAR ROTORS (Fig. 7.1, 7.2, and 7.4 a, b, c). These spectra reveal some features of resonance absorption and consequently may be termed *quasi-resonance spectra.* Indeed, the absorption band due to free rotors (Fig. 7.2c) is rather broad. We shall see in Part IV that the situation alters, if a strong conservative field is imposed.

For a *long lifetime limit* , i.e. at $y = \eta/\tau \to 0$, it follows from (7.25) that

$$\chi''(x) = G\sqrt{\pi}\, x \exp(-x^2), \tag{7.37a}$$

$$\chi'(x) = G\left[1 - 2x \exp(-x^2) \int_0^x \exp(t^2)\, dt\right]. \tag{7.37b}$$

The absorption maximum, given by

$$\max\left[x\chi''(x)\right] = \frac{G\sqrt{\pi}}{e}, \tag{7.37c}$$

is located at the frequency

$$x_L = 1, \text{ that is, } \omega_L = \frac{1}{\eta} = \sqrt{\frac{2k_BT}{I}}. \tag{7.37d}$$

As the collision frequency increases, the maximal absorption decreases, shifting towards lower frequencies at F_G and towards higher frequencies at F_B . The other distinctions between these models are follows:

1. The static susceptibility χ_S is equal to G at F_G, whereas at F_B it decreases from G to zero, when y increases.

2. Substitution F_B for F_G leads to a more rapid falloff of the dispersion curve in the Poley region ($x > 0.5$), while at low frequencies ($x < 0.5$), in contrast, absorption increases. Nevertheless, at F_G, as well at F_B, we have no Debye relaxation, since the only loss maximum is placed in the vicinity of the absorption maximum. Indeed, the Debye relaxation arises only, if some localization of molecules appears, e.g. due to precession of molecular tops (see Sec.7.3).

3. For large y substitution of F_B for F_G causes an essential change in the form of the $\chi''(\chi')$ dependence, cf. Fig. 7.4 b with 7.4 c, although it does not affect the overall area of the integrated absorption Π.

THE SPECTRA OF SPATIAL ROTORS (Fig. 7.3 and 7.4 d, e, f). The spectra $\chi^*(x)$ of planar and spatial rotors differ. Particularly,

a. If y $\to$ 0, then, in accord with Eq. 7.28,

$$\chi''(x) = G\pi x^2 \exp(-x^2); \tag{7.38}$$

$$\max\left[x\chi''(x)\right] = G\pi\left[\frac{3}{2}\right]^{\frac{3}{2}} \exp\left(-\frac{3}{2}\right) \quad \text{at } x_L = \sqrt{\frac{3}{2}}. \tag{7.39}$$

So, the loss / absorption maxima are for a linear molecule approximately twice that for a planar rotator, since the integrated absorption is twice as great. The position x_L of maximal absorption is shifted to higher frequencies in the ratio $\sqrt{3/2}:1$.

b. The minimum of the $\chi'(x)$ dependence is deeper for a spatial, than for a planar rotator.

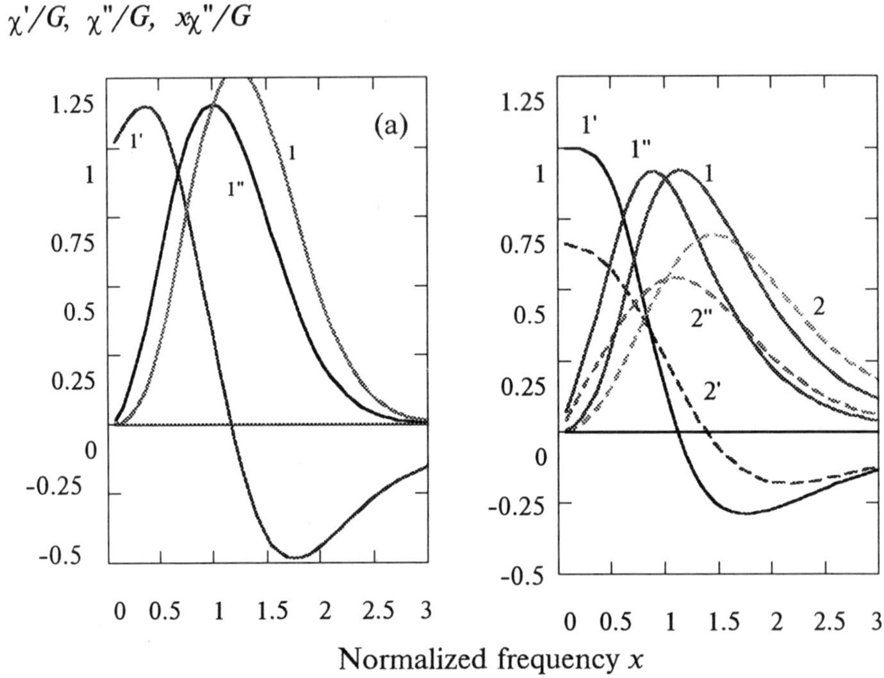

Fig. 7.3 **a, b:** The capture to this Figure see on the next page.

c. For sufficiently long lifetimes the $\chi'(x)$ curve passes through maximum for rotation in space, while for planar rotation this curve is decreasing at small x.

The dependence of absorption on the lifetime τ (or on the collision frequency $y = \eta/\tau$) is weak in both (planar / spatial) configurations.

* * *

The simplest ED model, just considered, reveals some characteristic features of polar dielectrics. However, it is hardly possible to apply this model for typical polar fluids. We shall study this problem in other chapters of the book.

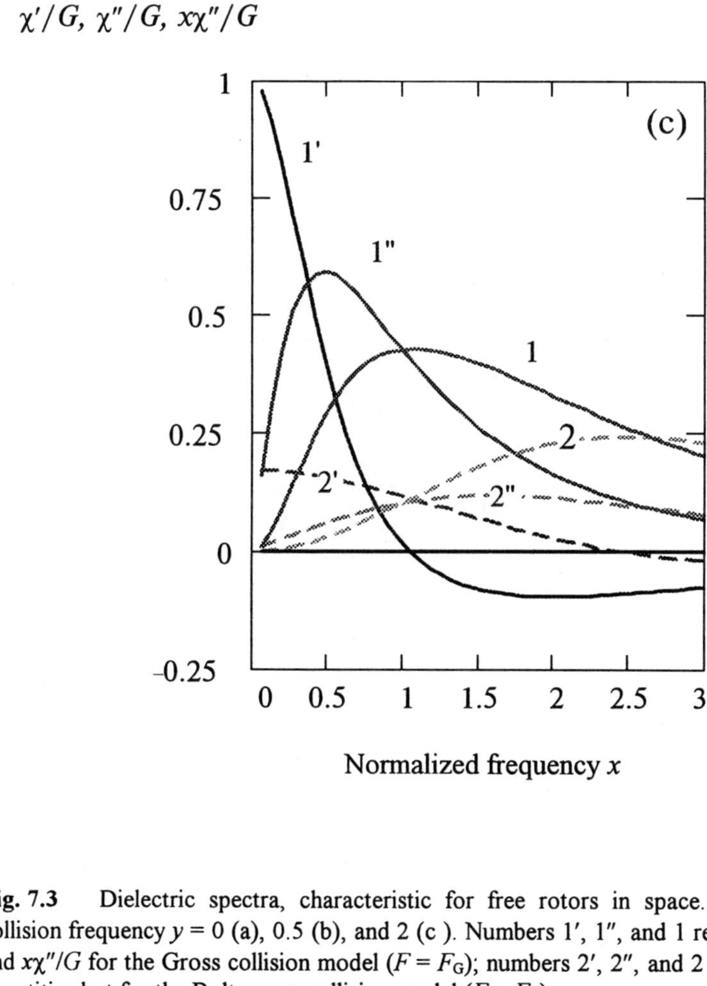

Fig. 7.3 Dielectric spectra, characteristic for free rotors in space. The normalized collision frequency $y = 0$ (a), 0.5 (b), and 2 (c). Numbers 1′, 1″, and 1 refer to χ'/G, χ''/G and $x\chi''/G$ for the Gross collision model ($F = F_G$); numbers 2′, 2″, and 2 refer to the same quantities but for the Boltzmann collision model ($F = F_B$).

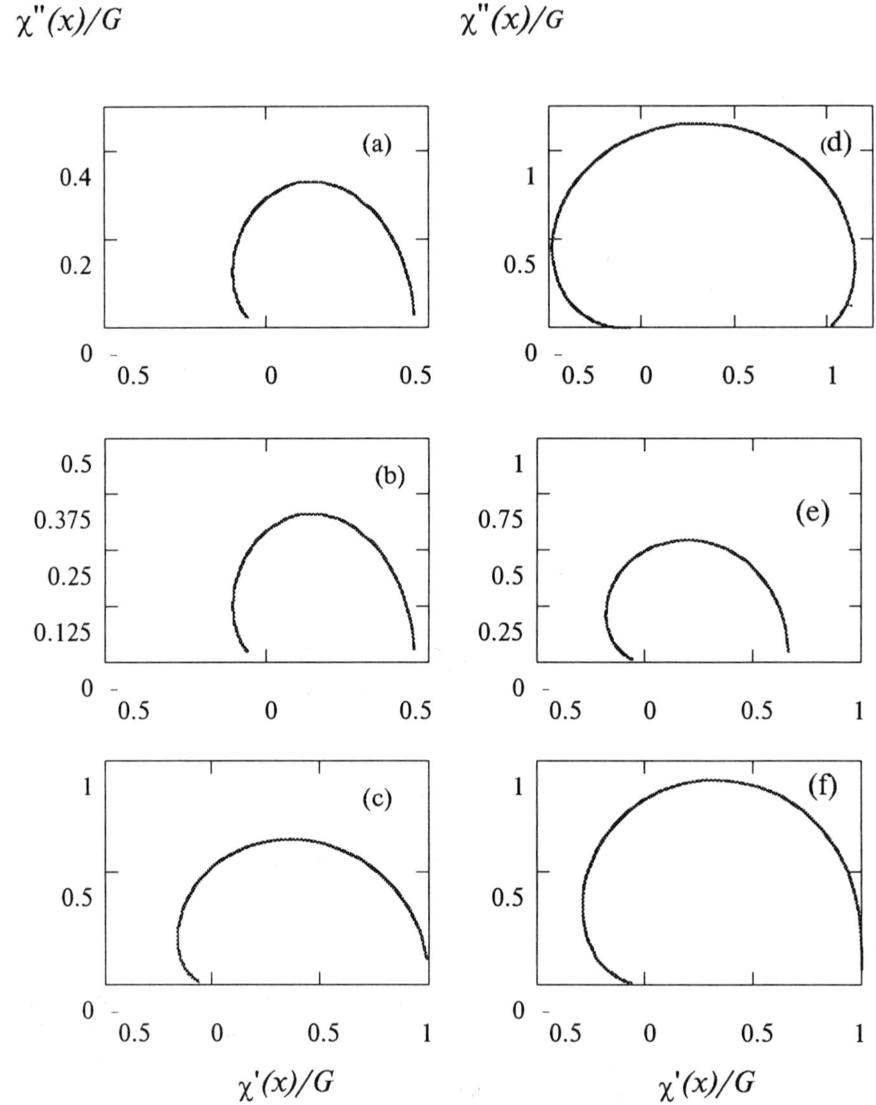

Fig. 7.4 The Cole-Cole diagram for planar (a, b, c) and spatial (d, e, f) rotators, calculated for $y = 0$ (a, d); for $y = 0.5$ at F_B (b, e) and at F_G (c, f).

7.2 Ensemble of Symmetric Top Molecules

7.2.1 The Spectral Function

The Reader is recommended to look previously through Sec. 3.4.6 and 3.4.7.

THE LONGITUDINAL SF $L_{\|}(z)$. Using Fig. 3.4a, we express the projection of vector μ on the Z axis by means of angles $\bar{\theta}$, δ. Let ϑ be angle on the plane $X'Y'$ between the projection $\mu_{X'Y'}$ and the axis Y'. We write the law of motion in the form $\vartheta = (\varphi + \varphi_0)\, p$, as describing precession of a free top. Comparing Eq. 3.111 and 5.104, we find Fourier amplitudes:

$$a_0 = \cos\delta \cos\bar{\theta};\ a_1 = \sin\delta \sin\bar{\theta};\ a_n = 0 \quad \text{for } n > 2. \tag{7.40}$$

Then the sum $S_{\|}$ in (5.108) is given by:

$$S_{\|} = \sum_{n=1}^{\infty} \frac{n^2 a_n^2}{p^2 n^2 - z^2} = \frac{\sin^2\delta \sin^2\bar{\theta}}{p^2 - z^2}. \tag{7.41}$$

We find the SF $L_{\|}(z)$ by taking account of the mean value $\langle q_{\|}^2 \rangle = 1/3$, of Eq. 3.110 for l, m, of Eq. 3.112 for the Hamiltonian h and of Eq. 3.121 for the Maxwellian distribution W:

$$L_{\|}(z) = 3\pi \sqrt{\frac{1+b}{\pi^3}} \int_0^{\pi/2} \sin^3\delta \, d\delta$$

$$\times \int_0^{\infty} \exp\left(-p^2\left[1 + b\cos^2\delta\right]\right) \frac{2p^4 dp}{p^2 - z^2} \int_0^{\pi/2} \sin^3\bar{\theta}\, d\bar{\theta}.$$

The integral over $\bar{\theta}$ yields $\int_0^{\pi/2} \sin^3\bar{\theta}\, d\bar{\theta} = 2/3$. Finally, the SF is found as a double integral over precession angle δ and over fundamental frequency p of a top:

$$L_{\|}(z) = 4\sqrt{\frac{1+b}{\pi}} \int_0^{\infty} \frac{p^4 dp}{p^2 - z^2} \int_0^{\pi/2} \exp\left[-p^2\left(1 + b\cos^2\delta\right)\right] \sin^3\delta \, d\delta \tag{7.42}$$

THE TRANSVERSE SF $L_\perp(z)$. The same expression should, of course, be valid for ⊥-orientation (for $\boldsymbol{E}_m \parallel Y$, see Fig. 3.3a). To show this, we express the angular displacement $\psi(\varphi)$ of the μ_{XY} projection relative to the axis Y in terms of angles δ, $\bar{\theta}$, $\vartheta(\varphi)$. Let us compare formula (5.104a) for $q_\perp(\varphi)$ at $\psi_0 = 0$ with expression of the same quantity in terms of the projection of $\boldsymbol{\mu}$-vector on the axes Y' and Z':

$$q_\perp(\varphi) = \sin\bar{\theta}\cos\psi = (\mu_{Y'})_Y + (\mu_{Z'})_Y = \sin\delta\cos\bar{\theta}\cos\vartheta - \cos\delta\sin\bar{\theta}\ . \tag{7.43}$$

To obtain a more general expression, in which ϑ is substituted by $\vartheta + \vartheta_0$, we write:

$$q_\perp(\varphi) = \sin\delta\ \cos\bar{\theta}\ \cos(\vartheta + \vartheta_0) - \cos\delta\ \sin\bar{\theta}\ , \tag{7.44a}$$

so that

$$q_\perp(0) = \sin\delta\ \cos\bar{\theta}\ \cos\vartheta_0 - \cos\delta\ \sin\bar{\theta}\ . \tag{7.44b}$$

It is seen in Fig. 3.4a, that during the period $\varphi = \Phi = 2\pi/p$ of a complete rotation of the vector $\boldsymbol{\mu}$ around vector $\boldsymbol{p}$ the latter performs a similar rotation around the axis Z. On the contrary, in Fig. 3.4b it is seen that vector $\boldsymbol{\mu}$ is oriented in such a way, that it does not revolve around Z. Thus, we have two specific cases:

$$\textbf{(a)}.\ \lambda = \frac{1}{2}\psi\left(\varphi = \frac{2\pi}{p}\right) = \pi \text{ at } \delta > \bar{\theta}\ ;\quad \textbf{(b)}.\ \lambda = 0 \text{ at } \delta < \bar{\theta} \tag{7.45}$$

Now we turn to Eq. 5.117 and 7.44. Noting that the average over ψ_0 is equivalent to the average over ϑ_0, we write

$$\frac{1}{2}\mathfrak{I} = \langle q_\perp(\varphi)\, q_\perp(0)\rangle_{\psi_0} \equiv \langle q_\perp(\varphi)\, q_\perp(0)\rangle_{\vartheta_0}$$

$$= \frac{1}{2}\sin^2\delta\ \cos^2\bar{\theta}\ \cos p\varphi + \cos^2\delta\ \sin^2\bar{\theta}\,. \tag{7.46}$$

Since $\Phi = 2\pi/p$, Eq.5.117 yields

$$\mathfrak{I} = \sum_{n=-\infty}^{\infty} b_n^2 \cos\left(n - \frac{\lambda}{\pi}\right) p\varphi\ . \tag{7.47}$$

(a). For $\lambda = \pi$, comparing (7.46) and (7.47), we find, that we must retain terms, corresponding to $n = 1$ and $n = 2$:

$$b_1^2 = 2\cos^2\delta \sin^2\overline{\theta} \text{ for } n = 1;\; b_2^2 = \sin^2\delta \cos^2\overline{\theta} \text{ for } n = 2.$$

The term b_1^2 gives no contribution to the integral (5.121), while the term b_2^2 yields

$$\left\langle q_\perp^2 \right\rangle L_\perp(z) = \pi \int Wp \frac{\sin^2\delta \cos^2\overline{\theta}}{p^2 - z^2} dh\, dl\, dm \; . \tag{7.48}$$

(b). For $\lambda = 0$ we must retain three terms, corresponding to $n = -1$, $n = 0$, and $n = 1$:

$$b_0^2 = 2\cos^2\delta \sin^2\overline{\theta};\; b_{-1}^2 + b_1^2 = \sin^2\delta \cos^2\overline{\theta} \; . \tag{7.49}$$

The term with $n = 0$ gives no contribution to (5.121), while the other two yield the same result (7.48), as for the case **(a)**.

Comparing (7.48) with Eq. 7.41 and noting that $\left\langle q_\parallel^2 \right\rangle = \left\langle q_\perp^2 \right\rangle = 1/3$, we conclude that the transverse SF $L_\perp(z)$ coincides in both **(a)** and **(b)** cases with the longitudinal SF $L_\parallel(z)$, as it should be. Our statement is proved. Hence, in ensemble of free rotating tops the SF is given by

$$L_\parallel(z) = L_\perp(z) = L(z,\, b)$$

$$= 4\sqrt{\frac{1+b}{\pi}} \int_0^\infty \frac{\exp(-p^2)\, p^4\, dp}{p^2 - z^2} \int_0^{\frac{\pi}{2}} \exp\left(-p^2 b \cos^2\delta\right) d\delta \tag{7.50a}$$

$$= 4\sqrt{\frac{1+b}{\pi}} \int_0^\infty \frac{\exp(-p^2)\, p^4\, dp}{p^2 - z^2} \int_0^1 \exp\left(-p^2 b \gamma^2\right)\left(1-\gamma^2\right) d\gamma \, . \tag{7.50b}$$

We remark that the integral over γ, which we denote $f(u)$, can be expressed in terms of the probability function:

$$f(u) \equiv \int_0^1 \exp\left(-p^2 b\gamma^2\right)\left(1-\gamma^2\right) d\gamma = \frac{1}{2u^2}\left[\exp(-u^2) + \frac{\sqrt{\pi}}{2u}\operatorname{erf}(u)\right], \tag{7.51}$$

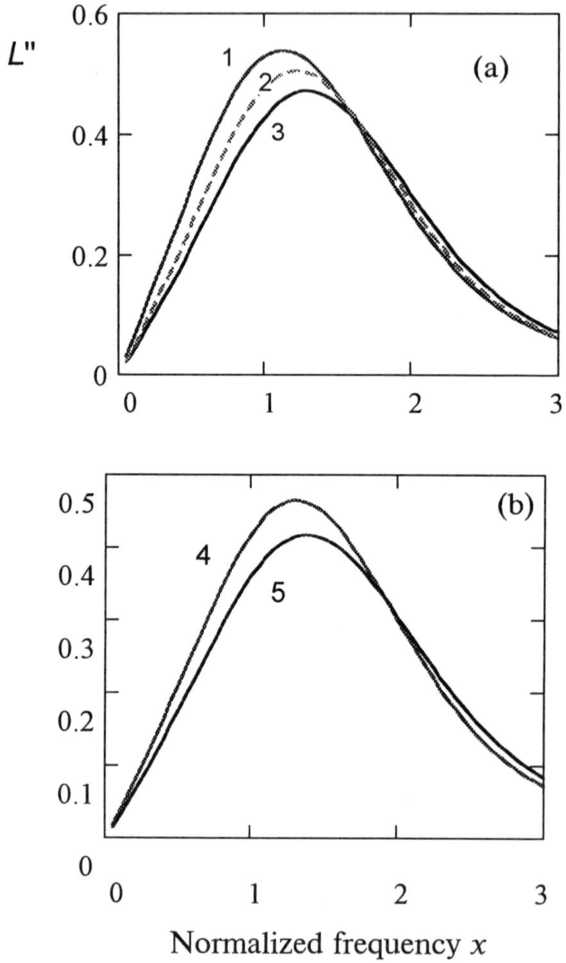

Fig. 7.5 The imaginary part of the spectral function vs. normalized radiation frequency, calculated for a prolate (a) and oblate (b) molecular tops. $y = 0.5$. For curves 1, 2, 3, 4, and 5 the formfactor b is taken respectively 10, 1, 0, - 0.1, and - 0.5.

$$u = p\sqrt{b}\,,\ \mathrm{erf}(u) = \frac{2}{\sqrt{\pi}} \int_0^u \exp(-t^2)\,dt\ . \tag{7.52}$$

Then

$$L(z,b) = 4\sqrt{\frac{1+b}{\pi}} \int_0^\infty f(p\sqrt{b})\,\frac{\exp(-p^2)\,p^4\,dp}{p^2 - z^2}\ . \tag{7.53}$$

The frequency dependence of $L''(x)$ versus x is illustrated by Fig. 7.5. The dielectric loss is described by a typical quasi-resonance curve. Transition from prolate to oblate tops results in a small but noticeable shift to higher frequencies (in the Poley region) of maximal loss. For a linear molecule ($b = \infty$) the corresponding dependence is presented by curves 2" in Fig.7.3.

Following (Gaiduk and Kalmykov 1987, p.112), the SF can be expressed in terms of the probability integral $w(z)$, Eq. 5.100, 5.101, 7.25, 7.29:

$$L(z,b) = 1 + \frac{L_{\mathrm{pl}}}{b} + \frac{i\pi}{2}\left(2z^2 - \frac{1}{b}\right)\Phi\mathrm{i}(z,b)\,, \tag{7.54}$$

$$L_{\mathrm{pl}}(Z) = 1 + i\sqrt{\pi}\,Z\,w(Z);\ Z = z\sqrt{1+b}\,; \tag{7.55}$$

$$\Phi\mathrm{i}(Z,\,b) = \frac{i}{\pi}\int_{-\infty}^{\infty} \exp(-t^2)\,\frac{\mathrm{erf}\left(\sqrt{b}\,t\right)}{z-t}\,dt \quad .(7.56a)$$

$$= z\sqrt{\frac{b}{\pi}}\int_{-1}^{1} w\left[z\sqrt{1+\frac{b}{4}\left(1+t^2\right)}\right] - \frac{2i}{\pi}\ln\left(\sqrt{b} + \sqrt{1+b}\right). \tag{7.56b}$$

Now, in view of Eq.3.88a for the formfactor b, we write down formulas for the spectral functions for some limiting cases.

(A). LINEAR MOLECULE. In this case of an *extremely prolate* top we put $I_c \to 0$ and $b \to \infty$. Then in (7.51) we take $f(u) \to \dfrac{\sqrt{\pi}}{2u} = \dfrac{\sqrt{\pi}}{2p\sqrt{b}}$; Eq. 7.53 yields the same result, as (7.11)-(7.13):

$$L(z) = 4\sqrt{\frac{b}{\pi}}\sqrt{\frac{\pi}{b}}\int_0^\infty f(p\sqrt{b})\,\frac{\exp(-p^2)\,p^4\,dp}{p^2 - z^2}\ , \tag{7.57}$$

(B). SPHERICAL TOP. Putting $I = I_c$ and $b = I/I_c - 1 = 0$, we obtain from (7.50b):

$$L(z) = \frac{8}{3\sqrt{\pi}} \int_0^{\infty} \frac{\exp(-p^2)\, p^4\, dp}{p^2 - z^2} = \frac{2}{3} + \frac{8z^2}{3\sqrt{\pi}} \int_0^{\infty} \frac{\exp(-p^2)\, p^2\, dp}{p^2 - z^2}$$

$$= \frac{2}{3}\left\{1 + 2z^2\left[1 + i\sqrt{\pi}\, z w(z)\right]\right\} = \frac{2}{3}\left\{1 + 2z^2 L_{\rm pl}(z)\right\}. \tag{7.58}$$

(C). DISK. Putting $b = -1/2$ into (7.50b), we have

$$L(z) = 4\sqrt{\frac{1}{2\pi}} \int_0^{\infty} \frac{\exp(-p^2)\, p^4\, dp}{p^2 - z^2} \int_0^1 \exp\left(\frac{p^2 \gamma^2}{2}\right)\left(1 - \gamma^2\right) d\gamma \tag{7.59}$$

7.2.2 The Long Lifetime Limit and The Debye Relaxation for y << 1

SPECTRAL FUNCTIONS. Taking account of the limit (5.176), we obtain from Eq.7.53 formula for $L''(x) = \chi''/G$ in terms of elementary functions:

$$L''(x,b) = \frac{\sqrt{\pi(1+b)}}{2b} \exp(-x^2)\left[2x\exp\left(-bx^2\right) + \sqrt{\frac{\pi}{b}}\left(2bx^2 - 1\right)\operatorname{erf}\left(\sqrt{bx}\right)\right]. \tag{7.60}$$

The real part L' can be found with help of the Kramers-Kronig relation (5.171b). In the latter we substitute χ' and χ'' for L' and L'', respectively. In view of (7.60) we have for y << 1:

$$L'(z,b) = \frac{\chi'}{G} = \frac{1}{b}\sqrt{\frac{1+b}{\pi}} \int_0^{\infty} \left[2p\exp(-bp^2) + \sqrt{\frac{\pi}{b}}\left(2bp^2 - 1\right)\operatorname{erf}\left(p\sqrt{b}\right)\right] \tag{7.61}$$

In the case of a *linear molecule* we can use Eq.7.34-7.35.

In the case of a *spherical top*, using Eq.7.21-7.25, we have directly from Eq.7.57:

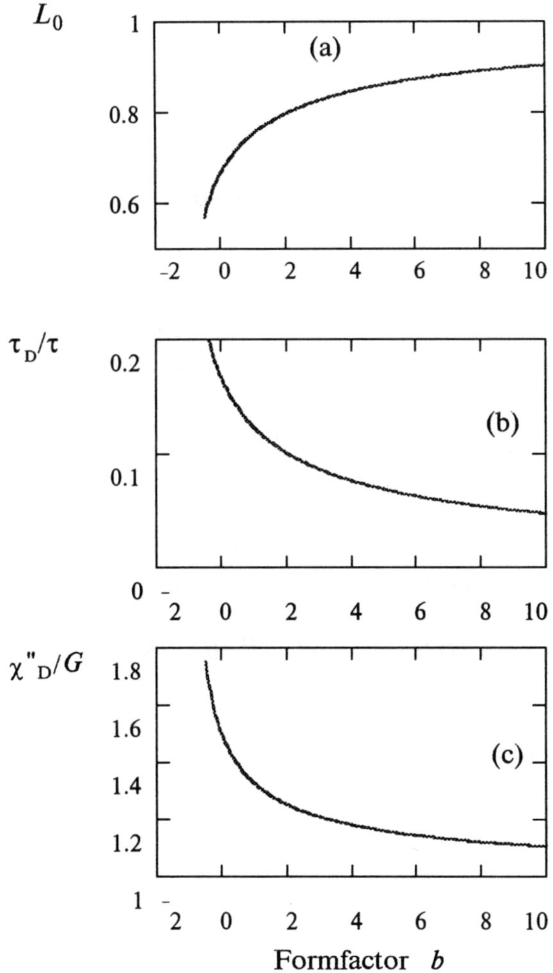

Fig. 7.6 The static spectral function (a), relaxation time / lifetime ratio (b) and maximum Debye loss (c) vs. formfactor b of a top, calculated for a long lifetime limit ($y = 0$).

$$L''(x) = \frac{4}{3}\sqrt{\pi}\, x^3 \exp(-x^2), \tag{7.62a}$$

$$L'(x) = \frac{2}{3}\sqrt{\pi}\left[1 + 2x - 4x^3 D(x)\right], \tag{7.62b}$$

where $D(x)$ is defined in (7.24).

PARAMETERS OF THE DEBYE RELAXATION. Putting $z = 0$ in Eq. 7.50a, we take integral over p. For a long lifetime limit we find the static SF in the form

$$L_0 = \sqrt{1+b}\int_0^1 \frac{\left(1-\beta^2\right)d\beta}{\left(1+b\beta^2\right)^{3/2}}$$

$$= \frac{\sqrt{1+b}}{b}\left[\sqrt{1+b} - \frac{\ln\left(\sqrt{b}+\sqrt{1+b}\right)}{\sqrt{b}}\right]. \tag{7.63}$$

Let us consider the Gross collision model ($F = F_G$). For the same limit $y = 0$, in view of Table 6.2, the ratio τ_D / τ is given by

$$\tau_D / \tau = \frac{1}{L_0} \quad \text{for } y = \eta/\tau \ll 1. \tag{7.64}$$

The L_0 value, given by Eq. 7.63, determines also the Debye maximum χ'', which is proportional to 1 - L_0:

$$\frac{\chi_D''}{G} = \frac{1}{2} - \frac{\sqrt{1+b}}{2b}\left[\sqrt{1+b} - \frac{1}{\sqrt{b}}\ln\left(\sqrt{b}+\sqrt{1+b}\right)\right]. \tag{7.65}$$

In Fig.7.6 we present the dependencies of L_0, τ_D/τ and χ''/G on the formfactor b. Let τ be constant. Then the relaxation time τ_D decreases with the increase of b (Fig.7.6b). Similar behavior has the b-dependence of the Debye loss (Fig. 7.6c).

For $b \to \infty$ Eq.7.63 yields $\tau_D/\tau = 1$, while the χ_D'' value tends to zero. The only loss maximum appears in the infrared region. Thus, *the typical Debye spectrum, separated from the resonance absorption band, does not appear in an ensemble of linear molecules.*

For a *spherical top*, $b = 0$, we find from Eq.7.62-7.64:

$$L_0 = \frac{2}{3}; \frac{\tau_D}{\tau} = \frac{3}{2}; \frac{\chi_D''}{G} = \frac{1}{6}. \tag{7.66}$$

For a *disk* we have from the same equations:

$$L_0 = \frac{\pi}{2} - 1 \approx 0.57; \frac{\tau_D}{\tau} = \frac{1}{\frac{\pi}{2}-1} \approx 1.75; \frac{\chi_D''}{G} = 1 - \frac{\pi}{4} \approx 0.215. \tag{7.67}$$

Thus, the ratio τ_D/τ and Debye loss maximum increase, when molecular top becomes more oblate; hence, precession of such top and localization of its trajectory becomes more pronounced. This result reinforces the conclusions of Sec. 6.5.2 and agrees with results of further numerical calculations (see Fig. 7.7).

7.2.3 Dielectric Spectra

Let x_L denote normalized frequency, at which normalized absorption $x\chi''(x)$ attains its maximum. Transition from a planar to a spatial rotor and after that - to spherical and oblate tops causes the increase of x_L. For a long lifetime τ limit (i.e. for $y = 0$) the x_L values are presented in Table 7.1.

For a planar rotor, linear molecule, and a spherical top these x_L values are obtained from Eq. 7.37a, 7.38 and 7.62a, respectively. For a disc x_L is calculated from Eq. 7.60. Hence, transition from prolate to oblate top results for $y = 0$ in the $\approx$ 24% increase of x_L, since $1.517 : \sqrt{3/2} \approx 1.24$.

Fig. 7.7 illustrates the shift of the frequency x_L with the decrease of b, as well as the effect of a shape of a top on the loss and absorption spectra $\chi''(x)$ and $x\chi''(x)$. We consider the self-consistent (at F_G) and Boltzmann (at F_B) field-induced distributions F. The smaller formfactor b, the larger is the shift of these dispersion curves into the Poley region. When y increases, the dielectric loss $\chi''(x)$ in the Poley region (near x_L) shifts to greater values of x at F_B and in the opposite direction at F_G.

Table 7.1 The frequency x_L of maximal absorption, Debye relaxation time / lifetime ratio τ_D / τ and Debye loss maximum χ''_D / G. Calculation for ensemble of free rotors at $\tau \to \infty$.

Para-meters	Rotation on plane	Rotation in space		
	Planar rotor	Prolate top $b \to \infty$	Spherical top $b = 0$	Disk $b = -1/2$
x_L	1	$\sqrt{3/2}$	$\sqrt{2}$	1.517
τ_D/τ	1	1	3/2	$\dfrac{1}{\frac{\pi}{2} - 1}$
χ_D''/G	$\to 0$	1/6		$1 - \dfrac{\pi}{4}$

Evolution of dielectric spectra, obliged to the change of the collision frequency y, is illustrated by Fig. 7.8. If the formfactor b is sufficiently small and the lifetime y is long ($y << 1$), then for the self-consistent collision model F_G, in contrast to ensemble of linear molecules, there appears a low-frequency loss maximum χ''_D, see Fig. 7.8 c, e. Formally, this χ''-maximum can be ascribed to non-zero mean μ_E-projection of a dipole moment onto radiation field amplitude E_m. To be more precise, the Debye loss χ''_D arises because the integral

$$\left\langle \left(\check{T}\right)^{-1} \int_0^{\check{T}} \mu_E(t)\,dt \right\rangle, \tag{7.68}$$

does not vanish. Here $\check{T}$ is the period of function $\mu_E(t)$.

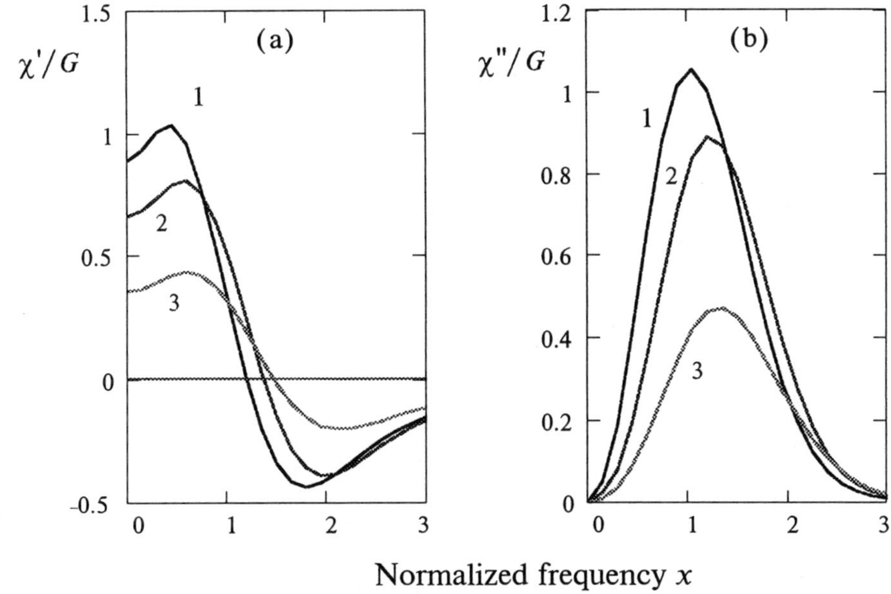

Fig. 7.7 The real (a, c, e) and imaginary (b, c, f) parts of the susceptibility, calculated for the Boltzmann (a - d) and Gross (e - f) collision models. The normalized collision frequency y = 0.05 (a, b) and 0.5 (c - f). Curves 1, 2, 3 correspond respectively to prolate (b = 9), spherical (b = 0) tops and disk (b = - 1/2).

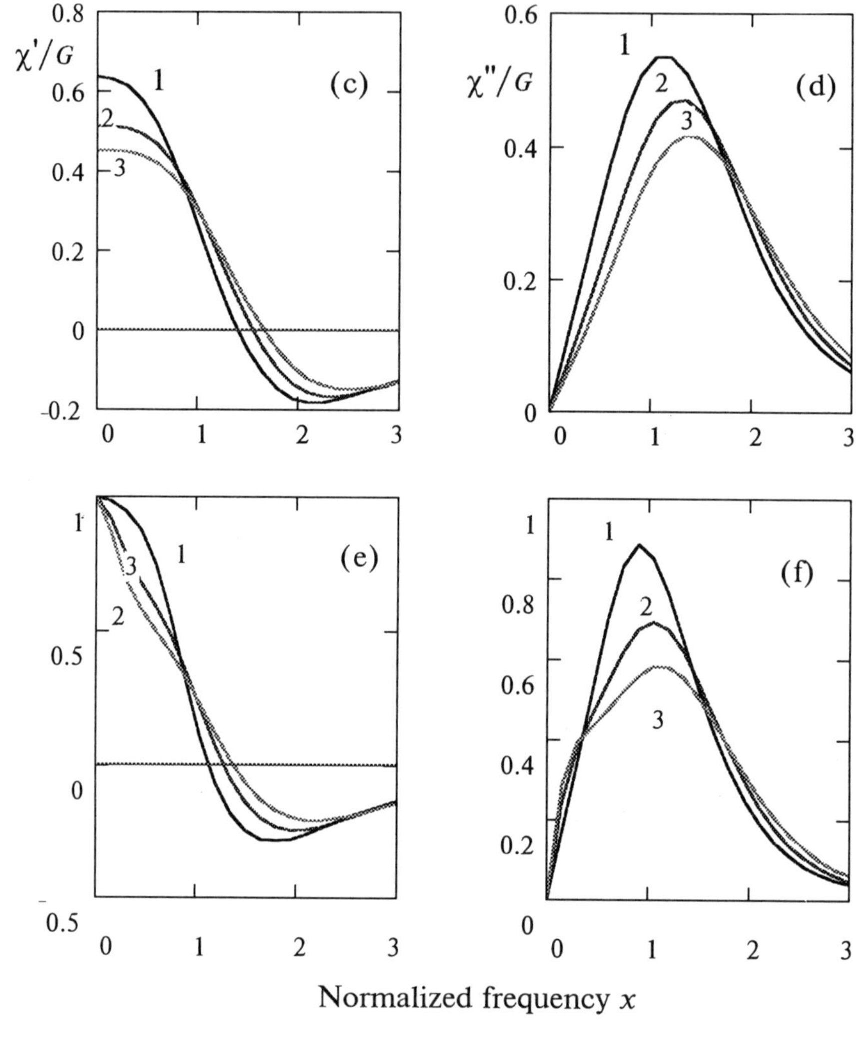

Fig. 7.7 c - f: The caption to Figure see on the previous page.

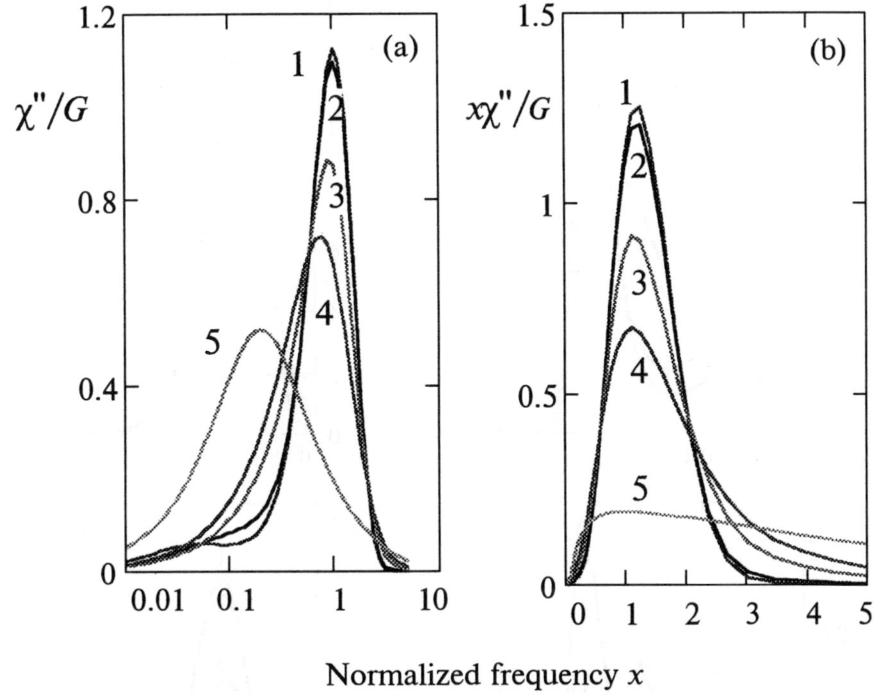

Fig. 7.8 Normalized loss (a, c, d) and normalized absorption (b, d, f), calculated for ensemble of prolate tops with b = 9 (a, b), for spherical tops (b = 0) and for disc (b = -1/2). The normalized collision frequency y = 0.05, 0.1, 0.5, 1, and 5 (curves from 1 to 5, respectively).

Precession of a molecular top about direction $\boldsymbol{p}$ of the steady state angular momentum results in some localization of a dipole moment vector $\mu(t)$. When time-varying field $\boldsymbol{E}(t)$ is switched off, the polarization of dielectric sample disappears after the relaxation time τ_D. The latter and the Debye loss χ_D'' increase, if the top becomes more oblate (see Table 7.1).

Frequent collisions ($y \geq 1$) are characterized by the only loss maximum.

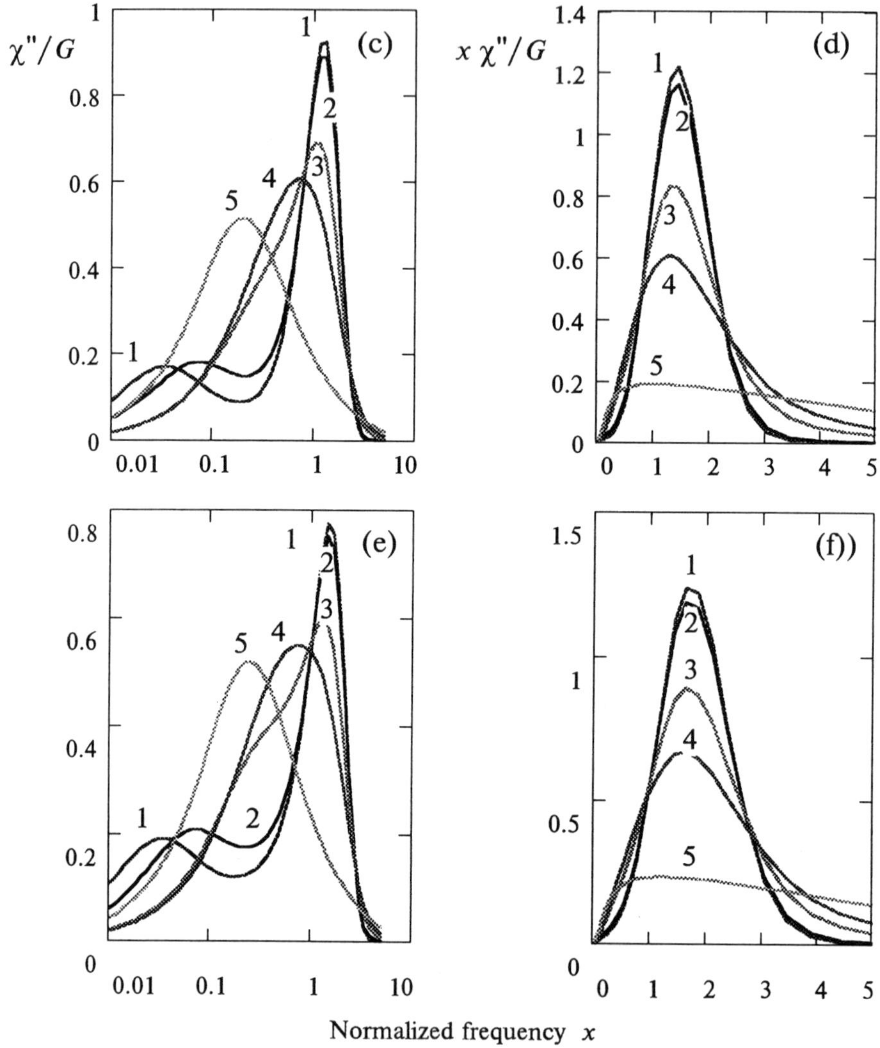

Fig. 7.8 **c - f:** The caption to Figure see on the previous page.

With the increase of y the latter shifts to lower frequencies x (see Fig. 7.8 c and e). The position of this maximum, being independent on the formfactor b, is determined by the Hubbard relation:

$$x_D = \frac{1}{y}, \text{ i.e. } \tau_D \tau = \eta^2 = \frac{I}{2k_B T} \text{ at } y >> 1. \tag{7.69}$$

However, it seems that such behavior is not encountered in polar liquids.

From Fig.7.8 we see that if $y << 1$ and if condition (7.60) approximately holds, the loss / absorption maxima, observed in the Poley region, depend weakly on the value of y.

In the next chapter we shall consider another molecular model, in which the localization of a dipole could be made pronounced, so that it may cause larger frequency shifts of dielectric spectra into Poley and Debye regions, than it is possible in an ensemble of free rotating tops.

CHAPTER 8
THE CONFINED ROTATOR (CR) MODELS

8.1 The Planar CR Model

8.1.1 Law of Motion and Steady State Distribution

THE POTENTIAL CONFIGURATION. We suggest, that dipoles of a polar medium librate because of presence of the rectangular conservative potential U (see Fig. 8.1). The elastically reflecting walls of the sector 2β present a rough model of neighboring molecules and are supposed to be at rest. The current rotor position is characterized by angle ϑ; we put $\vartheta = 0$, when the dipole moment μ is directed along the symmetry axis Z of the well, i.e. when μ coincides with the bisector. During the time interval t_v between strong collisions any dipole librates with a constant value $|\Omega|$ of angular velocity, while the sign of Ω changes after each reflection. Inside the sector 2β the potential energy of a dipole is set to be zero. Hence, the total energy H of a dipole is equal to its kinetic energy $I\Omega^2/2$.

We introduce the normalized Hamiltonian h and two basic direction cosines $q_{\|}(\varphi)$ and $q_{\perp}(\varphi)$:

$$h = \frac{H}{k_B T} = \eta^2\Omega^2;\ q_{\|}(\varphi) = \cos\vartheta(\varphi);\ q_{\perp}(\varphi) = \sin\vartheta(\varphi);\ \varphi = \frac{t}{\eta}. \quad (8.1)$$

The libration period $\check{T}$ and the normalized libration frequency $\hat{p}$ are functions of the current energy h of a dipole:

$$\check{T} = \frac{4\beta}{|\Omega|} = \frac{4\beta\eta}{\sqrt{h}}; \; \check{p} = 2\pi\frac{\eta}{\check{T}} = \frac{\sqrt{h}}{f}, \tag{8.2}$$

where the normalized libration amplitude

$$f = 2\beta/\pi. \tag{8.3}$$

A wide (Maxwellian) distribution W over energies h and similar distribution over libration frequencies $\check{p}$ are established at thermal equilibrium:

$$W = C\exp(-h) = C\exp\left(-f^2\check{p}^2\right). \tag{8.4}$$

We shall mostly consider the double well potential (DWP) configuration (Fig. 8.1a), in which the mean (over ensemble) direction cosine is equal to zero:

$$\langle q \rangle = 0. \tag{8.5}$$

Later we shall also briefly consider, how dielectric spectra change in the case of the single well potential (SWP), in which the only sector 2β is introduced (see Fig. 1.5a). Although the spectral functions $K(z)$, $L(z)$ are the same for both DWP (Fig. 8.1a) and SWP (Fig. 1.5a) configurations, the orientational susceptibilities $\chi^*(\omega)$ are different in these cases.

FOURIER AMPLITUDES AND SPECTRAL FUNCTIONS. Without loss of generality we set $\vartheta = 0$ at the instant $\varphi = 0$ (see Fig. 8.1b). If a dipole rotate from the left wall to the right one, the law of its motion is

$$\vartheta(\varphi) = \Omega t = \frac{\Omega\varphi}{\eta} = f\check{p}\varphi, \; \dot{\vartheta} = \sqrt{h} = f\check{p}, \; -\frac{\pi}{2\check{p}} \le \varphi \le \frac{\pi}{2\check{p}}. \tag{8.6a}$$

After the elastic reflection from the right wall the dipole will move to the left wall with the angular velocity, having the opposite sign. Then the law of motion become

$$\vartheta(\varphi) = \beta - f\check{p}\left(\varphi - \frac{\pi}{2\check{p}}\right); \; \dot{\vartheta} = -f\check{p}; \; \frac{\pi}{2\check{p}} \le \varphi \le \frac{3\pi}{2\check{p}}, \tag{8.6b}$$

and so on. The law of motion (8.6) is illustrated by Fig. 8.2. The period Φ of function $q_{\|}(\varphi)$ is given by

$$\Phi = \pi/\check{p} = \frac{\pi f}{\sqrt{h}} \,. \tag{8.7}$$

We take this quantity without superscript, $\check{\Phi} \to \Phi$. Indeed, in the CR model only subensemble of librators appears, since potential walls are supposed to be infinitely high. The period of $q_{\perp}(\varphi)$ function is twice that of $q_{\|}(\varphi)$. These time dependencies $q_{\perp}(\varphi)$, $q_{\|}(\varphi)$ are characterized by discontinuities at the instants of "weak" collisions of a dipole with walls, viz. at the instants

$$\varphi = \ldots - \frac{\pi}{2\check{p}}, \frac{\pi}{2\check{p}}, 2\frac{\pi}{2\check{p}}, \ldots \,.$$

However, this circumstance does not lead to any mathematical difficulty, related to description of dielectric spectra. We recall the Reader that "strong" collisions, that sometimes occur, disrupt a local-order potential configuration (the DWP, in our case).

Periodic functions $q_{\|}(\varphi)$ and $q_{\perp}(\varphi)$ are represented by Fourier series (5.81) over cosines and sines, respectively. Fourier amplitudes a_n, b_n are determined in accord with Eq. 5.82 by integrals over φ in the time interval [0, $\Phi/2$]:

$$\begin{Bmatrix} a_n \\ b_n \end{Bmatrix} = 2\,\frac{1 \pm (-1)^n}{\Phi} \int_0^{\Phi/2} \begin{Bmatrix} q_{\|}(\varphi)\cos \check{p} n\varphi \\ q_{\perp}(\varphi)\sin \check{p} n\varphi \end{Bmatrix} d\varphi, \; n = 0,\ 1,\ 2, \ldots. \tag{8.8}$$

Taking ϑ instead of φ as the variable and using Eq. 8.6a, 8.7, we have

$$\begin{Bmatrix} a_n \\ b_n \end{Bmatrix} = 2\,\frac{1 \pm (-1)^n}{2\beta} \int_0^{\beta} \begin{Bmatrix} \cos\vartheta \\ \sin\vartheta \end{Bmatrix} \begin{Bmatrix} \cos\dfrac{\pi n \vartheta}{2\beta} \\ \sin\dfrac{\pi n \vartheta}{2\beta} \end{Bmatrix} d\vartheta,$$

that is,

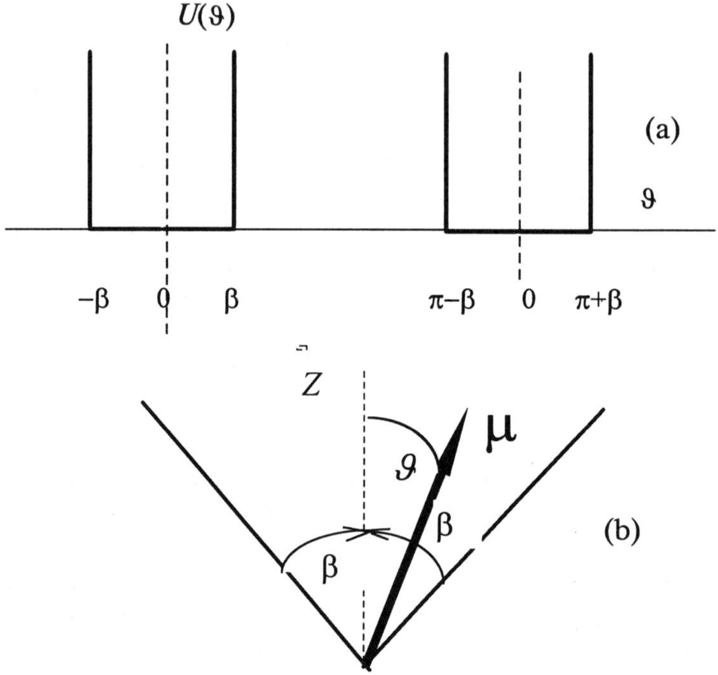

Fig. 8.1 The double well potential configuration (a) and geometric scheme, pertaining to the planar CR model (b).

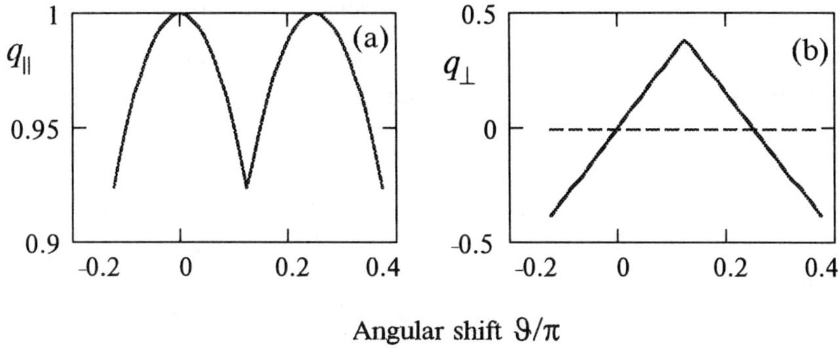

Fig. 8.2 The diagram, pertaining to the law of motion of a dipole.

$$a_n = \frac{2}{\beta}\int_0^{\beta} \cos\vartheta \cos\frac{\pi n\vartheta}{\beta} d\vartheta = \frac{1}{\beta}\left[\frac{\sin(\beta+\pi n)}{1+\frac{\pi n}{2\beta}} - \frac{\sin(\beta-\pi n)}{1-\frac{\pi n}{2\beta}}\right],\ n=0,2,4\ldots, \tag{8.9a}$$

$$b_n = \frac{2}{\beta}\int_0^{\beta} \sin\vartheta \sin\frac{\pi n\vartheta}{\beta} d\vartheta = \frac{1}{\beta}\left[\frac{\sin\left(\frac{\pi n}{2\beta}-1\right)}{\frac{\pi n}{2\beta}-1} - \frac{\sin\left(\frac{\pi n}{2\beta}+1\right)}{\frac{\pi n}{2\beta}+1}\right],\ n=1,3,5\ldots \quad . \tag{8.9b}$$

We need squares of these coefficients:

$$\begin{Bmatrix} a_n^2 \\ b_n^2 \end{Bmatrix} = \frac{4}{\beta^2}\begin{Bmatrix} \sin^2\beta \\ \cos^2\beta \end{Bmatrix}\left[\left(\frac{\pi n}{2\beta}\right)^2 - 1\right]^{-2} = 2s_n,\ \begin{Bmatrix} n=0,2,4\ldots \\ n=1,3,5\ldots \end{Bmatrix}, \tag{8.10}$$

$$s_n \equiv \frac{8f^2}{\pi^2}\frac{\sin^2\left[\frac{\pi}{2}(n-f)\right]}{\left(n^2-f^2\right)^2},\ n=1,2,3\ldots \quad . \tag{8.11}$$

Thus, s_n yield squares $b_n^2/2$ for the transverse orientation of alternating field (odd n) and the squares $a_n^2/2$ for the longitudinal orientation (even n).

Putting $u = 1$ into Eq. 5.83 and accounting for (8.7), we find the normalizing coefficient C of the distribution (8.4):

$$C^{-1} = \int_0^\infty \exp(-h)dh \int_0^\Phi d\varphi_0 = \int_0^\infty \exp(-h)\frac{\pi}{p}dh = 2\beta\int_0^\infty \exp(-u^2)du = 2\beta\sqrt{\pi} = f\sqrt{\pi^3},$$

so that

$$W(h) = \frac{\exp(-h)}{f\pi^{3/2}}. \tag{8.12}$$

Using Eq. 8.6a, 8.12, we calculate the mean squares

$$\left\{\begin{matrix}\langle q_\parallel^2\rangle \\ \langle q_\perp^2\rangle\end{matrix}\right\} = \frac{1}{f\pi^{3/2}}\int_0^\infty \exp(-h)dh \int_0^{\pi/\check{p}} \left\{\begin{matrix}\cos^2\left(f\check{p}\varphi_0\right) \\ \sin^2\left(f\check{p}\varphi_0\right)\end{matrix}\right\} d\varphi_0 = \frac{1}{2}\left(1 \pm \frac{\sin 2\beta}{2\beta}\right) \tag{8.12a}$$

We see that for any β

$$\langle q_\parallel^2\rangle + \langle q_\perp^2\rangle = 1, \tag{8.12b}$$

as it should be.

For very small libration amplitude β we have obvious relations:

$$\lim_{\beta\to 0}\langle q_\parallel^2\rangle = 1, \quad \lim_{\beta\to 0}\langle q_\perp^2\rangle = 0. \tag{8.12c}$$

8.1.2. Spectral Functions

Eq. 5.88 reduces to the following integral:

$$\int_0^\infty \Im(h)\exp(-h)dh, \tag{8.13}$$

$$\Im(h) = \frac{\pi}{4} 22 \sum_{n=1}^{\infty} \frac{s_n n^2 \frac{\sqrt{h}}{f}}{\pi^{3/2} f\left(\frac{n^2 h}{f} - z^2\right)} = \sqrt{\frac{h}{\pi}} \sum_{n=1}^{\infty} \frac{s_n}{h - z^2},$$

$$z_n = \frac{zf}{n} = x_n + iy_n,\ x_n = \frac{xf}{n},\ y_n = \frac{yf}{n}, \tag{8.14}$$

$\Im(h)$ refers to transverse (for odd n) and longitudinal (for even n) orientations of alternating field.

At last, we find from Eq. 5.88, 8.12 - 8.14 formulas for the longitudinal and transverse SFs $K_{\|}(z)$ and $K_{\perp}(z)$:

$$\frac{1}{3}\begin{Bmatrix} K_{\|}(z) \\ K_{\perp}(z) \end{Bmatrix} = \begin{Bmatrix} \langle q_{\|}^2 \rangle L_{\|}(z) \\ \langle q_{\perp}^2 \rangle L_{\perp}(z) \end{Bmatrix}$$

$$= \int_0^{\infty} \Im(h) \exp(-h) dh = \sum_{n=1}^{\infty} \begin{Bmatrix} s_{2n} L_F(z_{2n}) \\ s_{2n-1} L_F(z_{2n-1}) \end{Bmatrix}, \tag{8.15}$$

where $L_F(\bullet)$ denotes function (5.98)-(5.99), which already appeared in the theory of the planar ED model:

$$L_F(z_m) = \frac{2}{\sqrt{\pi}} \int_0^{\infty} \exp(-p^2) \frac{p^2\, dp}{p^2 - z_m^2} \tag{8.16a}$$

$$= 1 + i\sqrt{\pi} z_m w(z_m),\quad m = 2n \text{ or } m = 2n-1\ . \tag{8.16b}$$

Consequently, odd / even terms of the sum (8.15) are caused by transverse / longitudinal orientation of AC field.

For an *isotropic medium*, assuming $\langle q \rangle = 0$, we, in accord with Eq. 5.65, must take the sum over all n:

$$L(z) \equiv L^{\text{pl}}(z) = \langle q_{\|}^2 \rangle L_{\|}(z) + \langle q_{\perp}^2 \rangle L_{\perp}(z) = \sum_{n=1}^{\infty} s_n L_F(z_n) \tag{8.17a}$$

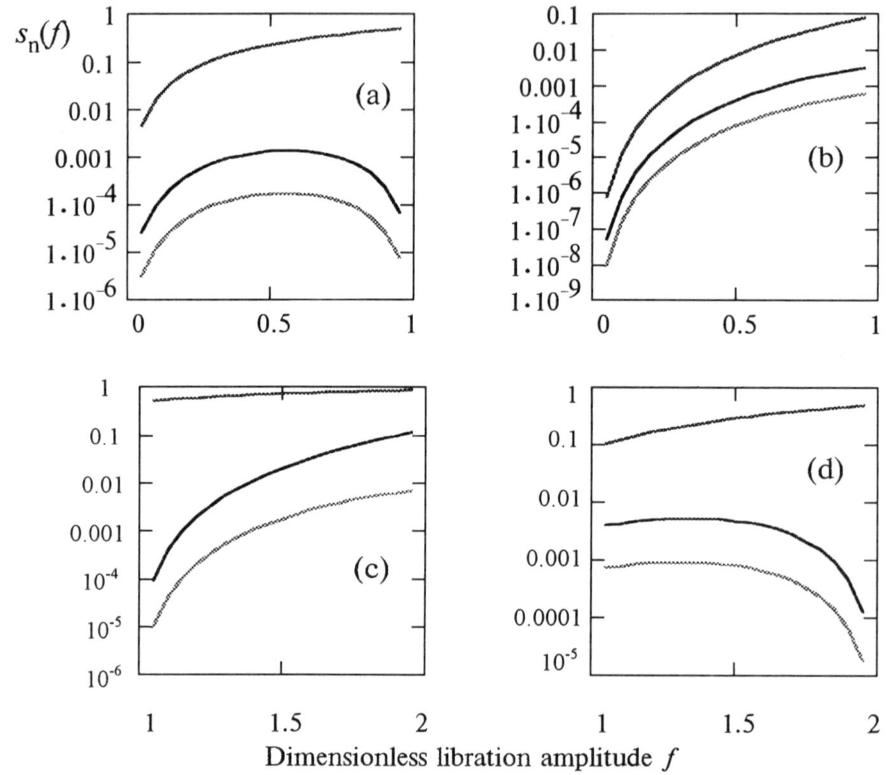

Fig. 8.3 The coefficients s_n vs. f for amplitudes β, lesser (a, b) and larger (c, d) than π/2. Odd (a, c) and even (b, d) values of s_n for n = 1, 3, 5 and 2, 4, 6, respectively (from top to bottom).

$$= \sum_{n=1}^{\infty} s_n \frac{2}{\sqrt{\pi}} \int_0^{\infty} \exp(-p^2) \frac{p^2\, dp}{p^2 - z_n^2}. \tag{8.17b}$$

Although the spectral function is expressed through the same complex function $w(u)$, as was employed in the ED model, the distinction between ED and CR models is considerable, since the *localization of molecules appears due to effect of a conservative potential well*, while such localization may appear in the ED model only due to precession of tops.

a. The main distinction consists in the *shift of the maximal absorption to higher frequencies*. Indeed, the argument ω of the SF for the ED model transforms into $f\omega$ for the CR model ($f < 1$).

b. An *infinite number* of ED-like spectral lines appears in the case of the CR model. However, this spectrum is usually contributed mainly by the first term of the series ($n = 1$), which is caused by interaction with dipoles of transverse-oriented radiation field. The dependence of coefficients s_n on the libration amplitude β is pictured in Fig. 8.3.

c. We shall see below that the transition from rotors in the ED model to librators in the CR model may *radically* change also the low frequency (Debye) spectrum.

Now we consider three particular cases.

LIBRATION IN A CIRCLE. If $\beta = \pi$, i.e. $f = 2$, the time dependence of the μ_Z projection is the same, as for the free rotation, since the reflection of a dipole from the potential wall alters the sign of the other projection (μ_Y). Hence, $L_{||}(z) \equiv L_{\perp}(z)$. Let us prove the validity of this conclusion. We have from Eq. 8.11, 8.12a, and 8.15:

$$s_2(f=2) = \lim_{f \to 2} s_2 = \frac{1}{2};\; s_n(f=2) = 0 \text{ for } n \neq 2; \tag{8.18a}$$

$$\langle q_{||}^2 \rangle = \frac{1}{2};\; L_{||}(z) = L_F(z) \text{ for } f = 2. \tag{8.18b}$$

LIBRATION IN A SEMICIRCLE. The same relation must hold at $\beta = \pi/2$ for the spectral function $L_{\perp}$. Indeed, after reflection of a dipole the time dependence $q_{\perp}(\varphi)$ does not alter. This dependence should be the same, as in the case of free rotation. By analogy with the previous calculation we have for $\beta = \pi/2$, i.e., for $f = 1$:

$$s_1(f=1) = \lim_{f \to 1} s_1 = \frac{1}{2};\; s_n(f=1) = 0 \text{ for } n \neq 1; \tag{8.19a}$$

$$\langle q_{\perp}^2 \rangle = \frac{1}{2};\; L_{\perp}(z) = L_F(z) \text{ for } f = 1. \tag{8.19b}$$

SMALL-AMPLITUDE LIBRATION IN A PLANE. At $\beta << 1$ we may put, in view of Eq. 8.12a, $\left\langle q_{\|}^2 \right\rangle \approx 1$; $\left\langle q_{\perp}^2 \right\rangle \approx \frac{\beta^2}{3}$ and may leave the first term in the series $\sum_n (\bullet)$ for odd and even n.

We have from (8.11), (8.15) for the *transverse susceptibility* ($n = 1$):

$$\left\langle q_{\perp}^2 \right\rangle L_{\perp}(z) \approx \quad \approx \frac{8f^2}{\pi^2} L_F(zf) \tag{8.20a}$$

and for the *longitudinal susceptibility* ($n = 2$)

$$\left\langle q_{\|}^2 \right\rangle L_{\|}(z) \approx \frac{8f^2}{\pi^2}\frac{\pi^2 f^2}{64} L_F(zf) \approx \frac{f^4}{8} L_F(zf). \tag{8.20b}$$

We remark that the ratio

$$\left\langle q_{\|}^2 \right\rangle L_{\|}(z) \Big/ \left\langle q_{\perp}^2 \right\rangle L_{\perp}(z) \approx \frac{\pi^2 f^2}{8^2} \approx \left(\frac{\beta}{4}\right)^2 <<1. \tag{8.21}$$

Thus, the transverse susceptibility is for a small libration amplitude much larger, than the longitudinal one. For this reason in the isotropic medium

$$L(z) \approx L_{\perp}(z) \approx \frac{8f^2}{\pi^2} L_F(z) \qquad \text{for } f << 1 . \tag{8.21a}$$

8.2 The Simplified Spatial CR Model

Let us consider the spectral function $L(z)$ of an isotropic medium. On comparing the planar and spatial *extended diffusion* models (see Sec. 7.1), it becomes evident that *formally* the distinction between these models results in the appearance in the integrand of an additional pre-exponential multiplier (see the accompanying tabulation), such that $\int_0^{\infty} W(p)dp = 1$, where $W(p)$ is given by Eq. 5.187.

The extended diffusion model of linear molecules	$L(z) = \int_0^\infty \frac{W(p)p^2\,dp}{p^2 - z^2}$
Planar ensemble	Spatial ensemble
$W(p) = \frac{2}{\sqrt{\pi}} \exp(-p^2)$	$W(p) = 2p \exp(-p^2)$

Consequently, for so chosen the $W(p)$ distribution we use the same normalization conditions for both systems.

We remark that for *translational* motion of a point mass m the difference in analogously distribution functions also results in the pre-exponential multiplier:

For *one-dimensional* motion: $W(\mathrm{v}) \propto \exp\left(-\frac{m\mathrm{v}^2}{2k_BT}\right)$,

For *three-dimensional* motion: $W(\mathrm{v}) \propto \mathrm{v}^2 \exp\left(-\frac{m\mathrm{v}^2}{2k_BT}\right)$.

In the planar confined-rotator model we have *the same distribution function* $W(p)$, *as in the ED model.* Hence, one may suppose that for an approximate transition to in-space libration we may change *only* the distribution function in the integrand of $L(z)$, as shown in the second tabulation:

The confined rotator model of linear molecules	$L(z) = \sum_{n=1}^{\infty} s_n \int_0^\infty \frac{W(p)p^2\,dp}{p^2 - z_n^2}$
Planar ensemble	Spatial ensemble
$W(p) = \frac{2}{\sqrt{\pi}} \exp(-p^2)$	$W(p) = 2p \exp(-p^2)$

With account of Eq. 7.11 - 7.13 we finally find that an approximate description of in-space librations is rendered by the formula

$$L(z) = \sum_{n=1}^{\infty} s_n \int_0^{\infty} W(p) \frac{p^2\,dp}{p^2 - z_n^2} = \sum_{n=1}^{\infty} s_n L_F(z_n) \tag{8.22a}$$

$$= \sum_{n=1}^{\infty} s_n \int_0^{\infty} \exp(-h) \frac{h dh}{h - z_n^2} = \sum_{n=1}^{\infty} s_n \left[1 + z_n^2 E_1(-z_n^2)\exp(-z_n^2)\right]. \tag{8.22b}$$

For a long lifetime limit we find the imaginary part of the SF by analogy with Eq. 7.26 - 7.28:

$$L''(x) = \sum_{n=1}^{\infty} s_n \left[1 + x_n^2 \left(-\operatorname{Ei}(x_n^2) + i\pi\right)\exp(-x_n^2)\right]; \; y \to 0. \tag{8.23}$$

Integrating imaginary parts over x in Eq. 8.17 and 8.22, we can easily prove that the Gordon sum rule holds both for planar and spatial confined rotators:

$$\Pi = \int_{-\infty}^{\infty} x\chi''(x)\,dx = \frac{1}{3}\begin{Bmatrix}1\\2\end{Bmatrix} \frac{\pi\mu^2 N}{If^2} \sum_n n^2 s_n = \frac{1}{3}\begin{Bmatrix}1\\2\end{Bmatrix} \frac{\pi\mu^2 N}{I} \text{ for } \begin{Bmatrix}\text{plane}\\\text{space}\end{Bmatrix} \tag{8.24}$$

We have applied here the identity (Gaiduk, Kalmykov 1987, p.201):

$$\sum_n s_n = f^2 . \tag{8.25}$$

8.3 Dielectric Spectra of an Isotropic Medium for the DWP configuration

In Fig. 8.4 we compare the loss frequency dependencies, calculated: (a) for the confined rotator and (b) for the extended diffusion models.

SMALL COLLISION FREQUENCY. The SF is calculated from Eq. 8.22 and the susceptibility from Eq. 7.33. At rather small y values ($y < 1$) two loss $\chi''(x)$ maxima appear (Fig. 8.4a, curves 1, 2, 3). The low-frequency peak χ''_D, arising at $x = x_D$, corresponds to the Debye relaxation time $\tau_D = \eta / x_D$. For $y \ll 1$ we have $x_D \ll 1$; the parameter x_D depends strongly on y, i.e. on the lifetime τ, since $\tau = \eta/y$. The high-frequency loss maximum χ''_L is positioned in the Poley absorption region, near the frequency x_L of maximal absorption.

The comparison with the case of free rotors (Fig. 8.4b, curves 1, 2, 3) shows significant distinction of the CR and ED models. In the latter the high frequency loss maximum, placed near x_L, is considerably greater and is positioned at lower frequency. Besides, the low-frequency maximum of $\chi''(x)$ does not appear, since in the ED model we have no localization of dipoles. This localization is controlled by the fitting parameter β of the CR model. Consequently, in the CR and ED models dielectric spectra considerably differ for $f << 1$.

HIGH COLLISION FREQUENCY. The two-humped loss curve $\chi''(x)$ transforms for $y = 1$ into curve 4, having one maximum (see Fig. 8.4a), its frequency x_D being an order of magnitude lower than for $y << 1$. For larger y (Fig. 8.4a, curves 5, 6, 7 corresponding to $y = 10, 100, 1000$, respectively) the loss maximum shifts to still lower frequencies, while the absorption maximum disappears.

The behavior of the ED model is similar (Fig. 8.4b, curves 5, 6, 7, corresponding to $y = 10, 100, 1000$, respectively).

The physical reason of such similarity consists in the following. The localization of dipoles is determined at $y >> 1$ just by frequent collisions, since the mean angular shift $\langle \vartheta \rangle$, on which a dipole rotates during the mean lifetime τ, becomes lesser than the sector of swinging:

$$\langle \vartheta \rangle << 2\beta \quad \text{for } y >> 1. \tag{8.26}$$

Such dielectric relaxation corresponds more to a picture of typical *Brownian motion.* It will be shown in the final Part V that very frequent collisions of Brownian-like type (with $y > 5$, to which curves 5, 6, 7 in Fig. 8.4 correspond) do not appear in polar liquids.

Stochastic processes determine for very rare (at $y < 1$) collisions the appearance of the low-frequency loss maximum χ''_D, as well as the quasi-resonance absorption in the Poley region. For extremely frequent collisions ($y >> 1$) the only low-frequency loss maximum appears, while the resonance absorption disappears.

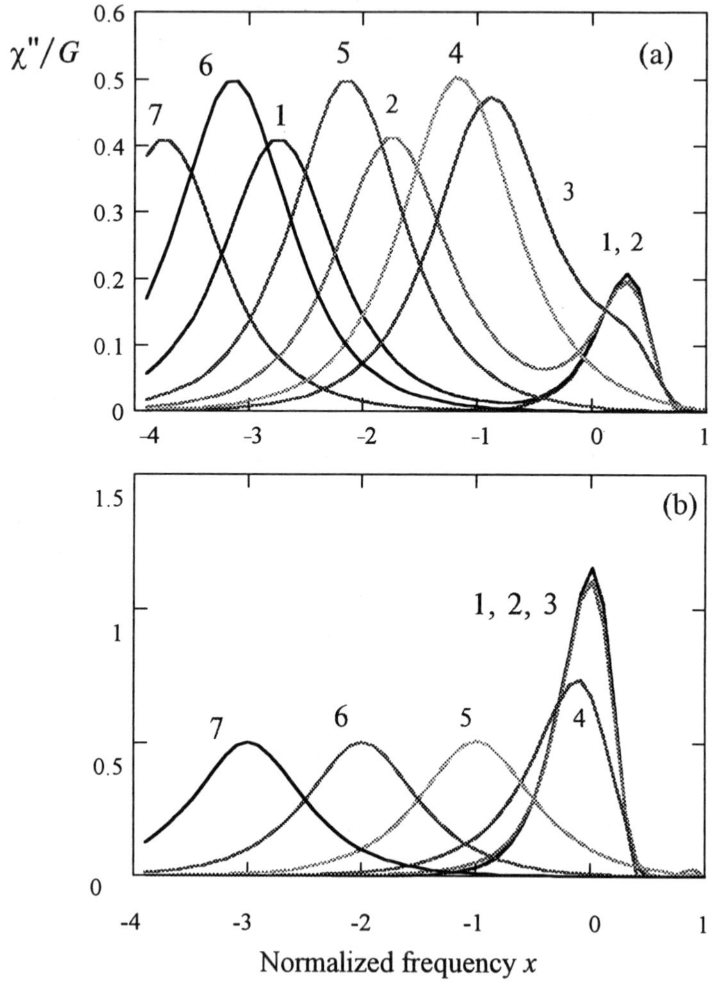

Fig. 8.4 Dielectric loss vs. x calculated for the spatial CR (a) and ED (b) models. The collision frequency y = 0.001, 0.01, 0.1, 1, 10, 100, and 1000 (curves from 1 to 7, respectively). The libration amplitude β of the CR model is taken equal to $\pi/4$. The Gross collision model.

We remark that the following asymptotic formula for $E_1(Z)$ function is applicable for $y > 5$ (Abramowitz and Stegun 1964, p.58):

$$Z\exp(Z)\,E_1(Z) \approx 1 - \frac{1}{Z} + \frac{2}{Z^2} - \frac{2\cdot 3}{Z^3} + \ldots \quad . \tag{8.27a}$$

Consequently, if $y > 5$, then for the spatial CR model we can take the SF in the form:

$$L(z) \approx 1 - Z\exp(Z)\,E_1(Z) \approx \frac{s_1}{Z}\left(1 - \frac{2}{Z} + \frac{2\cdot 3}{Z^2} - \frac{2\cdot 3\cdot 4}{Z^3} + \ldots\right); \tag{8.27b}$$

$$Z = -(fz)^2 = -f^2(x+iy)^2. \tag{8.27c}$$

THE EFFECT OF THE LIBRATION AMPLITUDE β. Fig. 8.5 illustrates typical (for $y < 1$) frequency dependencies of $\chi'(x)$, $\chi''(x)$, $x\chi''(x)$, relating to the spatial motion in the confined rotator model with induced distribution F_G.

A. The resonance frequency x_L is approximately determined by the equation

$$fx_L = \sqrt{\frac{3}{2}} \quad \text{for } y \ll 1, \tag{8.28}$$

analogous to Eq. 7.39 for the ED model. The only distinction is related to the multiplier $f = 2\beta/\pi$. For smaller f the absorption peak $x\chi''(x)$ shifts to higher frequencies (Fig. 8.5a). This shift ensures the constancy of the integrated absorption. With the increase of the collision frequency y the absorption line broadens (Fig. 8.5b). For small y the maximal absorption depends weakly on the lifetime τ (or on y). A similar phenomenon was already described for free rotating dipoles (see Fig. 8.4a and 8.4b).

The frequency x_L of the absorption maximum is $1/f$ times higher for the CR model, than for the ED model. Due to this property one can apply the CR model to polar liquids. Indeed, in a liquid the absorption peak usually is considerably shifted relatively to the peak of the rotational spectrum, observed for the same molecules. (8.29)

B. Real part $\chi'(x)$ of the susceptibility has the minimum, positioned near the absorption peak frequency x_L (see Fig. 8.5 e, f). This property also suggests existence of resonance interaction of far infrared radiation with polar medium.

C. The Debye loss maximum shifts to lower frequencies with the decrease of β and y. Indeed, localization of librators becomes more pronounced for smaller libration amplitude β and for lower collision frequency y.

THE COMPARISON OF PLANAR AND SPATIAL ENSEMBLES OF CONFINED ROTATORS (Fig. 8.6). For the former ensemble we double the SF $L(z)$ given by Eq. 8.17 and use Eq. 7.33 for $\chi^*(x)$:

$$\frac{\chi^*(x)}{G} = \frac{z\,2L^{\text{pl}}(z)}{x + iy2L^{\text{pl}}(z)}, \tag{8.30}$$

$$L^{\text{pl}}(z) = \sum_{n=1}^{\infty} s_n \frac{2}{\sqrt{\pi}} \int_0^{\infty} \frac{p^2 \exp(-p^2)\, dp}{p^2 - (fz/n)^2}\,. \tag{8.31}$$

We remind the Reader that Eq. 8.30 yields the true value of the integrated absorption and of static susceptibility χ_S. (For this Gross collision model we set the Kirkwood correlation factor $g = 1$).

Since the planar and spatial SFs are determined by *different* complex functions, $w(z)$ and $E_1(z)$, the positions x_L of the absorption peak slightly differ. Unlike Eq. 8.28, in the planar CR model we have

$$fx_L = 1 \quad \text{for } y \ll 1. \tag{8.32}$$

The position x_D of the low-frequency Debye maximum also differs (see Fig. 8.6):

$$x_D\big|_{\text{pl}} > x_D\big|_{\text{sp}}, \text{ that is, } \tau_D\big|_{\text{pl}} < \tau_D\big|_{\text{sp}}. \tag{8.33}$$

However, frequency dependencies of the complex susceptibility are similar in both ensembles (planar and spatial) (see Fig. 8.6).

8.4 The Peak Absorption and Debye Relaxation Time as Functions of the Lifetime

The possible application of the CR model, as well as of any other one, is related to the possibility to fit free parameters, aimed for the coincidence of calculated and experimental dielectric spectra. Namely,

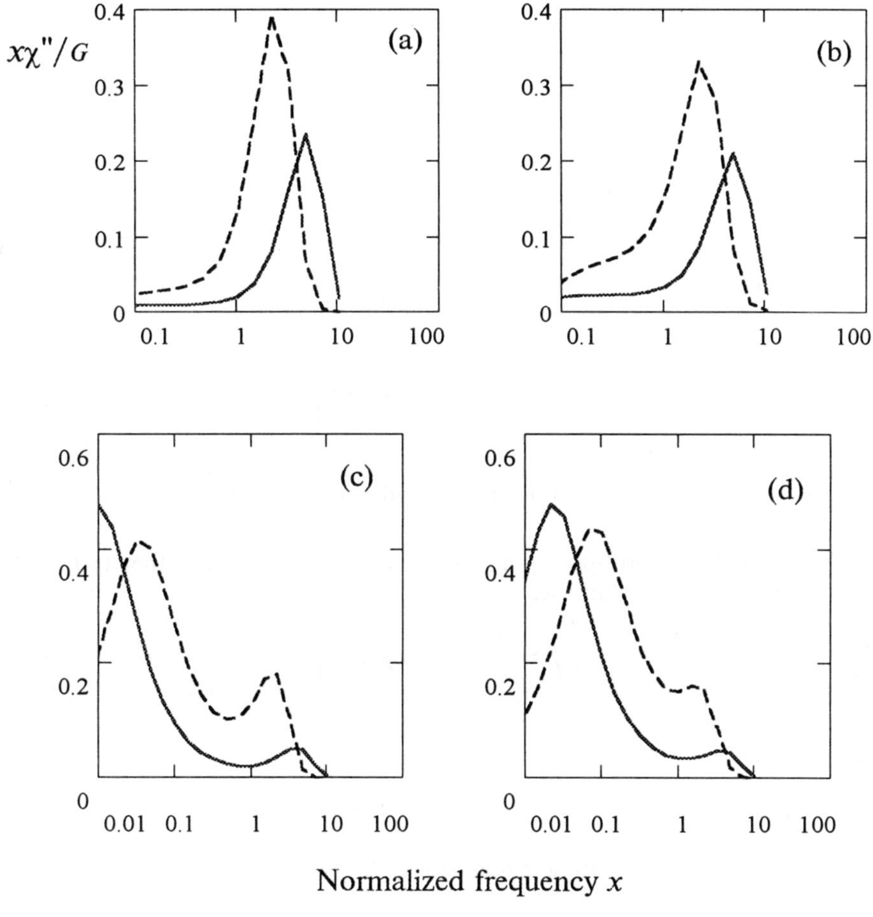

Fig. 8.5 Normalized absorption (a, b), loss (c, d), and real part of the susceptibility (e, f) vs. x. Solid lines for $\beta = \pi/8$ and dashed lines for $\beta = \pi/4$. The collision frequency $y = 0.2$ (to the left) and 0.5 (to the right). Calculated for the spatial confined rotator model at $F = F_G$.

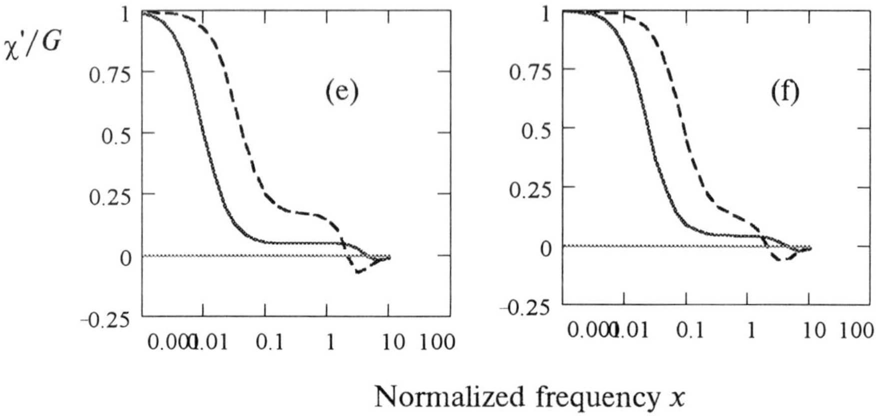

Fig. 8.5 **e, f:** The caption to this Figure see on the previous page.

(i) The peak absorption α_L and its frequency ω_L, or, in other terms, the dimensionless quantities

$$\gamma_L = \frac{1}{G}\max\left[x\chi''(x)\right], \tag{8.34}$$

$$x_L = \eta\omega_L. \tag{8.35}$$

(ii) The frequency $\omega_D = \dfrac{1}{\tau_D}$ of the Debye loss peak ε''_D.

The corresponding dimensionless quantities are given by

$$\frac{\chi''_D}{G} = \frac{1}{G}\max\left[\chi''(x)\right] \qquad \text{for } x << 1, \tag{8.36}$$

$$x_D = \eta\omega_D = \eta / \tau_D. \tag{8.37}$$

In the case of an isotropic fluid the position x_L of the absorption maximum can be estimated at rare collisions (i.e. for $y < 1/2$) from Eq. 8.28:

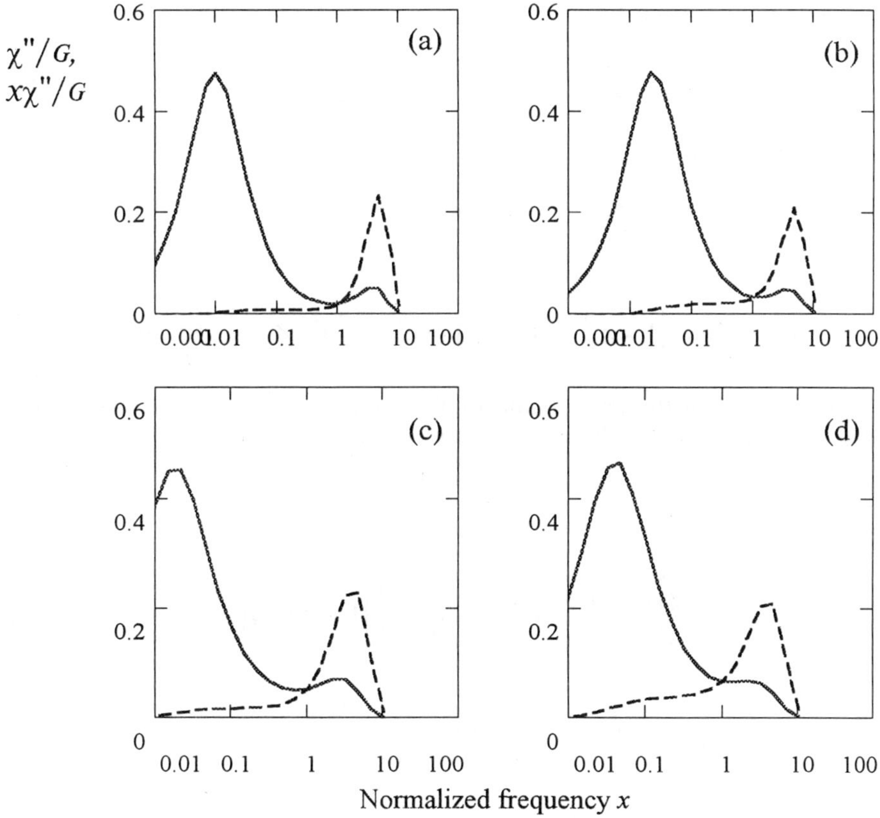

Fig. 8.6 The comparison of absorption (dashed curves) and loss (solid curves) frequency dependencies for spatial (a, b) and planar (c, d) confined rotators (for the latter - with the doubled spectral function). The libration amplitude $\beta = \pi/8$. The collision frequency $y = 0.2$ (to the left) and 0.5 (to the right). The Gross collision model.

$$x_L \approx \frac{1}{f}\sqrt{\frac{3}{2}} = \frac{\pi}{2\beta}\sqrt{\frac{3}{2}} \,. \tag{8.38}$$

Then, as it follows from Eq. 7.33 and 8.22,

$$\gamma_L = \frac{1}{G}\max\left[x\chi''(x)\right] \approx \frac{z_L}{x_L + iyL(z_L)}, \tag{8.39}$$

where $z_L = x_L + iy_L$,

$$L(z_L) = \sum_n s_n \int_0^\infty \frac{h\exp(-h)\,dh}{h-(xf/n)^2} \tag{8.40}$$

$$= \sum_n s_n\left\{1+Z_L^2\, E_1\left(-Z_L^2\right)\exp\left(-Z_L^2\right)\right\};\ Z_L \equiv \frac{z_L f}{n} \tag{8.41}$$

Accounting for the last column of Table 6.2, we can estimate the Debye-relaxation parameters for the chosen Gross collision model:

$$\frac{\tau_D}{\eta} \approx \frac{1}{yL(iy)}, \tag{8.42}$$

$$\frac{\chi_D''}{G} = \frac{1}{G}\max\left[\chi''(x)\right] \approx \frac{1}{2}\left[1 - L(iy)\right], \tag{8.43}$$

$$L(iy) = \sum_n s_n \int_0^\infty \frac{h\exp(-h)dh}{h+(xf/n)^2} \tag{8.44}$$

$$= \sum_n s_n\left\{1-Y^2\, E_1\left(Y^2\right)\exp\left(Y^2\right)\right\};\ Y \equiv \frac{yf}{n}, \tag{8.45}$$

where we assume that $x_D = \eta/\tau_D << 1$.

For calculation of the static SF (8.44) we may use Eq. 7.31.

8.4.1 The Estimated Spectral Parameters as Functions of the Normalized Lifetime τ/η

We take as the variable

$$\tau/\eta = y^{-1} . \tag{8.46}$$

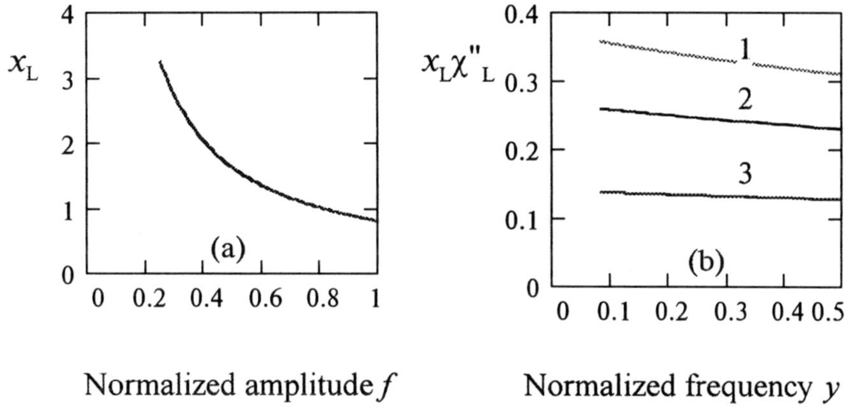

Fig. 8.7 Estimation of the absorption peak frequency x_L vs. f (a) and of the absorption peak $x_L\chi_L''$ vs. y for $f = 0.2$ (1), 0.4 (2), and 0.6 (3) (b).

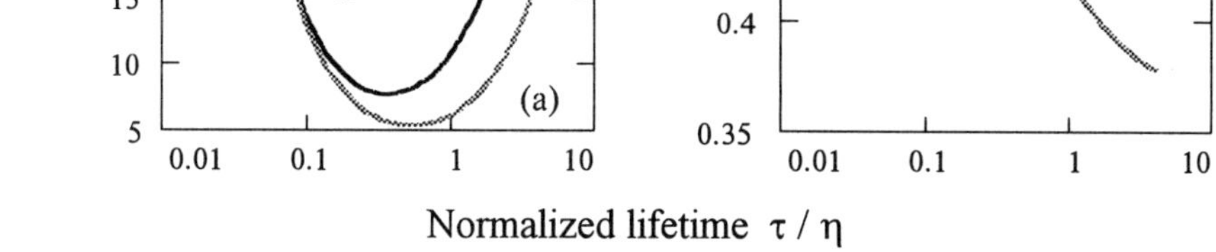

Fig. 8.8 Estimation of the Debye relaxation time (a) and the peak loss (b) vs. dimensionless lifetime τ/η. $f = 0.2$, 0.4, and 0.6 (from top to bottom).

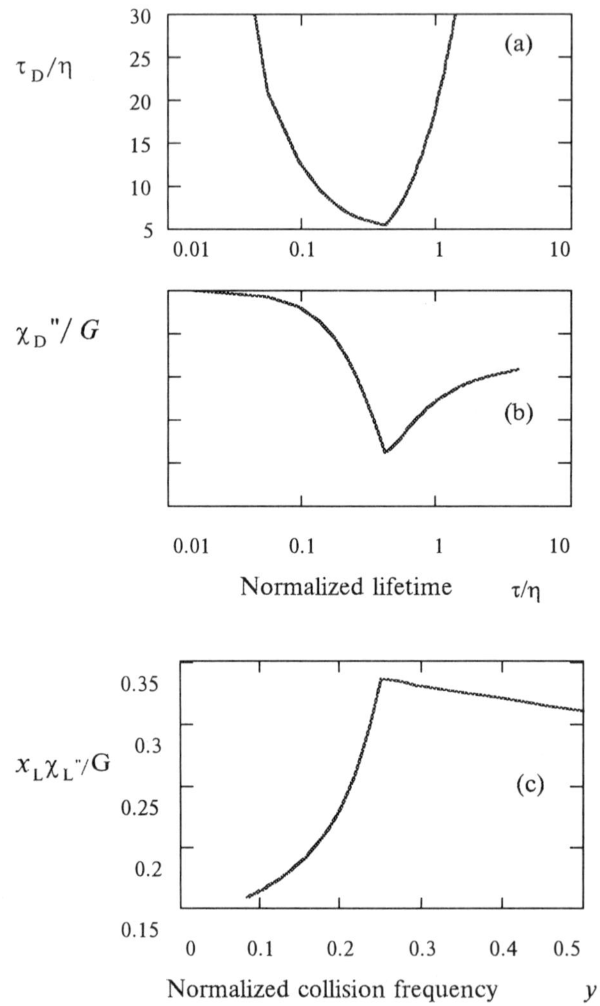

Fig. 8.9 Estimation of the relaxation time (a) and the loss peak (b) on the dimensionless lifetime τ/η; absorption peak vs. collision frequency (c). The case of τ-dependent amplitude β.

THE CASE OF CONSTANT LIBRATION AMPLITUDE β. At first, we assume that β = const, i.e. f is independent on τ. The x_L vs. f plot is shown in Fig. 8.7a; this estimation is applicable for $y < 0.5$. The correspondent dependence on y of the dimensionless absorption peak is shown in Fig. 8.7b. Both plots confirm the aforementioned reasoning (Sec. 8.3): the frequency x_L considerably decreases with the rise of the libration amplitude f, while the absorption peak decreases slowly with f. The new feature is seen in Fig. 8.8a, viz. the *minimum* of the relaxation time τ_D / η, calculated as a function of the ratio (8.46). One may regard the decrease of the lifetime as caused by the increase of temperature T. It seems natural, that in a liquid the relaxation time decreases with τ as well. Such correlation corresponds to the right wing of the estimated

$$\tau_D / \eta \quad \text{dependence on} \quad \tau / \eta = y^{-1} \quad . \tag{8.47}$$

The same reasoning may suggest that *the left wing of dependence* (8.47) *cannot be realized for the same molecular model.* Consequently, we may suppose that the lifetime

$$\tau_{cr} = \eta / y_{cr} \quad , \tag{8.48}$$

which corresponds to

$$\left[\tau_D / \eta\right]_{min} \quad , \tag{8.49}$$

is some *characteristic point* of a liquid.

In Part V we shall show that for a number of polar liquids this minimum approximately corresponds to the *critical temperature* T_{cr}. The critical values of τ/η, y, and τ_D /η are given in the left part of Table 8.1. The left wing of the curve (8.47) has no physical sense, since the phase state of a liquid changes in this point, consequently, one should employ some new model (not CR!), to describe the dielectric behavior of this new state of matter. From Fig. 8.8b we see that the loss maximum increases, when collisions become more frequent.

THE CASE OF VARIABLE LIBRATION AMPLITUDE β. It is reasonable to suppose that the libration amplitude β should also depend on the ratio (8.47), viz. f should increase, when τ/η is decreasing, e.g. when temperature T rises. For a qualitative illustration we take the rational dependence

$$f(\tau / \eta) = \frac{(\tau / \eta)a}{(\tau / \eta) - (\tau / \eta)_{cr} + b} \tag{8.50}$$

or, equivalently,

$$f(y) = \frac{a}{1-\left[b-(\tau/\eta)_{cr}\right] y}. \tag{8.51}$$

Table 8.1 The estimated parameters of the critical point

	Rigorous calculation			For $f(\frac{\tau}{\eta})$ from Eq. 8.52		
f	$\left(\frac{\tau}{\eta}\right)_{cr}$	y_{cr}	$\left(\frac{\tau_D}{\tau}\right)_{min}$	$\left(\frac{\tau}{\eta}\right)_{cr}$	y_{cr}	$\left(\frac{\tau_D}{\tau}\right)_{min}$
0.2	0.14	7.4	15.0			
0.4	0.30	3.5	7.7	0.42	2.4	5.5
0.8	0.42	2.4	5.4			

As an example we take $a = 0.2$, $f_{cr} = 0.6$. In view of Table 8.1 we find: $(\tau/\eta)_{cr} = 0.416$. Then Eq. 8.51 yields $b = 0.139$, so that

$$f(\tau/\eta) = \frac{(\tau/\eta)\,0.2}{(\tau/\eta) \ -0.277}. \tag{8.52}$$

For the τ-dependent amplitude f, taken from this expression, the results of calculation are presented in the right part of Table 8.1. We see that the minimum of the Debye relaxation time appears again (Fig. 8.9a) but the loss peak vs. lifetime dependence alters (cf. Fig. 8.9b with Fig. 8.8b), having now its minimum near the critical point. The absorption peak also changes its dependence on the collision frequency y (cf. Fig. 8.9c with Fig. 8.7b), going through maximum at some intermediate y-value, viz. at $y \approx 0.25$. We suggest that such behavior of dielectric spectra is more adequate than that found for the constant β case. Further consideration we postpone till Part V.

8.5 The Cone Confined Rotator (CCR) Model

8.5.1 Geometry

In this model we regard an axisymmetric potential *U*, viz. the potential box with an elastic reflecting *conical* wall. All calculations are made for the ensemble of linear molecules. One may regard that each dipole «starts» free rotation in a plane (see Fig. 8.10); the end of the vector μ rotates through an angle 2λ, passing successively points **A"** (on the cone surface), **R** (the greatest distance from it) and **A'** (again on the cone surface). Having being reflected at the point **A'**, the dipole rotates through the angle 2λ in some other plane, touches the cone surface, and so on. It is seen from the Fig. 8.10 that $\angle$OA'A" $= \pi/2 - \lambda = \angle$OA'B, so that

$$2\lambda = \angle \text{A'OA''} = \angle \text{A'OB}. \tag{8.53}$$

Consequently, we have a particular case of free rotation between two successive «weak» reflections, that was considered in Sec. 7.1. Therefore, we can describe rotation of a dipole in the sector A'OA" or A'OB, if we allow the vector μ in Fig. 3.4a to coincide with its projection $\mu_{X'Y'}$ on the plane of rotation and then draw confining radii in the plane X'Y', that are oriented to the *Y'* axis by the angle $\pm\alpha$.

Let $\vartheta(\varphi)$ be the angular shift of a dipole μ from the bisector OR, where $\vartheta = 0$ corresponds to $\varphi = 0$. In ensemble of linear molecules, unlike that of tops, the angular momentum vector $\boldsymbol{k} = \boldsymbol{p}$ is normal to the vector μ. Then, setting $\delta = \pi/2$ and $\varphi_0 = 0$ in Eq. 3.111a, we have

$$q_{\|}(\varphi) = \cos\theta(\varphi) = \sin\bar{\theta}\cos p\varphi = \sin\bar{\theta}\cos\vartheta(\varphi) \ , \tag{8.54}$$

where

$$\vartheta(\varphi) = p\varphi \tag{8.55}$$

is the law of free rotation in the sector under consideration, $\theta(\varphi)$ and $\bar{\theta}$ are angles between axis *Z* and dipole moment μ and between *Z* and angular momentum $\boldsymbol{k}$, respectively (see Fig. 8.10). From Eq. 7.43 we have

$$q_{\perp}(\varphi) = \sin\theta(\varphi)\cos\psi(\varphi) = \cos\bar{\theta}\cos\vartheta(\varphi) \ ; \tag{8.56}$$

the angle ψ in the plane A'OA is shown in Fig. 8.10. Combining Eq. 8.54 and 8.56, we obtain relations

$$\cos^2\theta + [1 - \cos^2\theta] \cos^2\psi = \cos^2\theta \sin^2\psi + \cos^2\psi = \cos^2\vartheta,$$

that is,

$$1 - \cos^2\theta \sin^2\psi - \cos^2\psi = \sin^2\psi \sin^2\theta = \sin^2\vartheta,$$

$$\sin\psi \sin\theta = \sin\vartheta. \tag{8.57}$$

For $m = k \cos\delta = 0$ we find, in accordance with the Eq. 7.1, the angular momentum

$$k = p = \sqrt{h}\,, \tag{8.58}$$

its projection on the Z axis being

$$l = k\cos\bar{\theta} = \sqrt{h}\,\cos\bar{\theta}\ . \tag{8.59}$$

When a dipole touches the surface of a cone, the angles ϑ, ψ and θ reach their maximum values

$$\vartheta_{max} = \alpha,\ \psi_{max} = \lambda,\ \theta_{max} = \beta\ , \tag{8.60}$$

where β is the opening of a cone (see Fig. 8.10). Inserting these values into Eq. 8.54, 8.56, 8.57, we find:

$$\sin\alpha = \sin\beta\,\sin\lambda;\ \cos\alpha = \frac{\cos\beta}{\sin\bar{\theta}};\ \cos\bar{\theta} = \frac{\sin\beta\,\cos\lambda}{\cos\alpha}\ . \tag{8.61}$$

Let Φ be the period of $q_{\|}(\varphi)$:

$$\Phi = 2\alpha\,/\sqrt{h}\ . \tag{8.62}$$

During Φ the angle ϑ alters in the range

$$\vartheta \in [-\alpha, \alpha]\ . \tag{8.63}$$

The angular frequency is given by

$$p = \frac{2\pi}{\Phi} = \frac{\pi}{\alpha}\sqrt{h}\ . \tag{8.64}$$

$$h \in [0, \infty]\ . \tag{8.65}$$

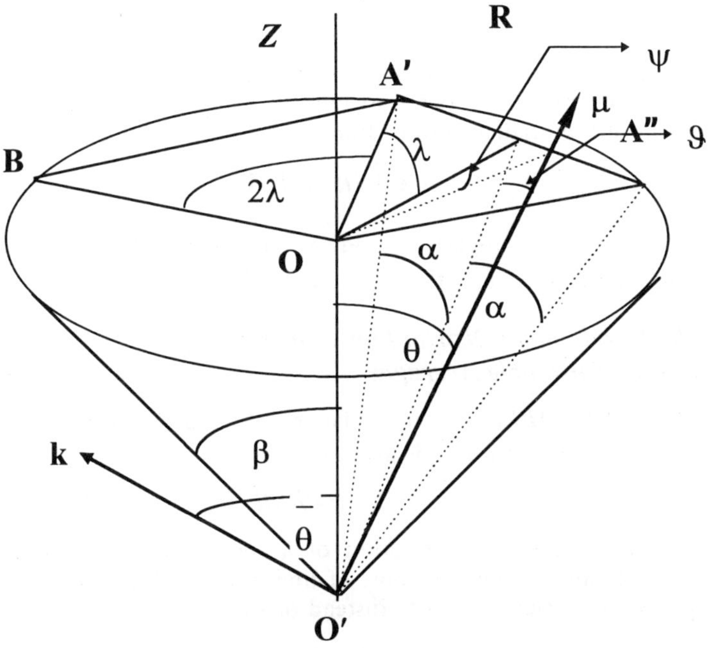

Fig. 8.10 Cone-confined rotator model: geometric scheme.

At fixed energy h the continuum of trajectories is determined by intervals

$$\lambda \in \left[0, \frac{\pi}{2}\right], \alpha \in [0, \beta], \bar{\theta} \in \left[\frac{\pi}{2} - \beta, \frac{\pi}{2}\right] . \tag{8.66}$$

where the least quantities correspond to the infinitesimal sector (near the cone surface) and the maximal ones - to motion in a diametric cross-section. The Z-projection of the angular momentum varies in the interval

$$l \in \left[\sqrt{h}\, \sin\beta, 0\right]. \tag{8.67}$$

We remark that angles α and λ should not be confused with the absorption coefficient α and the wavelength λ!

8.5.2 *The Steady State Distribution Function and the Ensemble Averages*

THE PHASE VARIABLES. In accordance with the definition (see Sec. 5.6.3) the phase-volume element $d\Gamma'$ is given by

$$d\Gamma' = dh\; dl. \tag{8.68a}$$

Differential dm_0, related to rotation of a top around its symmetry axis, is omitted, this holds for our ensemble of linear molecules. Noting Eq. 8.59 and omitting the minus sign, we write instead of Eq. 8.68:

$$d\Gamma' = dh \sqrt{h} \sin\bar{\theta}\, d\bar{\theta} . \tag{8.68b}$$

In view of relations (8.61) angles λ, α and $\bar{\theta}$ are not independent one from another. In accordance with the ranges (8.66) it is useful to take λ as the variable in calculating ensemble averages, since the limits of integration over λ do not depend on β : $\lambda \in (0, \pi/2)$. For this transformation one may use relations

$$\frac{d\alpha}{d\lambda} = \frac{\sin\beta \cos\lambda}{\sqrt{1 - \sin^2\beta \sin^2\lambda}} = \frac{\sin\beta \cos\lambda}{\cos\alpha} = \cos\bar{\theta} , \tag{8.69}$$

which follow from Eq. 8.61.

In view, that the steady-state distribution function is

$$W = C \exp(-h), \tag{8.70}$$

the ensemble average of any quantity $u(\Gamma)$, independent on initial angle ψ_0 on XY plane, can be found as follows:

$$\langle u \rangle = \int_{\Gamma} W(h) u(\Gamma) d\Gamma$$

$$= C \int_0^{\infty} \exp(-h) dh \int_{\frac{\pi}{2}-\beta}^{\frac{\pi}{2}} \sqrt{h} \sin\bar{\theta}\, d\bar{\theta} \int_{-\frac{\alpha}{\sqrt{h}}}^{\frac{\alpha}{\sqrt{h}}} u(\Gamma) d\varphi \tag{8.71a}$$

$$= C \int_0^{\infty} \exp(-h) dh \int_{\frac{\pi}{2}-\beta}^{\frac{\pi}{2}} \sqrt{h} \sin\bar{\theta}\, d\bar{\theta} \int_{-\alpha}^{\alpha} u(\Gamma) d\vartheta \;. \tag{8.71b}$$

Here $\vartheta \equiv \alpha/\sqrt{h}$. If we take λ instead of $\bar{\theta}$ as an independent variable, then

$$\bar{\theta}(\lambda) = \operatorname{acos}\frac{\sin\beta\cos\lambda}{\sqrt{1-\sin^2\beta\sin^2\lambda}} \;. \tag{8.72}$$

THE DISTRIBUTION FUNCTION. In order to find the normalization constant C, we put

$$u = 1. \tag{8.73}$$

Then we have from Eq. 8.70, 8.71:

$$\frac{1}{C} = 2\int_0^{\frac{\pi}{2}} \cos^2\bar{\theta}\, d\lambda = 2\int_0^{\frac{\pi}{2}} \frac{\sin^2\beta\cos^2\lambda}{1-\sin^2\beta\sin^2\lambda} d\lambda = \pi\left(1-\cos\beta\right), \tag{8.74}$$

$$W = \frac{\exp(-h)}{\pi\left(1-\cos\beta\right)} . \tag{8.75}$$

The same result is obtained, if one takes

$$\theta, \dot{\theta}, l \tag{8.76}$$

as phase variables. Then we have

$$\frac{1}{C} = \int_0^\infty d\dot{\theta} \int_0^{\Phi} dl \int_{-\beta}^{\beta} \exp\left(-\dot{\theta}^2 - \frac{l^2}{\sin^2\theta} \right) d\theta \tag{8.77a}$$

$$= \sqrt{\pi}\,\frac{\sqrt{\pi}}{2}\, 2\int_0^\beta \sin\theta\, d\theta = \pi(1 - \cos\beta), \tag{8.77b}$$

just as in Eq. 8.74.

CALCULATION OF $\langle q_\parallel^2 \rangle$ *AND* $\langle q_\perp^2 \rangle$. Quite analogously we find:

$$\langle q_\parallel^2 \rangle = \pi C \int_0^\beta \cos^2\theta \sin\theta\, d\theta = \pi C\,\frac{1-\cos^3\beta}{3}$$

$$\frac{1-\cos^3\beta}{3(1-\cos\beta)} = \frac{1+\cos\beta+\cos^2\beta}{3}; \tag{8.78}$$

$$\langle q_\parallel \rangle = \pi C\,\frac{\left(1-\cos^2\beta\right)}{2} = \frac{1+\cos\beta}{2}. \tag{8.79a}$$

The last result is valid for the single well potential (SWP). For the double well potential (DWP)

$$\langle q_\parallel \rangle = 0. \tag{8.79b}$$

For a very *narrow cone*, for an extremely open one (a *needle*) and for a *hemisphere* we have, respectively:

$$\langle q_\parallel^2 \rangle \to 1, \text{ if } \beta \to 0; \langle q_\parallel^2 \rangle \to \frac{1}{3}, \text{ if } \beta \to \pi; \langle q_\parallel^2 \rangle \to \frac{1}{3}, \text{ if } \beta \to \frac{\pi}{2} \tag{8.80}$$

The first result is evident, since all dipoles are oriented along its symmetry axis, while the last two mean that dipoles behave like free rotors.

CALCULATION OF $\langle q_\perp^2 \rangle$. Eq. 3.66 are valid now, averaging them, we have

$$\left\langle q_{\perp}^{2}\right\rangle = \frac{1-\left\langle q_{\parallel}^{2}\right\rangle}{2} = \frac{2-\cos\beta-\cos^{2}\beta}{6} = \frac{(1-\cos\beta)(2+\cos\beta)}{6} \,. \quad (8.81)$$

For the three particular cases, mentioned above, this expression yields (cf. with Eq. 8.80):

$$\left\langle q_{\perp}^{2}\right\rangle \to 0, \text{ if } \beta\to 0; \left\langle q_{\perp}^{2}\right\rangle \to \frac{1}{3}, \text{ if } \beta\to\pi; \left\langle q_{\perp}^{2}\right\rangle \to \frac{1}{3}, \text{ if } \beta\to\frac{\pi}{2} \,. \quad (8.82)$$

The first result agrees and the last two coincide with those, given by Eq. 8.80.

However, we shall see, that for the particular cases of a hemisphere and a needle the CCR model does *not* transform to the model of free rotation (to the ED model of linear molecules).

THE DISPERSION OF THE DIRECTION COSINE. First, we consider the *double well potential,* where $\left\langle q_{\parallel}\right\rangle = 0$. We have from Eq. 8.78, 8.79b:

$$\sigma_{\parallel} = 3\Delta_{\parallel} = 3\left\langle q_{\parallel}^{2}\right\rangle = 1 + \cos\beta + \cos^{2}\beta \text{ (for DWP).} \quad (8.83)$$

For the *single well potential,* where $\left\langle q_{\parallel}\right\rangle \neq 0$, we have from Eq. 8.78, 79a:

$$\sigma_{\parallel} = 3\Delta_{\parallel} = 3\left(\left\langle q_{\parallel}^{2}\right\rangle - \left\langle q_{\parallel}\right\rangle^{2}\right) = \frac{(1-\cos\beta)^{2}}{4} \text{ (for SWP).} \quad (8.84)$$

For our three particular cases

$$\sigma_{\parallel} = 0, \text{ for extremely narrow cone } (\beta = 0), \quad (8.85a)$$

$$\sigma_{\parallel} = 1/4, \text{ for a hemisphere } (\beta = \pi/2), \quad (8.85b)$$

$$\sigma_{\parallel} = 1, \text{ for a needle } (\beta = \pi). \quad (8.85c)$$

For both aforementioned potential configurations (DWP and SWP) the $\Delta_{\perp}$ value coincides with $\left\langle q_{\perp}^{2}\right\rangle$, see Eq. 8.81. In accordance with (3.66 a, b) $\left\langle q_{\perp}\right\rangle = 0$, so that

$$\sigma_{\perp} = 3\Delta_{\perp} = 3\left(\left\langle q_{\perp}^{2}\right\rangle - \left\langle q_{\perp}\right\rangle^{2}\right) = \frac{(1-\cos\beta)(2+\cos\beta)}{4} \,. \quad (8.86)$$

8.5.3 The Spectral Functions

THE LONGITUDINAL FOURIER AMPLITUDE a_n. To obtain the coefficient a_n (see Eq. 5.104) we take into account Eq. 8.54, 8.62, 8.64:

$$\int_0^{\frac{2\alpha}{\sqrt{h}}} q_{\|}(\varphi)\cos pn\varphi \, d\varphi = \frac{1}{\sqrt{h}} \int_{-\alpha}^{\alpha} q_{\|} \cos pn\vartheta \, d\vartheta = \frac{a_n \Phi}{2}$$

Hence,

$$a_n = \frac{2}{\Phi\sqrt{h}} \sin\bar{\theta} \int_{-\alpha}^{\alpha} \cos\vartheta \cos\frac{\pi n\vartheta}{\alpha} d\vartheta = \frac{\sin\bar{\theta}}{\alpha} \int_{-\alpha}^{\alpha} \cos\vartheta \cos\frac{\pi n\vartheta}{\alpha} d\vartheta$$

$$(-1)^n\left(\frac{1}{\pi n+\alpha} - \frac{1}{\pi n-\alpha}\right) \sin\bar{\theta} \sin\alpha$$

$$-2\alpha(-1)^n \frac{\sin\bar{\theta}\sin\alpha}{\pi^2 n^2 - \alpha^2} = -2\alpha(-1)^n \frac{\sin\bar{\theta}\sin\beta\sin\lambda}{\pi^2 n^2 - \alpha^2}. \qquad (8.87)$$

THE TRANSVERSAL FOURIER AMPLITUDE b_n. Taking into account definition (3.81) and Eq. 8.56, we may write:

$$q_{\perp}(\varphi) = \sin\theta(\varphi)\cos\psi(\varphi) = \cos\bar{\theta}\cos\vartheta(\varphi) = \rho(\varphi)\cos\psi(\varphi). \qquad (8.88)$$

It follows from Eq. 5.120 and 8.55, that

$$b_n = \frac{1}{\Phi} \int_0^{\Phi} \rho(\varphi)\cos\left(\tilde{\psi} + pn\varphi\right)d\varphi = \frac{1}{2\alpha} \int_{-\vartheta}^{\vartheta} \rho \cos\left(\tilde{\psi} + pn\varphi\right)d\vartheta, \qquad (8.89)$$

where in accordance with Eq. 5.110 and 8.64 the periodic function

$$\tilde{\psi}(\varphi) = \psi - \frac{p\lambda}{\pi}\varphi = \psi - \frac{\lambda}{\alpha}\vartheta \ .$$

Then

$$\tilde{\psi}(\varphi) + pn\varphi = \psi + \frac{n\vartheta - \lambda}{\alpha} \ . \qquad (8.90)$$

With account of Eq. 8.57, 8.88 - 8. 90 we have:

$$\rho\cos\left[\tilde{\psi}(\varphi)+pn\varphi\right] = \cos\bar{\theta}\cos\vartheta\cos\frac{n\vartheta-\lambda}{\alpha} - \sin\theta\sin\frac{n\vartheta-\lambda}{\alpha}$$

$$\cos\bar{\theta}\cos\vartheta\cos\frac{(\pi n-\lambda)\vartheta}{\alpha} - \sin\theta\sin\frac{(\pi n-\lambda)\vartheta}{\alpha}$$

$$\frac{1}{2}\left\{\left(1+\cos\bar{\theta}\right)\cos\frac{(\pi n+\alpha-\lambda)\vartheta}{\alpha} - \left(1-\cos\bar{\theta}\right)\cos\frac{(\pi n-\alpha-\lambda)\vartheta}{\alpha}\right\}. \quad (8.91)$$

Substituting into Eq. 8.89, we have after integration over ϑ:

$$b_n = \frac{(-1)^n}{2}\left\{\frac{1+\cos\bar{\theta}}{\pi n+\alpha-\lambda}\sin(\alpha-\lambda) - \frac{1-\cos\bar{\theta}}{\pi n-\alpha-\lambda}\sin(\alpha+\lambda)\right\} \quad (8.92)$$

We transform $\sin(\alpha\pm\lambda)$ with use of Eq. 8.61:

$$\sin(\alpha\mp\lambda) = \sin\alpha\cos\lambda \mp \cos\alpha\sin\lambda = \sin\beta\sin\lambda\frac{\cos\bar{\theta}\cos\alpha}{\sin\beta} \mp \cos\alpha\sin\lambda$$

$$= \cos\alpha\sin\lambda\left(\cos\bar{\theta}\mp 1\right);$$

$$b_n = (-1)^n\left\{\left(\frac{1}{\pi n-\alpha-\lambda} - \frac{1}{\pi n+\alpha-\lambda}\right)\sin^2\bar{\theta}\cos\alpha\sin\lambda\right\}$$

Finally, we find with account of Eq. 8.61:

$$b_n = (-1)^n\alpha\frac{\sin^2\bar{\theta}\cos\alpha\sin\lambda}{(\pi n-\lambda)^2} = (-1)^n\alpha\frac{\sin^2\bar{\theta}\sin\beta\sin\lambda}{(\pi n-\lambda)^2}. \quad (8.93)$$

THE SFs OF AN ANISOTROPIC MEDIUM. We substitute Fourier amplitudes (8.87), (8.93) into (5.108), (5.121). Using Eq. 8.64 and 8.75 and denoting $t=\sqrt{h}$, we obtain:

$$\frac{1}{3}\begin{Bmatrix} K_{\|} \\ K_{\perp} \end{Bmatrix} = \frac{2\pi}{1-\cos\beta}\int_0^{\infty} t^3 \exp\left(-t^2\right)dt \int_{\frac{\pi}{2}-\beta}^{\frac{\pi}{2}} \alpha \sin^3\bar{\theta}\sin^2\lambda \begin{Bmatrix} 4\,\Sigma_{\|}\,\sin^2\beta \\ \Sigma_{\perp}\,\cos^2\beta \end{Bmatrix} d\bar{\theta}, \tag{8.94}$$

$$\Sigma_{\|} \equiv \sum_{n=1}^{\infty} \frac{n^2}{\left(\pi^2 n^2 - \alpha^2\right)^2 \left[\left(\frac{t\pi n}{\alpha}\right)^2 - z^2\right]}, \tag{8.95a}$$

$$\Sigma_{\perp} \equiv \sum_{n=-\infty}^{\infty} \frac{\left(n - \frac{\lambda}{\pi}\right)^2}{\left((\pi n - \lambda)^2 - \alpha^2\right)^2 \left[\left(\frac{t(\pi n - \lambda)}{\alpha}\right)^2 - z^2\right]}, \tag{8.95b}$$

where α, λ are found from Eq. 8.61:

$$\alpha(\bar{\theta}) = \operatorname{acos}\left(\frac{\cos\beta}{\sin\bar{\theta}}\right);\ \lambda(\bar{\theta}) = \operatorname{acos}\left(\cot\beta\,\cot\bar{\theta}\right). \tag{8.96}$$

Substituting the variable of integration by $w = \cos\bar{\theta}$ and using again Eq. 8.61, we transform integrals (8.94) to simpler forms:

$$\begin{Bmatrix} K_{\|} \\ K_{\perp} \end{Bmatrix} = \frac{6\pi}{1-\cos\beta}\int_0^{\infty} t^3 \exp\left(-t^2\right)dt \int_0^{\sin\beta} \alpha\left(1 - w^2 - \cos^2\beta\right) \begin{Bmatrix} 4\,\Sigma_{\|} \\ \Sigma_{\perp}\,\cot^2\beta \end{Bmatrix} dw, \tag{8.97}$$

$$\alpha(w) = \operatorname{acos}\left(\frac{\cos\beta}{\sqrt{1-w^2}}\right);\ \lambda(w) = \operatorname{acos}\left(\frac{w\,\cot\beta}{\sqrt{1-w^2}}\right). \tag{8.98}$$

The longitudinal and transverse spectral functions $K_{||}(z)$ and $K_{\perp}(z)$ are thus expressed through double integrals (8.97) with $\Sigma_{||}$ and $\Sigma_{\perp}$ given by Eq. 8.95, 8.97, 8.98.

For the maximal development ($\beta = \pi$) the cone transforms into a needle, for which angle $\bar{\theta}$ between angular momentum $\boldsymbol{k}$ and the cone axis Z widens to $\bar{\theta} \in [-\pi/2, \pi/2]$.

The SF of macroscopically isotropic medium can be found with account of Eq. 6.33 from expression

$$L(z) = (1/3)\,(K_{||} + 2K_{\perp}), \tag{8.99}$$

where $K_{||}$ and $K_{\perp}$ are given by Eq. 8.97

The SFs of the cone-confined rotator model depend on the two parameters: the opening β of the cone and on the collision frequency $y = \eta/\tau$.

THE SFs IN THE LONG LIFETIME LIMIT. Let

$$u^2 = a^2 t^2, \quad F(h) = h\exp(-h), \tag{8.100}$$

$$B(h) = \frac{1}{a^2 h - z^2} = \frac{1}{2u}\left(\frac{1}{u - x - iy} + \frac{1}{u + x + iy}\right). \tag{8.101}$$

Then for $u \geq 0,\ x \geq 0$

$$\lim_{y \to 0} \operatorname{Im}\{B(h)\} = \frac{\pi}{2u}\,\delta(u - x); \tag{8.102}$$

$$\lim_{y \to 0} \int_0^{\infty} \operatorname{Im}\{B(h)\,F(h)\}\,dh = \frac{\pi}{a^2}\,F\left[\left(\frac{x}{a}\right)^2\right]. \tag{8.103}$$

Using this rule and noting, that in the CCR model

$$a_{||} = \frac{\pi n}{\alpha}, \; a_{\perp} = \frac{\pi n - \lambda}{\alpha}, \tag{8.104}$$

we can eliminate integration over t in Eq. 8.97 for the imaginary parts of the longitudinal and transverse spectral functions:

$$K_{||}'' = \frac{12}{1 - \cos\beta} \int_0^{\sin\beta} \alpha\left(1 - w^2 - \cos^2\beta\right) S_{||}\,dw\ , \tag{8.105}$$

$$K_{\perp}'' = \frac{3\cot^2\beta}{1-\cos\beta}\int_0^{\sin\beta}\alpha\left(1-w^2-\cos^2\beta\right)S_{\perp}\,dw\,, \tag{8.106}$$

$$S_{\parallel}(x,w) = \sum_{n=1}^{\infty}\frac{\left(\frac{x\alpha}{\pi n}\right)^2\exp\left[-\left(\frac{x\alpha}{\pi n}\right)^2\right]}{\left(\pi^2 n^2-\alpha^2\right)^2}\,, \tag{8.107}$$

$$S_{\perp}(x,w) = \sum_{n=-\infty}^{\infty}\frac{\left(\frac{x\alpha}{\pi n-\lambda}\right)^2\exp\left[-\left(\frac{x\alpha}{\pi n-\lambda}\right)^2\right]}{\left[(\pi n-\lambda)^2-\alpha^2\right]^2}\,, \tag{8.108}$$

$\alpha(w)$ and $\lambda(w)$ are defined by Eq. 8.98.

8.5.4 *Transition to Free Rotation*

If we take $\beta = \pi$, the conical surface is transformed into the needle, directed along the symmetry axis. In this case dipoles rotate freely and do not encounter any obstacle (this statement is not valid for dipoles, moving in planes that contain the Z axis). Thus, we might expect, that the SFs, given by Eq. 8.94 - 8.95 or 8.99, 8.100, reduce to $L_F(z)$, Eq. 7.11, 7.12, derived for the ED model. Indeed, a freely rotating dipole performs during the period $\varphi = \Phi$ a full cycle of rotation around the Z axis. Hence, the angle $\lambda = \pi$ and the sector, in which rotation takes place, transforms into a full circle, i.e. $\alpha = \pi$. Taking this into account, let us repeat the derivation of Eq. 8.87 and 8.93 at $\alpha = \beta = \lambda = \pi$:

$$a_n = \frac{\sin\bar{\theta}}{\pi}\int_{-\pi}^{\pi}\cos\vartheta\cos n\vartheta\,d\vartheta - = \begin{Bmatrix}\sin\bar{\theta},\ n=1\\ 0,\ \ n\neq 1\end{Bmatrix}, \tag{8.109}$$

$$b_n = \frac{1}{4\pi}\int_{-\pi}^{\pi}\left[\left(1+\cos\bar{\theta}\right)\cos n\vartheta - \left(1-\cos\bar{\theta}\right)\cos(n-2)\vartheta\right]d\vartheta\,,$$

$$= \pm\frac{1}{2}+\frac{1}{2}\cos\bar{\theta} \text{ for } n=\begin{Bmatrix}0\\2\end{Bmatrix};\ b_n = 0 \text{ for } n\neq 0,2\ . \tag{8.110}$$

Thus, noting that

$$W=\frac{\exp(-h)}{2\pi},\ p=\sqrt{h},\ \left\langle q_{\parallel}^2\right\rangle = \left\langle q_{\perp}^2\right\rangle = \frac{1}{3};\left\langle q_{\parallel}\right\rangle = \left\langle q_{\perp}\right\rangle = 0, \tag{8.111}$$

we find, that sums in Eq. 5.108, 5.121 reduce to

$$S_{\parallel}=\frac{a_1^2}{h-z^2}=\frac{\sin^2\bar{\theta}}{h-z^2},\ S_{\perp}=\frac{b_0^2+b_2^2}{h-z^2}=\frac{1}{2}\frac{1+\cos^2\bar{\theta}}{h-z^2} \tag{8.112}$$

With account of Eq. 8.122 we obtain from Eq. 5.108, 5.121 the required result for spatial rotors:

$$K_{\parallel}=\frac{3\pi}{2\pi}\int_0^{\infty}\exp(-h)\sqrt{h}\,dh\int_0^{\frac{\pi}{2}}\frac{\sqrt{h}}{h-z^2}\sin^3\bar{\theta}\,d\bar{\theta}\ ,$$

$$K_{\perp}=\frac{3\pi}{2\pi}\int_0^{\infty}\exp(-h)\sqrt{h}\,dh\int_0^{\frac{\pi}{2}}\frac{\sqrt{h}}{h-z^2}\,\frac{1}{2}\left(1+\cos^2\bar{\theta}\right)\sin\bar{\theta}\,d\bar{\theta}\,.$$

Finally, we have:

$$\begin{aligned} K_{\parallel} = K_{\perp} &= \frac{3}{2}\,\frac{2}{3}\int_0^{\infty}\frac{\exp(-h)\,h\,dh}{h-z^2} \\ &= 1+z^2\exp(-z^2)E_1(-z^2)\equiv L_F(z)\,. \end{aligned} \tag{8.113}$$

The same result is obtained for an isotropic medium, since we have no anisotropy in the case of free rotors:

$$L(z)=\frac{1}{3}\left(K_{\parallel}+2K_{\perp}\right)=L_F(z). \tag{8.114}$$

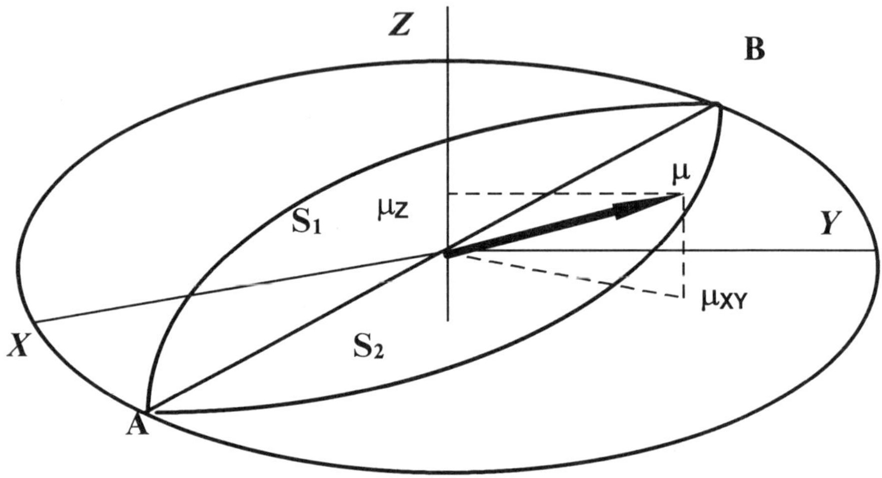

Fig. 8.11 The cone-confined rotator model: reflection of a dipole from a plane.

8.5.5 Rotation in Hemisphere

Let us obtain the SF $K_{\perp}(z)$ for $\beta = \pi/2$. The conical surface transforms into a reflecting plane, orthogonal to the Z axis. As shown in Fig. 8.11, a dipole rotates in plane S_1, then in S_2, then again in S_1, and so on. These planes cross at the line AB; one plane is a mirror reflection of another. The sector of rotation presents a semicircle ($\alpha = \pi/2$); $\lambda = \pi/2$, since angular distance from point A to B on plane XY is equal to π. By analogy with Eq. 8.110 we have:

$$b_n = \frac{1}{2\pi} \int_{-\pi}^{\pi} \left[\left(1+\cos\bar{\theta}\right)\cos 2n\vartheta - \left(1-\cos\bar{\theta}\right) \cos(n-1)2\vartheta \right] d\vartheta$$

$$= \pm\frac{1}{2}+\frac{1}{2}\cos\bar{\theta} \text{ for } n = \begin{Bmatrix} 0 \\ 1 \end{Bmatrix}; \; b_n = 0 \text{ for } n\neq 0,1 \;; \tag{8.115}$$

$$p = 2\sqrt{h},\, W = \frac{\exp(-h)}{\pi},\, \left\langle q_\perp^2 \right\rangle = \frac{1}{3}. \tag{8.116}$$

$$S_\perp = \frac{b_0^2\left(0-\frac{1}{2}\right)^2}{4h\left(0-\frac{1}{2}\right)^2 - z^2} + \frac{b_1^2\left(1-\frac{1}{2}\right)^2}{4h\left(1-\frac{1}{2}\right)^2 - z^2} = \frac{1+\cos^2\bar{\theta}}{8\left(h-z^2\right)}. \tag{8.117}$$

Substitution of these values into (5.121) yields:

$$K_\perp = L_\perp = 3\pi \int_0^\infty \frac{\exp(-h)\,2\sqrt{h}\;dh}{8\pi\left(h-z^2\right)} \int_0^{\frac{\pi}{2}} \sqrt{h}\left(1+\cos^2\bar{\theta}\right)\sin\bar{\theta}\;d\bar{\theta}$$

$$= \int_0^\infty \frac{\exp(-h)\,h\,dh}{h-z^2} \equiv L_F(z). \tag{8.118}$$

The longitudinal SF $K_{||}(z)$ does not obey this relation. This may be understood, if we turn our attention to Fig. 8.11. The law of motion $\mu_Y(t)$ does not alter after transition of a dipole to plane $\mathbf{S}_2$ from $\mathbf{S}_1$. Therefore, the $q_\perp(\varphi)$ dependence is the same for $\beta = \pi/2$, as for free motion. Since this statement is not true for the μ_Z projection, $K_{||}(z) \neq L\ (z)$ for $\beta = \pi/2$.

8.5.6 The Approximation of Small-Amplitude Librations

Let $2\beta << 1$. Then angular momenta of almost all dipoles are inclined to the symmetry axis Z at angles ca. $\pi/2$. *Sectors of rotation are displaced nearly in the diametric cross-section of a cone*, i.e. $\lambda \approx \pi/2$. Then, in accord with Eq. 8.61, we may put $\alpha = \beta$. Instead integration over $\bar{\theta}$ in Eq. 8.94 one may multiply integrand by the increment $\Delta\bar{\theta} = \pi/2 - (\pi/2 - \beta) = \beta$, with $\bar{\theta} = \lambda = \pi/2$ and $\alpha = \beta$. We may also leave only the first term in the sum $S_{||}$ for $K_{||}$, corresponding to $n = 1$. For $K_\perp$ we must leave in the sum $S_\perp$ the two first terms, corresponding to $n = 0$ and $n = 1$, since they have the same order of magnitude. Thus, the order of integrals is reduced:

$$K_{\|} \approx \frac{12\pi \sin^2\beta}{1-\cos\beta} \int_0^{\infty} \frac{\beta\beta\beta^2 \exp(-h)dh}{\pi^6\left[h - \left(\frac{\beta z}{\pi}\right)^2\right]} = \frac{12}{\pi^5} \frac{\beta^4 \sin^2\beta}{1-\cos\beta} L_F\left(\frac{\beta z}{\pi}\right) \quad (8.119)$$

$$K_{\perp} \approx \frac{3\pi \cos\beta}{1-\cos\beta} \int_0^{\infty} \frac{2\beta^2 16 \cdot 4\beta^2 \exp(-h)h\, dh}{4\pi^6\left[h - \left(\frac{2\beta z}{\pi}\right)^2\right]}$$

$$= \frac{96}{\pi^5} \frac{\beta^4 \cos^2\beta}{1-\cos\beta} L_F\left(\frac{2\beta z}{\pi}\right) \quad (8.120)$$

For an isotropic medium the spectral function is given by

$$L(z) = \frac{1}{3}\left(K_{\|} + 2K_{\perp}\right)$$

$$= \frac{4}{\pi^5} \frac{\beta^4}{1-\cos\beta}\left(\sin^2\beta\, L_F\left(\frac{\beta z}{\pi}\right) + 16\cos^2\beta\, L_F\left(\frac{2\beta z}{\pi}\right)\right). \quad (8.121)$$

Since the transversal SF is much larger than the longitudinal one, we may omit the latter. Then

$$L(z) \approx \frac{2}{3} K_{\perp} \approx \frac{128\beta^2}{\pi^5} L_F\left(\frac{2\beta z}{\pi}\right). \quad (8.122)$$

For relatively large libration amplitude, viz. $\beta \approx \pi/2$, the last expression closely resembles Eq. 7.11, 7.19 for the extended diffusion model, since $32/\pi^3 \approx 1.032$. Hence, we may suggest that the approximation (8.120) is applicable in a rather wide interval of angles β.

We can obtain an independent confirmation of this statement: Eq. 8.122 yields the value Π_{CCR} of integrated absorption, which differs from the exact value (π) only by 3%. To show this we write:

$$\int_0^\infty \operatorname{Im}\{L_F(z)\}\, x\, dx = \pi. \tag{8.123}$$

Then, taking $z' = (2\beta z/\pi) = x' + iy'$ as a new variable, we have for our approximation:

$$\Pi_{\mathrm{CCR}} = \int_0^\infty \operatorname{Im}\{L_F(z)\}\, x\, dx$$

$$\approx \frac{128\beta^2}{\pi^5}\frac{\pi^2}{4\beta^2}\int_0^\infty \operatorname{Im}\{L_F(z')\}\, x' dx' = \pi\frac{32}{\pi^3}. \tag{8.124}$$

The difference reduces to the multiplier $32/\pi^3 \approx 1.032$ instead of 1 for the exact Π_{CCR} value. Consequently, we may suppose, that formula (8.122) is applicable for approximate calculations in all practically interesting interval $[0, \pi/2]$ of libration amplitudes β.

8.5.7 Transition to the Planar Confined Rotator Model

To perform this transition we should take the trajectory of a dipole in the diametric cross-section of the cone. The conical surface, characterized by the solid angle 2β, is replaced by two elastic reflecting planes, that cross at the angle 2β, the symmetry axis being directed along the bisector of this angle. The angular momentum $\boldsymbol{k}$ is now orthogonal to the symmetry axis ($l = 0$); the angle $\bar{\theta} = \operatorname{acos}(l/k) = \pi/2$. Since the sector coincide with the cone angle ($\alpha \equiv \beta$), then, in accord with Eq. 8.61, $\lambda = \operatorname{asin}\dfrac{\sin\alpha}{\sin\beta} = \dfrac{\pi}{2}$. In our model we exclude the rotational subensemble. We introduce the libration frequency $\check{p}$, corresponding to the period $2\pi/\check{p}$ of function $q_\parallel(\varphi)$:

$$\check{p} = p/2 = \pi\sqrt{h}/(2\beta) = \sqrt{h}/f. \tag{8.125}$$

Our purpose is to prove that Fourier amplitudes a_n^2, b_n^2, calculated in Sec. 8.1.1 *directly* for the CR model, can also be obtained from the reduction of the CCR model. From Eq. 8.87, applied to $\lambda = \theta = \pi/2$ and $\alpha = \beta$, we have:

$$a_n^2 = \frac{4\beta^2 \sin^2 \beta^2}{\left(\pi n^2 - \beta^2\right)^2} \text{ for } n = 0,2,4\ldots \; . \tag{8.126}$$

These a_n^2 coincide with those, given by Eq. 8.10. Analogously, Eq. 8.93 reduces to

$$b_n^2 = \frac{16\beta^4 \cos^2 \beta}{\left[\pi(2n-1)^2 - 4\beta^2\right]^2} \text{ for } n = 1,3,5\ldots \quad .$$

In accordance with Eq. 5.141, derived for the planar CR model, the square of Fourier amplitudes is equal to the value

$$b_n^2 = \frac{16\beta^4 \cos^2 \beta}{\left[\pi(2n-1)^2 - 4\beta^2\right]^2} \text{ for } n = 1,3,5\ldots \quad . \tag{8.127}$$

(here multiplication by 4 is employed). These b_n^2 coincide with b_n^2 in Eq. 8.10.

8.5.8 Dielectric Behavior

Dielectric spectra, calculated with use of the exact double-integral expressions (8.95), (8.97), (8.98), are shown in Fig. 8.12 for the longitudinal (a, d) and transversal (b, e) orientations of alternating field, and for an isotropic medium (c, f). The evolution of spectra is stipulated by the increase of the opening β of a cone from $\pi/6$ (dashed curves) to $\pi/4$ (solid curves), the normalized collision frequency y being the same for all curves (y = 0.5). We employ the Gross collision model ($F = F_G$) and suggest that $\langle q \rangle = 0$, see Table 6.1, so that:

$$\frac{\chi^*}{G} = \frac{z\,K(z)}{x + \dfrac{iy}{\sigma} K(z)} \; . \tag{8.128}$$

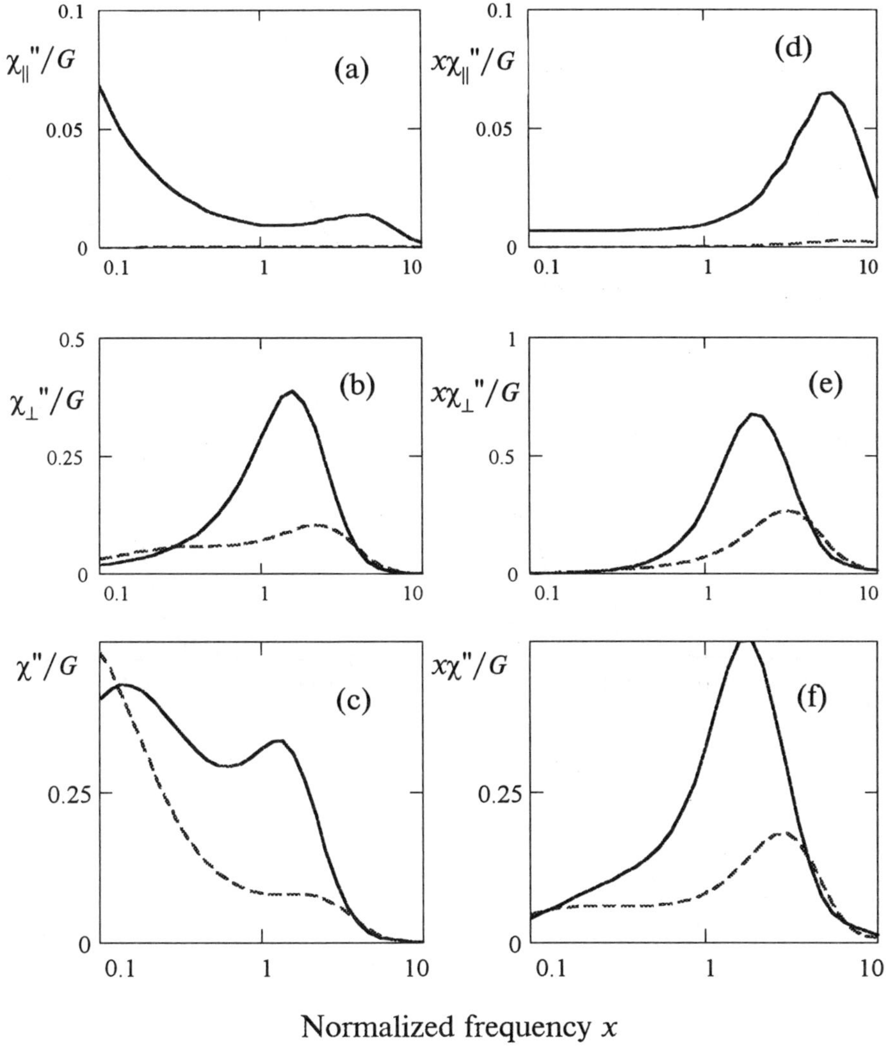

Fig. 8.12 Loss (left) and absorption (right) vs. x, calculated rigorously for the CCR model. Longitudinal (a, d), transverse (b, e) orientations and the case of an isotropic medium (c, f); $\beta = \pi/6$ (dashed curves) and $\pi/4$ (solid curves); $\langle q \rangle = 0$.

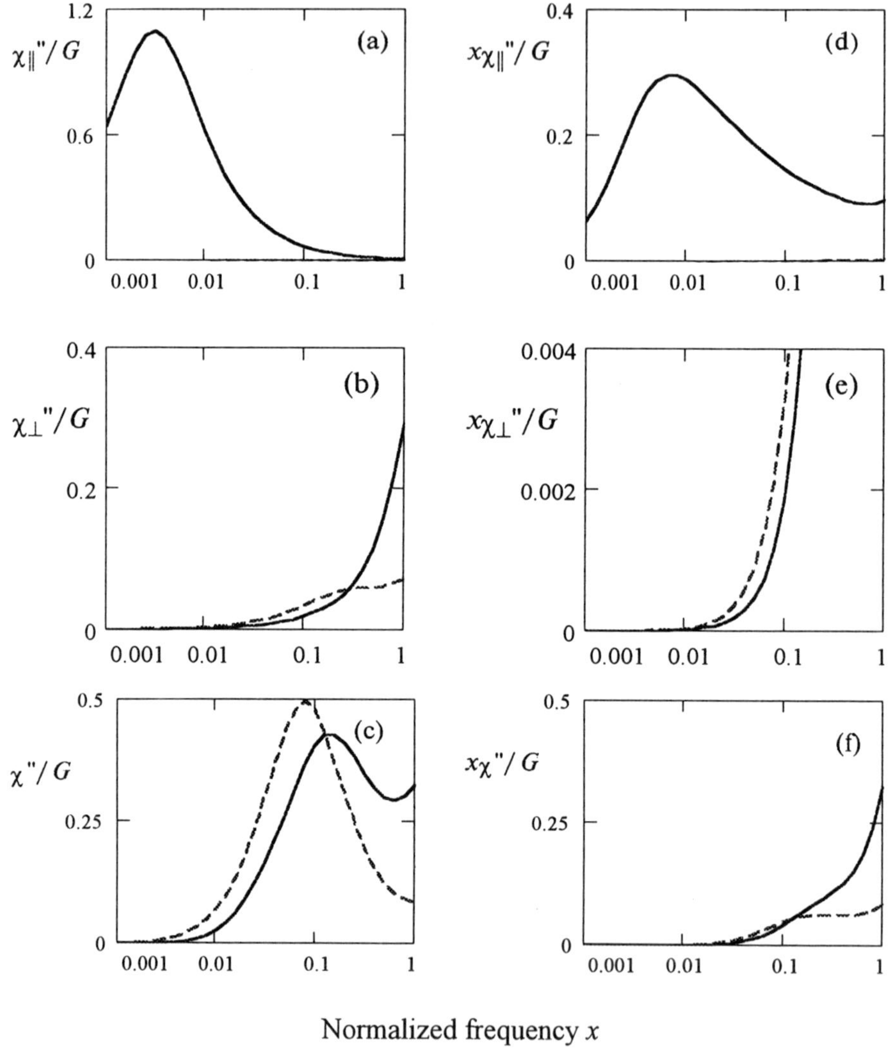

Fig. 8.13 The same, as in Fig. 8.12, but for the low frequency (Debye) range.

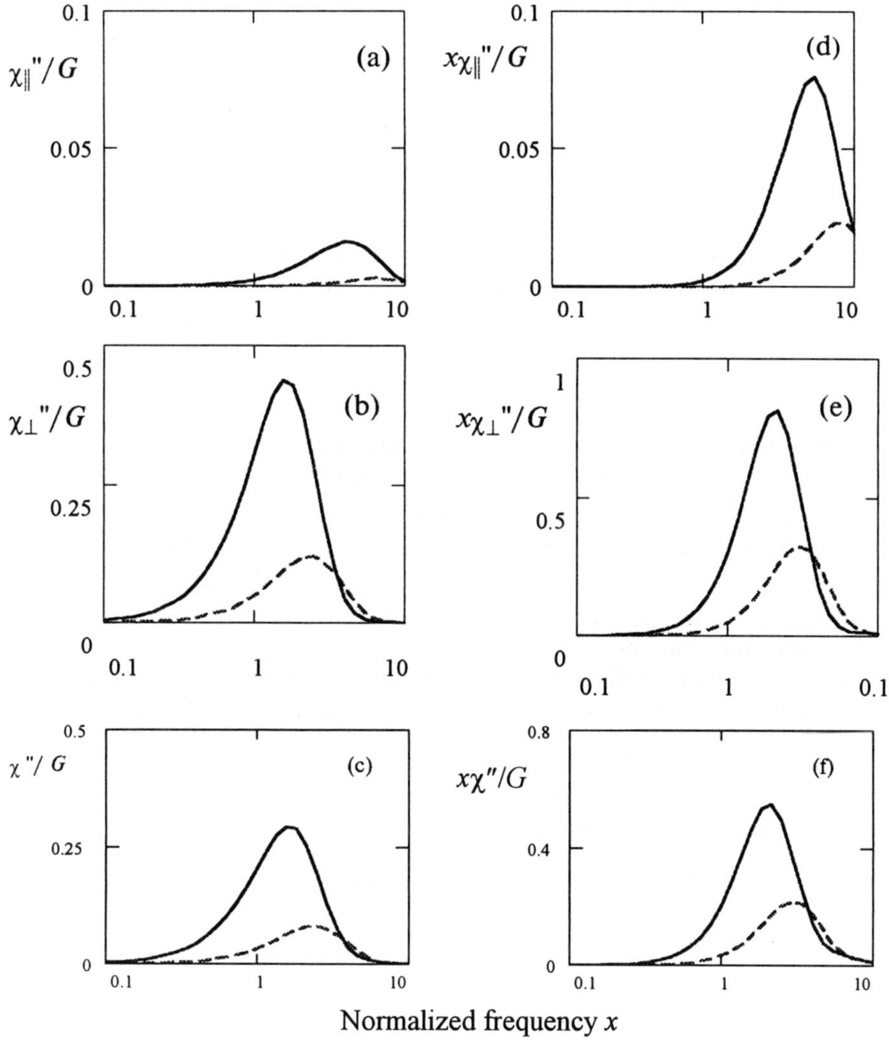

Fig. 8.14 The same, as in Fig. 8.12, but for the long lifetime limit.

For an *anisotropic* medium $K = K_{\|}$, $\sigma = \sigma_{\|}$ or $K_{\perp} = K_{\perp}$, $\sigma = \sigma_{\perp}$. The dispersion coefficients $\sigma_{\|}$ and $\sigma_{\perp}$ are found from Eq. 8.83 and 8.86, respectively.

For an *isotropic medium*, where $\sigma = 1$,

$$K(z) = L(z) = (1/3)\,(K_{\|} + 2K_{\perp})\,. \tag{8.129}$$

In Fig. 8.13 a, b, c the low-frequency dielectric spectra are pictured.

Analogous spectra are shown in Fig. 8.14 for the zero collision frequency ($y = 0$). The imaginary parts $K''(x)$ of the SFs are calculated now from ordinary integrals over w, see expressions (8.98), (8.105), (8.100). For $x > 0.5$ high- frequency loss / absorption spectra $\chi''(x)$ and $x\chi''(x)$ can be estimated by using expression

$$\chi''(x)\,/\,G \approx K''(x). \tag{8.130}$$

ANISOTROPIC ENSEMBLES OF CONFINED ROTORS. It is seen from Fig. 8.13, that Debye spectra are considerably shifted for $\perp$-orientation to the *high frequency region* compared with the case of $\|$-orientation, the maximum loss $\chi''_{D\|}$ being much larger, than $\chi''_{D\perp}$, cf. Fig. 8.13b and 8.13a. As is seen from Fig. 8.12 d, e, in the HF region the resonance loss / absorption maxima are on the contrary shifted for $\perp$-orientation to lower frequencies and are much larger, than for $\|$-orientation.

AN ISOTROPIC MEDIUM. For such medium dielectric spectra are as if "intermediate" between those for $\|$- and for $\perp$-orientations, cf. Fig. 8.12 c, f and 8.13 c, f.

8.5.9 The Comparison of Exact and Approximate Theories

In Fig. 8.15 we present dielectric spectra, calculated for the CCR model in the small β approximation. We use Eq. 8.119-8.122, 8.128, 8.129 with dispersions $\sigma_{\|}$ and $\sigma_{\perp}$ found from Eq. 8.83 and 8.86. For an isotropic medium we put $\sigma = 1$.

The comparison of Fig. 8.15 with Fig. 8.12, 8.13 shows, that a simplified theory, given in Sec. 8.5.6, is applicable for a *qualitative* description, at least for $\beta < \pi/4$.

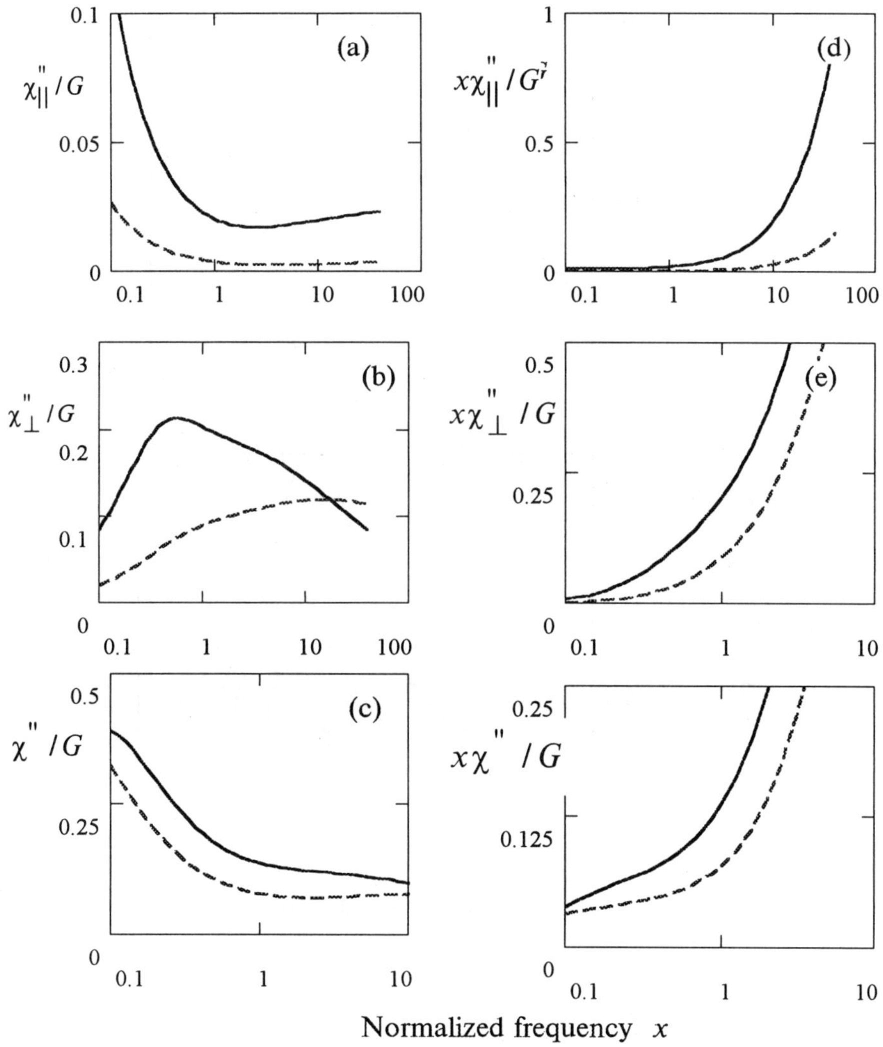

Fig. 8.15 The same as in Fig. 8.12 but for the small β approximation.

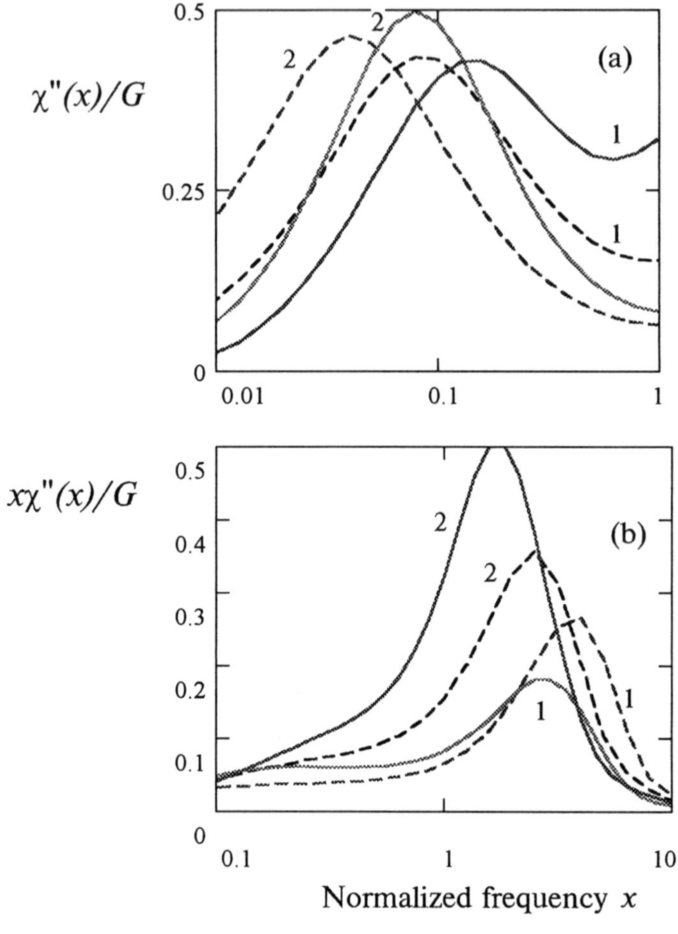

Fig. 8.16 The comparison of dielectric loss (a) and absorption (b), calculated for the CCR model (solid lines) and for spatial CR model (dashed lines). The libration amplitude $\beta = \pi/6$ and $\pi/4$ (curves 1 and 2, respectively). An isotropic medium, the double well potential configuration.

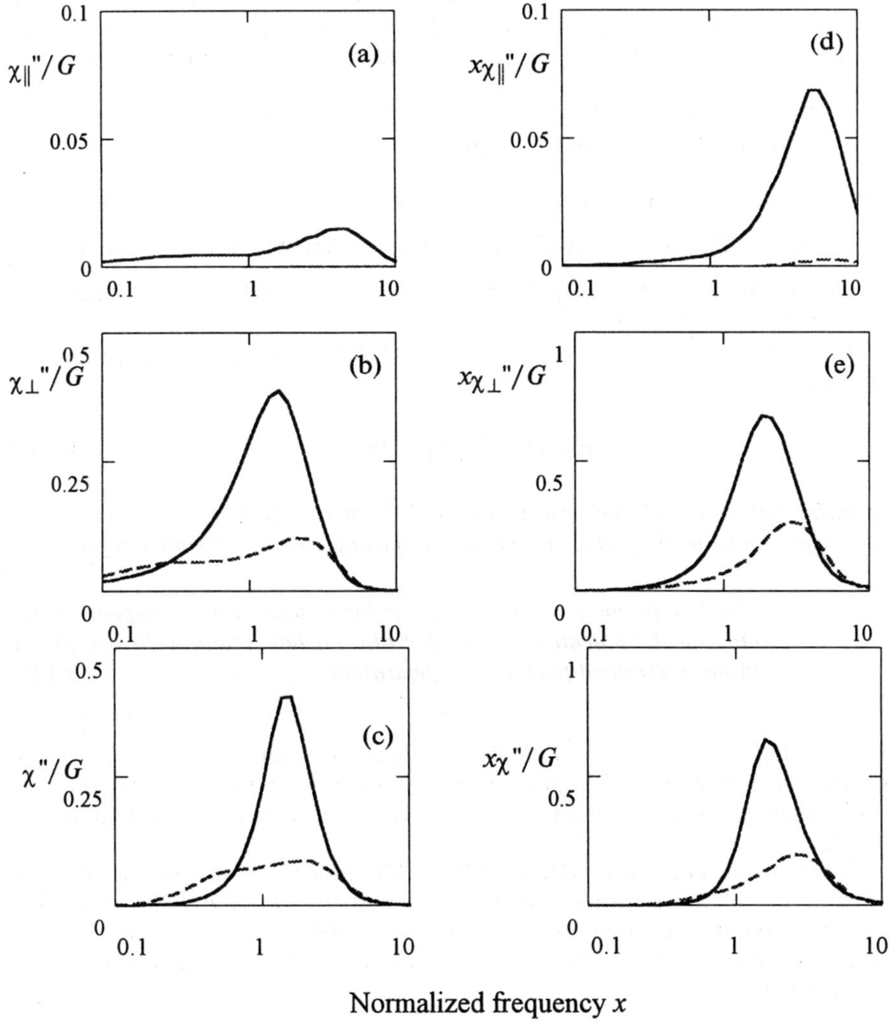

Fig. 8.17 The same, as in Fig. 8.12, but for the single well potential configuration, where $\langle q_\parallel \rangle \neq 0$.

On the other side, rather satisfactory result is obtained (see Fig. 8.16), if the simple spatial CR model is employed, where solid lines are drawn for the exact theory and dashed lines - for the CR model, described in Sec. 8.4.

8.6 The Cone Confined Rotator Model with a Single Potential Well: Dielectric Behavior

Dielectric spectra radically change, if it is assumed, that $\langle q \rangle \neq 0$. This is seen, if we compare Fig. 8.17 with Fig. 8.12, 8.13, calculated for $\langle q \rangle = 0$. In Fig. 8.17 we use the same SFs $K_{\|}(z)$, $K_{\perp}(z)$ and $\sigma_{\perp}$, as in Fig. 8.12, while $\sigma_{\|}$ is determined from Eq. 8.84.

For an isotropic medium we use Eq. 8.128, with $L(z)$, given by Eq. 8.129, and

$$\sigma = \bar{\sigma} = \frac{1}{3}\left(\sigma_{\|} + 2\sigma_{\perp}\right). \tag{8.131}$$

The main difference of the single potential from the double one is related to the *disappearance in the former of the low-frequency Debye relaxation region.*

The low-frequency dielectric behavior is thus strongly dependent not only on the potential profile, but also on how many wells are placed in the prescribed potential configuration. (8.132)

We shall see in 10, 11 and 12 chapters, that the aforementioned distinction between the SWP and DWP configurations retains also for the *field models,* that are characterized by the cosine and cosine squared potential profiles.

We remark in conclusion, that for the CCR model calculations were made here with use of the rigorous theory, elaborated for *librational spatial motion.* For other potential profiles we, in most cases, restrict our consideration by *planar motion with some corrections, in which the statistics of spatial motion is accounted for.*

CHAPTER 9

THE HYBRID CONFIHED ROTATOR / EXTENDED DIFFUSION MODEL (HM)

In this chapter we try to overcome the approximation of an infinitely deep potential well, used in the previous chapter. Roughly speaking, we employ the cone-confined rotator (CCR) model for the low-energy dipoles and the extended diffusion (ED) model for the rest molecules. These particles are termed L- and R-molecules and are marked respectively by superscripts $^{\vee}$ and $^{\circ}$. The boundary between them is determined by the normalized potential welldepth $u = U_0/(k_B T)$.

In the hybrid model (HM) we have *three* molecular parameters,

$$\beta, \tau, u, \tag{9.1}$$

viz., the libration amplitude β (the sector of the cone), the lifetime τ, and the depth u of the rectangular potential, shown in Fig. 1.5b. We remark that in the CR or CCR model only two parameters, β and τ, are employed. The theory of the ED and CCR models, considered in chapters 7 and 8, should be generalized in order to account for that only a part of all molecules participates now in libration or in free rotation.

LIBRATORS. L-molecules move inside a cone, shown in Fig. 8.10 and 9.1. The corresponding potential well has a flat bottom (see Fig. 1.5b). The time-varying angle θ between dipole moment μ and the symmetry axis Z does not exceed β, the opening of a cone. We also introduce angle Θ between Z and angular momentum vector (in Fig. 8.10 this angle was denoted $\bar{\theta}$); angle Θ is varied in the interval $\left[\frac{\pi}{2} - \beta, \frac{\pi}{2}\right]$, see Eq. 8.66. Normalized energy h of a dipole (it coincides with its kinetic energy) does not exceed the welldepth:

$$h \equiv H/(k_B T) \le u. \tag{9.2}$$

In this chapter we consider the single and double well potentials. In the first SWP case the only well exists in space, while in the last case both cones are supposed to be similar and to have the same symmetry axis Z (see Fig. 9.1). We assume that motion of librators in each cone is quite independent.

ROTATORS. R-dipoles, belonging to the second molecular group, have energy $h > u$. Normalized kinetic energy of a rotator

$$h_{\text{kin}} = h - u \tag{9.3}$$

may have any possible value, viz.

$$0 < h - u < \infty; \tag{9.4}$$

For angle Θ between Z and the angular-momentum vector $\boldsymbol{p}$ we have:

$$\Theta < \pi. \tag{9.5}$$

R-dipoles, strictly speaking, should be subdivided into two groups (**I** and **II**) of molecules: those, which fall *periodically* into the well (group **I**), and those, which rotate freely outside it (group **II**). Correspondingly, the normalized Z-projection l of the angular momentum should be given by:

$$\textbf{I.}\quad 0 \le l \le \sqrt{h-u}\,\sin\beta\,; \tag{9.6}$$

$$\textbf{II.}\quad \sqrt{h-u}\,\sin\beta \le l \le \sqrt{h-u}\ . \tag{9.7}$$

For simplicity we assume that *all R-molecules rotate freely:*

$$0 \le l \le \sqrt{h-u}\ . \tag{9.8}$$

THE LAW OF MOTION. Let us fix the time interval between two instances of elastic collisions of a dipole with the surface of a cone.

For *L-molecules* the time dependence $\theta(\varphi)$ during this interval is given by Eq. 8.54, 8.55, 8.58, 8.59 (we employ symbol $\vee$, so that $\vartheta \to \check{\vartheta}$, $\theta \to \check{\theta}$):

$$\check{\vartheta} = \operatorname{a\cos}\left[q_{\|}(\varphi)\right] - \operatorname{a\cos}\left[\sin\Theta\cos\check{\vartheta}(\varphi)\right]$$

$$= \operatorname{a\cos}\left[\sqrt{1-\frac{l^2}{h}}\cos\sqrt{h}(\varphi+\varphi_0)\right]. \tag{9.9}$$

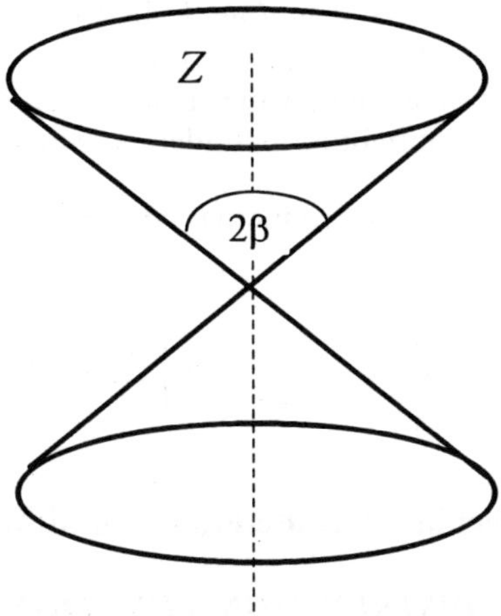

Fig. 9.1 The geometric scheme for the hybrid confined rotator model with the double well configuration of an intermolecular potential.

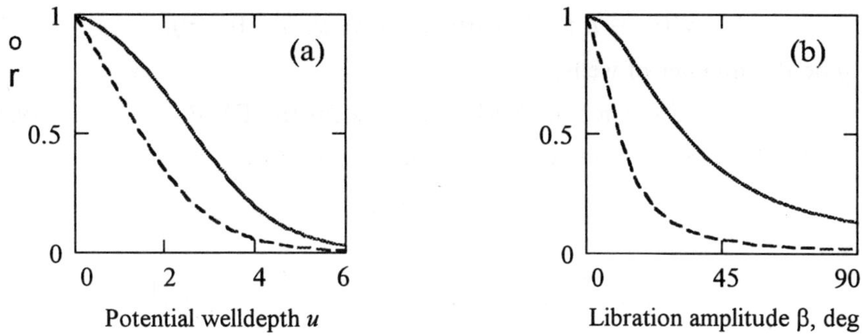

Fig. 9.2 The proportion of rotators vs. potential welldepth u (a) or libration amplitude β (b). In **a**: $\beta = \pi/8$ (solid line) and $\pi/4$ (dashed line); in **b**: $u = 2$ (solid line) and $u = 4$ (dashed line). The double well potential.

We recall the Reader that Θ is *spatial* angle between vector μ and axis Z of a cone and ϑ is the angular shift in *plane of rotation.*

For R-molecules similar angle-time dependence is obtained from Eq. 7.1 - 7.3. For the potential configuration under consideration (see Fig. 1.5b) we should substitute h by $h-u$. Introducing symbol °, so that $\vartheta \to \overset{\circ}{\vartheta}$, $\theta \to \overset{\circ}{\theta}$, we write:

$$\overset{\circ}{\theta}(\varphi) = a\cos\left[\sqrt{1-\frac{l^2}{h-u}}\ \cos\left[\sqrt{h-u}\ (\varphi+\varphi_0)\right]\right]. \tag{9.10}$$

9.1 The Mean Molecular Parameters at Equilibrium

THE STEADY-STATE DISTRIBUTION FUNCTION $W(h) = C\exp(-h)$.

We chose $\theta, \dot{\theta}, l$ as phase variables and take into account the profile $v(\theta)$ of the normalized potential:

$$v(\theta) = 0 \text{ for L-particles } (\theta \text{ inside the cone}),$$

$$v(\theta) = u \text{ for R-particles } (h > u,\ 0 \le \theta \le \pi).$$

Let j be the number of wells,

$$j = 1 \text{ for the SWP and } j = 2 \text{ for the DWP.} \tag{9.11}$$

The normalization constant C is found analogously to Eq. 8.74:

$$\int W\, d\Gamma = 1; \tag{9.12}$$

$$\frac{1}{C} = 2j \int_{-\infty}^{\infty} d\dot{\theta} \int_{0}^{\infty} dl \int_{0}^{\pi/j} d\theta\, \exp\left(-\dot{\theta}^2 - \frac{l^2}{\sin^2\theta} - v(\theta)\right)$$

$$= 2j\sqrt{\pi}\,\frac{\sqrt{\pi}}{2}\left\{\int_0^{\beta}\sin\theta\, d\theta + \exp(-u)\int_0^{\pi/j}\sin\theta\, d\theta\right\}$$

$$= j\pi\left[1-\cos\beta + \exp(-u)\left(\cos\beta + \delta_j\right)\right], \tag{9.13a}$$

where

$$\delta_j = \frac{1-(-1)^{-j}}{2}; \; \delta_j = 0 \text{ for } j = 2, \; \delta_j = 1 \text{ for } j = 1. \tag{9.13b}$$

For the main (DWP) configuration we find, taking $j = 2$:

$$C = \frac{1}{2\pi} \frac{1}{1-\cos\beta\left[1-\exp(-u)\right]} \, . \tag{9.14}$$

THE MEAN VALUES OF DIRECTION COSINES AND OF THEIR SQUARES. We employ again the variables $\theta, \dot{\theta}, l$. Taking account of Eq. 9.13, we obtain:

$$3\left\langle \begin{matrix} q_{\|}^2 \\ q_{\perp}^2 \end{matrix} \right\rangle = 3\int \left\langle \begin{matrix} \cos^2\theta \\ \dfrac{\sin^2\theta}{2} \end{matrix} \right\rangle W\, d\Gamma$$

$$= 3j\pi C \left\{ \int_0^{\beta} \langle \bullet \rangle \sin\theta\, d\theta + \exp(-u) \int_{\beta}^{\pi} \langle \bullet \rangle \sin\theta\, d\theta \right\}$$

$$= \frac{\left\{ \begin{matrix} 1-\cos^3\beta \\ \frac{1}{2}\left(2-3\cos\beta+\cos^3\beta\right) \end{matrix} \right\} + \exp(-u) \left\{ \begin{matrix} \delta_j + \cos^3\beta \\ \frac{1}{2}\left(2\delta_j + 3\cos\beta - \cos^3\beta\right) \end{matrix} \right\}}{1-\cos\beta + \exp(-u)\left(\delta_j + \cos\beta\right)}; \tag{9.15}$$

$$\langle q_{\|} \rangle = \pi C \left\{ \int_0^{\beta} \cos\theta \sin\theta\, d\theta + \exp(-u) \int_{\beta}^{\pi} \cos\theta \sin\theta\, d\theta \right\}$$

$$= \frac{\pi C}{2}\left[\sin^2\beta - \exp(-u)\right] \quad \text{for } j{=}1, \text{ i.e. for SWP;} \tag{9.16a}$$

$$\langle q_{\|} \rangle = 0 \quad \text{for } j{=}2, \text{ i.e. for DWP.} \tag{9.16b}$$

PROPORTIONS $\overset{\circ}{r}$ *and* $\overset{\vee}{r}$ *OF ROTATORS AND LIBRATORS.* Let us calculate two statistic integrals (which we denote st) for L- and R-ensembles:

$$\overset{\vee}{\mathrm{st}} = \int\limits_{\overset{\vee}{\Gamma}} \exp(-h)\, d\overset{\vee}{\Gamma} \; ; \; \overset{\circ}{\mathrm{st}} = \int\limits_{\overset{\circ}{\Gamma}} \exp(-h)\, d\overset{\circ}{\Gamma} \; ,$$

where $\overset{\vee}{\Gamma}$, $\overset{\circ}{\Gamma}$ are phase volumes of corresponding subensembles. We chose h, l and φ_0 as phase variables.

For *L-molecules*

$$0 \le h \le u, \; 0 \le l \le \sqrt{h}\, \sin\beta, \; 0 \le \varphi_0 \le \overset{\vee}{\Phi} . \tag{9.17a}$$

For *R-molecules*

$$u \le h < \infty, \; 0 \le l \le \sqrt{h-u} \; , 0 \le \varphi_0 \le \overset{\circ}{\Phi} , \tag{9.17b}$$

where

$$\overset{\vee}{\Phi} = \frac{2\alpha}{\sqrt{h}} \approx \frac{2\beta}{\sqrt{h}} \text{ and } \overset{\circ}{\Phi} = \frac{2\pi}{\sqrt{h-u}} \tag{9.18}$$

are periods of librational and rotational motion. Here the $\approx$ sign corresponds to the small β approximation (the latter is used mostly in this chapter). Ranges (9.17) of phase variables look "natural". However, they are *approximate;* more rigorous consideration will be given in Sec. 9.4.

To calculate statistic integral $\overset{\vee}{\mathrm{st}}$ for librators we should replace the upper limit $h = \infty$ in integral (8.71) by $h = u$ and should multiply the result by j. Therefore, we should multiply expression (8.74) by $j\,(1 - \cos\beta)$:

$$\overset{\vee}{\mathrm{st}} = j\pi\left[1 - \exp(-u)\right]\left(1 - \cos\beta\right) . \tag{9.19}$$

The second integral, $\overset{\circ}{\mathrm{st}}$, is given by

$$\overset{\circ}{\mathrm{st}} = \int_0^{\infty} \exp(-h)\, dh \int_0^{\sqrt{h-u}} dl \int_0^{\frac{2\pi}{\sqrt{h-u}}} d\varphi = 2\pi \exp(-u) \; . \tag{9.20}$$

We remark that the normalization constant is given by:

$$C = \frac{1}{\overset{\vee}{\mathrm{st}} + \overset{\circ}{\mathrm{st}}} . \tag{9.21}$$

It is easily seen that for the DWP (j = 2) we have from Eq. 9.19 - 9.21 exact expression (9.14) for C. For the SWP (j = 1), Eq. (9.19) - (9.21) yield for C expression, that differ from the rigorous one, Eq. 9.13, 9.14. However for $u >> 1$ this difference becomes negligible.

Thus, the proportions $\overset{\circ}{\mathrm{st}}$ and $\overset{\vee}{\mathrm{st}}$ of rotators and librators are given for the DWP configuration by

$$\overset{\circ}{r} = \frac{\overset{\circ}{\mathrm{st}}}{C} = \frac{\exp(-u)}{1 - \cos\beta[1 - \exp(-u)]}, \tag{9.22a}$$

$$\overset{\vee}{r} = \frac{\overset{\vee}{\mathrm{st}}}{C} = \frac{[1 - \exp(-u)](1 - \cos\beta)}{1 - \cos\beta[1 - \exp(-u)]} . \tag{9.22b}$$

Eq. 9.22 satisfy an obvious relation $\overset{\circ}{r} + \overset{\vee}{r} = 1$. The proportion $\overset{\circ}{r}$ of rotating molecules decreases (see Fig. 9.2), when the amplitude β and the potential u rise. *In the HM* $\overset{\circ}{r}$ *is relatively large even for a rather deep well.* For instance, if $u = 4$, then $\overset{\circ}{r} \approx 19.7\%$ and 6% for $\beta = \pi/8$ and $\pi/4$, respectively.

The weak dependence of $\overset{\circ}{r}$ on the barrier u can be explained as follows. The librational phase volume $\overset{\vee}{\Gamma}$ is *finite* and decreases as β^2 with $\beta \to 0$. Indeed, we can estimate $\overset{\vee}{\Gamma}$ from the assumption, that a dipole moves in a diametric cross-section of a cone. Then

$$\overset{\vee}{\Gamma} = j \int_0^u dh \int_0^{\sqrt{h}\sin\beta} dl \int_0^{\overset{\vee}{\Phi}} d\varphi = 2 j \alpha u \sin\beta.$$

Since in this assumption $\alpha \approx \beta$, $\check{\Gamma} \le 2\beta u \sin\beta$, the phase volume $\check{\Gamma}$ decreases with β.

For the rectangular potential well the population of rotators is by no means proportional to $\exp(-u)$, *as one might suppose for the first glance;* $\mathring{r}$ *decreases with the rise of* u *more slowly (especially for small* β*), because the phase volume* $\check{\Gamma}$ *of librators, Eq.* 9.18, *is finite and is proportional to* β *for small* β.

THE MEAN LIBRATIONAL / ROTATIONAL FREQUENCIES $\check{p}$ and $\mathring{p}$. We shall determine the first quantity, $\check{p}$, as the reciprocal of the libration period. For small β (viz. for $\beta < 1$) $\check{p}$ is approximately twice period $\check{\Phi}$ of function $\cos\check{\vartheta}(\varphi)$:

$$\check{p} \equiv \frac{2\pi}{2\check{\Phi}} = \frac{\pi}{\check{\Phi}} \tag{9.23}$$

By definition

$$\langle\check{p}\rangle = \frac{\int\limits_{\check{\Gamma}} W\check{p}\,d\Gamma}{\int\limits_{\check{\Gamma}} W\,d\Gamma} = \frac{\int\limits_{\check{\Gamma}} W\check{p}\,d\Gamma}{\check{r}} \equiv \frac{C}{\check{r}}\int\limits_{\check{\Gamma}} \exp(-h)\,\check{p}\,d\Gamma = \frac{\int\limits_{\check{\Gamma}} \exp(-h)\,\check{p}\,d\Gamma}{\check{st}} \;. \tag{9.24a}$$

The mean rotational frequency $\mathring{p}$ is found in analogous way:

$$\overset{\circ}{p} \equiv \frac{2\pi}{\overset{\circ}{\Phi}}; \left\langle \overset{\circ}{p} \right\rangle = \frac{\int\limits_{\overset{\circ}{\Gamma}} W \overset{\circ}{p} d\Gamma}{\int\limits_{\overset{\circ}{\Gamma}} W d\Gamma} = \frac{\int\limits_{\overset{\circ}{\Gamma}} W \overset{\circ}{p} d\Gamma}{\overset{\circ}{r}} \equiv \frac{C}{\overset{\circ}{r}} \int\limits_{\overset{\circ}{\Gamma}} \exp(-h) \overset{\circ}{p} d\Gamma \quad . \qquad (9.24b)$$

Noting Eq. 9.22, we obtain explicit expressions:

$$\left\langle \overset{\vee}{p} \right\rangle = \frac{jC}{\overset{\vee}{r}} \int\limits_0^u \exp(-h)\, dh \int\limits_0^{\sqrt{h}\sin\beta} dl \int\limits_0^{\overset{\vee}{\Phi}} \frac{\pi}{\overset{\vee}{\Phi}} d\varphi_0 = \frac{j\pi \sin\beta}{\overset{\vee}{\mathrm{st}}} \int\limits_0^u \sqrt{h} \exp(-h)\, dh$$

$$= \frac{\cot(\beta/2)}{1-\exp(-u)} \left[\frac{\sqrt{\pi}}{2} \operatorname{erf}(\sqrt{u}) - \sqrt{u}\exp(-u) \right]; \qquad (9.25)$$

$$\left\langle \overset{\circ}{p} \right\rangle = \frac{C}{\overset{\circ}{r}} \int\limits_u^\infty \exp(-h)\, dh \int\limits_0^{\sqrt{h-u}} dl \int\limits_0^{\overset{\circ}{\Phi}} \frac{2\pi}{\overset{\circ}{\Phi}} d\varphi_0$$

$$= \frac{1}{\overset{\circ}{\mathrm{st}}} \int\limits_u^\infty \sqrt{h-u}\, \exp(-h) dh = \frac{2\pi\exp(-u)}{2\pi\exp(-u)} \frac{\sqrt{\pi}}{2} = \frac{\sqrt{\pi}}{2} \qquad (9.26)$$

Thus, the mean rotational frequency is the same, as in the ED model, since it does not depend on the potential u. The librational frequency $\left\langle \overset{\vee}{p} \right\rangle$ increases with the rise of u and decreases with the rise of β (see Fig. 9.3 a, b);

$$\left\langle \overset{\vee}{p} \right\rangle = 0 \quad \text{for } u = 0; \left\langle \overset{\vee}{p} \right\rangle = \frac{\sqrt{\pi}}{2} \cot\frac{\beta}{2} \quad \text{for } u \gg 1. \qquad (9.27)$$

The most important property of the HM is related to the *considerable distinction between the mean frequencies* $\left\langle \overset{\circ}{p} \right\rangle$ and $\left\langle \overset{\vee}{p} \right\rangle$:

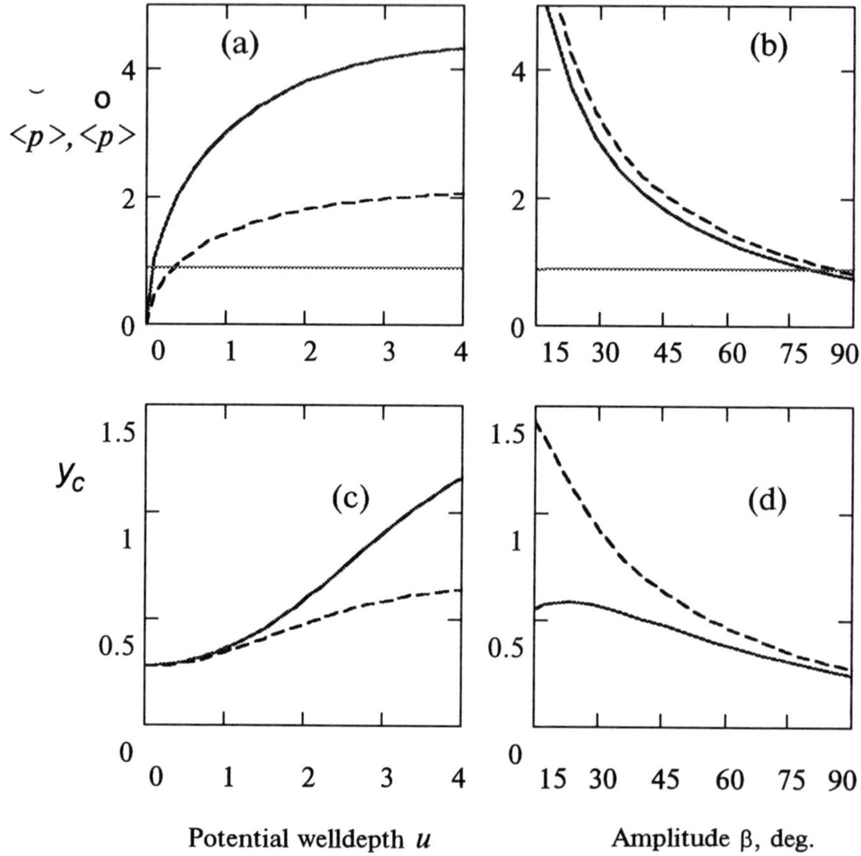

Fig. 9.3 The mean librational frequencies (a, b) and collision-time parameter (c, d) vs. potential welldepth (a, c) or libration amplitude (b, d). In **a, c:** $\beta = \pi/8$ (solid lines) and $\pi/4$ (dashed lines); in **b, d:** $u = 2$ (solid lines) and 4 (dashed lines). The mean rotational frequency in Fig. a, b is drawn by horizontal line.

$$\left\langle \overset{\circ}{p} \right\rangle \ll \left\langle \overset{\vee}{p} \right\rangle \text{ for } u \gg 1,\ \beta < 1. \qquad (9.28)$$

For instance, if $\beta = \pi/8$ and $u = 4$, then the $\left\langle \overset{\vee}{p} \right\rangle$-value is about 4 times larger than $\left\langle \overset{\circ}{p} \right\rangle$, cf. curves 1 and 2 in Fig. 9.3a. Consequently,

---

We may expect that two-humped absorption curve may appear in the case of a deep rectangular potential well, in which the low-frequency peak is due to R-molecules, rotating with energies above the potential u, and the high-frequency peak - due to librators, moving near the bottom of the well. (9.29)

THE COLLISION-TIME PARAMETER y_c. Now we shall calculate the mean normalized frequency of rotational motion as the sum of corresponding values, found for two subensembles:

$$\langle p\rangle = \left\langle \overset{\vee}{r} <\overset{\vee}{p}> + \overset{\circ}{r} <\overset{\circ}{p}> \right\rangle = C\left[\int_{\overset{\vee}{\Gamma}} \overset{\vee}{p}\, d\Gamma + \int_{\overset{\circ}{\Gamma}} \overset{\circ}{p}\, d\Gamma\right]$$

$$= C\left\{ j\int_0^u \exp(-h)\,dh \int_0^{\sqrt{h}\sin\beta} dl \int_0^{\overset{\vee}{\Phi}} \frac{\pi}{\overset{\vee}{\Phi}}\, d\varphi \right.$$

$$\left. + \int_u^\infty \exp(-h)\,dh \int_0^{\sqrt{h-u}} dl \int_0^{\overset{\circ}{\Phi}} \frac{2\pi}{\overset{\circ}{\Phi}}\,d\varphi \right\}$$

$$= C\left[j\pi\sin\beta \int_0^u \sqrt{h}\exp(-h)\,dh + 2\pi\int_u^\infty \sqrt{h-u}\,\exp(-h)\,dh\right]$$

$$= \frac{j\sin\beta\left[\dfrac{\sqrt{\pi}}{2}\,\mathrm{erf}\sqrt{u} - \sqrt{u}\,\exp(-u)\right] + \sqrt{\pi}\,\exp(-u)}{j\left[1 - \cos\beta + \left(\cos\beta + \delta_j\right)\exp(-u)\right]}. \tag{9.30}$$

For our most interesting case ($\delta = 0, j = 2$) we have:

$$\langle p\rangle = \frac{\sin\beta\left[\dfrac{\sqrt{\pi}}{2}\,\mathrm{erf}\sqrt{u} - \sqrt{u}\,\exp(-u)\right] + \dfrac{\sqrt{\pi}}{2}\exp(-u)}{1 - \left[1 - \exp(-u)\right]\cos\beta}. \tag{9.30a}$$

The reciprocal value, divided by π, we define as the *collision-time* parameter y_c:

$$y_c = \frac{\eta}{\tau_c} = \frac{1}{\pi}\left\langle \check{r} < \check{p} > + \mathring{r} < \mathring{p} > \right\rangle = \frac{\langle p\rangle}{\pi}. \tag{9.31}$$

Using this expression, we define time τ_c. The latter means the reciprocal semi-period of rotational / librational motion.

The average half-period of reorientation may serve as the measure of a collision time of particles in condensed media; this time is affected by the conservative potential U.

The parameter y_c is thus proportional to the reciprocal of the collision time. We shall see that strong collisions are usually so frequent in a condensed medium, that the mean collision frequency y exceeds y_c. In other terms, the mean lifetime τ is less than the time of collision τ_c, introduced just above. It follows from Fig. 9.3c that:

(a). y_c increases with the rise of u.

(b). For a rather deep well ($u >> 1$) and small β (20 - 45°) the y_c value falls into the interval 0.5 - 1 . Consequently,

---

We may expect that in the case of condensed medium the normalized collision frequency y_c *should exceed* 0.5 - 1. (9.32)

THE DISPERSION σ. In view of Eq. 9.14, 9.15 we have for the *DWP potential configuration*, where $j = 2$ and $\langle q \rangle = 0$:

$$\sigma_{\|} = \frac{(1-\cos\beta)\left(1+\cos\beta+\cos^2\beta\right)+\exp(-u)\cos^3\beta}{1-\left[1-\exp(-u)\right]\cos\beta}, \tag{9.33a}$$

$$\sigma_{\perp} = \frac{1}{2}\,\frac{(1-\cos\beta)\left(2-\cos\beta-\cos^2\beta\right)+\exp(-u)\cos\beta\left(3-\cos^2\beta\right)}{1-\left[1-\exp(-u)\right]\cos\beta}. \tag{9.33b}$$

For the *SWP configuration*, where $\langle q \rangle \neq 0$, we have:

$$\sigma_{\|} = \frac{\left[(1-\cos\beta)^2 - \exp(-u)(1+\cos\beta)^2\right]^2 + 16\exp(-u)\sin^2\beta}{4\left[1-\cos\beta+(1+\cos\beta)\exp(-u)\right]^2}, \tag{9.34a}$$

$$\sigma_{\perp} = \frac{(1-\cos\beta)^2(2+\cos\beta)+\exp(-u)(1+\cos\beta)^2(2-\cos\beta)}{2\left[1-\cos\beta+(1+\cos\beta)\exp(-u)\right]}. \tag{9.34b}$$

For a particular case $u = 0$, $\beta = \pi$ expressions (9.33)-(9.34) yield the same result: $\sigma_{\|} = \sigma_{\perp} = 1$.

9.2 The Spectral Functions *K(z)* and *L(z)*

We can find the SFs of the hybrid model by using the following rules:

$$K(z) = \overset{\vee}{K}(z) + \overset{\circ}{K}(z), \tag{9.35}$$

$$\overset{\vee}{K}(z) = jC\int_0^u \overset{\vee}{f}(\bullet)\, h\exp(-h)\,dh,\quad \overset{\circ}{K}(z) = C\int_u^\infty \overset{\circ}{f}(\bullet)\, h\exp(-h)\,dh \tag{9.36}$$

instead of

$$K(z) = C_{\mathrm{CCR}}\int_0^\infty \overset{\vee}{f}(\bullet)\, h\exp(-h)\,dh \quad \text{for the CCR model,} \tag{9.37a}$$

$$K(z) = C_{\text{ED}} \int_0^{\infty} \overset{\circ}{f}(\bullet)\, h \exp(-h)\, dh \quad \text{for the ED model,} \tag{9.37b}$$

where

$$\frac{C}{C_{\text{CCR}}} = \gamma\,(1 - \cos\beta); \; \frac{C}{C_{\text{ED}}} = \gamma \; . \tag{9.38}$$

For the DWP configuration, in accordance with Eq.9.14,

$$\gamma = 2\pi C = \frac{1}{1 - \cos\beta\left[1 - \exp(-u)\right]} \, . \tag{9.39}$$

Further we consider for simplicity the small β approximation (see Sec.8.5.6), which however is applicable, at least for a qualitative investigation, also for $\beta < \pi/2$.

To find the *SF of L-molecules* we, in accordance with the first relation in Eq. 9.36, use the following substitutions:

$$\int_0^{\infty} \frac{h \exp(-h)\, dh}{h + \check{Z}} \quad \rightarrow \quad \int_0^{u} \frac{h \exp(-h)\, dh}{h + \check{Z}} \, , \tag{9.40}$$

$$\check{Z} = -\left(\frac{\beta z}{\pi}\right)^2 \text{ for } K_{\|}(z) \text{ and } \check{Z} = -\left(\frac{2\beta z}{\pi}\right)^2 \text{ for } K_{\perp}(z) \; .$$

Thus, in the DWP-configuration case we may write expressions for longitudinal and transverse SFs of librators. Taking account of Eq. 8.119, 8.120, 9.38 - 9.40, we have:

$$\check{K}_{\|}(z) = \frac{12}{\pi^5}\, \gamma\, \beta^4 \sin^2\beta \int_0^{u} \frac{h \exp(-h)\, dh}{h - \left(\frac{\beta z}{\pi}\right)^2} \, , \tag{9.41}$$

$$\check{K}_{\perp}(z) = \frac{96}{\pi^5}\, \gamma\, \beta^4 \cos^2\beta \int_0^{u} \frac{h \exp(-h)\, dh}{h - \left(\frac{2\beta z}{\pi}\right)^2} \, . \tag{9.42}$$

If $u = \infty$, $\gamma \to \dfrac{1}{1-\cos\beta}$. Then Eq. 9.41 and 9.42 reduce to Eq. 8.119 and 8.120, respectively, as it should be.

To find the *SF of R-molecules*, we account for Eq. 7.11 *for the ED model,* where also $p = \sqrt{h}$, $\overset{\circ}{K}(z) = \overset{\circ}{K}_{\parallel}(z) = \overset{\circ}{K}_{\perp}(z)$;

$$\overset{\circ}{K}(z) = \gamma \exp(-u)\; L_F(z) = \gamma \exp(-u) \int_0^{\infty} \frac{h \exp(-h)\, dh}{h - z^2}; \qquad (9.43)$$

$$\frac{1}{3} \overset{\circ}{K}_{\parallel}(z) = \frac{1}{2\pi} \int_0^{\infty} p \exp(-h)\; dh \int_0^{\frac{\pi}{2}} p \sin\Theta\; \pi\, \frac{\sin^2\Theta\, d\Theta}{p^2 - z^2} = \frac{L_F(z)}{3}. \qquad (9.44)$$

For the case of *the hybrid model* we replace, by analogy with Eq. 9.26, the angular momentum p by $\sqrt{h-u}$, the lower limit 0 by u and the constant $\dfrac{1}{2\pi}$ by $C = \dfrac{\gamma}{2\pi}$. Then

$$\frac{1}{3} \overset{\circ}{K}_{\parallel}(z) = \frac{\gamma}{2\pi} \int_u^{\infty} \sqrt{h-u}\, \exp(-h)\; dh \int_0^{\frac{\pi}{2}} \sqrt{h-u}\;\; \pi\, \frac{\sin^3\Theta\, d\Theta}{(h-u)^2 - z^2}\;.$$

Substituting the variable of integration h by $h + u$, we get Eq. 9.43. Noting Eq. 9.22, we remark that

$$\overset{\circ}{K}_{\parallel}(z) = \overset{\circ}{r}\, L_F(z). \qquad (9.45)$$

The SF of R-molecules is equal to their proportion $\overset{\circ}{r}$ *multiplied by the SF L(z) of the ED model. Consequently, dielectric behavior of this molecular subensemble is the same, as that of free rotors, since the law of motion of R-molecules is not affected by the potential well, although their population decreases with the rise of u.* (9.46)

This property corresponds to the independence of the mean rotational frequency $\left\langle \overset{\circ}{p} \right\rangle$ on the parameter u, see Eq. 9.26. For $u = 0$ expression (9.45) reduces to Eq. 7.11.

The SF of an isotropic medium is given by

$$L(z) = \frac{1}{3}\left[\overset{\vee}{K}_{\parallel}(z) + 2\overset{\vee}{K}_{\perp}(z)\right] + \overset{\circ}{K}(z)\ . \tag{9.47}$$

9.3 Evolution of Dielectric Spectra with the Rise of the Rectangular Potential

We consider the double well potential (DWP) configuration, that is characterized by two wells with the adjacent apexes of cones and with common symmetry axis (see Fig.9.1), so that

$$\langle q \rangle = 0,\ \Delta \equiv \left\langle q^2 \right\rangle \quad \text{and} \quad \sigma = 3\Delta \equiv 3\left\langle q^2 \right\rangle \quad \text{for DWP.} \tag{9.48}$$

Taking the Gross collision model, we have for the *longitudinal* orientation of alternating field:

$$\frac{\overset{*}{\chi}_{\parallel}(x)}{G} = \frac{z\left[\overset{\vee}{K}_{\parallel}(z) + \overset{\circ}{K}(z)\right]}{x + iy\dfrac{\overset{\vee}{K}_{\parallel}(z) + \overset{\circ}{K}(z)}{\sigma_{\parallel}}}\ . \tag{9.49a}$$

For the *transverse* orientation

$$\frac{\overset{*}{\chi}_{\perp}(x)}{G} = \frac{z\left[\overset{\vee}{K}_{\perp}(z) + \overset{\circ}{K}(z)\right]}{x + iy\dfrac{\overset{\vee}{K}_{\perp}(z) + \overset{\circ}{K}(z)}{\sigma_{\perp}}}\ . \tag{9.49b}$$

For an *isotropic medium* the SF $L(z)$ is given by Eq. 9.47. Then we can calculate the complex susceptibility from the expression

$$\frac{\chi^*(x)}{G} = \frac{z\,L(z)}{x + iy\,L(z)} \qquad \text{for DWP;} \tag{9.50}$$

$$\sigma_{\|} = 3\Delta_{\|} = 3\left\langle q_{\|}^2 \right\rangle; \ \sigma_{\perp} = 3\Delta_{\perp} = 3\left\langle q_{\perp}^2 \right\rangle . \tag{9.51}$$

The frequency dependence of normalized absorption $x\chi''/G$ is illustrated for the Poley absorption region by Fig. 9.4. Fig. 9.5 shows the loss spectrum $\chi''(x)/G$ in all orientational-polarization region. In Fig. 9.4a we plot spectra for two limiting cases, viz. related to the ED model ($u = 0$, curve 1) and to the CCR model ($u \to \infty$, curves 2, 3, 4 respectively for $\|$-orientation, isotropic ensemble, and $\perp$-orientation). In Fig. 9.4 b, c, d we show plots for the hybrid model. If $u = 1$ (curves 5), absorption spectra, calculated for the HM, are close to those, obtained for the ED model (curve 1 in Fig. 9.4a). For large u ($u = 6$, curves 7) the spectra are close to those, calculated for the CCR model. For $u = 4$ (curves 6) we have some intermediate case. Thus,

---

In the hybrid model the maximal absorption frequency x_L strongly depends on the barrier u. In the ED model x_L is placed near $x = 1$, while in the CCR model it depends only on the amplitude β. (9.52)

For $u = 1$ the loss curve $\chi''(x)$ is close in the hybrid model to that in the ED model (cf. curves 5 in Fig. 9.5 b, c, d with curve 1 in Fig. 9.5a). The low-frequency (Debye) relaxation does not appear in this shallow-well case. However, for *deeper* potentials (for $u = 4$ and especially for $u = 6$) the Debye loss maximum arises. Note, that for all u values the loss maximum is placed in the Poley region for $\perp$-orientation (near $x = 1$, Fig. 9.5 d), while in the basic CCR model this maximum is shifted to much lower x (near $x = 0.01$, curve 3 in Fig. 9.5 a). In Fig. 9.6 we picture the u-dependence of the Debye relaxation time, estimated for an isotropic medium from the expression

$$\tau_D = \frac{\eta}{y\,L(iy)} , \tag{9.53}$$

given in Table 6.2. For smaller β value (curve 1) the decrease of τ_D with the decrease of potential u is more pronounced, than for larger amplitude β (curve 2).

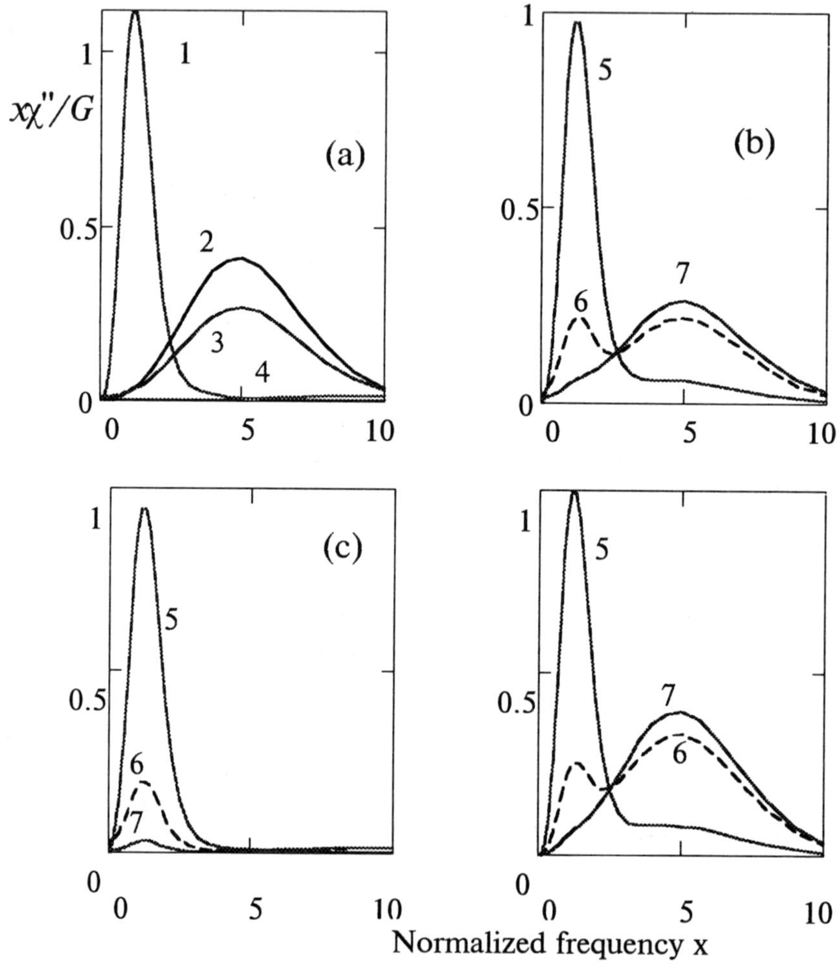

Fig. 9.4 Normalized absorption vs. x, calculated for $y = 0.2$, $\beta = \pi/8$; the double well potential configuration. **a:** for the ED model (upper curve 1) and for the CCR model, ⊥-orientation (curve 2), isotropic medium (curve 3), ||-orientation (lower curve 4, near the abscissa axis); **b, c, d:** for the hybrid model; an isotropic ensemble (b), ||-orientation (c), ⊥-orientation (d); $u = 1$ (curves 5), $u = 4$ (curves 6), and $u = 6$ (curves 7).

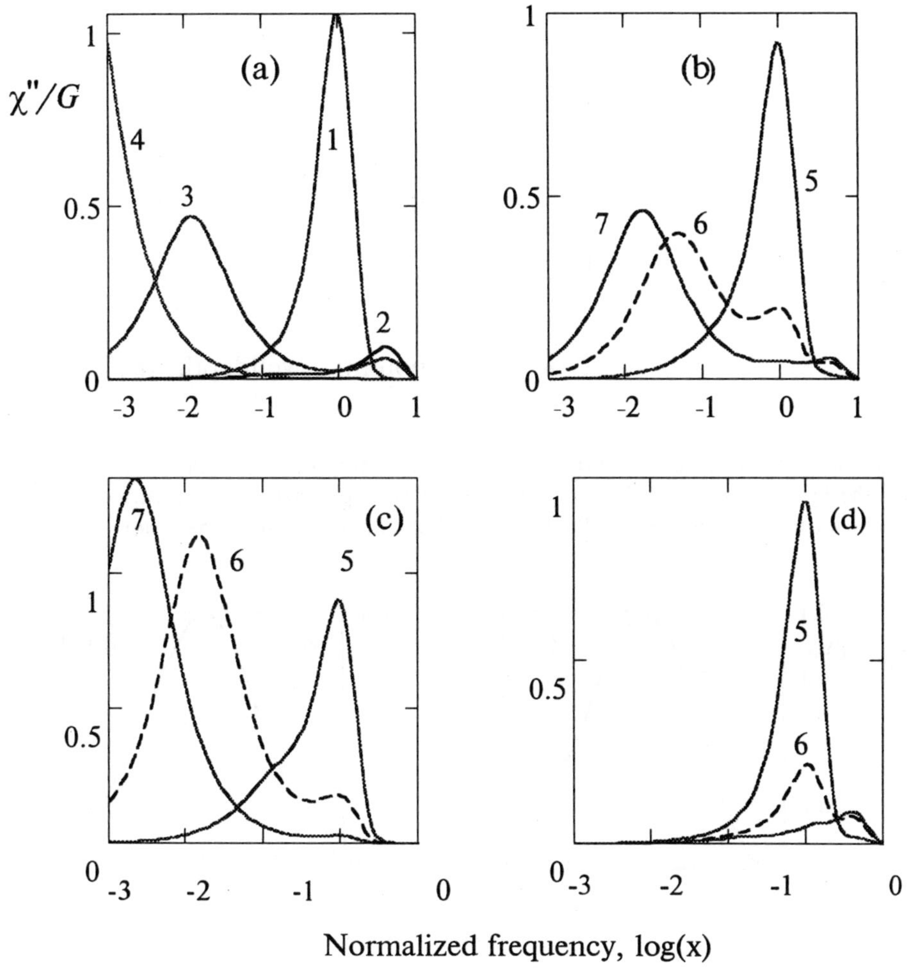

Fig. 9.5 Normalized loss vs. x dependence. All notations as in Fig. 9.4.

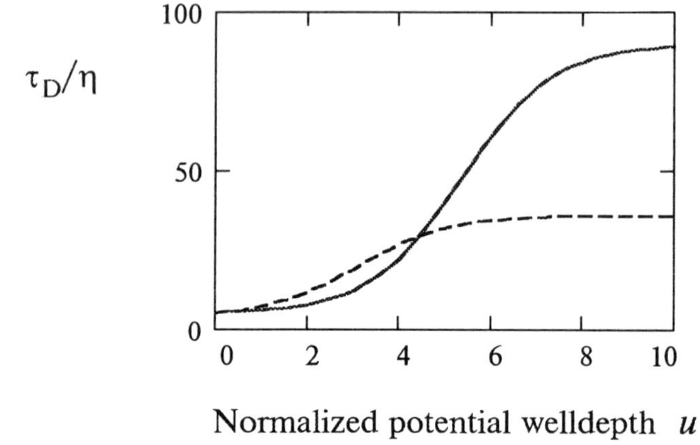

Fig. 9.6 The dependence of the relaxation time on the potential welldepth u. Estimated for the hybrid model (DWP configuration). An isotropic medium. Normalized collision frequency $y = 0.2$, libration amplitude $\beta = \pi/8$ (solid line) and $\pi/4$ (dashed line).

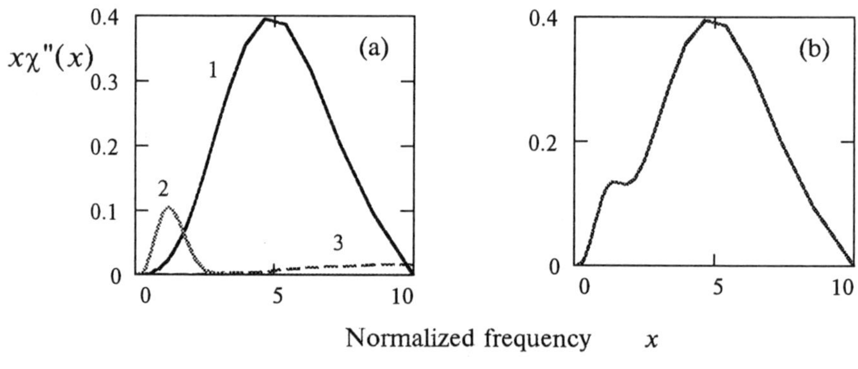

Fig. 9.7 The frequency dependence of the normalized absorption in a rare collision limit. The hybrid model, $\beta = \pi/8$ and $u = 5$. **a:** The contribution to absorption due to subensembles of transverse librators (1), of rotors (2) and of longitudinal librators (3). **b:** The normalized absorption in an isotropic medium.

THE TWO-HUMPED ABSORPTION CURVE. The new interesting phenomenon is related to two subensembles, in which molecules rotate / librate with different frequencies (see estimations 9.25, 9.26 and Fig. 9.3 a, b). For low ($u = 1$) and deep ($u = 6$) potentials absorption curves 5, 7 in Fig. 9.4 have the only maximum. However, for an intermediate potential welldepth, say for $u = 4$ (see curves 6 in Fig. 9.4 b, d) a two-humped absorption arises: now the contributions of librators and rotators become commensurable in some part of the Poley absorption region. This case is typical for liquid water and for many aqueous systems (see Chap. 15).

THE LIMITING LINES IN THE HYBRID MODEL. We consider the long lifetime limit ($y = 0$), when absorption due to each molecular subensemble, is characterized by the minimal bandwidth. Taking account of Eq. 8.100 - 8.103, we find that, if $y = 0$, then

$$\lim_{y \to 0} \left\{ \mathrm{Im} \int_0^\infty \frac{f(h)\, dh}{h - a^2 z^2} \right\} = \pi \, \mathrm{Im}\left[f(a^2 x^2) \right];$$

$$\lim_{y \to 0} \left\{ \mathrm{Im} \int_0^u \frac{f(h)\, dh}{h - a^2 z^2} \right\} = \pi \left\{ \begin{array}{c} \mathrm{Im}\left[f(a^2 x^2), \ x \le \dfrac{\sqrt{u}}{a} \right] \\ 0, \ x \ge \dfrac{\sqrt{u}}{a} \end{array} \right\}.$$

Using this expressions, we obtain from Eq. 9.41 - 9.43:

$$\check{K}_{\|}''(x) = \frac{12}{\pi^4} \gamma \beta^4 \sin^2 \beta \, (\beta x / \pi)^2 \exp\left[-(\beta x / \pi)^2 \right] \text{ for } x \le \frac{\pi \sqrt{u}}{\beta}; \quad (9.54a)$$

$$\check{K}_{\|}''(x) = 0 \text{ for } x > \frac{\pi \sqrt{u}}{2\beta};$$

$$\check{K}_{\perp}''(x) = \frac{96}{\pi^4} \gamma \beta^4 \cos^2 \beta \, (2\beta x / \pi)^2 \exp\left[-(\beta x / \pi)^2 \right] \text{ for } x \le \frac{\pi \sqrt{u}}{2\beta}; \quad (9.54b)$$

$$\check{K}_{\perp}''(x) = 0 \text{ for } x > \frac{\pi \sqrt{u}}{2\beta};$$

$$\overset{\circ}{K}''(z) = \pi\gamma \exp(-u) x^2 \exp(-x^2). \tag{9.54c}$$

Thus, the absorption curve due to librators *is cut at some maximal frequency* while for rotators the similar curve is continuous.

In Fig. 9.7a we take the u-value, which is about that we employ for liquid water ($u = 5$). A two-humped absorption appears in an isotropic medium, that resembles curve 6 in Fig. 9.4b.

9.4 The Hybrid Model 2 (HM2)

Here we replace expressions (9.41)-(9.43) for the spectral function of the hybrid model by taking as a basis Eq. 8.113 that are applicable for *any* β. We shall find dielectric response due to low-energy dipoles that reflect from the walls of a conical surface. For definiteness we shall consider only the case of two opposite wells, i.e. the DWP configuration.

9.4.1 The Spectral Function of Librators: a Rigorous Description

INTERVALS OF INTEGRATION OVER ENERGY h AND ANGULAR MOMENTUM l. In view of Eq. 3.53 and Fig. 8.8 the normalized kinetic energy $\dot{\theta}^2$ due to azimuthal deviation is related to the normalized energy h and to angular momentum projection l on the symmetry axis Z of a cone:

$$\dot{\theta}^2 = h - \frac{l^2}{\sin^2\theta} \quad \text{for } \theta < \beta, \tag{9.55a}$$

$$\dot{\theta}^2 = h - \frac{l^2}{\sin^2\theta} - u \quad \text{for } \theta \geq \beta, \tag{9.55b}$$

θ being angle between Z and the dipole moment vector μ; $0 \leq \theta \leq \beta$ in our case of librators.

For $\theta < \beta$, i.e. for $U(\theta) \equiv 0$, the minimal energy is given by

$$h_{\min} = \frac{l^2}{\sin^2\theta}. \tag{9.56}$$

Otherwise it follows from Eq. 9.55a, that if it were $h < h_{min}$, then the kinetic energy $\dot{\theta}^2$ will be negative for any $\theta < \beta$.

On the other side, the largest energy of a librator is

$$h_{max} = \frac{l^2}{\sin^2\theta} + u \ , \tag{9.57}$$

since in view of Eq. 9.55b only dipoles with energy $h < h_{max}$ will reflect from the surface of a cone.

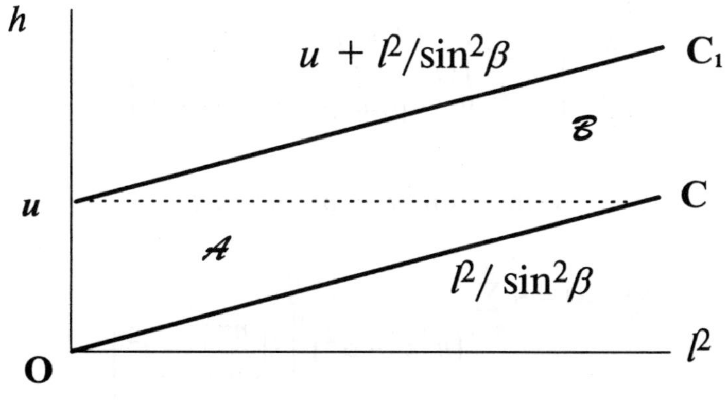

Fig. 9.8 The form of the phase region, occupied by librators.

Thus, for this refined hybrid model (that we term HM2, viz. the hybrid 2 model) the intervals of integration over l and h may be chosen as follows:

$$\frac{l^2}{\sin^2\theta} \le h \le \frac{l^2}{\sin^2\theta} + u; 0 \le l^2 \le \infty \ . \tag{9.58}$$

The phase diagram of this ensemble is pictured in Fig. 9.8. For the sake of clearness we draw l^2 (instead of l) as an abscissa axis. The lower border of this region corresponds to the minimal energy h (9.56), the upper border - to the maximal h.

THE SPECTRAL FUNCTION. The transition from the "pure" cone confined model to the hybrid model obeys the rule (9.36) with γ, given by Eq.9.39, where we put

$$j = 2,\ jC = 2\gamma\,(1 - \cos\beta)\, C_{CCR} = \frac{1 - \cos\beta}{E};$$
$$E \equiv 1 - \left[1 - \exp(-u)\right]\cos\beta \tag{9.59}$$

Thus, substituting the multiplier $\dfrac{1}{1-\cos\beta}$ in Eq. 8.94 by $\dfrac{1}{E}$ and accounting for the h-interval in (9.58), we have:

$$\begin{Bmatrix} \check{K}_{\|} \\ \check{K}_{\perp} \end{Bmatrix} = \frac{3\pi}{E}\int_0^{\infty} dl \int_{(l/\sin\beta)^2}^{(l/\sin\beta)^2+u} \begin{Bmatrix} f_{\|}\sin^2\beta, \\ f_{\perp}\cos^2\beta \end{Bmatrix}$$
$$\times\, \alpha \sin^2\bar{\theta}\sin^2\lambda\, \exp(-h)\sqrt{h}\, dh. \tag{9.60}$$

$$f_{\|}(h,\alpha) = 4\sum_{n=1}^{\infty} \frac{n^2}{\left(n^2\pi^2 - \alpha^2\right)^2 \left[h\left(\frac{n\pi}{\alpha}\right)^2 - z^2\right]}, \tag{9.61a}$$

$$f_{\perp}(h,\alpha,\lambda) = \sum_{n=-\infty}^{\infty} \frac{\left(n - \frac{\lambda}{\pi}\right)^2}{\left((n\pi - \lambda)^2 - \alpha^2\right)^2 \left[h\left(\frac{(n\pi - \lambda)}{\alpha}\right)^2 - z^2\right]}. \tag{9.61b}$$

Now we introduce variables

$$t = \frac{l}{\sin\beta} \text{ and } g = h - \left(\frac{l}{\sin\beta}\right)^2 = h - t^2 \tag{9.62}$$

instead of l and h, respectively. In view of relations (8.59), (8.61) we express after some trigonometry multiplier Λ as function of variables t, g:

$$\sin^2\bar\theta = 1-\frac{l^2}{h} = 1-\frac{l^2}{g + l^2/\sin^2\beta} = \frac{l^2\cos^2\beta+g\sin^2\beta}{g\sin^2\beta+l^2},$$

$$\cos^2\alpha = \cos^2\beta\ \frac{l^2+g\sin^2\beta}{g\sin^2\beta+l^2\cos^2\beta},$$

$$\sin^2\alpha = \frac{g\sin^4\beta\cos^2\beta}{g\sin^2\beta+l^2\cos^2\beta},\ \sin^2\lambda = \frac{g\sin^2\beta\cos^2\beta}{g\sin^2\beta+l^2\cos^2\beta},$$

$$\sin^2\bar\theta\sin^2\lambda = 1-\frac{h}{l^2\sin^2\beta} = \frac{g\sin^2\beta}{g\sin^2\beta+l^2} = \frac{g}{g+t^2}. \tag{9.63}$$

We recall the Reader that $\bar\theta$ is angle between Z cone axis and angular momentum vector; $\bar\theta$ should not be confused with θ in Eq. 9.55 !.

Thus,

$$\Lambda(g,t) = \alpha\sqrt{g+t^2}\sin^2\bar\theta\sin^2\lambda = \frac{g\,\alpha(g,t)}{\sqrt{g+t^2}}, \tag{9.64a}$$

$$\alpha(g,t) = \mathrm{a}\sin\frac{\sqrt{g}\,\sin\beta}{\sqrt{g+t^2\cos^2\beta}},\ \lambda(g,t) = \mathrm{a}\sin\frac{\sqrt{g}}{\sqrt{g+t^2\cos^2\beta}}. \tag{9.64b}$$

Finally, we get from Eq. 9.62 double integrals with limits, that do not depend on variables of integration:

$$\begin{Bmatrix}\check{K}_\parallel \\ \check{K}_\perp\end{Bmatrix} = \frac{3\pi}{E}\sin\beta\int_0^\infty \exp(-t^2)dt\int_0^u \begin{Bmatrix}F_\parallel \\ F_\perp\end{Bmatrix}\exp(-g)\,dg, \tag{9.65}$$

$$F_\parallel = 4\,\Lambda(g,t)\sin^2\beta\sum_{n=1}^{\infty}\frac{n^2}{\left(n^2\pi^2-\alpha^2\right)^2\left[\left(g+t^2\right)(n\pi/\alpha)^2-z^2\right]}, \tag{9.66a}$$

$$F_{\perp} = \Lambda(g,t)\cos^2\beta \sum_{n=-\infty}^{\infty} \frac{\left(n-\dfrac{\lambda}{\pi}\right)^2}{\left((n\pi-\lambda)^2-\alpha^2\right)^2\left[\left(g+t^2\right)\left(\dfrac{(n\pi-\lambda)}{\alpha}\right)^2-z^2\right]} \tag{9.66b}$$

If the potential barrier is extremely deep, the upper limit in inner integral may be substituted by ∞ and constant E - by $\left(1-\cos\beta\right)^{-1}$.

To simplify calculations one may substitute some variables by their mean values. Let us calculate these averages.

9.4.2 Librational Subensemble: Averaging of some Parameters

We shall calculate mean values of l^2, α, λ and mean proportion $\check{r}$ of librators. We introduce, just like in Sec. 9.1, statistic integral

$$\check{\mathrm{st}} = \int_{\check{\Gamma}} \exp(-h)\, d\check{\Gamma}\ . \tag{9.67}$$

The mean (over subensemble) value of any function $f(\Gamma)$ is given by

$$\left\langle f(\check{\Gamma})\right\rangle \equiv \frac{1}{\check{\mathrm{st}}}\int_{\check{\Gamma}} \exp(-h)\, f(\Gamma) d\Gamma\,. \tag{9.68}$$

Let $\check{\Gamma}$ be formed by two regions, $\mathcal{A}$ and $\mathcal{B}$, see Fig. 9.8:

$$0 \le h \le u;\quad 0 \le l \le l_{\max};\ l_{\max}(h) = \sqrt{u}\sin\beta,\quad \{\text{in } \mathcal{A}\}, \tag{9.69a}$$

$$\left|\cos\bar{\theta}\right|_{\min} = \frac{l_{\min}(h)}{\sqrt{h}} = \sqrt{1-\frac{u}{h}}\sin\beta, \ \left|\cos\bar{\theta}\right|_{\max} = \sin\beta \ . \tag{9.69b}$$

By analogy with Eq.8.71a, where we put $u(\Gamma) = 0$, we find from Eq. 9.91-9.94:

$$\overset{\vee}{\mathrm{st}} = 2j\left(\mathfrak{I}_A + \mathfrak{I}_B\right) , \tag{9.70}$$

$$\mathfrak{I}_A = \int_0^u \exp(-h)\,dh \int_0^{\sin\beta} \alpha\left(\cos\bar{\theta}\right) d\cos\bar{\theta} = \left[1-\exp(-u)\right]\mathfrak{I}_C , \tag{9.71a}$$

$$\mathfrak{I}_B = \int_u^\infty \exp(-h)\,dh \int_{\sqrt{1-\frac{u}{h}}\sin\beta}^{\sin\beta} \alpha\left(\cos\bar{\theta}\right) d\cos\bar{\theta} , \tag{9.71b}$$

$$\mathfrak{I}_C = \int_0^{\sin\beta} \alpha\left(\cos\bar{\theta}\right) d\cos\bar{\theta} = \int_0^{\frac{\pi}{2}} \cos^2\bar{\theta}\, d\lambda = \frac{\pi}{2}\,(1-\cos\beta). \tag{9.71c}$$

Taking account of relations

$$\alpha\left(\cos\bar{\theta}\right) = \mathrm{a}\cos\left(\frac{\cos\beta}{\sqrt{1-\cos^2\bar{\theta}}}\right) , \tag{9.72}$$

$$\alpha\left(\sqrt{1-\frac{u}{h}}\sin\beta\right) = \mathrm{a}\cos\left(\frac{\cos\beta}{\sqrt{\cos^2\beta + \dfrac{u\sin^2\beta}{h}}}\right) , \tag{9.73}$$

we take the second integral in (9.70) by parts:

$$\Im_B = \left[\exp(-h) \int_{\sqrt{1-\frac{u}{h}}\,\sin\beta}^{\sin\beta} \alpha\left(\cos\bar{\theta}\right)\, d\cos\bar{\theta} \right]_{h=\infty}^{h=u}$$

$$- \int_u^{\infty} \exp(-h)\alpha\left(\sqrt{1-\frac{u}{h}}\,\sin\beta\right) \frac{u\sin\beta\, dh}{2h^2\sqrt{1-\frac{u}{h}}}$$

$$= \Im_C \exp(-u) - \frac{u\sin\beta}{2} \int_u^{\infty} a\cos\left[\frac{\cos\beta}{\sqrt{\cos^2\beta + \frac{u\sin^2\beta}{h}}}\right] \frac{\exp(-h)\,dh}{h\sqrt{h(h-u)}}. \quad (9.74)$$

Finally, changing variable h by $t^2 + u$, we have from Eq. 9.70-9.72:

$$\frac{\overset{\vee}{\mathrm{st}}}{\pi j} = 1 - \cos\beta - \frac{2}{\pi} u \exp(-u) \sin\beta \int_0^{\infty} a\cos\left[\frac{\sqrt{u+t^2}\,\cos\beta}{\sqrt{t^2\cos^2\beta + u}}\right] \frac{\exp(-t^2)\,dt}{\left(u+t^2\right)^{3/2}}. \tag{9.75}$$

For the double potential configuration, in which $j = 2$, the proportion $\overset{\vee}{r}$ of librators is simply

$$\overset{\vee}{r} = \frac{\overset{\vee}{\mathrm{st}}}{2\pi E} = \frac{\overset{\vee}{\mathrm{st}}}{2\pi\left(1 - \cos^2\beta\,[1 - \exp(-u)]\right)}, \tag{9.76}$$

where $\overset{\vee}{\mathrm{st}}/(2\pi)$ is given by the r.h.s. of Eq. 9.75.

THE MEAN SQUARE $\left\langle l^2 \right\rangle$ *OF THE ANGULAR MOMENTUM PROJECTION.* We employ the same procedure, as in previous subsection. Since $l^2 = h\cos^2\bar{\theta}$, we have:

$$\langle l^2 \rangle = \frac{1}{\overset{\vee}{\mathrm{st}}} \int\limits_{\overset{\vee}{\Gamma}} l^2(\overset{\vee}{\Gamma}) \exp(-h)\, d\overset{\vee}{\Gamma},$$

so that

$$\langle l^2 \rangle \overset{\vee}{\mathrm{st}} = j\left(\Re_B + \Re_C\right), \tag{9.77a}$$

$$\Re_C = 2\int_0^u \exp(-h)\, h\, dh \int_0^{\sin\beta} \alpha(\cos\bar\theta) \cos^2\bar\theta\, d\cos\bar\theta, \tag{9.77b}$$

$$\Re_B = 2\int_u^\infty \exp(-h)\, h\, dh \int_{\sqrt{1-\frac{u}{h}}\sin\beta}^{\sin\beta} \alpha(\cos\bar\theta) \cos^2\bar\theta\, d\cos\bar\theta. \tag{9.77c}$$

The inner integral in (9.77b) does not depend on h and can be found with use of integrals 2.563.1 and 3.615.1 in (Gradshtein and Rydzyk 1963, pp. 166, 382):

$$\Im_\theta = 2\int_0^{\sin\beta} \alpha(\cos\bar\theta)\cos^2\bar\theta\, d\cos\bar\theta$$

$$= \frac{2}{3}\int_0^{\frac{\pi}{2}} \left[\frac{\sin^2\beta\cos^2\lambda}{1-\sin^2\beta\sin^2\lambda}\right]^2 d\lambda = \frac{\pi}{6}(1-\cos\beta)^2(2+\cos\beta). \tag{9.78}$$

Indeed, in view of the identity

$$\sin^4\beta\cos^4\lambda \equiv \left(1-\sin^2\beta\sin^2\lambda\right)^2 - 2\cos^2\beta\left(1-\sin^2\beta\sin^2\lambda\right) + \cos^4\beta,$$

Eq. 9.106 may be reduced to simpler form:

$$\Im_\theta = \frac{2}{3}\int_0^{\frac{\pi}{2}} \left[1 - \frac{2\cos^2\beta}{\left(1-\sin^2\beta\sin^2\lambda\right)^2} + \frac{\cos^4\beta}{1-\sin^2\beta\sin^2\lambda}\right] d\lambda\ .$$

Since

$$\int_0^{\frac{\pi}{2}} \frac{d\lambda}{\left(1-\sin^2\beta\sin^2\lambda\right)^2} = \frac{2-\sin^2\beta}{2\cos^2\beta}\int_0^{\frac{\pi}{2}} \frac{d\lambda}{1-\sin^2\beta\sin^2\lambda} \quad ,$$

we find:

$$\Im_\theta = \frac{2}{3}\left[\frac{\pi}{2} - \frac{\cos^2\beta}{2}\left(2+\sin^2\beta\right)\int_0^{\frac{\pi}{2}} \frac{d\lambda}{1-\sin^2\beta\,\sin^2\lambda}\right]$$

$$= \frac{2\pi}{3\cdot 2}\left[1 - \frac{\left(2+\sin^2\beta\right)\cos\beta}{2}\right] = \frac{2\pi}{3\cdot 4}\left(1-\cos^2\beta\right)^2\left(2+\cos\beta\right) .$$

We have got Eq. 9.78. We remark that the ratio $\Im_\theta \big/ \overset{\vee}{\mathrm{st}}$ is equal to the mean value $\left\langle q_\perp^2 \right\rangle_{\mathrm{CCR}}$ in the CCR model, cf. (9.78) with Eq. 8.81b.

Taking integral $\Re_B$ by parts, we have from Eq. 9.77:

$$\overset{\vee}{\mathrm{st}}\left\langle l^2\right\rangle \Big/ j = \Im_\theta\left[1-(u+1)\exp(-u)\right] + \Bigg[\, 2(h+1)\exp(-h)$$

$$\times \int_{\sqrt{1-\frac{u}{h}}\sin\beta}^{\sin\beta} \alpha\left(\cos\bar\theta\right)\cos^2\bar\theta\; d\cos\bar\theta \;\Bigg]_{h=\infty}^{h=u}$$

$$- \int_u^\infty (h+1)u\sin^3\beta\left(1-\frac{u}{h}\right)\frac{\alpha\left(\sqrt{1-\frac{u}{h}}\,\sin\beta\right)}{h\sqrt{h(h-u)}}\,\exp(-h)\,dh .$$

The non-integral term [•] reduces to $\Im_\theta\ (u+1)\exp(-u)$. Hence,

$$\check{\mathrm{st}}\left\langle l^2\right\rangle = \frac{\pi j}{6}\left[(1-\cos\beta)^2(2+\cos\beta) - \frac{12u}{\pi}\sin^3\beta\,\exp(-u)\right]$$

$$\times\int_0^\infty \mathrm{acos}\left[\frac{\sqrt{u+t^2}\,\cos\beta}{\sqrt{t^2\cos^2\beta+u}}\right]\frac{u+t^2+1}{\left(u+t^2\right)^{5/2}}\,t^2\exp(-t^2)\,dt\,. \qquad (9.79)$$

Substituting $\check{\mathrm{st}}$ from Eq. 9.75, we have finally:

$$\left\langle l^2\right\rangle = \frac{1}{6}\Bigg[\ (1-\cos\beta)^2(2+\cos\beta) - \frac{12u}{\pi}\exp(-u)(1+\cos\beta)\sin\beta$$

$$\times\int_0^\infty \mathrm{acos}\left[\frac{\sqrt{u+t^2}\,\cos\beta}{\sqrt{t^2\cos^2\beta+u}}\right]\frac{u+t^2+1}{\left(u+t^2\right)^{5/2}}\,t^2\exp(-t^2)\,dt\ \Bigg]$$

$$\times\left\{1-\frac{2u}{\pi}\cot\frac{\beta}{2}\,\exp(-u)\int_0^\infty \mathrm{acos}\left[\frac{\sqrt{u+t^2}\,\cos\beta}{\sqrt{t^2\cos^2\beta+u}}\right]\frac{\exp(-t^2)\,dt}{\left(u+t^2\right)^{3/2}}\ \right\}^{-1}. \qquad (9.80)$$

THE MEAN ANGLES $\langle\alpha\rangle$ *AND* $\langle\lambda\rangle$. The only distinction from the calculation of integral (9.70) is that now we have in the integrands an additional multiplier $f(\cos\bar{\theta})$, where $f = \alpha(\cos\bar{\theta})$ or $f = \lambda(\cos\bar{\theta})$.

Let $X = \cos(\bar{\theta})$. Then, taking account of Eq.9.72 and of relation $\lambda = \mathrm{a}\sin\left(\frac{\sin\alpha}{\sin\beta}\right)$, we may write:

$$\alpha(X) = \mathrm{acos}\left(\frac{\cos\beta}{\sqrt{1-X^2}}\right),\ \lambda(X) = \mathrm{asin}\left(\frac{1}{\sin\beta}\sqrt{\frac{1-X^2-\cos^2\beta}{1-X^2}}\right). \qquad (9.81)$$

By analogy with Eq. 9.70 - 9.77 we have:

$$\langle f\rangle\check{\mathrm{st}} = j\left(\mathfrak{I}_1+\mathfrak{I}_2\right), \qquad (9.81a)$$

$$\Im_1 = 2\int_0^u \exp(-h)\,dh \int_0^{\sin\beta} \alpha(X)\,f(X)\,dX = 2\left[1-\exp(-u)\right]\Im_f\,, \quad (9.81b)$$

$$\Im_f = \int_0^{\sin\beta} \alpha(X)\,f(X)\,dX\,, \quad (9.81c)$$

$$\Im_2 = 2\int_u^{\infty} \exp(-h)\,dh \int_{\sqrt{1-\frac{u}{h}}\sin\beta}^{\sin\beta} \alpha(X)\,f(X)\,dX \quad (9.81d)$$

$$= 2\left[\exp(-h) \int_{\sqrt{1-\frac{u}{h}}\sin\beta}^{\sin\beta} \alpha(X)\,f(X)\,dX\right]_{h=\infty}^{h=u} - 2\int_u^{\infty} \exp(-h)\,\alpha(Y)\,f(Y)\frac{u\sin\beta\,dh}{h\sqrt{h-u}}\,, \quad (9.82)$$

where $Y \equiv \sqrt{1-\frac{u}{h}}\sin\beta$, so that

$$\alpha(Y) = \mathrm{acos}\left(\frac{\sqrt{h}\cos\beta}{\sqrt{h\cos^2\beta+u\sin^2\beta}}\right), \quad \alpha(Y) = \mathrm{asin}\sqrt{\frac{u}{h\cos^2\beta+u\sin^2\beta}}\,. \quad (9.83)$$

The non-integral and integral parts of Eq.9.82 yield respectively

$$2\exp(-u)\,\Im_f \quad \text{and} \quad -2u\sin\beta\exp(-u)\int_0^{\infty} \alpha_1(t)\,f(t)\frac{\exp\left(-t^2\right)dt}{\left(u+t^2\right)^{3/2}}\,,$$

where for $h \to t^2 + u$,

$$\alpha_1(t) = \mathrm{acos}\left(\frac{\sqrt{u+t^2}\cos\beta}{\sqrt{t^2\cos^2\beta+u}}\right), \quad \alpha_2(t) = \mathrm{asin}\sqrt{\frac{u}{t^2\cos^2\beta+u}}\,. \quad (9.84)$$

Finally,

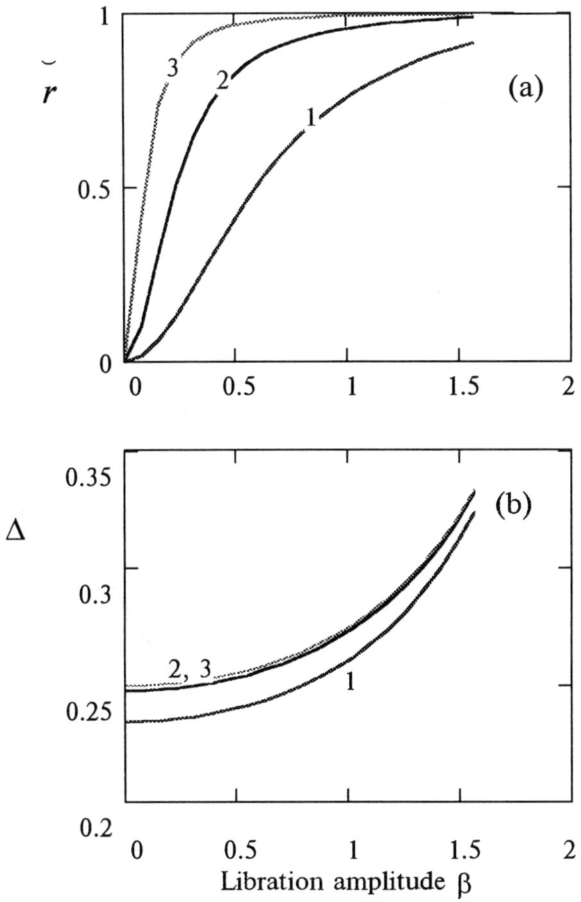

Fig. 9.9 The proportion of librators (a); mean values of normalized precession angular momentum Δ (b) and angles α (c) and λ (d) vs. angle β of a potential conical well. Line A in Fig. c is the ratio $\langle\alpha\rangle/\beta$. Curves 1, 2, 3 for the normalized potential welldepth u = 1.8, 3.6 and 5.4. Horizontal lines in Fig. c and d mark the ordinate value, equal to $\pi/2$.

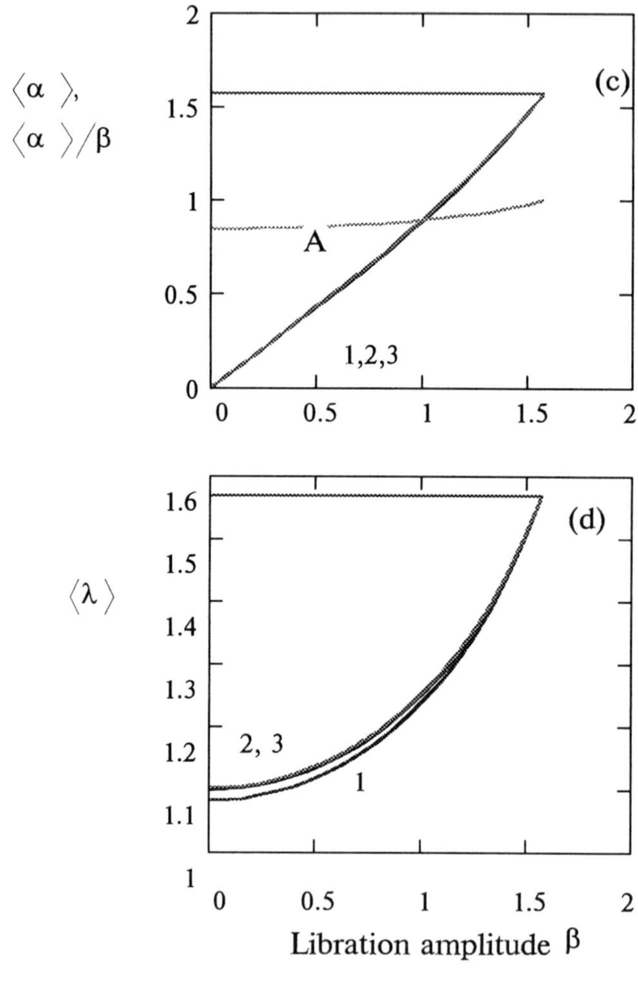

Fig. 9.9 **c, d:** The caption to this Figure see on the previous page.

$$\begin{Bmatrix} \langle\alpha\rangle \\ \langle\lambda\rangle \end{Bmatrix} = \frac{2j}{\check{\mathrm{st}}} \left[\Im_f - \sin\beta u \exp(-u) \int_0^\infty \alpha_1(t) \begin{Bmatrix} \alpha_1(t) \\ \lambda_1(t) \end{Bmatrix} \frac{\exp(-t^2)dt}{\left(u+t^2\right)^{3/2}} \right], \quad (9.85)$$

where $\check{\mathrm{st}}$ is given by Eq. 9.57, $\Im_f$ by Eq. 9.81c and $\alpha_1(t)$, $\lambda_1(t)$ by Eq. 9.84.

ESTIMATION OF THE AVERAGED PARAMETERS (Fig. 9.9). When cone angle β increases with u being constant, the proportion $\check{r}$ of librators also increases (see Fig. 9.9a); this proportion also increases with the rise of u. Such behavior is typical for the hybrid model (see previous sections).

The normalized square of the precession angular momentum, $\Delta = l^2/\sin^2\beta$, slightly increases, when β rises (Fig. 9.9b). However, Δ is not negligibly small.

The mean libration amplitude $\langle\alpha\rangle$ is almost equal to the cone angle β, see Fig. 9.9c. This property means that librations in a diameter plane contribute mainly to this parameter.

The mean "projection" $\langle\lambda\rangle$ of angle $\langle\alpha\rangle$ on horizontal plane is an increasing function of β. For the low β limit the $\langle\lambda\rangle$-value appreciably differs from the value $\lambda = \pi/2$, that was set in the first version of the hybrid model. For large β (viz. for $\beta \to \pi/2$) both averaged angles, $\langle\alpha\rangle$ and $\langle\lambda\rangle$, tend to $\pi/2$. This result corresponds to an obvious picture of reorientation process. The dependence of Δ and especially of the mean angles $\langle\lambda\rangle$ and $\langle\alpha\rangle$ on the potential barrier u is very weak.

9.4.3 Another Representation for the Spectral Function of Librators

Here we shall transform Eq. 9.60-9.61 to another form, that probably is more convenient for numerical integration. Choosing again h and l as variables, we change the order of integration comparing with that, used in Sec.9.4.1. It follows from Fig. 9.8 that the region of integration may be represented as the *difference of two semi-infinite triangles.*

We remark that $h = \left[\frac{l(h)}{\sin\beta}\right]^2$ corresponds to the border OC of the *outer triangle*. Thus, we should take the following intervals of integration:

$$h \in [0, \infty], \;\; l(h) \in [0, \sqrt{h}\,\sin\beta] \;. \tag{9.86a}$$

On the border uC_1 of the *inner triangle*, where $h = \left[\frac{l(h)}{\sin\beta}\right]^2 + u$, we put

$$h \in [u, \infty], \;\; l(h) \in [0, \sqrt{h-u}\,\sin\beta] \;. \tag{9.86b}$$

Thus, we write instead of Eq. 9.60:

$$\begin{Bmatrix} \check{K}_{\|} \\ \check{K}_{\perp} \end{Bmatrix} = \frac{3\pi}{E}\left(\int_0^\infty dh \int_0^{\sqrt{h}\sin\beta} dl \;\; \int_u^\infty dh \int_0^{\sqrt{h-u}\sin\beta} dl\right)$$

$$\times \begin{Bmatrix} f_{\|}\sin^2\beta \\ f_{\perp}\cos^2\beta \end{Bmatrix} \alpha \sin^2\bar{\theta}\,\sin^2\lambda \sqrt{h}\exp(-h)dh \,. \tag{9.87}$$

Now we replace the variable l by $v = \cos\bar{\theta} = l/\sqrt{h}$. Since

$$\alpha(v) = \mathrm{acos}\left(\frac{\cos\beta}{\sqrt{1-v^2}}\right), \;\; \lambda(v) = \mathrm{asin}\left(\frac{1}{\sin\beta}\sqrt{\frac{\sin^2\beta - v^2}{1-v^2}}\right), \tag{9.88a}$$

$$\sin^2\bar{\theta}\,\sin^2\lambda = \frac{\sin^2\beta - v^2}{\sin^2\beta}, \tag{9.88b}$$

we reduce Eq.9.87 to the following double integrals:

$$\begin{Bmatrix} \overset{\vee}{K}_{\|} \\ \overset{\vee}{K}_{\perp} \end{Bmatrix} = \frac{3\pi}{E\sin^2\beta}\left(\int_0^u h\exp(-h)dh\begin{Bmatrix} a_{\|}(h) \\ a_{\perp}(h) \end{Bmatrix}dh + \int_u^{\infty} h\exp(-h)\begin{Bmatrix} a_{\|}(h)-b_{\|}(h) \\ a_{\perp}(h)-b_{\perp}(h) \end{Bmatrix}dh\right), \tag{9.89a}$$

where

$$\begin{Bmatrix} a_{\|}(h) \\ a_{\perp}(h) \end{Bmatrix} = \int_0^{\sin\beta}\left(\sin^2\beta - \mathrm{v}^2\right)\alpha \begin{Bmatrix} f_{\|}(h,\alpha)\sin^2\beta \\ f_{\perp}(h,\alpha,\lambda)\cos^2\beta \end{Bmatrix} d\mathrm{v}, \tag{9.89b}$$

$$\begin{Bmatrix} b_{\|}(h) \\ b_{\perp}(h) \end{Bmatrix} = \int_0^{\sqrt{1-\frac{u}{h}}\sin\beta}\left(\sin^2\beta - \mathrm{v}^2\right)\alpha \begin{Bmatrix} f_{\|}(h,\alpha)\sin^2\beta \\ f_{\perp}(h,\alpha,\lambda)\cos^2\beta \end{Bmatrix} d\mathrm{v}, \tag{9.89c}$$

$\alpha = \alpha(\mathrm{v})$ and functions $f_{\|}(h,\alpha)$, $f_{\perp}(h,\alpha,\lambda)$ are determined by Eq. 9.61.

In the inner integral the interval of integration is finite; this is an important distinction between Eq. 9.89 and 9.65. In Eq. 9.89 this interval is not larger than 1.

9.4.4 *Approximate Representation of the SF in the Form of Ordinary Integrals*

We may transform approximately double integrals over h and v to the product of corresponding ordinary integrals. If we set $f(h,\mathrm{v}) \equiv f(h)V(\mathrm{v})$, then:

$$\int \exp(-h)\,dh \int f(h,\mathrm{v}) \approx \int f_1(h)\exp(-h)\,dh \int V(\mathrm{v})d\mathrm{v}\ .$$

As we have seen in Sec.9.4.2, two variables, α and λ, do not differ much from β and $\pi/2$, respectively. Consequently,

(a) In a first approximation one may replace α by β and λ by $\pi/2$ in series (9.61);

(b) In the *integrands* of Eq. (9.89b) and (9.89c) one may replace variable $\alpha(\mathrm{v})$ by its mean value $\langle\alpha\rangle$.

Then

$$\left\{\begin{matrix}\check{K}_{\|}\\ \check{K}_{\perp}\end{matrix}\right\} \approx \frac{3\pi\langle\alpha\rangle}{E\sin^2\beta}\left[\; V(\sin\beta)\int_0^u h\exp(-h)dh\left\{\begin{matrix}F_{\|}(h)\\ F_{\perp}(h)\end{matrix}\right\}dh\right.$$

$$\left. + \int_u^\infty h\exp(-h)\left\{\begin{matrix}F_{\|}(h)\\ F_{\perp}(h)\end{matrix}\right\}\left\{V(\sin\beta) - V\left(\sqrt{1-\frac{u}{h}}\,\sin\beta\right)\right\}dh\;\right], \tag{9.90}$$

where

$$V(\mathrm{v}) \equiv \int_0^{\mathrm{v}}\left(\sin^2\beta - \mathrm{v}^2\right)d\mathrm{v} = \mathrm{v}\left[\sin^2\beta - \frac{1}{3}\mathrm{v}^2\right], \tag{9.91}$$

so that

$$\frac{V(\sin\beta)}{\sin^2\beta} = \frac{2}{3}\sin\beta; \tag{9.92a}$$

$$\frac{V(\sin\beta) - V\left(\sqrt{1-\frac{u}{h}}\,\sin\beta\right)}{\sin^2\beta} = \frac{2}{3}\sin\beta\, S_1(h); \tag{9.92b}$$

$$S_1(h) = 1 - \sqrt{1-\frac{u}{h}}\left(1+\frac{u}{2h}\right). \tag{9.92c}$$

By substituting values $\alpha = \beta$ and $\lambda = \pi/2$ one may transform series $F_{\|}$, $F_{\perp}$ to the form

$$\left\{\begin{matrix}F_{\|}(h)\\ 2F_{\perp}(h)\end{matrix}\right\} = \left\{\begin{matrix}f_{\|}(h)\sin^2\beta\\ 2f_{\perp}(h)\cos^2\beta\end{matrix}\right\} = \frac{2}{\pi^2}\left\{\begin{matrix}\sum\limits_{n=2,4,\ldots}^{\infty}\\ \sum\limits_{n=1,3,\ldots}^{\infty}\end{matrix}\right\}\frac{s_n}{h-\left(\frac{2fz}{n}\right)^2}, \begin{Bmatrix}(9.93\text{a})\\ (9.93\text{b})\end{Bmatrix}$$

$$s_n = 8\left(\frac{f}{\pi}\right)^2 \frac{\sin^2\left[\frac{\pi}{2}(n-f)\right]}{(n^2-f^2)^2}, \tag{9.94}$$

that we already met in the *confined rotator model.* Indeed, it follows from Eq. 9.61a that

$$f_{\|}(h,\beta)\sin^2\beta = \frac{1}{\pi^4}\sum_{n=1}^{\infty}\frac{(2\beta/\pi)^2\sin^2\beta}{\left[n^2-(\beta/\pi)^2\right]^2}\frac{1}{h-\left(\frac{2\beta z}{\pi n}\right)^2}. \tag{9.95}$$

This expression reduces to Eq. 9.93a.

On the other side, in series Eq. 9.61b the 0 and 1, -1 and 2, -2 and 3, ... terms coincide. Thus,

$$f_{\perp}(h,\beta,\frac{\pi}{2})\cos^2\beta = \frac{2}{\pi^4}\sum_{n=1}^{\infty}\frac{(\beta/\pi)^2\cos^2\beta}{\left[\left(n-\frac{1}{2}\right)^2-\frac{\beta^2}{\pi^2}\right]^2}\frac{1}{h-\frac{(\beta z/\pi)^2}{\left(n-\frac{1}{2}\right)^2}}, \tag{9.96}$$

that reduces to Eq. 9.93b.

THE SF L(z) OF AN ISOTROPIC MEDIUM. Inserting Eq. 9.92, 9.93 into Eq.9.90, we derive simple expression for the spectral function $L(z)$ of *librators* in form of ordinary integrals:

$$\check{L}(z) = \frac{1}{3}\left(2\check{K}_{\perp}(z) + \check{K}_{\|}(z)\right) \tag{9.97a}$$

$$= D(u,\beta)\left\{\int_0^u dh\, h\exp(-h)\sum_{n=1}^{\infty}\frac{s_n}{h-(fz/n)^2}\right\}$$

$$+\int_u^{\infty} dh\, h\, S_1(h)\exp(-h)\sum_{n=1}^{\infty}\frac{s_n}{h-(fz/n)^2}, \tag{9.97b}$$

where

$$D(u,\beta) = \frac{4\langle\alpha(u,\beta)\rangle \sin\beta}{3\pi\, E(u,\beta)}, \quad E(u,\beta) = 1 - \left[1 - \exp(-u)\right]\cos\beta \ , \quad (9.97c)$$

$\langle\alpha(u,\beta)\rangle$ is given by Eq. 9.85. We remind the Reader that $\langle\alpha(u,\beta)\rangle$ is rather close to β and practically do not depend on u.

A. THE COMPARISON WITH THE FIRST VERSION OF THE HYBRID MODEL. **(a).** In the HM2 model we have series of integrals instead of two integrals for longitudinal and transverse orientations of an alternating field and **(b)** we have an additional integral in the interval $h \in [u, \infty]$.

B. THE COMPARISON WITH THE CONFINED ROTATOR MODEL. The second integral in Eq. 9.97b vanishes for $u \to \infty$, and we get just the form (8.22) of the CR model with the coefficient $D(u,\beta)$, that differs (but not much) from 1 (note that in the CR model $D = 1$). We remark that, if $u = \infty$, $E(u, \beta) = 1 - \cos\beta$, then

$$D = \frac{4}{3\pi}\langle\alpha\rangle \cot\left(\frac{\beta}{2}\right). \quad (9.98)$$

C. LIBRATION IN A HEMISPHERE. If $\beta = \pi/2$, $E(u, \beta) = 1$, $f = 1$, $\langle\alpha\rangle = \beta$,

$$D = 2/3 \text{ for } \beta = \pi/2. \quad (9.99)$$

Since all odd terms in series (9.97b), except the first, disappear, odd terms (9.94) reduce to

$$s_1 = \frac{8}{\pi^2}\frac{\sin^2[(\pi/2)(1-f)]}{4(1-f)^2} = \frac{1}{2}; \ s_3 = s_5 = \ldots = 0.$$

Returning to the transversal SF $\check{K}_{\perp}(z)$, we have from Eq. 9.90:

$$2\check{K}_{\perp}(z) = \int_0^u \frac{h\exp(-h)\,dh}{h - z^2} + \int_u^\infty \frac{h\exp(-h)\,S_1(h)\,dh}{h - z^2}. \quad (9.100)$$

For $u \to \infty$ this expression coincides with that, given by the CR model.

Eq. 9.97 is applicable, at least for a qualitative study, in all region of angles β, i.e. for $\beta \in [0, \pi/2]$.

9.4.5 *Rigorous Theory for the Large β Case*

If β is near $\pi/2$, we may obtain a simple analytical expressions of series (9.61). *When calculating these series* we take

$$\alpha \approx \lambda \approx \beta \approx \pi / 2. \tag{9.101}$$

We also employ the suggestion **(b)** of Sec. 9.4.4, i.e.

$$\int \alpha(v) \ldots dv \approx \langle a \rangle \int \ldots dv \quad .$$

Further derivations differ for $f_{\parallel}$ and $f_{\perp}$ series.

THE LONGITUDINAL ORIENTATION. For $\alpha = \beta = \pi/2$ we may represent the n th term of the sum $hf_{\parallel}(h, \pi/2)$, Eq. 9.61a, in the form

$$\frac{4}{\pi^4} \frac{n^2}{\left(n^2 - \frac{1}{4}\right)^2 \left(4n^2 - \frac{z^2}{h}\right)}$$

$$\frac{16}{\pi^4 \left(1 - \frac{z^2}{h}\right)} \frac{1}{\left(4n^2 - 1\right)^2} - \frac{z^2 / h}{\left(4n^2 - 1\right)\left(4n^2 - \frac{z^2}{h}\right)}$$

$$= \frac{16h}{\pi^4 \left(h - z^2\right)} \left[\frac{1}{\left(4n^2 - 1\right)^2} - \frac{z^2 / h}{\left(4n^2 - 1\right)\left(1 - \frac{z^2}{h}\right)} + \frac{z^2 / h}{\left(4n^2 - \frac{z^2}{h}\right)\left(1 - \frac{z^2}{h}\right)} \right].$$

Thus,

$$hf_{\parallel}\left(h, \frac{\pi}{2}\right) = \frac{16h}{\pi^4 \left(h - z^2\right)} \left[\sum_{n=1}^{\infty} \frac{1}{\left(4n^2 - 1\right)^2} \right.$$

$$- \frac{z^2}{h-z^2}\sum_{n=1}^{\infty}\frac{1}{4n^2-1} + \frac{z^2}{4\left(h-z^2\right)}\sum_{n=1}^{\infty}\frac{1}{n^2-\frac{z^2}{4h}} \Bigg].$$

The first and second series in [•] are equal to $(\pi^2 - 8)/16$ and 1/2, respectively. Since, see e.g. (Prudnikov, Brychkov and Marychev 1981),

$$\sum_{n=1}^{\infty}\frac{1}{n^2-p^2} = \frac{1}{2p^2}\left(1-\pi p \cot \pi p\right), \tag{9.102}$$

the last expression reduces to

$$hf_{\|} = \frac{h\,S_2(h)}{\pi^2\left(h-z^2\right)}, \tag{9.103}$$

$$S_2(h) = 1 - \frac{4z\sqrt{h}}{\pi}\,\frac{\cot\left(\frac{\pi z}{2\sqrt{h}}\right)}{h-z^2}. \tag{9.104}$$

Thus,

$$\check{K}_{\|}(z) = \frac{3\langle\alpha\rangle}{\pi E}\Bigg\{ V(\sin\beta)\int_0^u \frac{h\exp(-h)S_2(h)}{h-z^2}dh$$

$$+ \int_u^{\infty}\frac{h\exp(-h)S_2(h)}{h-z^2}\left[V(\sin\beta) - V\left(\sqrt{1-\frac{u}{h}}\,\sin\beta\right)\right]dh \Bigg\}.$$

Putting $\beta = \pi/2$ into Eq. 9.90, we finally have:

$$\check{K}_{\|}(z) = \int_0^u \frac{h\exp(-h)S_2(h)}{h-z^2}dh + \int_u^{\infty}\frac{h\exp(-h)S_1(h)S_2(h)}{h-z^2}dh. \tag{9.105}$$

THE TRANSVERSE ORIENTATION. If Eq. 9.101 holds, only 0 th and 1 th terms of series (9.61b) should be retained. Let us transform their denominators. Putting $\mathrm{v} = \cos\overline{\theta}$, we write

$$\lambda = a\sin X,\ \alpha = \mathrm{acos}Y,\ X = \frac{1}{\sin\beta}\sqrt{\frac{\sin^2\beta - \mathrm{v}^2}{1-\mathrm{v}^2}},\ Y = \frac{\cos\beta}{\sqrt{1-\mathrm{v}^2}},$$

$$(\pi n - \lambda)^2 - \alpha^2 = (\pi n - \lambda - \alpha)\,(\pi n - \lambda + \alpha)\ ,$$

$$\alpha \mp \lambda = a\cos Y \mp a\sin X = a\cos Y \pm a\cos X \mp \frac{\pi}{2}\ .$$

Noting that

$$XY = \frac{\cos\beta\,\sqrt{\sin^2\beta - \mathrm{v}^2}}{\left(1-\mathrm{v}^2\right)\sin^2\beta},\ \sqrt{\left(1-X^2\right)\left(1-Y^2\right)} = \frac{\mathrm{v}\ \cot\beta\,\sqrt{\sin^2\beta - \mathrm{v}^2}}{\left(1-\mathrm{v}^2\right)\sin^2\beta}\ ,$$

employing standard formulas for (acos $Y \pm$ acos X) and choosing signs in these with account that α and λ are nearly equal, one can easily obtain:

$$\alpha \pm \lambda = \mp\, \mathrm{a}\sin\left(A_\pm \pm \begin{Bmatrix}0\\ \pi\end{Bmatrix}\right);\ A_\pm = \frac{\cot\beta}{1\pm\mathrm{v}}\sqrt{\sin^2\beta - \mathrm{v}^2}\ . \qquad (9.106)$$

For β near $\pi/2$ we may replace asin(•) by its argument. Then $\cot^2\beta$ multiplier in $f_\perp\cot^2\beta$ divide out (this was the purpose of our transformations).

Summing terms with $n = 0$ and $n = 1$, we finally derive:

$$f_\perp\cot^2\beta = \frac{1+\mathrm{v}^2}{2\pi^2\left(h-z^2\right)\left(\sin^2\beta - \mathrm{v}^2\right)};\ \beta \approx \frac{\pi}{2}\ . \qquad (9.107)$$

Substitution Eq. 9.137 into (9.118) yields:

$$\check{K}_\perp(z) = \frac{3\langle\alpha\rangle}{2\pi\,\sin^2\beta\ E}\left\{\ V_1(\sin\beta)\int_0^u \frac{h\exp(-h))}{h-z^2}dh\right.$$

$$\left. + \int_u^\infty \frac{h\exp(-h)}{h-z^2}\left[V_1(\sin\beta) - V_1\left(\sqrt{1-\frac{u}{h}}\ \sin\beta\right)\right]dh\ \right\}, \qquad (9.108a)$$

where

$$V_1(\mathrm{v}) = \int_0^{\mathrm{v}} \left(1+\mathrm{v}^2\right) d\mathrm{v} = \mathrm{v} + \frac{\mathrm{v}^2}{3}. \tag{9.108b}$$

For $\beta = \pi/2$ we have:

$$\frac{3\langle\alpha\rangle}{2\pi \sin^2\beta\, E} = \frac{3}{4}, V_1(\sin\beta) = \frac{4}{3}, V_1(\sin\beta) - V_1\left(\sqrt{1-\frac{u}{h}}\sin\beta\right) = \frac{4}{3} S_3(h),$$

where

$$S_3(h) = 1 - \sqrt{1-\frac{u}{h}}\left(1-\frac{u}{4h}\right). \tag{9.109}$$

Then

$$\overset{\vee}{K}_{\perp}(z) = \int_0^u \frac{h \exp(-h)\, dh}{h - z^2} + \int_u^\infty \frac{h \exp(-h) S_3(h)\, dh}{h - z^2}. \tag{9.110}$$

Approximate expression (9.100) resembles rigorous formula (9.110), where the first integral is twice of that in (9.100).

Eq. 9.105, 9.110 yield the solution of our problem for large cone angle β. Resonant denominator $(h - z^2)$ appears in both expressions. It is corrected for the longitudinal orientation by $S_2(h)$ function. The contribution $\overset{\vee}{L}(z)$ of *librators* to the total SF $L(z)$ is determined by Eq. 9.97a with $\overset{\vee}{K}_{\parallel}$ and $\overset{\vee}{K}_{\perp}$ from (9.105) and (9.110), respectively:

$$\overset{\vee}{L}(z) = \frac{1}{3}\int_0^u \frac{h \exp(-h)}{h - z^2}\left(2+S_2(h)\right)dh + \frac{1}{3}\int_u^\infty \frac{h \exp(-h)}{h - z^2} S_3(h)\left(2+S_1(h)\right) dh\ , \tag{9.111}$$

where $S_1(h)$, $S_2(h)$, $S_3(h)$ are respectively given by Eq. 9.92c, 9.104 and 9.109.

One can regard Eq. 9.105, 9.110, 9.111 to be the generalization of the ED model for the subensemble of dipoles performing almost free librations above (or beneath) a nearly plane conical surface. In both approaches (Sec. 9.4.4 and 9.4.5) the cone welldepth u was supposed to be arbitrary.

9.4.6 Evolution of Dielectric Spectra due to the Rise of the Cone Angle

PARAMETERS AND THE CALCULATION SCHEME. We shall take for example four values of the cone angle β:

$$\beta_{\mathrm{m}} = (1/8)\ \pi m,\ m = 1, 2, 3, 4. \tag{9.112a}$$

The other parameters, i.e. u, y, and the Kirkwood correlation factor, g, will be chosen as follows:

$$u_m = 4.5 - \frac{3.5}{3}(m-1),\ y_m = 0.8 + \frac{2}{3}(m-1),\ g_m = 2 - \frac{1}{3}(m-1). \tag{9.112b}$$

These parameters and others, determined by them, are presented in Table 9.1.

We have an idea to estimate qualitatively the effect of *temperature* on dielectric behavior. Indeed, when T rises, the potential barrier u and Kirkwood factor g decrease, while the libration amplitude β and collision frequency y increase.

We shall apply the *isothermal* collision model to an isotropic polar medium. For the SF $\check{L}(z)$ of librators we employ simplified expression (9.97). For $\beta \approx \pi/2$ the latter will be compared with the rigorous one, Eq. 9.111.

For the SF of *rotators* we take, by analogy with Eq. 9.45, L_F^{sp} of the spatial extended diffusion model *multiplied* by the proportion $\mathring{r} = 1 - \check{r}$ of rotators; $\check{r}$ is estimated from Eq. 9.103 and $L_F^{\mathrm{sp}}(z)$ is given by

$$L_F^{\mathrm{sp}}(z) = 2\int_0^{\infty} \frac{p^3 \exp(-p^2)}{p^2 - z^2}\, dp\ . \tag{9.113}$$

APPLICATION OF THE SIMPLIFIED SF $L(z)$, DETERMINED BY THE AVERAGED PARAMETERS $\alpha \approx \langle\alpha\rangle \approx \beta$ and $\lambda \approx \pi/2$ (Fig. 9.10). The frequency dependence of the normalized absorption is shown in Fig. 9.10 a, b, c for $\beta = \pi/8$, $\pi/4$ and $3\pi/8$, respectively. The loss vs. frequency dependence is shown in Fig.9.12d. When β increases:

(a) The two-humped absorption curve transforms into a curve with a single maximum.

(b) The absorption maximum shifts to *lower* frequency and the loss maximum - to *opposite* direction.

Thus, we employ the following formulas for the total SF $L(z)$ and the complex susceptibility $\chi^*(x)$:

$$\check{L}(z) = \frac{1}{3}\left[2\check{K}_{\perp}(z) + \check{K}_{\parallel}(z)\right], \tag{9.114a}$$

$$L(z) = \check{L}(z) + \left(1 - \check{r}\right) L_F^{\text{sp}}(z), \tag{9.114b}$$

$$\frac{\chi^*(x)}{G} = \frac{z\, L(z)}{gx + iy(1+gxz)L(z)}. \tag{9.115}$$

Properties (**a, b**) resemble those, found in the previous version of the hybrid model, cf. Fig. 9.10 and Fig. 9.11. However, as compared with the first variant, we have in the HM2 version:

(c) The contribution of free rotors to the total absorption is lesser for a given barrier *u*, since the proportion $\check{r}$ is lower in the HM2 model (see Table 9.1).

(d) The Debye loss maximum is shifted in the HM2 model to *higher frequencies* and depend strongly on the amplitude β, cf. Fig. 9.10d with 9.11d.

Distinctions between two HM models are founded on the account in the HM2 model of contribution to spectra of *"peripheral" librators*. These travel near the surface of the conical well, far from its diametric cross-section. In the previous HM version the contribution to spectra of such particles was neglected, since the $|\, l \,| \to 0$ approximation was used. This distinction is accompanied by the change of corresponding statistical parameters, since the phase volume, employed in the HM2 model (see Fig. 9.8), differs from that, taken in the HM model. In particular, we repeat, the proportion $\check{r}$ of librators are different in both versions.

Table 9.1 Parameters used in calculation of dielectric spectra illustrated by Fig. 9.12 - 9.15

Independent parameters					The steady-state functions				
m	1	2	3	4	m	1	2	3	4
β	$\frac{\pi}{8}$	$\frac{\pi}{4}$	$\frac{3\pi}{8}$	$\frac{\pi}{2}$	r%[a)]	12.3	9.5	12.3	22.2
β	0.4	0.8	1.1	π	r%[b)]	12.9	11.2	17.3	36.8
u	4.5	3.3	2.2	1.0	Δ	0.25	0.26	0.28	0.31
g	2.0	1.7	1.3	1.0	$\langle\alpha\rangle$	0.34	0.69	1.08	$\frac{\pi}{2}$
y	0.8	1.5	2.1	2.8	$\langle\lambda\rangle$	1.12	1.19	1.31	$\frac{\pi}{2}$
Notes	a) for the HM2 model; b) for the HM model								

DIELECTRIC BEHAVIOUR FOR MAXIMUM CONICAL ANGLE β. The rigorous absorption / loss spectra are shown in Fig. 9.12 by solid lines. In Fig. 9.12 a, b these spectra are compared with those (see dashed curves), calculated for the model of free rotation, where $u = 0$. The existence of a well in the HM2 model leads to a wider absorption band (Fig. 9.12a) and to the low-frequency shift of a loss curve (Fig. 9.12b). However, these distinctions are not pronounced, since the absorption maximum is placed near $x = 1$ (the latter property is typical for the ED model). For the simplified HM2 model, described in Sec.9.4.4, calculated absorption / loss curves (dashed curves in Fig. 9.12 c, d) are lower than those obtained rigorously.

We remark that

---

if we insert into Eq. 9.115 *the SF* (9.114a), *found from the simplified HM2 model, Eq.* 9.97, *and multiplied by the coefficient A (independent of frequency), the resultant spectra will be close to those, described by the rigorous SF, Eq.* 9.111*:*

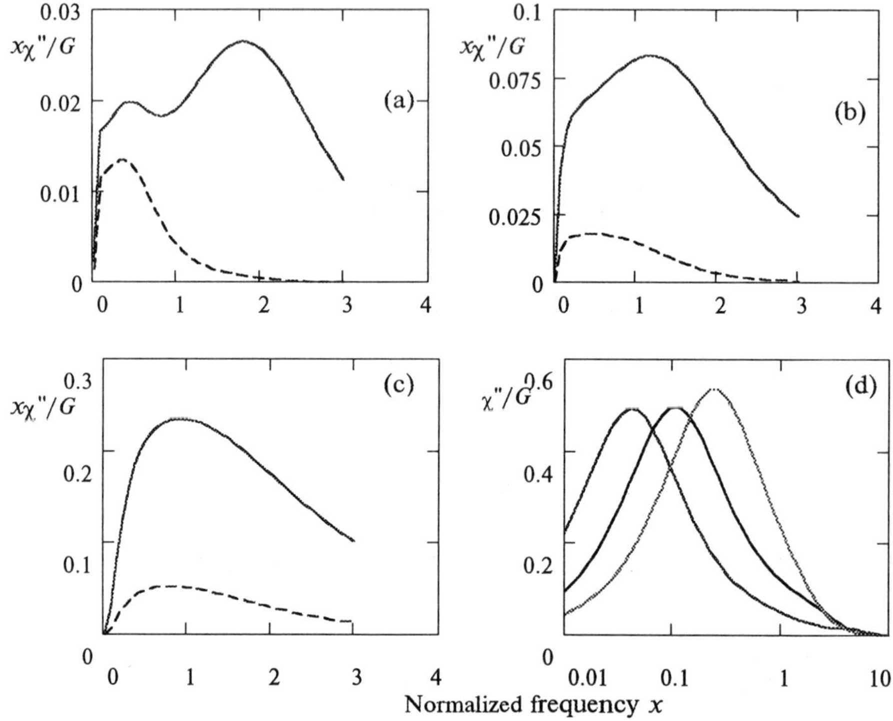

Fig. 9.10 Dielectric spectra, calculated for the hybrid 2 model with approximations, described in Sec. 9.44 (solid curves).

a, b, c: Transformation of two-humped absorption spectrum into frequency dependence, having one maximum; $\beta = \pi/8$ (a), $\pi/4$ (b) and $3\pi/8$ (c). The contribution of the subensemble of free rotors is shown by dashed curves.

d: The loss spectrum. calculated for $\beta = \pi/8$, $\pi/4$ and $3\pi/8$ (from left to right). Other parameters of the model are given in Table 9.1.

$$\underbrace{A \bullet \check{L}(z)}_{\text{from Eq. 9.97}} \text{ is } ca. \text{ equivalent to } \underbrace{\check{L}(z)}_{\text{from Eq. 9.111}} \text{ for } \beta = \frac{\pi}{2}. \tag{9.116}$$

If $u = 1$, $y = 2.8$ and $g = 1$, coefficient $A = 1.6$.

We may try to apply this idea for all angles β. Indeed, we may replace the constant A by a β-depending coefficient

$$B(\beta) = [1 + (A\text{-}1) \sin^2\beta]. \tag{9.117}$$

$B(\beta)$ is near 1 for small β and is near A for $\beta \to \pi/2$;

$$\underbrace{B(\beta) \bullet \check{L}(z)}_{\text{From Eq. 9.97}} \quad \text{is ca. equivalent to a rigorous description from Eq. 118 for } \beta.$$

Note, that we should determine the A-coefficient for each $\{u, y\}$-set of parameters of the HM2 model by comparing rigorous and approximate values of $\check{L}$ found for $\beta = \pi/2$. We regard that the HM2 model could be useful for study of fine dielectric phenomena in polar fluids.

Returning to the last example, we see from Fig. 9.13a, that absorption in an isotropic medium is contributed mainly by "transverse librators", to which the transverse part $\check{L}_{\perp}(z) = (1/3)\, \check{K}_{\perp}(z)$ of the total SF $\check{L}(z)$ corresponds. The least contribution to the absorption near its maximum is due to the longitudinal part $\check{L}_{\parallel}(z) = (1/3)\, \check{K}_{\parallel}(z)$ of $\check{L}(z)$.

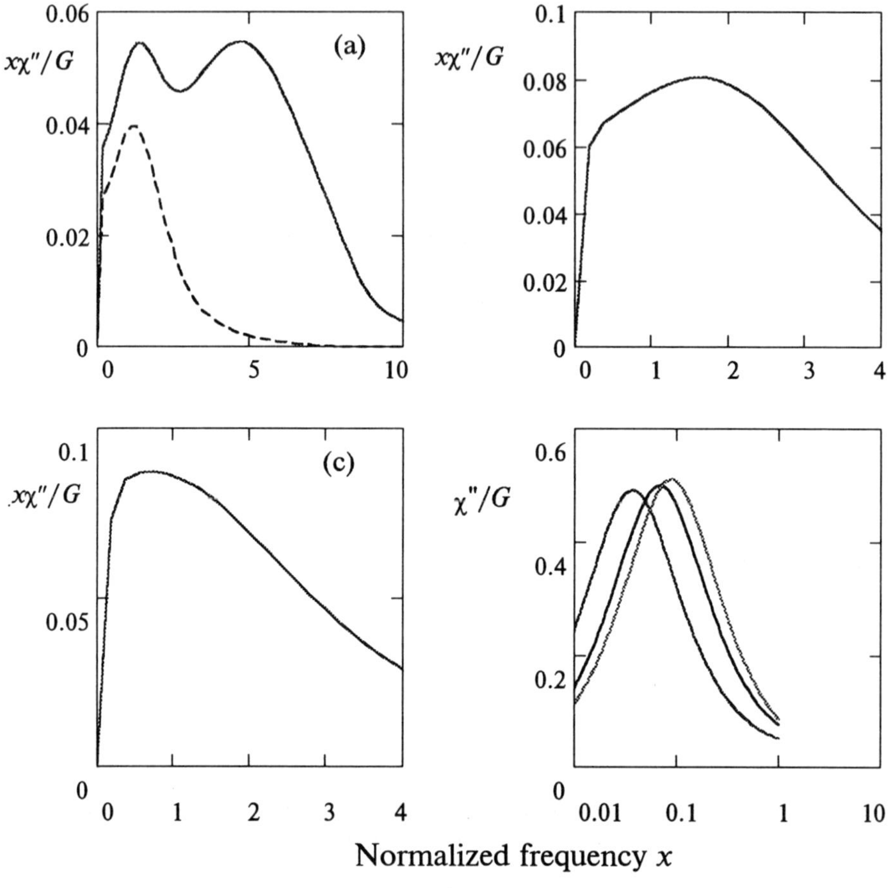

Fig. 9.11 The same as in Fig. 9.10 but calculated for the first version of the hybrid model.

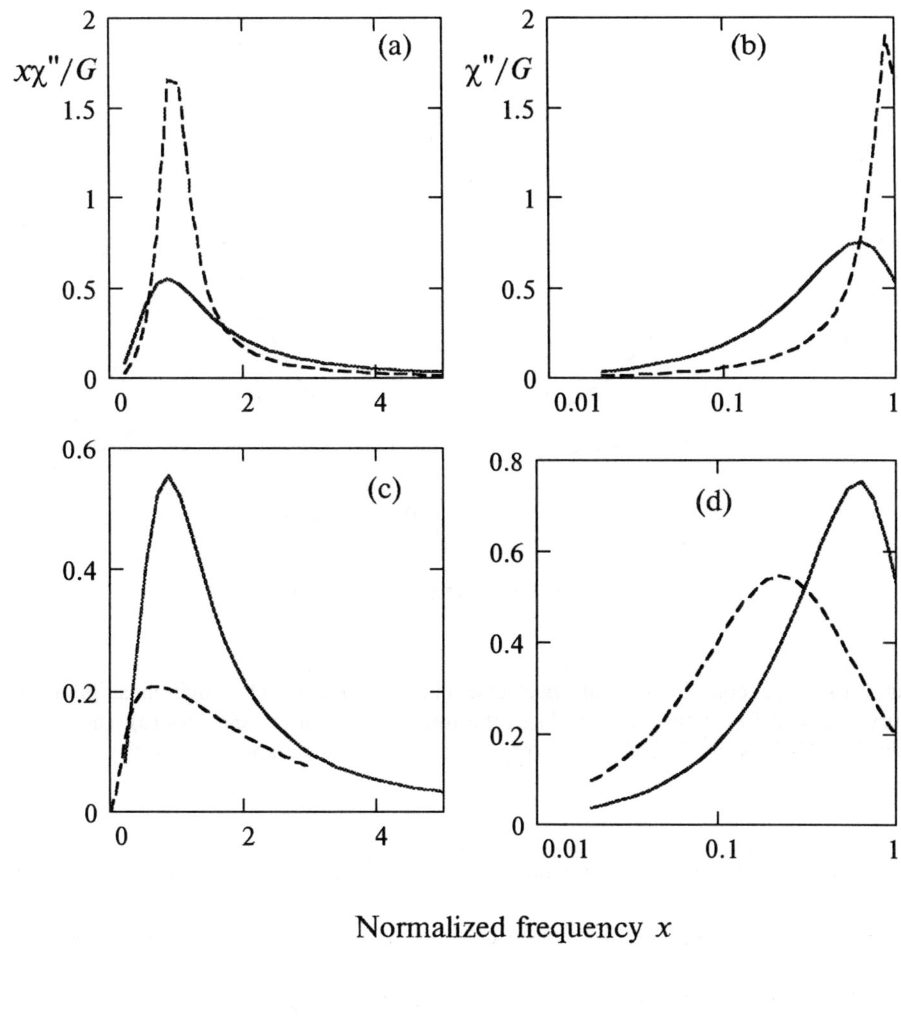

Fig. 9.12. Dielectric behavior for libration in hemisphere ($\beta = \pi/2$): normalized absorption (a, c) and imaginary part of the complex susceptibility (b, d) vs. frequency, calculated for $u = 1$, $y = 2.8$, $g = 1$. Rigorous theory of hybrid 2 model (solid lines); the extended diffusion model (dashed lines in Figs. a and b); the simplified hybrid 2 model (dashed lines in Figs. c and d).

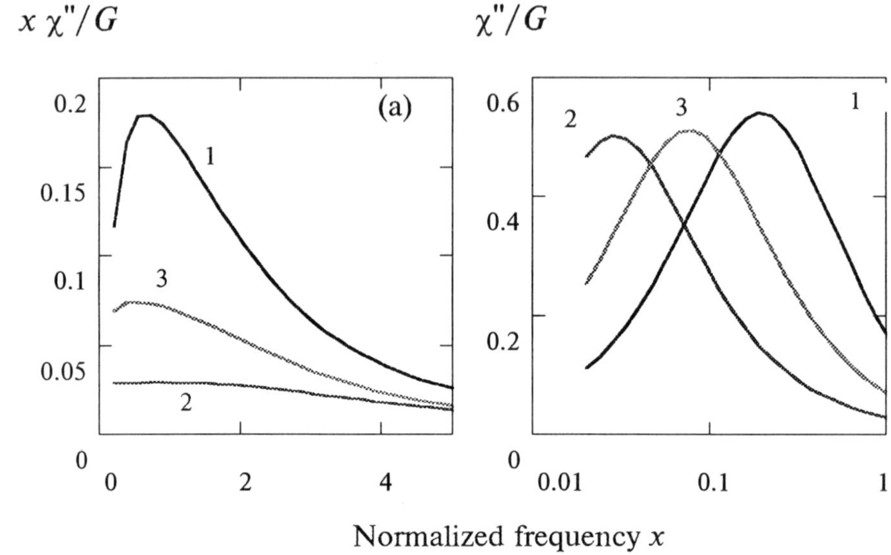

Fig. 9.13 The contributions of transverse librators (curves 1), longitudinal librators (curves 2) and free rotors (curves 3) to the total absorption (a) and loss (b), shown by solid lines in Figs. 9.12a and 9.12b, respectively. Calculated rigorously for the hybrid 2 model.

The contributions of these parts, $\check{L}_{\|}(z)$, $\check{L}_{\perp}(z)$, as well as that of free rotors, $\mathring{L}(z) = \mathring{K}(z)$, is illustrated by Fig. 9.13 b for the region of maximum loss. The maxima of three curves, $\check{\chi}''_{\perp}(x)$, $\check{\chi}''_{\|}(x)$, $\mathring{\chi}''(x)$ are nearly equal, but they are placed on different frequencies. The sum of three loss curves yields the total loss dependence $\chi''(x)$, shown by solid line in Fig. 9.12b. The frequency x_D of the total maximum χ''_D is *much higher,* than those, obtained for individual curves $\check{\chi}''_\perp(x)$, $\check{\chi}''_\parallel(x)$, $\mathring{\chi}''(x)$, shown in Fig.9.13.

These estimations were made with use of Eq. 9.115, in which the numerator was represented as the sum of three terms:

$$\chi^*(x) = \check{\chi}_\perp(x) + \check{\chi}_\parallel(x) + \mathring{\chi}(x)$$

$$= \frac{Gz\left\{\check{L}_\perp(z) + \check{L}_\parallel(z) + \mathring{L}(z)\right\}}{gx + iy(1+gxz)\left\{\check{L}_\perp(z) + \check{L}_\parallel(z) + \mathring{L}(z)\right\}}. \tag{9.118}$$

PART FOUR

FIELD MODELS

The concept of the intermolecular potential, employed here, is quite different as compared with that, considered in the previous Part. Indeed, the «walls» of the well lose their «mechanical» clearness, since *instantaneous reflections* of dipoles from the walls of the rectangular box are replaced by *continuous interaction* of reorienting dipoles with non-homogeneous potential field. As a result, the libration amplitude β becomes dependent on the energy h of a dipole: the greater h, the greater β. We shall show, that dielectric behavior substantially depends on the steepness of the potential and on its depth U_0.

First, we consider in Chap. 10 the *parabolic potential.* The latter presents a rather good approximation of any potential curve near the potential minimum. It will be seen, that simple Lorentz-type formulas describe the spectral function / dielectric susceptibility due to polar librators. It seems, that the parabolic potential is *too steep*: because of this property it cannot be successfully applied to classical fluids.

Next, we consider in Chap. 11 the *cosine potential profile* that corresponds to the constant field (CF), imposed on polar medium. A non-zero dipole moment, induced by conservative external field, appears in this system. The dielectric-response theory becomes in this case rather complicated, even if we restrict our consideration by planar rotation. Such theory will be given below. The examination of the steady state distributions and the values of averaged parameters yield valuable preliminary information about dielectric properties of this molecular ensemble. We shall see, that the decrease of the potential steepness, which occurs, when the angular shift ϑ of a dipole increases, leads to the radical distinction with the previous system of Chap. 10, where the potential steepness does not depend on ϑ.

For simplicity all numerical calculations will be made in Chap. 11 on the basis of the *quasi-harmonic approximation* (QHA). In the latter one should take rather simple ordinary integrals with the Lorentz-like integrands. The CF model is the system, that is characterized by a *single* potential maximum. We shall show, that because of this property the low-frequency dielectric behavior, specific for isotropic fluids, is rather unusual as described in frames of this model.

In Chap. 12 we consider another example, the *cosine-squared* (CS) *potential profile*, for which the low-frequency spectra agree *qualitatively better* (than on the basis of the CF model) with such spectra, observed in isotropic polar liquids. We remark, that the induced dipole moment of the medium vanish in the CS potential field. Moreover, since the *potential curve is more flat near the potential bottom*, than that, used in Chap 11, dielectric properties substantially change.

Still more radical step is made in the last 13 th Chapter of this Part. We consider the *hat model*, in which the form of the conservative potential can be varied. In distinction with the 11 th and 12 th chapters we assume, that the rather steep increase of the potential follows the flat bottom. It seems, that this potential profile is more adequate for the description of dielectric spectra in fluids, than simple cosine or cosine squared profiles.

In all four chapters almost all calculations are made for *planar motion*. This drawback restricts the applicability of the theory, so that one may consider this Part as an outline of more correct theory, that should deal with the *spatial rotation* of molecules. In this respect the physical understanding, gained here, of the effects of the potential steepness on dielectric behavior may become very useful.

CHAPTER 10
THE ELASTIC BOND (EB) APPROXIMATION

In Sec. 3.5 we have found the law of motion of a dipole, librating in a parabolic potential well

$$\frac{U(\vartheta)}{k_B T} = p_m^2 \begin{Bmatrix} \vartheta^2 \\ \theta^2 \end{Bmatrix} + \text{const}, \qquad \text{for} \begin{Bmatrix} p \\ s \end{Bmatrix}, \tag{10.1}$$

here and further "p" and "s" refer to motion in a plane and in space, respectively. We have also calculated the steady state distribution function

$$W = \frac{p_m}{\pi} \begin{Bmatrix} 1 \\ 2p_m \end{Bmatrix} \exp(-h), \qquad \text{for} \begin{Bmatrix} p \\ s \end{Bmatrix}, \tag{10.2}$$

and a number of mean values

$$\langle q_{\|}^2 \rangle, \langle q_{\perp}^2 \rangle, \Delta \equiv \langle q_{\|}^2 \rangle - \langle q_{\|} \rangle^2, \langle \beta \rangle, \tag{10.3}$$

where $q_{\|}(\varphi)$ and $q_{\perp}(\varphi)$ are normalized projections μ_E/μ of the dipole moment vector $\boldsymbol{\mu}$ for the longitudinal ($\| Z$) and transverse ($\perp Z$) orientations of alternating field $\boldsymbol{E}(t)$ relative symmetry axis Z of the conservative potential U; $\beta(h)$ is the libration amplitude of the angular shift $\vartheta(t)$ or $\theta(t)$; $2p_m$ is the normalized angular frequency of function $q_{\|}(\varphi)$:

$$q_{\|}(\varphi) = \begin{Bmatrix} \cos\vartheta \\ \cos\theta \end{Bmatrix} \approx \bar{q}_{\|} + \text{const} \bullet \cos 2p_m(\varphi + \varphi_0), \quad \text{for} \begin{Bmatrix} p \\ s \end{Bmatrix}, \tag{10.4}$$

$\Phi \equiv \pi / p_m$ being the period of this function.

We shall find the spectral functions (SFs) for both ensembles.

10.1 The Spectral Function of the Planar Ensemble

We recall to the Reader that in the case of the parabolic potential only a librational subensemble exists (denoted by the symbol $^{\vee}$). Comparing Eq. 3.132 with series (5.81), we find that $\check{p} \equiv p_m$ and that only three Fourier amplitudes a_0, a_1 and b_1 do not vanish:

$$a_0 = 1 - \frac{\beta^2}{4} = 1 - \frac{h}{4p_m^2},\ a_1 = \frac{\beta^2}{4} = \frac{h}{4p_m^2},\ b_1 = \beta = \frac{\sqrt{h}}{p_m}. \tag{10.5}$$

Substituting into Eq. 5.86, we find

$$\frac{K_\parallel}{3} \equiv \left\langle q_\parallel^2 \right\rangle L_\parallel(z)$$

$$= \frac{\pi}{2}\frac{p_m}{\pi}\int_0^\infty \exp(-h)\, p_m \left(\frac{h}{4p_m^2}\right)^2 \frac{1}{4p_m^2}\frac{4\,dh}{4p_m^2 - z^2} = \frac{1}{4p_m^2}\frac{h}{4p_m^2 - z^2}, \tag{10.6}$$

$$\frac{K_\perp}{3} \equiv \left\langle q_\perp^2 \right\rangle L_\perp(z) = \frac{\pi}{2}\frac{p_m}{\pi}\int_0^\infty \exp(-h)\, p_m \left(\frac{h}{p_m}\right)^2 \frac{dh}{p_m^2 - z^2} = \frac{h}{2\left(p_m^2 - z^2\right)}. \tag{10.7}$$

Hence, it is the *Lorentz line* $(a^2 - z^2)^{-1}$, that describes dielectric dispersion for the case of the parabolic potential, the resonance frequency $x_\parallel \approx 2p_m$ of the longitudinal SF being twice that of the transverse SF (in the latter $x_\perp \approx p_m$.

To find the SF in an isotropic medium we take the sum of two terms:

$$L(z) = \frac{1}{3}\left(K_\parallel + K_\perp\right) = \frac{1}{2}\frac{1}{p_m^2 - z^2} + \frac{1}{4p_m^2}\frac{1}{4p_m^2 - z^2} \tag{10.8}$$

If $p_m >> 1$, the main peak is obliged to the transverse SF:

$$L(z) \approx\ = \frac{1}{2}\frac{1}{p_m^2 - z^2}. \tag{10.8a}$$

10.2 The Spectral Function of the Spatial Ensemble

In general expressions of Sec.5.6.1 we must put

$$p \equiv 2p_m, \tag{10.8b}$$

since the first harmonic (for $n = 1$) in series (5.104) is proportional to $\cos 2p_m(\varphi + \varphi_0)$, see Eq. 3.145.

10.2.1 The Longitudinal SF

Comparing Eq. 3.145 with series (5.104), we find, that only two Fourier amplitudes a_0 and a_1 do not vanish. To calculate the SF $K_{\|}(z)$ we need only a_1^2:

$$a_1^2 = \frac{h^2 - 4p_m^2 l^2}{16p_m^4} \tag{10.9}$$

From Eq. 5.108 we have:

$$\frac{1}{3}K_{\|}(z) = \pi\frac{2p_m^2}{\pi}\int_0^\infty dh\, \exp(-h) 2p_m \int_0^{\frac{h}{2p_m}} \frac{h - 4p_m^2 l^2}{4p_m^2 - z^2}\frac{dl}{16p_m^4}\ . \tag{10.10}$$

Since

$$\int_0^{\frac{h}{2p_m}} \left(h - 4p_m^2 l^2\right) dl = \frac{h^3}{3p_m} \quad \text{and} \quad \int_0^\infty h^3 \exp(-h)\, dh = 6,$$

we find:

$$\frac{1}{3}K_{\|}(z) = \pi\frac{2p_m^2}{\pi}\frac{2p_m}{16p_m^4}\frac{1}{4p_m^2 - z^2} = \frac{1}{2p_m^2}\frac{1}{4p_m^2 - z^2}. \tag{10.11}$$

This quantity is two times larger than the SF (10.6) of the planar ensemble.

10.2.2 The Transverse SF

We remark that $\lambda = \pi/2$, see Eq. 3.146a. We may avoid tedious transformations by the following way. Taking into account: **(a)** the result

(10.7), obtained for a planar ensemble, and **(b)** the similarity of its SF (10.6) to the spatial SF (10.11), we may suggest that

---

The sum $S_\perp$, defined by (5.121b), contains only terms with $n = 0$ and $n = 1$. (10.12)

The validity of this statement will be proved later. Consequently, since $p \to 2p_{\mathrm{m}}$, we find

$$S_\perp = \frac{b_0^2 + b_1^2}{p_m^2 - z^2} . \tag{10.13}$$

Substituting Eq. 3.150a into Eq. 5.121b, we have:

$$\int_0^{\pi/p_m} \left\langle \sin^2\left[\theta(\varphi+\varphi_0)\right] \cos^2\left[\psi(\varphi+\varphi_0)+\psi_0\right] \right\rangle_{\psi_0} d\varphi_0$$

$$\approx \frac{1}{2}\int_0^{\pi/p_m} \theta^2(\varphi)\,d\varphi = \frac{1}{2}\frac{h}{2p_m^2}\frac{\pi}{p_m} = \frac{\pi}{2p_m}\sum_{n=-\infty}^{\infty} b_n^2 = \frac{\pi\left(b_0^2+b_1^2\right)}{2p_m} . \tag{10.14}$$

Combining with Eq. 10.13, we obtain

$$S_\perp = \frac{h}{8p_m^2}\frac{1}{p_m^2 - z^2} . \tag{10.15}$$

Substitution of $S_\perp$ into Eq. 5.121 yields:

$$\frac{K_\perp(z)}{3} = 4p_m^3\int_0^\infty dh\,\exp(-h)\int_0^{\frac{h}{2p_m}} \frac{h\,dl}{p_m^2 - z^2}\frac{1}{8p_m^2} = \frac{1}{2\left(p_m^2 - z^2\right)} . \tag{10.16}$$

This quantity coincides with that, given by Eq. 10.7 for the planar ensemble.

10.2.3 The Spectral Function L(z) of an Isotropic Medium

Since

$$L(z) = \frac{2K_\perp(z) + K_\parallel(z)}{3} = \frac{1}{p_m^2 - z^2} + \frac{1}{2p_m^2}\frac{1}{4p_m^2 - z^2}\,, \qquad (10.17)$$

the spatial SF $L(z)$ and correspondent integrated absorption are twice than those in the planar ensemble, cf. with Eq. 10.8.

10.2.4 Detailed Calculation of the Fourier Amplitudes b_n

We have found above the sum of these amplitudes, Eq. 10.14. In order to calculate each amplitude b_n, we return to the law of motion (3.144). Putting $\dot{\theta} = 0$, we find the extreme values

$$\left(\theta^2\right)_{\min} = \left(\bar{\theta} - \delta\right)^2 \text{ and } \left(\theta^2\right)_{\max} = \left(\bar{\theta} + \delta\right)^2\,,$$

angles $\bar{\theta}$ and δ are shown in Fig. 3.3b:

$$\left(\bar{\theta} \pm \delta\right)^2 = \frac{1}{2p_m^2}\left(h \pm \sqrt{h^2 - 4p_m^2 l^2}\right), \qquad (10.18)$$

the first and the second terms on the r.h.s. of Eq. 10.18 are, respectively, the constant and the amplitude of the $\theta^2(\varphi)$ dependence. Using Eq. 10.18, we express h and l through angles $\bar{\theta}$ and δ:

$$h = 2p_m^2\left(\bar{\theta} + \delta\right)^2,\ \sqrt{h^2 - 4p_m^2 l^2} = 2p_m^2\,\bar{\theta}\,\delta,\ l = p_m\left(\bar{\theta}^2 - \delta^2\right). \qquad (10.19)$$

Then Eq. 3.144 may be represented as

$$\theta^2(\varphi) = \bar{\theta}^2 + \delta^2 - 2\bar{\theta}\delta\cos\left(2p_m\varphi\right), \qquad (10.20)$$

so that

$$q_\parallel(\varphi) \approx 1 - \frac{\bar{\theta}^2}{2} = 1 - \frac{\bar{\theta}^2 + \delta^2}{2} + \bar{\theta}\,\delta\cos\left(2p_m\varphi\right). \qquad (10.21)$$

It follows from Eq. 10.18-10.20, that:

1. The angular shift θ is constant, when the end of the vector μ rotates along the mean line $\theta = \bar{\theta}$ about the Z axis (see Fig. 3.3b). In view of Eq. 10.18 such motion occurs, if $l = h/(2p_m)$. Thus,

$$\left(\bar{\theta} + \delta\right)^2 = \bar{\theta}^2 = \frac{h}{2p_m^2} \ . \tag{10.22}$$

2. The minimum value of θ^2 is zero. Noting that $l = 0$, we find from (10.18)(10.19), that we have libration in a plane, passing through the Z axis. The libration amplitude may be found from Eq. 10.18:

$$\beta^2 = \left(\bar{\theta} \pm \delta\right)^2 = 4\bar{\theta}^2 = \frac{h}{p_m^2} \ . \ (???) \tag{10.23}$$

It follows from the comparison of (10.23) with (10.22), that the squared libration amplitude β is two times larger, than the mean squared shift $\bar{\theta}^2$ found at the maximal value $l = h/(2\,p_m^2\ )$.

Now we shall express the time dependence $\psi(\varphi)$ in terms of $\bar{\theta}$ and δ, where ψ is the angular shift of the μ vector around the Z axis. From Eq. 3.85, 10.18 - 10.19 we have

$$\psi \approx l\int_0^{\varphi}\frac{d\varphi'}{\theta^2(\varphi')} = l\int_0^{\varphi}\frac{d\varphi'}{\bar{\theta}^2 + \delta^2 - 2\bar{\theta}\delta\cos\left(2p_m\varphi'\right)}$$

$$= \frac{l}{p_m\left(\bar{\theta}^2 - \delta^2\right)}\operatorname{atan}\left(\frac{\left(\bar{\theta}^2 - \delta^2\right)\tan\left(p_m\varphi\right)}{\left(\bar{\theta} - \delta\right)^2}\right) = \operatorname{atan}\left(\frac{\left(\bar{\theta} + \delta\right)\tan\left(p_m\varphi\right)}{\bar{\theta} - \delta}\right). \tag{10.24}$$

Then we find from Eq. 5.120, 10.20 and 3.146a:

$$\xi_n(\varphi) = \sin\theta\,\cos\left(\psi(\varphi) - \frac{\lambda}{\pi}2p_m\varphi + 2p_m n\varphi\right)$$

$$= \sin\theta\cos\left(\psi + \left[n - \frac{1}{2}\right]2p_m\varphi + 2p_m n\varphi\right). \tag{10.25}$$

Using Eq. 10.24, we obtain after some trigonometry:

$$\sin^2\theta\cos^2\psi = \sin^2\theta\frac{\left(\bar{\theta} - \delta\right)^2\cos^2 p_m\varphi}{\left(\bar{\theta} - \delta\right)^2\cos^2 p_m\varphi + \left(\bar{\theta} + \delta\right)^2\sin^2 p_m\varphi}$$

$$= \frac{(\bar{\theta}-\delta)^2 \sin^2\theta \cos^2 p_m\varphi}{\bar{\theta}^2+\delta^2 - 2\bar{\theta}\,\delta\cos 2p_m\varphi}, \tag{10.26}$$

$$\sin\theta\cos\psi = (\bar{\theta} - \delta)\cos p_m\varphi;\ \sin\theta\sin\psi = (\bar{\theta}+\delta)\sin p_m\varphi, \tag{10.27}$$

$$\sin\theta\cos(\psi - p_m\varphi) = (\bar{\theta} - \delta)\cos^2 p_m\varphi + (\bar{\theta}+\delta)\sin^2 p_m\varphi$$

$$= \theta-\delta\cos 2p_m\varphi, \tag{10.28}$$

$$\sin\theta\sin(\psi - p_m\varphi) = (\bar{\theta}+\delta)\cos p_m\varphi \sin p_m\varphi$$

$$-(\bar{\theta} - \delta)\cos p_m\varphi \sin p_m\varphi = \delta\sin 2p_m\varphi. \tag{10.29}$$

Finally,

$$\xi_n(\varphi) = (\bar{\theta} - \delta\cos 2p_m\varphi)\cos 2p_m n\varphi - \delta\sin 2p_m\varphi\sin 2p_m n\varphi$$

$$= \bar{\theta}\cos 2p_m n\varphi - \delta\cos 2p_m n\varphi - \delta\sin 2p_m(n-1)n\varphi. \tag{10.30}$$

In accordance with Eq. 5.120a we now calculate expressions for $b_n = \frac{p_m}{\pi}\int\limits_0^{\pi/p_m} \xi_n(\varphi)\, d\varphi$:

$$b_0 = \bar{\theta},\ b_1 = -\delta,\ b_n = 0 \quad \text{for } n \neq 0, 1. \tag{10.31}$$

We find, at last, that the sum $S_\perp$ in Eq. 5.121 reduces to

$$S_\perp = \frac{b_0^2 + b_1^2}{p_m^2 - z^2} = \frac{h}{8p_m^2\left(p_m^2 - z^2\right)}. \tag{10.32}$$

Thus, we have obtained Eq. 10.15, the statement (10.12) is proved.

10.3 The Complex Susceptibility of Linear Molecules Subjected to the Influence of the Parabolic Potential

In Table 10.1 we collect formulas for the statistical averages, spectral functions and complex susceptibility. For the latter we employ the Gross collision model (see Table 6.1 for $F = F_G$):

Table 10.1 Formulas for the elastic-bond approximation: the Gross collision model ($p_m >> 1$)

No-tations	$\begin{Bmatrix} \text{SWP} \\ \text{DWP} \end{Bmatrix} \rightarrow \begin{Bmatrix} \text{single} \\ \text{double} \end{Bmatrix}$ well potential with $\begin{Bmatrix} \langle q \rangle \neq 0 \\ \langle q \rangle = 0 \end{Bmatrix}$ $\begin{Bmatrix} p \\ s \end{Bmatrix} \rightarrow \begin{Bmatrix} \text{libration on place} \\ \text{libration in space} \end{Bmatrix}$			
A. STATISTICAL AVERAGES				
$\langle q_{\perp}^2 \rangle$	$\dfrac{1}{2p_m^2}$	$\langle q_{\parallel}^2 \rangle$	$1 - \dfrac{1}{p_m^2}\begin{Bmatrix} 1/2 \\ 1 \end{Bmatrix} \approx 1$	$\begin{Bmatrix} p \\ s \end{Bmatrix}$
$\langle q_{\parallel} \rangle$	$1 - \dfrac{1}{p_m^2}\begin{Bmatrix} 1/4 \\ 1/2 \end{Bmatrix} \approx 1$ for SWP	$\langle q_{\parallel} \rangle$	0 for DWP	$\begin{Bmatrix} p \\ s \end{Bmatrix}$
Δ	$\dfrac{1}{p_m^4}\begin{Bmatrix} 1/8 \\ 1/4 \end{Bmatrix}$ for SWP	Δ	$1 - \dfrac{1}{p_m^2}\begin{Bmatrix} 1/2 \\ 1 \end{Bmatrix} \approx 1$ for DWP	$\begin{Bmatrix} p \\ s \end{Bmatrix}$
$\bar{\Delta} = \dfrac{1}{2p_m^2} + \dfrac{1}{8p_m^4} \approx \dfrac{1}{2p_m^2}$ for SWP				{p}
$\bar{\Delta} = \dfrac{1}{p_m^2} + \dfrac{1}{4p_m^4} \approx \dfrac{1}{p_m^2}$ for SWP				{s}
$\bar{\Delta} = 1$ for DWP				{s,p}

Continuation of Table 10.1

B. SPECTRAL FUNCTIONS			
$\frac{1}{3}K_{\parallel}$	$\begin{Bmatrix}4\\2\end{Bmatrix}\frac{p_m^2}{4p_m^2-z^2},\begin{Bmatrix}p\\s\end{Bmatrix}$	$\frac{1}{3}K_{\perp}$	$\frac{1}{2(p_m^2-z^2)},\ \{s,p\}$
L	$\begin{Bmatrix}1/2\\1\end{Bmatrix}\left[\frac{1}{p_m^2-z^2}+\frac{1}{2p_m^2\left(4p_m^2-z^2\right)}\right]\approx\begin{Bmatrix}1/2\\1\end{Bmatrix}\frac{1}{p_m^2-z^2},\begin{Bmatrix}p\\s\end{Bmatrix}$		

C. COMPLEX SUSCEPTIBILITY		
$\chi_{\parallel}^{*}$	$\frac{3Gz}{\left(4p_m^2-z^2\right)\begin{Bmatrix}4\\2\end{Bmatrix}xp_m^2+2\,iyp_m^2}$ for SWP	$\begin{Bmatrix}p\\s\end{Bmatrix}$
$\chi_{\parallel}^{*}$	$\frac{3Gz}{\left(4p_m^2-z^2\right)\begin{Bmatrix}4\\2\end{Bmatrix}xp_m^2+2\,iy}$ for DWP	$\begin{Bmatrix}p\\s\end{Bmatrix}$
$\chi_{\perp}^{*}$	$\frac{3G}{2\left(p_m^2-xz\right)}$ for SWP, DWP	$\{p,s\}$
χ^{*}	$\frac{3G}{\left(p_m^2-xz\right)\begin{Bmatrix}1/2\\1\end{Bmatrix}}$ for SWP	$\begin{Bmatrix}p\\s\end{Bmatrix}$
χ^{*}	$\frac{3G\begin{Bmatrix}1/2\\1\end{Bmatrix}}{\left(p_m^2-z^2\right)x+iy}$ for DWP	$\begin{Bmatrix}p\\s\end{Bmatrix}$

$$\frac{\chi^*}{3G} = \frac{z(1/3)K(z)}{x + \frac{iy}{\Delta}(1/3)K(z)} = \frac{x + iy}{\frac{x}{\langle q^2 \rangle L} + \frac{iy}{\Delta}}. \quad (10.33)$$

For an *anisotropic* medium we have to calculate

$$\chi_{\|}^*,\ (1/3)K_{\|}(z) \equiv \langle q_{\|}^2 \rangle L_{\|},\ \Delta \equiv \langle q_{\|}^2 \rangle - \langle q_{\|} \rangle^2 ; \quad (10.34)$$

$$\chi_{\perp}^*,\ (1/3)K_{\perp}(z) \equiv \langle q_{\|}^2 \rangle L_{\perp},\ \Delta \equiv \langle q_{\perp}^2 \rangle . \quad (10.35)$$

For an isotropic medium we take in Eq. 10.33

$$L(z) = \frac{1}{3}\left(K_{\|}(z) + \begin{Bmatrix} 1 \\ 2 \end{Bmatrix} K_{\perp}(z) \right),\ \bar{\Delta} \equiv \Delta + \begin{Bmatrix} 1 \\ 2 \end{Bmatrix} \langle q_{\perp}^2 \rangle \text{ for } \begin{Bmatrix} p \\ s \end{Bmatrix}. \quad (10.36)$$

The results of calculations are shown in Fig. 10.1 - 10.3; we take $p_m = 4$ and two values of collision frequency y ($y = 0.1$, solid lines; $y = 0.3$, dashed lines). We consider the single and double well potentials (SWP and DWP). Assuming small angular deviations θ, we take potential profiles $U(\theta)$ in the form

$$U(\theta) = (1/2)\ U_0\ (1 - \cos\theta) \text{ for SWP, } U(\theta) = U_0\ (1 - \cos^2\theta) \text{ for DWP.}$$

In the first (SWP) case the potential $U(\theta)$ corresponds to the constant field (CF) model. In the DWP configuration two equilibrium states are introduced, viz. those, corresponding to $\theta = 0$ and $\theta = \pi$, both being coincident with the symmetry axis Z. The mirror symmetry of the potential in addition to the axial one is a characteristic feature of the DWP configuration, whereas in the SWP one only axial symmetry of $U(\theta)$ exists.

The following conditions should be satisfied for the corresponding potential wells: $|\theta| \ll 1$ and $|\pi - \theta| \ll 1$. Dipoles, librating in two wells in the case of the DWP configuration, are considered to be *statistically* independent; in our approximation the law of motion in each well is the same as that, found for the SWP configuration. However, statistical averages, spectra and relaxation times may differ.

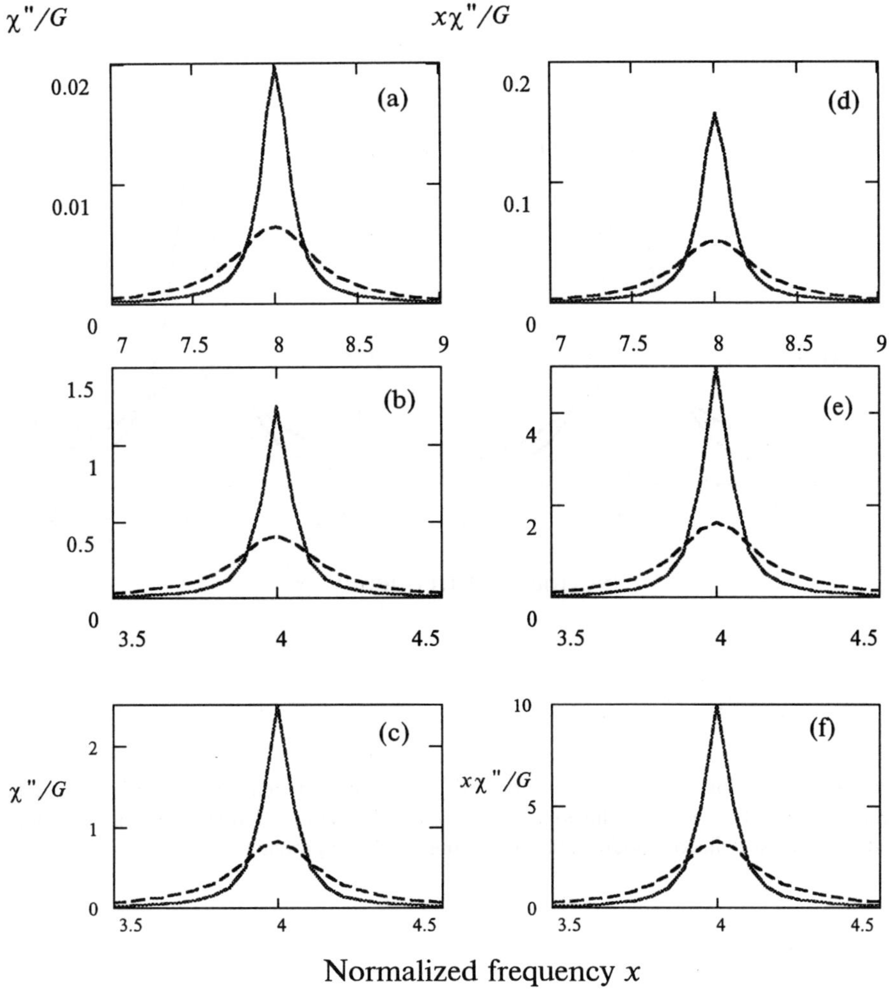

Fig. 10.1 The loss (a, b, c) and absorption (d, e, f) frequency dependencies for the single potential well: the parabolic potential, Gross collision model. y = 0.1 (solid lines) and 0.3 (dashed lines). The longitudinal (a, d), transverse (b, e) orientations of alternating field and an isotropic medium (c, f).

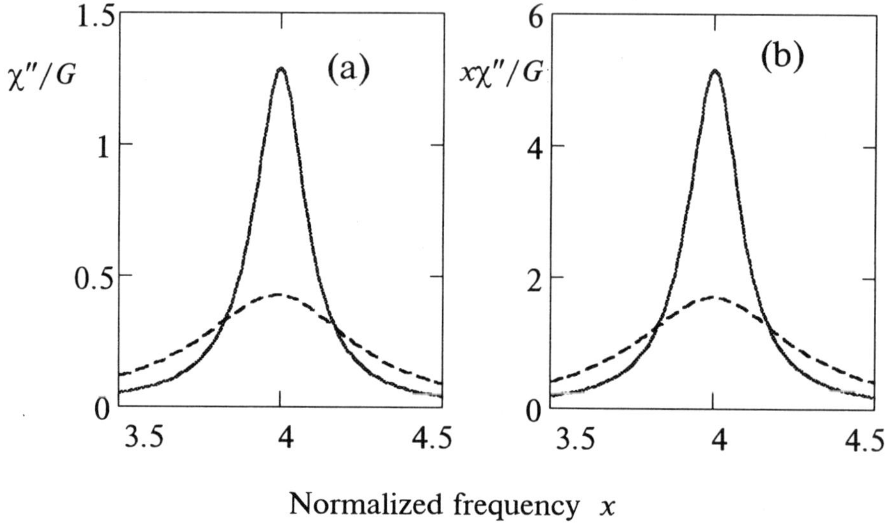

Fig. 10.2 The loss (a) and absorption (b) frequency dependencies for the double potential well, $y = 1$ (solid lines) and 0.3 (dashed lines). Calculated for an isotropic medium / Gross collision model. The case of the parabolic potential.

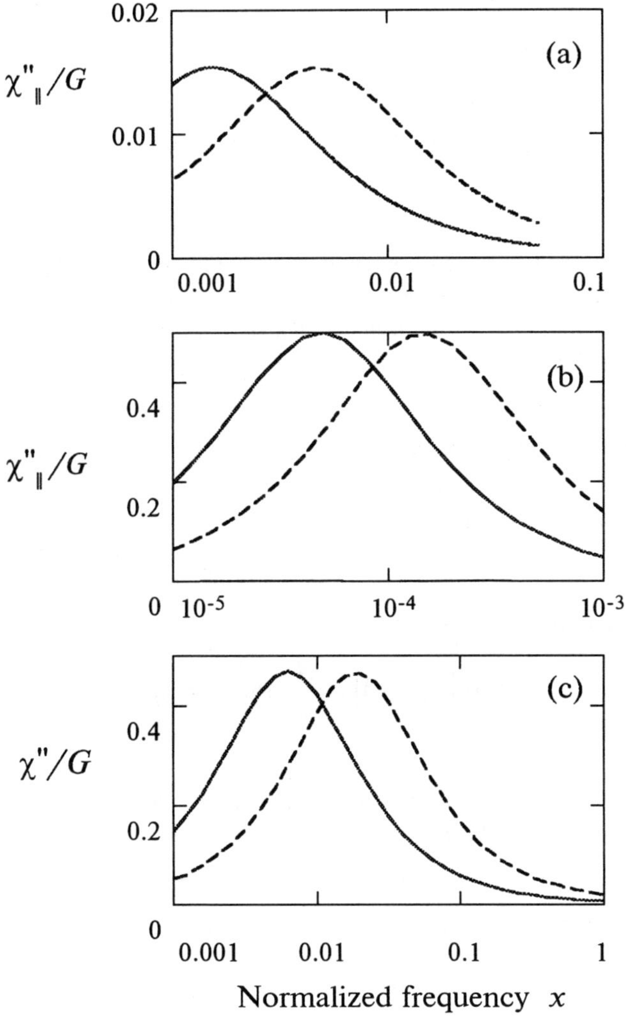

Fig. 10.3 The low-frequency (Debye) loss spectra calculated for the single (a) and double (b, c) well potentials. The longitudinal (a, b) susceptibility and that, relevant to an isotropic medium (c). $y = 0.1$ (solid lines) and 0.3 (dashed lines); $p_m = 4$.

10.4 The Poley Absorption

The latter is practically determined by the resonance terms

$$4p_m^2 - z^2, \; p_m^2 - z^2 \text{ or } p_m^2 - xz$$

in denominators of expressions for the susceptibility, see Table 10.1, Sec. C. Due to these denominators

---

Absorption $x\chi''(x)$ have peaks at the normalized frequency $x_{\|} \approx 2p_m$ and $x_{\perp} \approx p_m$ respectively for longitudinal and transverse orientations of alternating field. For an isotropic medium the main absorption peak is placed near $x \approx p_m$. (10.37)

Indeed, the frequency dependence

$$x\chi''(x) = Ax \operatorname{Im}\left(\frac{1}{p_m^2 - xz}\right) \equiv A \frac{x^2 y}{\left(p_m^2 - x^2\right)^2 + x^2 y^2} \tag{10.38}$$

goes through maximum exactly at $x = p_m$:

$$\max\left[x\chi''(x)\right] = \frac{A}{y} \quad \text{for } x = p_m. \tag{10.39}$$

The value of this maximum increases infinitely, when the collision frequency y tends to zero, (10.40)

---

the integrated absorption being, of course, independent on y.

For other formulas, given in Table 10.1, Sec. C, the property (10.37) is approximately valid, if the collision frequency y is not large:

$$x\chi''(x) = \max, \quad \text{for } x = \begin{Bmatrix} x_{\|} \\ x_{\perp} \end{Bmatrix} \approx \sqrt{\begin{Bmatrix} 4 \\ 1 \end{Bmatrix} p_m^2 - y^2}\,, \text{ if } y << p_m. \tag{10.41}$$

This statement is confirmed by Fig.10.1, 10.2. We consider only the *spatial* ensemble of librators, since for rotation in plane the form of dispersion curves remains practically the same (Sec. C of Table 10.1).

Note, that the maximal absorption is much larger for the transverse orientation, than for the longitudinal one:

$$\max\,[x\chi_{\perp}''(x)] >> \max\,[x\chi_{\parallel}''(x)] \quad \text{for } p_m >> 1. \qquad (10.41a)$$

In the small y limit the loss frequency dependence reduces to the δ-function:

$$\chi_{\parallel}''(x) \propto \delta\left(x^2 - 4p_m^2\right),\ \chi_{\perp}''(x) \propto \delta\left(x^2 - p_m^2\right) \ \text{for } y \to 0, \qquad (10.42)$$

$\chi_{\parallel}''(x)$ being $\approx$ $\chi_{\perp}''(x)$ in the case of an isotropic medium.

For the SWP and DWP configurations dielectric spectra are very close: they practically coincide in the case of an anisotropic media and slightly differ in the case of an isotropic medium (cf. Fig. 10.1 c, f with Fig. 10.2).

10.5 The Loss Frequency Dependence for the Single Well Potential

The imaginary part of the susceptibility

$$\chi''(x) = A\,\mathrm{Im}\left(\frac{1}{p_m^2 - xz}\right) \equiv A\,\frac{xy}{\left(p_m^2 - x^2\right)^2 + x^2y^2} \qquad (10.43)$$

attains its maximum at

$$x_m = \frac{1}{\sqrt{6}}\left[\sqrt{\left(2p_m^2 - y^2\right) + 12p_m^4} + 2p_m^2 - y^2\right]^{1/2}. \qquad (10.44)$$

It follows from Eq. 10.44, that the frequency x_m is commensurable with p_m for small and moderate y values. For instance,

$$\text{If } y \to 0,\ x_m = p_m\ , \qquad (10.45)$$

$$\text{If } y \to 2\,p_m^2,\ x_m = \frac{p_m}{3^{1/4}}. \qquad (10.46)$$

The loss-peak frequency x_m becomes small only at very frequent collisions, i.e. if

$$y >> p_m. \qquad (10.47)$$

Consequently, noting Table 10.1, Sec. C, we conclude, that

The low-frequency (Debye) dielectric relaxation does not appear, if one considers the transverse/isotropic susceptibility $\chi_{\perp}^{*}(x)$ *or* $\chi^{*}(x)$ *for dipoles, librating in the parabolic potential well.* (10.48)

We have seen in Sec.8.6 that such situation is relevant also to the confined rotator model.

We remark that the longitudinal susceptibility behaves in another fashion, since $\chi_{\|}^{*}(x)$ exhibits the low frequency dispersion (see Fig. 10.3 a).

10.6 The Loss Frequency Dependence for the Double Well Potential

The transverse susceptibility is the same as that, found for the SWP-case, i.e. the Debye relaxation does not appear for transverse susceptibility $\chi_{\perp}^{*}(x)$. But the situation changes, if one considers the susceptibilities $\chi_{\|}^{*}(x)$ and $\chi^{*}(x)$ (Fig. 10.3 b, c). The loss maximum $\chi''_{D\|}$ is shifted to small x-region comparing with its position relevant to the longitudinal SWP susceptibility.

We can estimate the relaxation time, using formulas, presented in Table 6.2 (the Gross collision model, $F = F_G$, is considered now):

$$\frac{\tau_{D\|}}{\tau} \approx \frac{\sigma_{\|}}{K_{\|}(iy)} = \frac{A}{(1/3)\,K_{\|}(iy)} = \frac{\Delta}{\left\langle q_{\|}^{2}\right\rangle L_{\|}(iy)}. \tag{10.49}$$

Taking into account Table 10.1, Sec. C, we have:

$$\frac{\tau_{D\|}}{\tau} \approx \begin{Bmatrix} 16 \\ 8 \end{Bmatrix} p_m^4 \left(1 + \frac{y^2}{4p_m^2}\right) \quad \text{for} \quad \begin{Bmatrix} p \\ s \end{Bmatrix}, \text{ DWP.} \tag{10.50}$$

Consequently, the low frequency loss peak is placed on the frequency

$$x_{D\|} \approx \frac{\eta}{\tau_{D\|}} \approx y\left[\begin{Bmatrix}16\\8\end{Bmatrix} p_m^4\left(1+\frac{y^2}{4p_m^2}\right)\right]^{-1} \quad \text{for} \quad \begin{Bmatrix}p\\s\end{Bmatrix}, \text{ DWP.} \quad (10.51)$$

The comparison of Fig. 10.3b with Fig. 10.3a shows, that the longitudinal relaxation time is much larger for the DWP, than that for the SWP. This property is caused by the difference in the number of the potential wells.

For an isotropic medium the relaxation time is lesser, than the value (10.50):

$$\frac{\tau_D}{\tau} \approx \frac{\eta}{\tau_{D\|}} \approx \begin{Bmatrix}2\\1\end{Bmatrix} p_m^2\left(1+\frac{y}{p_m^2}\right) \quad \text{for} \quad \begin{Bmatrix}p\\s\end{Bmatrix}, \text{ DWP.} \quad (10.52)$$

The corresponding relaxation time-lifetime dependence,

$$\frac{\tau_D}{\tau} \approx \begin{Bmatrix}2\\1\end{Bmatrix} p_m^2\left[\tau+\frac{\eta^2}{\tau p_m^2}\right], \quad (10.53)$$

has the point of minimum:

$$\min\{\tau_D(\tau)\} \approx \begin{Bmatrix}4\\2\end{Bmatrix} \eta\, p_m \quad \text{at} \quad \tau_{\min} = \frac{\eta}{p_m}. \quad (10.54)$$

This $\tau_{\min}$ corresponds to very frequent collisions. In this case $\tau_{\min} << \eta$, where η is much longer time scale, corresponding to the mean period of isolated free rotating polar molecule.

CHAPTER 11
THE CONSTANT FIELD (CF) MODEL

11.1 The Steady State Law of Motion

EQUATION OF MOTION. In this and next chapters we consider a planar motion of a dipole. We denote ϑ as the angular shift from the symmetry axis Z of the potential $U(\vartheta)$. The conservative electric field $\boldsymbol{E}_0$ is constant in the CF model by definition. The corresponding conservative potential $U(\vartheta)$ has the cosine profile:

$$U(\vartheta) = \mu E_0 \,(1 - \cos\vartheta) = (U_0/2)\,(1 - \cos\vartheta), \tag{11.1}$$

where $U_0 = 2\mu E_0$ is the potential welldepth (we assume that $U = 0$ at the bottom of the well, where $\vartheta = 0$). Without loss of generality the constant term $U_0/2 = \mu E_0$ in (11.1) may be excluded:

$$U(\vartheta) = \mu E_0 \cos\vartheta = (U_0/2)\cos\vartheta, \tag{11.2}$$

For planar motion we put $\dot{\psi} = l = 0$. Substituting θ by ϑ in equation of motion (3.45) and using dimensionless variables, we have

$$h = \frac{H}{k_B T} = \dot{\vartheta}^2 + \frac{U(\vartheta)}{k_B T} = \dot{\vartheta}^2 - 2p^2\cos\vartheta, \tag{11.3}$$

where $\dot{\vartheta} \equiv d\vartheta / d\varphi$ and the field parameter p is defined as

$$p = \sqrt{\frac{\mu E_0}{2k_B T}} = \frac{1}{2}\sqrt{\frac{U_0}{k_B T}}\,. \tag{11.4}$$

The longitudinal directional cosine $q_{\|} = \cos\vartheta$; $\dot{q}_{\|} = -\dot{\vartheta}\sin\vartheta$; we may rewrite equation of motion (11.3) in terms of the $q_{\|}(\varphi)$ function:

$$h = \frac{\dot{q}_{\|}^{2}}{1-q_{\|}^{2}} - 2p^{2}q_{\|}\,; \tag{11.5}$$

$$\frac{dq_{\|}}{d\varphi} = \sqrt{\left(h+p^{2}q_{\|}\right)\left(1-q_{\|}^{2}\right)}\,. \tag{11.6}$$

This equation can be solved in terms of elliptic integrals, while the time dependence $q_{\|}(\varphi)$ - in terms of Jacobi elliptic functions, h being an arbitrary constant, and the other one (constant φ_0) is additional to time φ (see further Eq. 11.11).

The *type* of molecular rotation depends on the ratio h/p^2.

A. LIBRATORS. The minimal energy h, corresponding to the constant (over time φ) orientation of a dipole μ along $\boldsymbol{E}_0$, can be found from Eq. 11.5, if we take $\dot{q}_{\|}$ and set $\vartheta = 0$, i.e. $q_{\|} = 1$:

$$h_{\min} = h\left\{\dot{q}_{\|} - 0,\ q_{\|} = 1\right\} = -2p^{2}, \text{ so that } H_{\min} = -\mu E_0. \tag{11.7}$$

Dipoles with this h value possess only potential energy (it is minimal). If $h > h_{\min}$, the zero velocity ($\dot{q}_{\|} = 0$) corresponds to the instant, when the direction of angular velocity of a dipole changes (in other words, when the «reflection» occurs of a dipole inside the potential well). We have again from Eq. 11.5:

$$h\left\{\dot{q}_{\|} = 0,\right\} = -2p^{2}\cos\vartheta_{\max}, = -2p^{2}q_{\|}\left\{\dot{q}_{\|} = 0\right\}. \tag{11.8}$$

Therefore, the libration amplitude β is given by

$$\beta(h) = \vartheta_{\max}(h) = \operatorname{acos}\left(-\frac{h}{2p^2}\right), \tag{11.9}$$

$h = 2p^2$ being the *threshold* value of the energy, viz. for $h > 2p^2$ librations vanish. Since in accordance with the definition $|q_\parallel| < 1$, the finite energy interval, corresponding to the *librational state*, is given by

$$-2p^2 \le h \le 2p^2, \{ \text{librational state} \}. \tag{11.10}$$

HINDERED ROTATORS. For $h > 2p^2$ reflections of a dipole do not occur. The law of motion of a dipole resembles that of a fast rotating pendulum. The larger energy h, the closer the law of motion of a dipole to the law of a free rotation, when velocity $\dot{\vartheta} = \sqrt{h}$ is constant.

THE LAW OF MOTION. We denote $q_\parallel = x$. The $\varphi(x)$ dependence can be found from Eq. 11.6 as integral

$$\sqrt{2}\, p(\varphi + \varphi_0) = \int_1^{q_\parallel} \frac{dx}{\sqrt{(1-x)\left(x + \frac{h}{2p^2}\right)(1+x)}}, \quad , \tag{11.11}$$

wherein the lower limit $q_\parallel = \cos\vartheta = 1$ corresponds to $\vartheta = 0$, viz. to the orientation along the symmetry axis Z.

Now we write down the standard integral (Prudnikov, Brychkov and Marychev 1981, p. 61, No. 1.2.27.1):

$$-\frac{\sqrt{a-c}}{2} \int_a^{q_\parallel} \frac{dx}{\sqrt{(a-x)(x-b)(x-c)}} \equiv u = F(\gamma, k)), \tag{11.12}$$

$$u = -\sqrt{\frac{a-c}{2}}\, p(\varphi + \varphi_0); \tag{11.13}$$

$$a > x \ge b > c; \tag{11.14}$$

$$k = \sqrt{\frac{a-b}{a-c}};\ \sin^2\gamma = \operatorname{sn}^2(u,k) = \frac{a-x}{a-b}; \tag{11.15}$$

$F(\gamma, k)$ is the elliptic integral of the first kind; k is its modulus; $\mathrm{sn}(u, k)$ and (see further) $\mathrm{cn}(u, k)$, $\mathrm{dn}(u, k)$ are Jacobi elliptic functions, having the argument u and the same modulus k. Later we shall also use the parameter

$$m = k^2. \tag{11.16}$$

In the A class (of librators) in view of inequality (11.14) and Eq. 11.8 we have for the lower value of $x \equiv q_{||}$:

$$\min\left(q_{||}\right) = \cos\theta_{\max} = -h / p^2 . \tag{11.17}$$

So we put

$$b = -h/\left(2p^2\right), a = 1, c = -1. \tag{11.18}$$

Substituting these parameters into Eq. 11.12 - 11.15 and using symbol $^{\vee}$ for librators, we express $k \equiv \check{k}$ and $q_{||} \equiv \check{q}_{||}$ respectively as functions of h and φ:

$$k = \sqrt{\frac{1}{2}}\sqrt{1+\frac{h}{2p^2}}, u = p(\varphi + \varphi_0); \tag{11.19}$$

$$\check{q}_{||}(\varphi) = 1 - 2\check{k}^2 \mathrm{sn}^2(u,\check{k}) = 2\,\mathrm{dn}^2\left[p(\varphi + \varphi_0), \check{k}\right] - 1, \tag{11.20}$$

where we employ identities

$$\mathrm{dn}^2 u + k^2 \sin^2\gamma \equiv \mathrm{dn}^2 u + k^2 \mathrm{sn}^2 u \equiv 1, \quad ; \tag{11.21}$$

where γ denotes argument of elliptic integral. In Eq. 11.21 and in many others, that follow further, we omit designation of modulus k:

$$\mathrm{dn}\, u \equiv \mathrm{dn}\,(u, k);\ \mathrm{sn}\, u \equiv \mathrm{sn}\,(u, k);\ \mathrm{cn}\, u \equiv \mathrm{cn}\,(u, k). \tag{11.22}$$

The direction cosine for $\perp$-orientation is easily found as

$$\check{q}_{\perp}(\varphi) = \sqrt{1 - \check{q}_{||}^2(\varphi)} = \sqrt{4\,\mathrm{dn}^2 u + 4\,\mathrm{dn}^4 u}$$

$$= 2\check{k}\,\mathrm{sn}\left[p(\varphi+\varphi_0),\,\check{k}\right]\mathrm{dn}\left[p(\varphi+\varphi_0),\,\check{k}\right]. \tag{11.23}$$

In the B class (of hindered rotors), where $h/(2p^2) > 1$, the minimal value of $q_{\|}$ is - 1; then we put in integral (11.12):

$$a = 1,\ b = -1,\ c = -\frac{h}{2p^2}. \tag{11.24}$$

We employ symbol ° and have from Eq. 11.13 - 11.16:

$$\mathring{k} = \sqrt{\frac{2}{1+\dfrac{h}{2p^2}}} = \frac{1}{\check{k}}\ , \tag{11.25}$$

$$u = -\frac{1}{\sqrt{2}}\sqrt{\frac{1}{1+\dfrac{h}{2p^2}}}\ p(\varphi+\varphi_0) = -\frac{p(\varphi+\varphi_0)}{\mathring{k}}, \tag{11.26}$$

$$\mathring{q}_{\|}(\varphi) = 1-2\,\mathrm{sn}^2(u,\,\mathring{k}) \equiv 2\,\mathrm{cn}^2\left(\frac{p(\varphi+\varphi_0)}{\mathring{k}},\,\mathring{k}\right) - 1\ , \tag{11.27}$$

where we have used identity

$$\mathrm{cn}^2 u + \mathrm{sn}^2 u = 1. \tag{11.28}$$

By analogy with Eq. 11.23 we write:

$$\mathring{q}_{\perp}(\varphi) = \sqrt{1-\mathring{q}_{\|}^{\,2}(\varphi)} = \sqrt{4\,\mathrm{cn}^2 u - 4\,\mathrm{cn}^4 u}$$

$$= 2\,\mathrm{sn}\left[\frac{p(\varphi+\varphi_0)}{\mathring{k}},\,\mathring{k}\right]\mathrm{cn}\left[\frac{p(\varphi+\varphi_0)}{\mathring{k}},\,\mathring{k}\right]. \tag{11.29}$$

THE ENERGY DEPENDENT PERIOD OF REORIENTATION. Let $\check{p}$ be the normalized libration frequency of a dipole around the symmetry axis of the potential (11.1); $\check{p}$ is two times lesser than the fundamental frequency of $\check{q}_{\|}(\varphi)$ function. The period of sn(u) is $4\mathbf{K}(m)$, and that of sn(u) and dn(u) is $2\,\mathbf{K}(m)$, where

$$\mathbf{K}(m) \equiv F\left(\frac{\pi}{2}\middle| m\right) \tag{11.30}$$

is complete elliptic integral of the first kind.

In the A class, in accordance with definition of frequency $\check{p}$, we have:

$$\check{p} = \frac{2\pi p}{2 \cdot 2\mathbf{K}(\check{m})} = \frac{\pi p}{2\mathbf{K}(\check{m})}, \tag{11.31}$$

$$\check{m} = \check{k}^2 \equiv \frac{4p^2}{2p^2 + h}. \tag{11.32}$$

In the B class the period of

$$f(\varphi) = \mathrm{sn}\left[\frac{p(\varphi + \varphi_0)}{\mathring{k}}, \mathring{k}\right]$$

is equal to $\dfrac{4\mathbf{K}(\mathring{m})\sqrt{\mathring{m}}}{p}$; the period $\mathring{q}_{\perp}(\varphi)$ q (v) in (11.29) is half of the latter. Thus the frequency of full rotation is given by

$$\mathring{p} = \frac{2\pi p}{2\mathbf{K}(\mathring{m})\sqrt{\mathring{m}}} \tag{11.33}$$

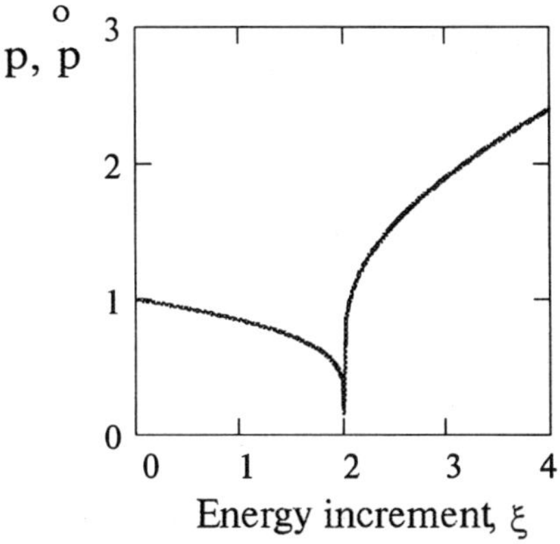

Fig. 11.1 The fundamental frequency of librational (1) and hindered-rotational (2) motion vs. energy increment

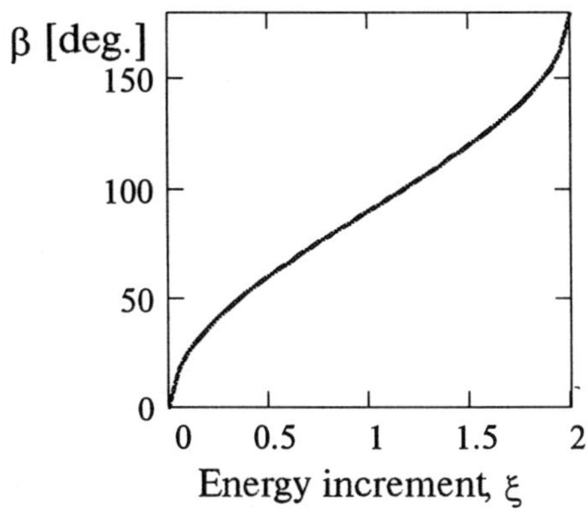

Fig.11.2 Libration amplitude β vs. normalized energy increment.

$$\mathring{m} = \mathring{k}^2 = \frac{4p^2}{2p^2 + h} \tag{11.34}$$

In Fig. 11.1 we picture frequencies $\check{p}$ and $\mathring{p}$ as functions of the energy increment

$$\xi = \frac{h - h_{\text{min}}}{h_{\text{th}}} = \frac{h + 2p^2}{2p^2} \equiv \frac{H + \mu E_0}{\mu E_0}, \tag{11.35}$$

where $h_{\text{th}} = 2p^2$ is the threshold energy of a dipole. The dependence on ξ of the libration amplitude β is presented in Fig. 11.2. We remark that β, $\check{m}$ and $\mathring{m}$ depend on ξ as

$$\beta = a\cos(1 - \xi),\ \check{m} = \frac{\xi}{2},\ \mathring{m} = \frac{2}{\xi}. \tag{11.36}$$

If energy is minimal ($\xi = 0$) the librational frequency $\check{p}$ is maximal. A dipole, being oriented along $\boldsymbol{E}_0$ field, performs librations with an infinitesimal amplitude β:

$$\max\left(\check{p}\right) = p,\ \max\left(\check{\Omega}\right) = \max\left(\frac{\check{p}}{\eta}\right) = \sqrt{\frac{\mu E_0}{I}}, \tag{11.37}$$

$$\beta\left(h_{\min}\right) = \beta\left(\xi = 0\right) = 0. \tag{11.38}$$

Since $\check{m}\left(\xi = 0\right) = 0$, these librations are harmonic, so that sn$u$ and cn$u$ reduce respectively to sin$u$ and cos$u$.

When ξ (or h) rises, the frequency $\check{p}$ decreases and drops to zero at $\xi = 2$ (or $h = h_{\text{th}} = 2p^2$). The libration amplitude β becomes maximal and equal to π at this threshold energy. For $h > h_{\text{th}}$ librations become forbidden; rotational frequency $\mathring{p}$ sharply increases in the small region near h_{th}. For $h >> 1$ we have an asymptotic dependence $\mathring{p} \approx \sqrt{h}$, corresponding to free rotors. In the vicinity of h_{th} both parameters, $\check{m}$ and $\mathring{m}$, tend to 1, Jacobi elliptic functions reduce to hyperbolic functions:

$$q_{\|}(\varphi) \to 2\operatorname{sech}^2(p\varphi) - 1;\ q_{\|}(\varphi) = 2\operatorname{sech}^2(p\varphi)\sinh(p\varphi)\ . \tag{11.39}$$

At $\varphi \to \infty$ vector $\boldsymbol{\mu}$ asymptotically tends to the orientation, opposite to that of $\boldsymbol{E}_0$ vector, since $q_{\perp} \to 0$ and $q_{\|} \to \infty$ in this limit.

11.2 Distribution Functions

THE STEADY STATE ENERGY DISTRIBUTION $W(h) = C\exp(-h)$.

To find normalizing constant C we chose the pair ϑ, $\dot{\vartheta}$ of canonical variables. In accordance with Eq. 11.3

$$\frac{1}{C} = \int_{-\infty}^{\infty} \exp\left(-\dot{\vartheta}^2\right) d\dot{\vartheta} \int_{-\pi}^{\pi} \exp\left(2p^2 \cos\vartheta\right) d\vartheta . \tag{11.40}$$

Integrals over $\dot{\vartheta}$ and ϑ yield respectively $\sqrt{\pi}$ and zero-order modified Bessel function, since

$$I_n(z) = \frac{1}{\pi} \int_0^{\pi} \exp(z \cos\vartheta) \cos n\vartheta \, d\vartheta . \tag{11.41}$$

Hence,

$$C = \frac{1}{2\pi\sqrt{\pi}\, I_0\left(2p^2\right)} . \tag{11.42}$$

It is instructive to write down another expression for the same constant C by choosing h and φ_0 as a pair of canonical variables, , where $h \in [h_{\min}, \Phi]$, Φ being the period of $q_{\|}(\varphi + \varphi_0)$ function. We remark that

$$h = \dot{\vartheta}_0^2 + u(\vartheta_0); \; \varphi_0 = \int_0^{\vartheta_0} \frac{d\vartheta_0}{\dot{\vartheta}_0} . \tag{11.43}$$

We may transform integrals over $\dot{\vartheta}_0$ and ϑ_0 to those over h, φ_0 by calculating the Jacobian

$$D = \frac{\partial(h, \varphi_0)}{\partial\left(\dot{\vartheta}_0, \vartheta_0\right)} = \begin{vmatrix} 2\dot{\vartheta}_0 & \dfrac{\partial u}{\partial \dot{\vartheta}_0} \\ 0 & \dfrac{1}{\dot{\vartheta}_0} \end{vmatrix} = 2 . \tag{11.44}$$

We note that interval $[0, \Phi]$ of integration over φ_0 corresponds to $\vartheta \in [-\pi, \pi]$. On the other side, integral over h corresponds to $\dot{\vartheta} \in [0, \infty]$. Thus, in view of Eq. 11.40

$$\frac{1}{C} = \int_{-2p^2}^{\infty} \exp(-h)\, dh \int_{0}^{\Phi} d\varphi_0 \,. \tag{11.45}$$

The normalized distribution function

$$W = \frac{\exp(-h)}{2\pi\sqrt{\pi}\; I_0\left(2p^2\right)} \tag{11.46}$$

can be employed for calculation of any averaged quantity $\langle F \rangle$ in two following fashions:

$$\langle F \rangle = \frac{1}{2\pi\sqrt{\pi}\; I_0\left(2p^2\right)} \int_{-\infty}^{\infty} \exp\left(-\dot{\vartheta}^2\right) d\dot{\vartheta} \int_{-\pi}^{\pi} F\left(\vartheta, \dot{\vartheta}\right) \exp\left(2p^2 \cos\vartheta\right) d\vartheta \tag{11.47}$$

or equivalently

$$\langle F \rangle = \frac{1}{2\pi\sqrt{\pi}\; I_0\left(2p^2\right)} \int_{h_{\min}}^{\infty} \exp(-h)\, dh \int_{0}^{\Phi} F\left(h, \varphi_0\right)\, d\varphi_0 \tag{11.48}$$

DISTRIBUTIONS OVER FREQUENCIES OF PERIODIC MOTION. We put $F(\bullet) \equiv 1$ in Eq. 11.48. Then the last formula reduces to the normalization condition

$$\int_{h_{\min}}^{\infty} W(h)\, \Phi(h)\, dh = 1, \tag{11.49a}$$

which we rewrite as the sum of two integrals over frequencies $\left\{ \overset{\vee}{p}, \overset{\circ}{p} \right\}$ of periodic motion, the variable h being substituted by the corresponding frequency, as it was done in Eq. 5.179.

Thus, Eq. 11.49 reduces to

$$C \int_{-2p^2}^{2p^2} \exp\left[-h\left(\overset{\vee}{p}\right)\right] \frac{\overset{\vee}{\Phi}}{\partial \overset{\vee}{p} / \partial h} d\overset{\vee}{p} + C \int_{2p^2}^{\infty} \exp\left[-h\left(\overset{\circ}{p}\right)\right] \frac{\overset{\circ}{\Phi}}{\partial \overset{\circ}{p} / \partial h} d\overset{\circ}{p}\,, \tag{11.50}$$

where C is given by Eq. 11.42. Then we have for functions $\check{f}\left(\check{p}\right)$ and $\mathring{f}\left(\mathring{p}\right)$, defined by Eq. 5.187:

$$\check{f}\left(\check{p}\right) = -\frac{\exp\left[-h\left(\check{p}\right)\right]}{2\sqrt{\pi}\,\check{p}\,I_0\left(2p^2\right)}\frac{1}{\partial\check{p}/\partial h};\quad \mathring{f}\left(\mathring{p}\right) = \frac{\exp\left[-h\left(\mathring{p}\right)\right]}{\sqrt{\pi}\,\mathring{p}\,I_0\left(2p^2\right)}\frac{1}{\partial\mathring{p}/\partial h}. \tag{11.51}$$

To find explicit expressions, we remark that derivative $d\mathbf{K}/dk$ is given by

$$\frac{d\mathbf{K}}{dk} = \frac{\mathbf{E} - m'\mathbf{K}}{m'\sqrt{m}} \tag{11.52}$$

$\mathbf{E}(m)$ is complete elliptic integral of the second kind;

$$m' \equiv 1 - m = 1 - k^2;\ \text{we employ also}\ k' \equiv \sqrt{1-k^2} \equiv \sqrt{m'}\,. \tag{11.53}$$

Thus, with account of Eq. 11.31 we have:

$$\frac{\partial\check{p}}{\partial h} = \frac{\pi p}{2}\frac{\partial}{\partial h}\frac{1}{\mathbf{K}\left[\check{m}(h)\right]};\quad \check{m}(h) = \frac{1}{2} + \frac{h}{4p^2}; \tag{11.54}$$

$$\frac{\partial\check{p}}{\partial h} = \frac{\pi p}{2}\frac{1}{\mathbf{K}^2}\frac{\mathbf{E} - m'\mathbf{K}}{m'\sqrt{m}}\frac{1}{2\sqrt{m}}\frac{\partial\check{m}}{\partial h} = -\frac{\mathbf{E} - m'\mathbf{K}}{4mm'}\frac{\mathbf{K}}{\pi p}\left(\frac{\check{p}}{p}\right)^2. \tag{11.55}$$

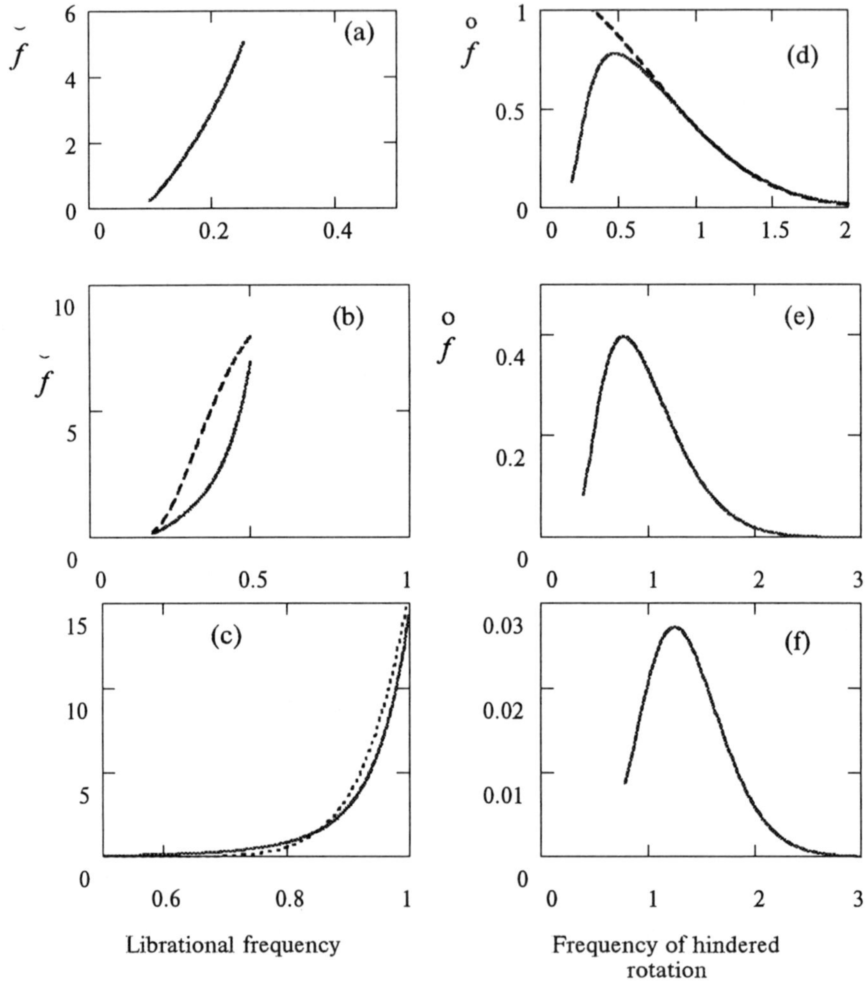

Fig.11.3 Distributions functions over normalized frequencies of a periodic librational (a, b, c) and hindered-rotational (d, e, f) motions. The field parameter p = 0.25 (a, c), 0.5 (b, e), and 1 (c, f). Dashed lines in Fig. b, c - asymptotic dependence (11.64a) and in Fig. d - the Maxwellian distribution (11.59).

Finally, noting, that $h(\overset{\vee}{m}) = 4p^2\left(\overset{\vee}{m}-\frac{1}{2}\right)$, we derive:

$$\overset{\vee}{f}\left(\overset{\vee}{p}\right) = \frac{2\sqrt{\pi}}{I_0\left(2p^2\right)}\left(\frac{p}{\overset{\vee}{p}}\right)^3 \frac{m\,m'\exp\left(-4p^2 m+2p^2\right)}{\mathbf{E}(m)-m'\mathbf{K}(m)} \quad , \tag{11.56a}$$

$$p(\overset{\vee}{m}) = \frac{\pi p}{2\mathbf{K}(m)}; \quad m \equiv \overset{\vee}{m} \in [0, 1]. \tag{11.56b}$$

With account of Eq. 11.34, 11.52 we find distribution $\overset{\circ}{f}\left(\overset{\circ}{p}\right)$ for hindered rotators:

$$\frac{\partial \overset{\circ}{p}}{\partial h} = \pi p \frac{\partial}{\partial h}\frac{1}{\overset{\circ}{k}\mathbf{K}} = -\pi p\left(\frac{1}{\overset{\circ}{k}\mathbf{K}}\right)^2 \frac{\mathbf{E}}{m'}\frac{\partial \overset{\circ}{m}}{\partial h}\frac{1}{2\overset{\circ}{k}} = \frac{\overset{\circ}{p}{}^2}{\pi p}\frac{\overset{\circ}{m}{}^2}{4p^2}\frac{\mathbf{E}}{2\overset{\circ}{m}'\sqrt{\overset{\circ}{m}}};$$

Since $h(\overset{\circ}{m}) = \exp\left(2p^2 - \frac{4p^2}{\overset{\circ}{m}}\right)$,

$$\overset{\circ}{f}\left(\overset{\circ}{p}\right) = \frac{8\sqrt{\pi}}{I_0\left(2p^2\right)}\left(\frac{p}{\overset{\circ}{p}}\right)^3 \frac{(1-m)\exp\left(2p^2-\frac{4p^2}{m}\right)}{m\sqrt{m}\,\mathbf{E}(m)} \tag{11.57a}$$

$$p(\overset{\circ}{m}) = \frac{\pi p}{\sqrt{m}\,\mathbf{K}(m)}; \quad m \equiv \overset{\circ}{m} \in [0, 1]. \tag{11.57b}$$

THE ZERO FIELD LIMITING CASE. Let $p \to 0$. Then

$$I_0\left(2p^2\right)\to 1,\ \overset{\circ}{m}\to 0,\ \mathbf{E}\left(\overset{\circ}{m}\right)\to\frac{\pi}{2},\ \overset{\circ}{p}\to\frac{2p}{\sqrt{\overset{\circ}{m}}},\ \overset{\circ}{m}\sqrt{\overset{\circ}{m}}\to\left(\frac{2p}{\overset{\circ}{p}}\right)^3 \quad ; \qquad (11.58)$$

$$\overset{\circ}{f}\left(\overset{\circ}{p}\right)\to\frac{2}{\sqrt{\pi}}\exp\left(-\overset{\circ}{p}{}^2\right) \qquad (11.59)$$

We have got Maxwellian distribution. Indeed, it should be established in the case of vanishing conservative field. In this limit $\overset{\vee}{f}\left(\overset{\vee}{p}\right)\to 0$.

THE STRONG-FIELD LIMITING CASE. For $p > 1$ the asymptotic formula

$$I_n(z)\approx\frac{\exp(z)}{\sqrt{2\pi z}}\left(1-\frac{4n^2-1^2}{1!8z}+\frac{\left(4n^2-1^2\right)\left(4n^2-3^2\right)}{1!(8z)^2}-\ldots\right) \qquad (11.60)$$

can be employed for Bessel functions. Thus,

$$I_0\left(2p^2\right)\approx\frac{\exp\left(2p^2\right)}{2\sqrt{\pi}\,p} \qquad (11.61)$$

Since $m=\frac{1}{2}+\frac{h}{4p^2}$ is rather small, we can use expansions

$$\mathbf{K}\left(\overset{\vee}{m}\right)\approx\frac{\pi}{2}\left(1+\frac{\overset{\vee}{m}}{4}\right);\ \mathbf{E}\left(\overset{\vee}{m}\right)\approx\frac{\pi}{2}\left(1-\frac{\overset{\vee}{m}}{4}\right), \qquad (11.62)$$

so that

$$\mathbf{E}\left(\overset{\vee}{m}\right)-\overset{\vee}{m}{}'\,\mathbf{K}\left(\overset{\vee}{m}\right)\approx\frac{\pi}{2}\left(1-\frac{\overset{\vee}{m}}{4}-\left[1+\frac{\overset{\vee}{m}}{4}\right]\left[1-\overset{\vee}{m}\right]\right)\approx\frac{\pi\overset{\vee}{m}}{4}.$$

As a first approximation, we find from Eq. 11.57b:

$$\mathbf{K} \approx \frac{\pi}{2}, \check{m} \approx \frac{4\left(p-\check{p}\right)}{p} \tag{11.63}$$

At last, for $p \leq \check{p}$ we find:

$$\check{f}\left(\check{p}\right) \approx 16p\left(\frac{p}{\check{p}}\right)^3 \exp\left[-16p\left(p-\check{p}\right)\right] \approx 16p\exp\left[-16p\left(p-\check{p}\right)\right].. \tag{11.64a}$$

If $p \to \infty$, $f\left(\mathring{p}\right) \to 0$, and

$$\check{f}\left(\check{p}\right) \to \delta\left(\check{p}-p\right) \text{ at } \check{p} < p, \quad \check{f}\left(\check{p}\right) \equiv 0 \text{ at } \check{p} \geq p, \tag{11.64b}$$

since $\check{f}\left(\check{p}\right)$ vanishes in all points, except $p \to \check{p}$, while the integral

$$\int_0^p \check{f}(\check{p})\, d\check{p} \approx 16p \int_0^p e^{-16p\left(p-\check{p}\right)} d\check{p} = 1 - e^{-16\check{p}^2} \to 1. \tag{11.64c}$$

For numerical calculations we may use further the following representations of the ***K***(m) and ***E***(m) functions (Abramowitz and Stegun 1964, Sec. 17.3.33 and 17.3.35):

$K(m) = a_0 + a_1 m' + a_2 m'' + \left(b_0 + b_1 m' + b_2 m''\right)\ln\frac{1}{m'}$	$E(m) = 1 + a_1 m' + a_2 m'' + \left(b_1 m' + b_2 m''\right)\ln\frac{1}{m'}$
$a_0 = 1.38629$, $b_0 = 0.5$	$a_1 = 0.46302$, $b_1 = 0.24527$
$a_1 = 0.11197$, $b_1 = 0.12135$	$a_2 = 0.10778$, $b_2 = 0.04125$
$a_2 = 0.07253$, $b_2 = 0.028873$	

(11.65)

where m' is defined by Eq. 11.53.

THE $\check{f}(\check{p})$ *AND* $\mathring{f}(\mathring{p})$ *PLOTS.* These distribution functions (DF) are shown respectively in Fig. 11.3 a, b, c and 11.3 d, e, f. The DF $\mathring{f}(\mathring{p})$ of rotators goes through a maximum. In the limit $p \to 0$ it reduces to monotonically falling Maxwellian distribution (11.59). For $p > 1$ curve $\mathring{f}(\mathring{p})$ vanishes.

The DF $\check{f}(\check{p})$ increases with the rise of $\check{p}$ and falls to zero at $\check{p} = p$; for $p > 1$ this curve becomes very narrow and close to one-side δ-function (11.64b). Specific features of distributions $\check{f}(\check{p})$ and $\mathring{f}(\mathring{p})$ stipulate those of dielectric spectra (see Sec. 11.6). For example, at $p > 1$ one may neglect the contribution of hindered rotors to spectra, since almost all dipoles librate harmonically with frequencies $\check{p}$ close to p.

11.3 The Field Dependence of the Steady State Parameters

11.3.1 Statistical Weights of Librators and Hindered Rotators

The proportion $\check{r}$ of the A-subensemble is given by

$$\check{r} = \int_{-2p^2}^{2p^2} W(h)\check{\Phi}(h)\, dh = C \int_{-2p^2}^{2p^2} \check{\Phi}(h) \exp(-h)\, dh\,. \tag{11.66}$$

We remark that

$$dh = 4p^2\, d\check{m},\ \exp\left[-h\left(\check{p}\right)\right] = \exp\left(2p^2\right) \exp\left(-4p^2 \check{m}\right). \tag{11.67}$$

Substituting $\check{\Phi} = \pi / \check{p}$, we have with account of Eq. 11.42:

$$\check{r} = \frac{4p\exp\left(2p^2\right)}{\pi\sqrt{\pi}\,I_0\left(2p^2\right)}\int_0^1 \exp\left(-4p^2 m\right)\mathbf{K}(m)\,dm. \tag{11.68}$$

For the other subensemble we have by analogy:

$$\mathring{\Phi}(h) = \frac{2\pi}{\mathring{p}},\ \mathring{r} = \int_{2p^2}^{\infty} W(h)\mathring{\Phi}(h)dh = C\int_{2p^2}^{\infty}\mathring{\Phi}(h)\exp(-h)\,dh\,. \tag{11.69}$$

Since

$$dh = -\frac{4p^2}{\mathring{m}^2}d\mathring{m},\ \exp\left[-h\left(\mathring{m}\right)\right] = \exp\left(2p^2\right)\exp\left(-\frac{4p^2}{\mathring{m}}\right), \tag{11.70}$$

we have:

$$\mathring{r} = \frac{4p\exp\left(2p^2\right)}{\pi\sqrt{\pi}\,I_0\left(2p^2\right)}\int_0^1 \exp\left(-\frac{4p^2}{m}\right)\mathbf{K}(m)\,dm. \tag{11.71}$$

For $p \to 0$ the proportion of librators tends to

$$\mathring{r} \approx \frac{8p}{\pi\sqrt{\pi}}\,. \tag{11.72}$$

Indeed, in view of standard integral

$$\Im(m) = \int k\,\mathbf{K}\,dk = \mathbf{E} - \left(1-k^2\right)\mathbf{K} \equiv \mathbf{E}(m) - m'\mathbf{K}(m)\quad; \tag{11.73}$$

$$\mathbf{E}(0) = \mathbf{K}(0) = \frac{\pi}{2}\,;\ \mathbf{E}(1) = 1;\ \Im(1) - \Im(0) = \int_0^1 k\,\mathbf{K}\,dk \approx 1.$$

Then we get the result (11.72).

In the other limit $p > 1$ one can put $\mathbf{K} \approx \pi/2$ in the integrand of Eq. 11.68 and use approximation (11.61). Then

$$\check{r} \approx \frac{4p\exp\left(2p^2\right)2\sqrt{\pi}\,p}{\pi\sqrt{\pi}\,I_0\left(2p^2\right)4p^2}\,\frac{\pi}{2}\int_0^{4p^2}\exp(-t)dt \approx 1,$$

as it should be, since almost all particles are librators.

These considerations are confirmed by plots shown in Fig. 11.4a. For $p < 1$ both $\check{r}(p)$ and $\mathring{r}(p)$ dependencies are practically linear. On the other side, we may expect that the contribution of hindered rotors to dielectric spectra should be extremely low for $p > 1$.

We remark that the sum of expressions (11.68) and (11.71), found numerically, amounts to 1.

11.3.2 The Mean Libration Amplitude

In view of Eq. 11.9 and 11.32 we find:

$$\beta(h) = \operatorname{acos}\left(-\frac{h}{2p^2}\right) = \operatorname{acos}\left(1 - 2\check{m}\right). \tag{11.74a}$$

Since $1 - \cos\beta = 2\sin^2(\beta/2) = 2\check{m}$, we can write the equivalent $\beta\left(\check{m}\right)$ representation:

$$\beta = 2\operatorname{asin}\left(\sqrt{\check{m}}\right) = 2\operatorname{asin}\left(\check{k}\right). \tag{11.74b}$$

The mean value of this quantity is given by

$$\bar{\beta} = \int_{-2p^2}^{2p^2} \beta\check{\Phi}W\,dh \bigg/ \int_{-2p^2}^{2p^2} \check{\Phi}W\,dh, \tag{11.75a}$$

or, in the equivalent form [see Eq. 11.68)],

$$\bar{\beta} = \frac{1}{\check{r}} \int_{-2p^2}^{2p^2} \beta\,\check{\Phi}(h)W(h)\,dh. \tag{11.75b}$$

Noting that

$$\check{\Phi} = \frac{2\mathbf{K}}{p}, \quad \check{m} = \frac{1}{2} + \frac{h}{4p^2}, \quad dh = 4p^2\,d\check{m}, \tag{11.76a}$$

$$\overset{\circ}{\Phi} = \frac{2\mathbf{K}}{pk}, \quad \overset{\circ}{m} = \frac{1}{\dfrac{1}{2} + \dfrac{h}{4p^2}}, \quad dh = -\,4p^2\,\frac{d\overset{\circ}{m}}{\overset{\circ}{m}{}^2} \tag{11.76b}$$

and omitting symbol $\vee$, we finally derive:

$$\bar{\beta} = \frac{\int_0^1 \operatorname{a}\cos(1-2m)\ \mathbf{K}(m)\ \exp\left(-4p^2 m\right) dm}{\int_0^1 \mathbf{K}(m)\ \exp\left(-4p^2 m\right) dm}. \tag{11.77}$$

The $\bar{\beta}$ vs. p dependence is shown in Fig. 11.4 b (curve 1). The libration amplitude decreases with the rise of conservative field. For $p > 1$ the approximation

$$\bar{\beta} \approx \frac{\sqrt{\pi}}{2p}\ \operatorname{erf}(2p)\ -\ \pi\ \exp\left(-4p^2\right) \quad \text{for } p > 1 \tag{11.78}$$

holds (see curve 2), which is close to simple expression (3.140), viz. $\bar{\beta} = \sqrt{\pi}/(2p)$, obtained for the elastic bond approximation. In order to derive Eq. 11.78 we put in Eq. 11.77

$$\mathbf{K} = \frac{\pi}{2},\ \bar{\beta} \approx 2 \int_{m=0}^{m=1} \exp\left(-4p^2 m\right) \operatorname{a}\sin(k)\ d\left(4p^2k^2\right)$$

$$= 2\left\{-\frac{\pi}{2}\exp\left(-4p^2\right) + \int_0^1 \frac{\exp\left(-4p^2k^2\right) dk}{\sqrt{1-2k^2}}\right\}$$

Since $\exp\left(-4p^2k^2\right)$ rapidly decreases with the rise of k, we may substitute multiplier $\left(1-2k^2\right)^{-1/2}$ by 1. Then the last expression yields Eq. 11.78.

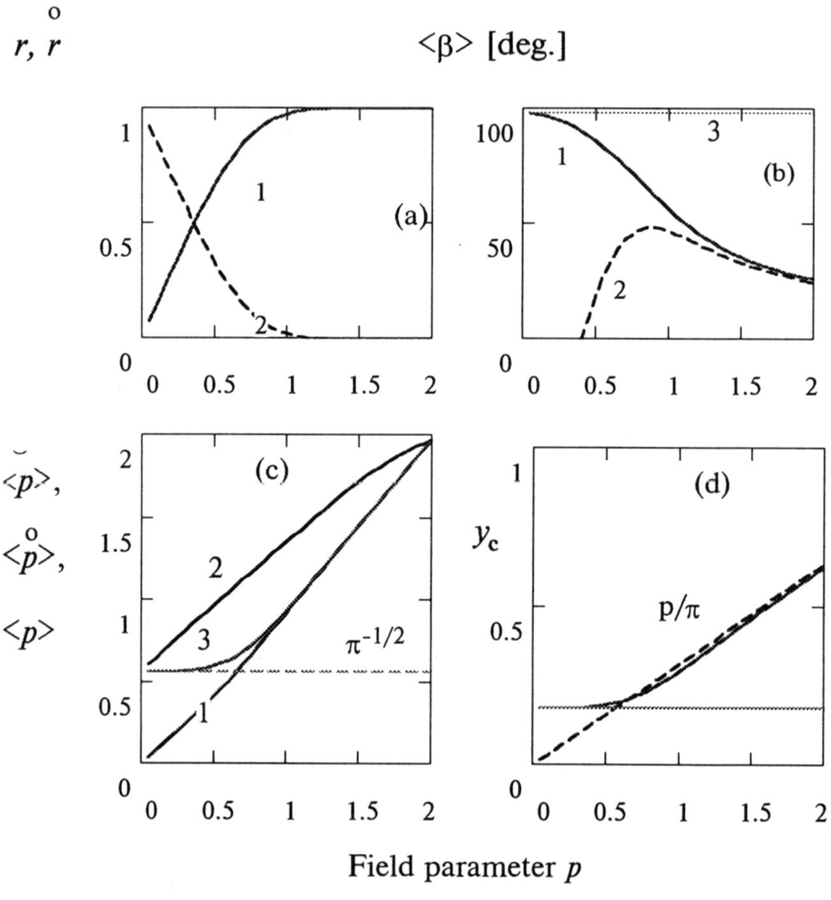

Fig. 11.4 **a:** The proportions of librators (1) and of rotators (2); **b:** the mean values of the libration amplitude β (1); curve 2 is the asymptotic dependence (11.78) and line 3 is the value (11.79); **c:** the mean fundamental frequency of librators (1), hindered rotators (2) and of the whole ensemble (3); **d:** the effective collision time parameter y_c (solid line).

In the other limiting case ($p \to 0$) we take $\check{r}$ from Eq. 11.72 and substitute $I_0\left(2p^2\right)$ and exponential term $\exp\left(-4p^2 m\right)$ in the integrand by 1. Then Eq. 11.77 reduces to

$$\bar{\beta} = 2\int_0^1 k\mathbf{K}\,\mathrm{asin}(k)dk = \pi\left(1-\frac{\pi}{2\mathbf{K}^2\left(1/\sqrt{2}\right)}\right). \tag{11.79}$$

One finds that $\bar{\beta} \approx \pi/2$, see curve 3 in Fig. 11.4b.

To derive last expression we should take integral by parts with account of Eq. 11.73:

$$\frac{\bar{\beta}}{2} = \left[\left(\mathbf{E}-k'^2\,\mathbf{K}\right)\mathrm{asin}\,k\,\right]_{k=0}^{k=1} - \int_0^1\left(\mathbf{E}-k'^2\,\mathbf{K}\right)\frac{dk}{k'} = \frac{\pi}{2} - \int_0^1\left(\mathbf{E}-\mathbf{K}+k^2\mathbf{K}\right)\frac{dk}{k'}. \tag{11.80}$$

In order to calculate integral

$$\mathfrak{I} = \int_0^1 \frac{k^2\,\mathbf{K}\,dk}{k'} \tag{11.81}$$

we: **(a)** note, that derivative of elliptic integral **K** is given by

$$\frac{d\mathbf{K}}{dk} \equiv \frac{\mathbf{E}-k'^2\,\mathbf{K}}{k\,k'^2} \tag{11.82}$$

and **(b)** write two equivalent expressions for $\mathfrak{R} = \int_0^1 kk'\frac{d\mathbf{K}}{dk}dk$:

$$\mathfrak{R} = \int_0^1 \frac{\mathbf{E}-k'^2\,\mathbf{K}}{k'}dk = \int_0^1 \frac{\mathbf{E}-\mathbf{K}+k^2\mathbf{K}}{k'}dk\ ; \tag{11.83}$$

$$\mathfrak{R} = \left[kk'\,\mathbf{K}\right]_{k=0}^{k=1} - \int_0^1 \frac{\left(k'^2-k^2\right)\mathbf{K}}{k'}dk = -\int_0^1 \frac{\left(1-2k^2\right)\mathbf{K}}{k'}dk. \tag{11.84}$$

Combining (11.83) and (11.84), we derive: $\Im = \int_0^1 \frac{\mathbf{E}dk}{k'}$ and transform Eq. 11.80 to expression

$$\frac{\bar{\beta}}{2} = \frac{\pi}{2} - \int_0^1 \frac{2\mathbf{E} - \mathbf{K}}{k'} dk \ , \tag{11.85}$$

for which following standard integrals are known:

$$\int_0^1 \frac{\mathbf{E}}{k'} dk = \frac{1}{2}\mathbf{K}^2\left(\frac{1}{\sqrt{2}}\right) + \frac{\pi^2}{8\mathbf{K}^2\left(\frac{1}{\sqrt{2}}\right)} \ ; \int_0^1 \frac{\mathbf{K}}{k'} dk = \mathbf{K}^2\left(\frac{1}{\sqrt{2}}\right) \ . \tag{11.86}$$

As a result, we get Eq. 11.79 from Eq. 11.85.

13.3.3 The Mean Fundamental Frequencies of a Periodic Motion

We define the aforementioned frequencies $\left\langle \overset{\vee}{p} \right\rangle$ and $\left\langle \overset{\circ}{p} \right\rangle$ as

$$\left\langle \overset{\vee}{p} \right\rangle = \frac{\int_{-2p^2}^{2p^2} W(h)\overset{\vee}{p}\overset{\vee}{\Phi}(h)dh}{\int_{-2p^2}^{2p^2} W(h)\overset{\vee}{\Phi}(h)dh} = \frac{\int_{-2p^2}^{2p^2} W(h)\overset{\vee}{p}\overset{\vee}{\Phi}(h)dh}{\overset{\vee}{r}}, \tag{11.87a}$$

$$\left\langle \overset{\circ}{p} \right\rangle = \frac{\int_{2p^2}^{\infty} W(h)\overset{\circ}{p}\overset{\circ}{\Phi}(h)dh}{\int_{2p^2}^{\infty} W(h)\overset{\circ}{\Phi}(h)dh} = \frac{\int_{2p^2}^{\infty} W(h)\overset{\circ}{p}\overset{\circ}{\Phi}(h)dh}{\overset{\circ}{r}}. \tag{11.87b}$$

In view of (11.42) and relations $\overset{\vee}{\Phi} = \pi/\overset{\vee}{p}$, $\overset{\circ}{\Phi} = 2\pi/\overset{\circ}{p}$ we derive following expressions for the fundamental frequencies:

$$\left\langle \overset{\vee}{p} \right\rangle = \frac{\mathrm{sh}\left(2p^2\right)}{\sqrt{\pi}\, I_0\left(2p^2\right)} \frac{1}{\overset{\vee}{r}}; \quad \left\langle \overset{\circ}{p} \right\rangle = \frac{\exp\left(2p^2\right)}{\sqrt{\pi}\, I_0\left(2p^2\right)} \frac{1}{\overset{\circ}{r}}, \tag{11.87c}$$

where the proportions $\overset{\vee}{r}$ and $\overset{\circ}{r}$ are given by Eq. 11.68 and 11.71. In particular cases of strong and weak conservative fields we have respectively

$$\left\langle \overset{\vee}{p} \right\rangle \approx p \text{ at } p \geq 1 \text{ and} \left\langle \overset{\vee}{p} \right\rangle \approx \frac{\pi p}{4}; \quad \left\langle \overset{\circ}{p} \right\rangle \approx \frac{1}{\sqrt{\pi}} \text{ at } p << 1. \tag{11.88}$$

We shall also define the mean value of the fundamental frequency for a whole ensemble of dipoles:

$$\langle p \rangle \approx \overset{\vee}{r} \left\langle \overset{\vee}{p} \right\rangle + \overset{\circ}{r} \left\langle \overset{\circ}{p} \right\rangle = \frac{\mathrm{ch}\left(2p^2\right)}{\sqrt{\pi}\, I_0\left(2p^2\right)}. \tag{11.89}$$

Dependencies (11.87)-(11.89) are shown in Fig. 11.4c. Frequencies $\left\langle \overset{\vee}{p} \right\rangle$ and $\left\langle \overset{\circ}{p} \right\rangle$ increase with the rise of the field parameter p; the mean librational frequency $\left\langle \overset{\vee}{p} \right\rangle$ (curve 1) is lower than that of hindered rotors (curve 2). We remark, that in the hybrid model we have for small libration angle β (i.e. for large p) quite another situation (see Fig. 9.3 a, b). Thus, we may expect that *dielectric behavior of both confined rotator and hybrid models should substantially differ.*

The mean frequency $\langle p \rangle$ (curve 3 in Fig. 11.4c) is lesser, than $\left\langle \overset{\circ}{p} \right\rangle$, and larger, than $\left\langle \overset{\vee}{p} \right\rangle$. If $p > 1$, then $\langle p \rangle \approx \left\langle \overset{\vee}{p} \right\rangle$, since the proportion $\overset{\circ}{r}$ of hindered rotors tends to zero.

11.3.4 The Reverse Value of the Collision Time

We define the collision time parameter y_c just as in Sec. 9.1, Eq. 9.31. In view of Eq. 11.89 we have

$$y_c = \frac{\mathrm{ch}\left(2p^2\right)}{\pi\sqrt{\pi}\, I_0\left(2p^2\right)}\;; \tag{11.90}$$

$$y_c \approx \frac{p}{\pi} \quad \text{at } p \geq 1; \;\; y_c \approx \frac{1}{\pi\sqrt{\pi}} \quad \text{at } p \to 0\,. \tag{11.91}$$

From Fig. 11.4c (solid line) we see that y_c is near 0.2 for small p and increases till ≈ 0.6 for $p = 2$. The last example corresponds, as we shall see in Part V, to the case of condensed polar media, where collisions are rather frequent.

11.3.5 The Mean Values of Direction Cosines

AN ANISOTROPIC MEDIUM. We find $\left\langle q^2 \right\rangle$ by analogy with normalization integral (11.40):

$$\begin{Bmatrix} \left\langle q_{\parallel}^2 \right\rangle \\ \left\langle q_{\perp}^2 \right\rangle \end{Bmatrix} = \frac{1}{2\pi\sqrt{\pi}\, I_0\left(2p^2\right)} \int_{-\infty}^{\infty} d\dot{\vartheta} \int_0^{2\pi} \exp\left(-\dot{\vartheta}^2 + 2p^2\cos\vartheta\right) \begin{Bmatrix} \cos^2\vartheta \\ \sin^2\vartheta \end{Bmatrix} d\vartheta$$

$$= \frac{1}{4\pi\, I_0\left(2p^2\right)} \int_0^{2\pi} \exp\left(2p^2\cos\vartheta\right) \{1 \pm \cos 2\vartheta\}\, d\vartheta\,. \tag{11.92}$$

In view of definition (11.41) we have

$$\begin{Bmatrix} \left\langle q_{\parallel}^2 \right\rangle \\ \left\langle q_{\perp}^2 \right\rangle \end{Bmatrix} = \frac{I_0\left(2p^2\right) \pm I_2\left(2p^2\right)}{2 I_0\left(2p^2\right)}\,. \tag{11.93}$$

Since

$$I_0(z) + I_2(z) = \frac{2 I_1(z)}{z}\,, \tag{11.94}$$

equivalently

$$\langle q_{\|}^2 \rangle = 1 - \frac{I_1(2p^2)}{2p^2 I_0(2p^2)}; \quad \langle q_{\perp}^2 \rangle = 1 - \langle q_{\|}^2 \rangle = \frac{I_1(2p^2)}{2p^2 I_0(2p^2)}. \tag{11.95}$$

To find the mean value $\langle q_{\|} \rangle = \langle \cos \vartheta \rangle$ we should substitute $\frac{1}{2}(1 \pm \cos 2\vartheta)$ in Eq. 11.92 by $\cos\vartheta$. With account of Eq. 11.41 we have:

$$\langle q_{\|} \rangle = I_1(2p^2) / I_0(2p^2). \tag{11.96}$$

The dispersion

$$\Delta = \langle q^2 \rangle - \langle q \rangle^2 \equiv \left\langle (q - \langle q \rangle)^2 \right\rangle \tag{11.97}$$

is given by the following combination of Bessel functions:

$$\Delta_{\|} = \frac{I_0(2p^2) + I_2(2p^2)}{2 I_0(2p^2)} - \left(\frac{I_1(2p^2)}{I_0(2p^2)} \right)^2 = \left(\frac{I_1(2p^2)}{I_0(2p^2)} \right) \left[\frac{I_1(2p^2)}{I_0(2p^2)} - \frac{1}{2p^2} \right]; \tag{11.98a}$$

$$\Delta_{\perp} = \langle q_{\perp} \rangle^2 = 1 - \frac{I_1(2p^2)}{2p^2 I_0(2p^2)}, \quad \text{since } \langle q_{\perp} \rangle = 0. \tag{11.98b}$$

Both terms in Eq. 11.98a are close one to another; it is convenient to obtain $\Delta_{\|}$ directly from the integral:

$$\Delta_{\|} = \frac{1}{2\pi\sqrt{\pi}\, I_0(2p^2)} \int_{-\infty}^{\infty} d\dot{\vartheta} \int_0^{2\pi} \exp\left(-\dot{\vartheta}^2 + 2p^2 \cos\vartheta \right) \left\{ \langle q_{\|} \rangle - \cos\vartheta \right\}^2 d\vartheta$$

$$= \frac{1}{\pi I_0(2p^2)} \int_0^{\pi} \exp(2p^2 \cos\vartheta) \left\{ \frac{I_1(2p^2)}{I_0(2p^2)} - \cos\vartheta \right\}^2 d\vartheta. \tag{11.98c}$$

Using expansions (11.60), one can derive asymptotic dependencies for the strong-field approximation ($p > 1$):

$$I_0\left(2p^2\right) \approx \frac{\exp\left(2p^2\right)}{2p\sqrt{\pi}}\left[1+\frac{1}{16p^2}+\frac{1}{\left(16p^2\right)^2 2}\right], \tag{11.99a}$$

$$I_1\left(2p^2\right) \approx \frac{\exp\left(2p^2\right)}{2p\sqrt{\pi}}\left[1-\frac{3}{16p^2}-\frac{15}{\left(16p^2\right)^2 2}\right], \tag{11.99b}$$

$$I_2\left(2p^2\right) \approx \frac{\exp\left(2p^2\right)}{2p\sqrt{\pi}}\left[1-\frac{15}{16p^2}+\frac{105}{\left(16p^2\right)^2 2}\right], \tag{11.99c}$$

so that

$$\left\langle q_\parallel^2\right\rangle \approx 1-\frac{1}{2p^2},\ \left\langle q_\perp^2\right\rangle \approx \frac{1}{2p^2},\ \Delta_\parallel \approx \frac{1}{8p^4}. \tag{11.100}$$

These expressions coincide (as it should be) with those, Eq. 3.137b, 3.138, calculated for the elastic bond model. For zero field ($p = 0$)

$$\left\langle q_\parallel^2\right\rangle = \left\langle q_\perp^2\right\rangle = 0.5,\ \left\langle q_\parallel\right\rangle = 0,\ \ \Delta_\parallel = 1\ . \tag{11.101}$$

In Fig. 11.5a we picture the p-dependencies of the mean squares of direction cosines for the longitudinal and transverse orientations of external field.

If $p < 1$, these quantities are rather close, while near $p = 2$ they considerably differ. For $p > 1$ the dispersion $\Delta_\parallel$ rapidly decreases with the rise of p; it becomes close to asymptotic function $(8p^4)^{-1}$, cf. curves 1 and 3 in Fig. 11.5b.

THE DISPERSION IN AN ISOTROPIC MEDIUM. Using Eq. 1.98 we take the mean (inside a "domain") value:

$$\bar{\Delta} = \Delta_\parallel + \Delta_\perp = 1 - \left\langle q_\parallel\right\rangle^2 = 1 - \left[\frac{I_1\left(2p^2\right)}{I_0\left(2p^2\right)}\right]^2; \tag{11.102a}$$

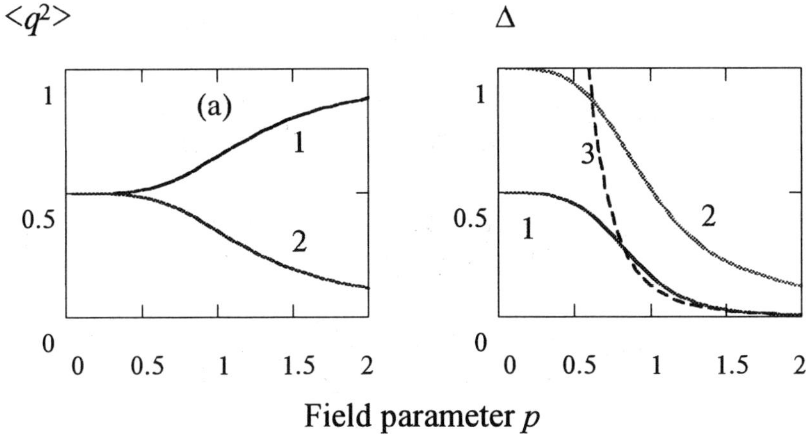

Fig. 11.5 The p-dependence of direction cosine. **a:** the mean values $\left\langle q_{\|}^{2}\right\rangle$ (curve 1) and $\left\langle q_{\perp}^{2}\right\rangle$ (curve 2); **b:** the dispersion Δ for the longitudinal orientation of the external field (curve 1) and for an isotropic medium (curve 2); the asymptotic dependence (11.103) for ||- orientation (curve 3).

$$\overline{\Delta} = 1 \text{ for } p = 0; \text{ and } \overline{\Delta} \approx \frac{1}{2p^2} \text{ for } p > 1. \qquad (11.102\text{b})$$

The decrease of $\overline{\Delta}$ with the rise of p is much slower than in an anisotropic medium (cf. curves 2 and 3 in Fig. 11.5b).

11.4 Rigorous Expressions for the Spectral Functions

11.4.1 Fourier Amplitudes a_n, b_n of Periodic Motion

At first, we write down useful relations for elliptic Jacobi functions and for some trigonometric series (Gradshtein and Rydzyk 1963, Sec. 8.146); (Prudnukov, Brychkov and Marychev 1981, Sec. 5.4.14); (Abramowitz and Stegun 1979, Sec. 16.23):

$$\mathrm{dn}^2 u = \frac{\mathbf{E}}{\mathbf{K}} + \frac{2\pi^2}{\mathbf{K}^2} \sum_{n=1}^{\infty} \frac{nQ^n}{1-Q^{2n}} \cos\frac{2n\,\pi u}{2\mathbf{K}}, \tag{11.103}$$

$$-\frac{d\,\mathrm{cn}u}{du} = \mathrm{sn}u\ \mathrm{dn}u = -\frac{d}{du}\frac{2\pi}{k\mathbf{K}} \sum_{n=1}^{\infty} \frac{Q^{n-1/2}}{1+Q^{2n-1}} \cos\frac{(2n-1)\,\pi u}{2\mathbf{K}}, \tag{11.104}$$

$$\mathrm{sn}u = \frac{2\pi}{k\mathbf{K}} \sum_{n=1}^{\infty} \frac{Q^{n-1/2}}{1-Q^{2n-1}} \sin\frac{(2n-1)\pi u}{2\mathbf{K}}, \tag{11.105}$$

$$\mathrm{dn}u = \frac{\pi}{2\mathbf{K}} + \frac{2\pi}{\mathbf{K}} \sum_{n=1}^{\infty} \frac{Q^n}{1+Q^{2n}} \cos\frac{2n\,\pi u}{2\mathbf{K}}, \tag{11.106}$$

$$\frac{d\,\mathrm{dn}u}{du} = -k^2\,\mathrm{sn}u\ \mathrm{cn}u\ , \tag{11.107}$$

where

$$k^2 = m,\ Q = \exp\left[-\pi\,\frac{K(m')}{K(m)}\right],\ m' = 1-m. \tag{11.108}$$

Let FA[*X*] denote extraction of Fourier amplitude from *X*. Using Eq. 11.103-11.104 and general representations (5.81) for motion in a plane, we find from Eq. 11.20, 11.23 amplitudes $\check{a}_{2n}, \check{b}_{2n-1}$ of *librating* particles (class A):

$$\check{a}_{2n} = 2\,\mathrm{FA}\left[\mathrm{dn}^2 u\right] = 4n\frac{\pi^2}{\mathbf{K}^2}\frac{Q^n}{1-Q^{2n}}\ , \tag{11.109}$$

$$\check{b}_{2n-1} = -2\check{k}\,\mathrm{FA}\left[\frac{d\,\mathrm{cn}(u)}{du}\right] = 2(2n-1)\,\frac{\pi^2}{\mathbf{K}^2}\frac{Q^{n-1/2}}{1+Q^{2n-1}}\ . \tag{11.110}$$

For *hindered rotators* we use representations (5.89), the law of motion (11.27), (11.29) and trigonometric series (11.103) and (11.106):

$$\overset{\circ}{a}_n = \text{FA}\left[1+\frac{2\,\text{dn}^2 u}{\overset{\circ}{k}^2}\right];\ \overset{\circ}{b}_n = \frac{2}{\overset{\circ}{k}}\text{FA}\left[1-\frac{d\,\text{dn}u/du}{\overset{\circ}{k}^2}\right];$$

$$\begin{Bmatrix}\overset{\circ}{a}_n\\ \overset{\circ}{b}_n\end{Bmatrix} = \left(\frac{2\pi}{\overset{\circ}{k}\mathbf{K}}\right)^2 \frac{nQ^n}{1-Q^{2n}} . \qquad (11.111)$$

Combining Eq. 11.109 - 11.110 and using Eq. 11.111, we calculate squares of amplitudes:

$$\begin{Bmatrix}\overset{\vee}{a}{}^2_{2n}\\ \overset{\vee}{b}{}^2_{2n-1}\end{Bmatrix} = 4\begin{Bmatrix}(2n)^2\\ (2n-1)^2\end{Bmatrix}\frac{\pi^4}{\mathbf{K}^4}Q^{\begin{Bmatrix}2n\\2n-1\end{Bmatrix}}\left[1-Q^{\begin{Bmatrix}2n\\2n-1\end{Bmatrix}}\right]^{-2}, \qquad (11.112)$$

$$\begin{Bmatrix}\overset{\circ}{a}{}^2_n\\ \overset{\circ}{b}{}^2_n\end{Bmatrix} = \left(\frac{2\pi}{\overset{\circ}{k}\mathbf{K}}\right)^4\left(\frac{nQ^n}{1-Q^{2n}}\right)^2 . \qquad (11.113)$$

In an anisotropic medium we represent the SF as the sum

$$K(z) = \overset{\vee}{K}(z) + \overset{\circ}{K}(z) , \qquad (11.114a)$$

while in an isotropic medium Eq. 11.114a reduces to the sum

$$L(z) = \overset{\vee}{L}(z) + \overset{\circ}{L}(z) \equiv \frac{\overset{\vee}{K}(z) + \overset{\circ}{K}(z)}{3} . \qquad (11.114b)$$

11.4.2 The Spectral Function of Librating Dipoles

AN ANISOTROPIC MEDIUM. Substituting Eq. 11.76a for dh and Eq. 11.112 into general expressions (5.86) and (5.87), we find the SF $\overset{\vee}{K}(z)$ of librators:

$$\frac{\check{K}_{\|}(z)}{3} = \left\langle q_{\|}^{2} \right\rangle \check{L}_{\|}(z) = \frac{2p^{3}\exp\left(2p^{2}\right)}{\sqrt{\pi}\, I_{0}\left(2p^{2}\right)}$$

$$\times \int_{0}^{1} dm \exp\left(-4p^{2}m\right)\left(\frac{\pi}{\mathbf{K}(m)}\right)^{5} \sum_{n=1}^{\infty} \frac{(2n)^{4}}{\left(1-Q^{2n}\right)^{2}} \frac{Q^{2n}}{\left[\dfrac{n\pi p}{\mathbf{K}(m)}\right]^{2} - z^{2}}, \tag{11.115a}$$

$$\frac{\check{K}_{\perp}(z)}{3} = \left\langle q_{\perp}^{2} \right\rangle \check{L}_{\perp}(z) = \frac{2p^{3}\exp\left(2p^{2}\right)}{\sqrt{\pi}\, I_{0}\left(2p^{2}\right)}$$

$$\times \int_{0}^{1} dm \exp\left(-4p^{2}m\right)\left(\frac{\pi}{\mathbf{K}(m)}\right)^{5} \sum_{n=1}^{\infty} \frac{(2n-1)^{4}}{\left(1+Q^{2n-1}\right)^{2}} \frac{Q^{2n-1}}{\left[\dfrac{(2n-1)\pi p}{2\mathbf{K}(m)}\right]^{2} - z^{2}}. \tag{11.115b}$$

If $p \to 0$, both SFs (11.115) vanish.

AN ISOTROPIC MEDIUM. In accord with Eq. 5.93

$$L(z) = \check{L}(z) = \left\langle q_{\|}^{2} \right\rangle \check{L}_{\|}(z) + \left\langle q_{\perp}^{2} \right\rangle \check{L}_{\perp}(z) \tag{11.116}$$

$$= \frac{\pi}{4} \int_{h_{\min}}^{h_{\text{th}}} dh\, W \check{p} \sum_{n=1}^{\infty} \frac{n^{2} A_{n}^{2}}{\check{p}^{2} n^{2} - z^{2}}, \tag{11.117}$$

$$A_{n}^{2} = \check{a}_{n}^{2} + \check{b}_{n}^{2} + (-1)^{n}\left(\check{a}_{n}^{2} - \check{b}_{n}^{2}\right). \tag{11.118}$$

Substituting (11.112) and (11.113), we have:

$$A_{n}^{2} = 4n^{2}\left(\frac{\pi}{\mathbf{K}}\right)^{4} Q^{n} \left\{ \frac{1}{\left(1-Q^{n}\right)^{2}} + \frac{1}{\left(1+Q^{n}\right)^{2}} \right.$$

$$+ (-1)^n \frac{1}{\left(1-Q^n\right)^2} - \frac{1}{\left(1+Q^n\right)^2} \Bigg\} = 8n^2 \left(\frac{\pi}{\mathbf{K}}\right)^4 Q^n \left[\frac{1+(-Q)^n}{1-Q^{2n}}\right]^2 . \qquad (11.119)$$

Substitution into integral (11.117) yields

$$\check{L}(z) = \frac{2p^3 \exp\left(2p^2\right)}{\sqrt{\pi}\, I_0\left(2p^2\right)}$$

$$\times \int_0^1 dm \exp\left(-4p^2 m\right) \left(\frac{\pi}{\mathbf{K}(m)}\right)^5 \sum_{n=1}^{\infty} \left[\frac{1+(-Q)^n}{1-Q^{2n}}\right]^2 \frac{n^4 Q^n}{\left[\dfrac{n\pi p}{2\mathbf{K}(m)}\right]^2 - z^2} . \qquad (11.120)$$

TRANSITION TO THE ELASTIC BOND MODEL. If $p > 1$, $\exp\left(-4p^2 m\right)$ in Eq. 5.115 is not extremely small only in the vicinity of small m values. Considering m to be small, we can:

(a). retain in the sum only the first term (for $n = 1$) and use the power series expansion of the $Q(m)$ function,

$$Q(m) \approx \frac{m}{16}\left(1 + \frac{m}{2} + \frac{21}{64} m^2 + \frac{31}{128} m^3 + \ldots\right); \qquad (11.121)$$

$$Q(m) \approx \frac{m}{16} \qquad \text{for } m \ll 1; \qquad (11.122)$$

(b). use asymptotic formula (11.61) for $I_0\left(2p^2\right)$ function;

(c). put $\mathbf{K}(m) = \mathbf{K}(0) = \pi/2$ in resonant denominator in (11.120). Then Eq. (11.115) reduce to

$$\frac{\check{K}_{\|}(z)}{3} = \frac{\int_0^{4p^2} t^2 \exp(-t)\, dt}{8p^2\left(4p^2 - z^2\right)}, \qquad (11.123a)$$

$$\frac{\check{K}_{\perp}(z)}{3} = \frac{\int\limits_0^{4p^2} t\exp(-t)\,dt}{2\left(p^2 - z^2\right)}. \tag{11.123b}$$

For large p we may replace the upper limit by ∞, thus we get results (10.11) and (10.16), obtained for the elastic bond (EB) model:

$$\frac{\check{K}_{\parallel}(z)}{3} = \frac{1}{4p^2\left(4p^2 - z^2\right)};\; \frac{\check{K}_{\perp}(z)}{3} = \frac{1}{2\left(p^2 - z^2\right)}. \tag{11.124}$$

11.4.3 The Spectral Function of Hindered Rotors

AN ANISOTROPIC MEDIUM. In accord with Eq. 5.90

$$\frac{1}{3}\begin{Bmatrix} \mathring{K}_{\parallel}(z) \\ \mathring{K}_{\perp}(z) \end{Bmatrix} = \begin{Bmatrix} \left\langle q_{\parallel}^2 \right\rangle \mathring{L}_{\parallel}(z) \\ \left\langle q_{\perp}^2 \right\rangle \mathring{L}_{\perp}(z) \end{Bmatrix}$$

$$= C\pi \int\limits_{2p^2}^{\infty} dh\, \mathring{p} \exp\left[-h\left(\mathring{p}\right)\right] \sum_{n=1}^{\infty} \begin{Bmatrix} a_n^2 \\ b_n^2 \end{Bmatrix} \frac{n^2}{\left(\mathring{p}n\right)^2 - z^2}. \tag{11.125}$$

Using Eq. 11.42 for C, (11.113) for squares of Fourier amplitudes and (11.70) for $\exp(-h)$ and dh, we get analogously to Eq. 11.115:

$$\frac{1}{3}\begin{Bmatrix} \mathring{K}_{\parallel}(z) \\ \mathring{K}_{\perp}(z) \end{Bmatrix} = \frac{2p^3 \exp\left(2p^2\right)}{\sqrt{\pi}\, I_0\left(2p^2\right)}$$

$$\times \int_0^1 \frac{dk}{k^8} \exp\left(-\frac{4p^2}{k^2}\right)\left(\frac{2\pi}{\mathbf{K}}\right)^5 \sum_{n=1}^{\infty} \left(\frac{n^2 Q^n}{1-Q^{2n}}\right)^2 \frac{1}{\left[\frac{\pi n p}{k\mathbf{K}}\right]^2 - z^2}. \quad (11.126)$$

It is convenient to change variable of integration: $k \to t = 2\pi / k$, so that

$$m = m(t) = \frac{4p^2}{t^2}\ ;\ \mathbf{K}(m) = \mathbf{K}\left(\frac{4p^2}{t^2}\right). \quad (11.127)$$

Then Eq. 11.126 transforms to

$$\frac{1}{3}\begin{Bmatrix} \overset{\circ}{K}_{\parallel} \\ \overset{\circ}{K}_{\perp} \end{Bmatrix} = \frac{\exp\left(2p^2\right)}{2\sqrt{\pi}\, I_0\left(2p^2\right)}$$

$$\times \int_{2p}^{\infty} dt\, t^6 \exp\left(-t^2\right)\left(\frac{\pi}{\mathbf{K}}\right)^5 \sum_{n=1}^{\infty} \left(\frac{n^2 Q^n}{p^2\left(1-Q^{2n}\right)}\right)^2 \frac{1}{\left[\frac{\pi n t}{2\mathbf{K}}\right]^2 - z^2}. \quad (11.128)$$

For strong conservative field ($p > 1$) these integrals are extremely small. Indeed, near the lower limit ($t = 2p$), where $\exp(-t^2)$ is still not small, modulus k and parameter m are near 1. Hence, the integrand in (11.128) vanishes because $[\mathbf{K}(m)]^{-5} \to 0$. Thus, for $p > 1$ only librators contribute to the dielectric susceptibility. This property corresponds to the $\overset{\circ}{f}\left(\overset{\circ}{p}\right)$ plot, Fig. 11.3f.

AN ISOTROPIC MEDIUM. Taking the sum of the aforecited expressions, we get:

$$\overset{\circ}{L} = \frac{1}{3}\left\{\overset{\circ}{K}_{\parallel} + \overset{\circ}{K}_{\perp}\right\} = \frac{\exp\left(2p^2\right)}{\sqrt{\pi}\, I_0\left(2p^2\right)}$$

$$\times \int_{2p}^{\infty} dt\, t^6 \exp\left(-t^2\right)\left(\frac{\pi}{\mathbf{K}}\right)^5 \sum_{n=1}^{\infty}\left(\frac{n^2 Q^n}{p^2\left(1-Q^{4n}\right)}\right)^2 \frac{1+Q^{4n}}{\left[\frac{\pi n t}{2\mathbf{K}}\right]^2 - z^2}. \quad (11.129)$$

TRANSITION TO THE EXTENDED DIFFUSION MODEL. If p is small, $\exp(-4p^2/k^2)$ in the integrand of Eq. 11.126 is not small. However *only small* k *values contribute the SFs*, since for k near 1 the integrand vanishes because of multiplier $(p/\mathbf{K})^5$. Thus, we may employ approximation (11.122). Retaining only the first term in sums over n, we have:

$$\sum_{n=1}^{\infty}\left(\frac{n^2 Q^n}{p^2\left(1-Q^{4n}\right)}\right)^2 \approx \left(\frac{Q}{p}\right)^2 = \left(\frac{m^2}{16p^2}\right)^2 = \left(\frac{1}{2t}\right)^4 \quad \text{for } p \to 0. \quad (11.130)$$

Neglecting Q^{2n} and Q^{4n} (in comparison with 1) and taking $\mathbf{K} = \pi/2$, we get from Eq. 11.128 and 11.129 the results (5.98) - (5.99), obtained for the planar ED model. Indeed,

$$\left\langle q_{\parallel}^2 \right\rangle = \left\langle q_{\perp}^2 \right\rangle = \frac{1}{2}; \ \exp\left(2p^2\right) = I_0\left(2p^2\right) = 1 \ \text{for } p \to 0; \quad (11.131)$$

$$\frac{\mathring{K}_{\parallel}}{3} = \frac{\mathring{K}_{\perp}}{3} = \frac{\mathring{L}_{\parallel}}{2} = \frac{\mathring{L}_{\perp}}{2} = \frac{1}{2\sqrt{\pi}} \int_0^{\infty} 2^5 t^6 \exp\left(-t^2\right) 2^{-4} t^{-4} \frac{dt}{t^2 - z^2}$$

$$= \frac{1}{\sqrt{\pi}} \int_0^{\infty} t^2 \exp\left(-t^2\right) \frac{dt}{t^2 - z^2} = \frac{L_F(z)}{2} = \frac{1}{2}\left[1 + i\sqrt{\pi}\, z\, w(z)\right]; \quad (11.132)$$

$$\mathring{L}(z) = \frac{1}{\sqrt{\pi}} \int_0^{\infty} 2^5 t^6 \exp\left(-t^2\right) 2^{-4} t^{-4} \frac{dt}{t^2 - z^2} L_F(z); \quad (11.133a)$$

$$\mathring{L}(z) = \mathring{L}_{\perp}(z) = \mathring{L}_{\parallel}(z) , \quad (11.133b)$$

as it should be in the case of an isotropic medium.

11.5 The Quasi-harmonic Approximation (QHA)

We have considered above the transition from the rigorous theory of the CF model to approximations, termed the EB and ED models, viz. respectively to very strong (parabolic) potential $U(\vartheta)$ and to zero one. The principal distinction between these two models and the QHA consists in the following. *In the QHA we take into account in the first approximation the dependence of the fundamental librational / hindered rotational frequencies on energy of dipoles.* Namely, we retain two terms in the power expansion (11.62):

$$K(m) \approx \frac{\pi}{2}\left(1+\frac{m}{4}\right). \qquad (11.134)$$

11.5.1 The Quasi-Harmonic Approximation for Librators

We use $t = 4p^2 m$ as variable of integration. Thus, multiplier $(\pi / \mathbf{K})^5$ in Eq. 11.120 is transformed as follows:

$$\left[\frac{\pi}{\mathbf{K}(m)}\right]^5 \approx \frac{2^5}{\left(1+\dfrac{m}{4}\right)^5} = \left(\frac{2}{\check{\gamma}}\right)^5 ; \check{\gamma} \equiv 1+\frac{t}{16p^2}. \qquad (11.135)$$

The *anharmonicity coefficient* $\check{\gamma}$, being proportional to the libration period, increases with the rise of energy of a dipole and / or with the decrease of constant field.

AN ANISOTROPIC MEDIUM. Retaining in sums over n, Eq. 11.115, only the first term, corresponding to $n = 1$, substituting $Q(m)$ function by the first term of its power expansion, Eq.11.122, we have

$$\sum_{n=1}^{\infty} \frac{(2n)^4}{\left(1-Q^{2n}\right)^2} \frac{Q^{2n}}{\left[\dfrac{n\pi p}{\mathbf{K}(m)}\right]^2 - z^2} \approx \frac{16Q^2}{4p^2 \check{\gamma}^{-2} - z^2} = \frac{16m^2 \check{\gamma}^2}{16^2\left(4p^2 - z^2 \check{\gamma}^2\right)}, \qquad (11.136)$$

$$\sum_{n=1}^{\infty} \frac{(2n-1)^4}{\left(1+Q^{2n-1}\right)^2} \frac{Q^{2n-1}}{\left[\dfrac{(2n-1)\pi p}{2\mathbf{K}(m)}\right]^2 - z^2} \approx \frac{(m/16)\overset{\vee}{\gamma}^2}{p^2 - z^2\overset{\vee}{\gamma}^2}. \quad (11.137)$$

Substituting m by $t/(4p^2)$, we finally derive:

$$\frac{\overset{\vee}{K}_{\|}(z)}{3} = \frac{\exp\left(2p^2\right)}{16p^2\sqrt{\pi}\, I_0\left(2p^2\right)} \int_0^{4p^2} \frac{t^2 \exp(-t)\, dt}{\overset{\vee}{\gamma}^3(t)\left[4p^2 - z^2\overset{\vee}{\gamma}^2(t)\right]}, \quad (11.138a)$$

$$\frac{\overset{\vee}{K}_{\perp}(z)}{3} = \frac{\exp\left(2p^2\right)}{4p\sqrt{\pi}\, I_0\left(2p^2\right)} \int_0^{4p^2} \frac{t \exp(-t)\, dt}{\overset{\vee}{\gamma}^3(t)\left[p^2 - z^2\overset{\vee}{\gamma}^2(t)\right]}. \quad (11.138b)$$

For $p \to \infty$ these expressions reduce to Eq. 11.123.

AN ISOTROPIC ENSEMBLE. In the QHA the SF $L(z)$ is given by

$$L(z) = \frac{\overset{\vee}{K}_{\|}(z) + \overset{\vee}{K}_{\perp}(z)}{3} = \frac{\exp\left(2p^2\right)}{4p\sqrt{\pi}\, I_0\left(2p^2\right)}$$

$$\times \int_0^{4p^2} \frac{t \exp(-t)}{\overset{\vee}{\gamma}^3(t)} \left(\frac{1}{p^2 - z^2\overset{\vee}{\gamma}^2(t)} + \frac{1}{4p^2 - z^2\overset{\vee}{\gamma}^2(t)} \right) dt. \quad (11.139)$$

11.5.2 The Quasi-Harmonic Approximation of Hindered Rotators

In this approximation we have instead of Eq. 11.135:

$$\frac{\pi}{\mathbf{K}(m)} \approx \frac{2^5}{\left(1+\frac{m}{4}\right)^5} = \frac{2^5}{\mathring{\gamma}^5(t)}, \tag{11.140}$$

$$\mathring{\gamma}(t) = 1 + \frac{m}{4} = 1 + \frac{p^2}{t^2} \tag{11.141}$$

The anharmonicity coefficient describes, quite analogously to Eq. 11.135, the increase of the period of hindered rotators with the rise of energy of particles, i.e. with the decrease of static field. We take two terms in *Q(m)* power expansion and retain only m th-order terms (comparing with 1). Then sums in Eq. 11.120 reduce to

$$\sum_{n=1}^{\infty} \frac{n^4 Q^{2n}}{p^4\left(1-Q^{2n}\right)^2} \frac{1}{\left[\frac{\pi n t}{2\mathbf{K}}\right]^2 - z^2} \approx \frac{Q^2}{p^4\left(1 \mp Q^2\right)^2\left[\left(t/\mathring{\gamma}\right)^2 - z^2\right]}$$

$$\approx \left(\frac{m}{16p^2}\right)^2 \frac{\left(1+\frac{m}{2}\right)^2}{\left(t/\mathring{\gamma}\right)^2 - z^2} = \frac{1}{16t^4} \frac{\left(1+2p^2/t^2\right)^2 \mathring{\gamma}^2}{t^2 - z^2\mathring{\gamma}^2}. \tag{11.142}$$

This result means, that hindered rotators behave like in an isotropic medium. Finally we derive

$$\frac{\mathring{K}_{\parallel}(z)}{3} = \frac{\mathring{K}_{\perp}(z)}{3} = \frac{\exp\left(2p^2\right)}{\sqrt{\pi}\, I_0\left(2p^2\right)} \int_{2p}^{\infty} dt\, t^2 \frac{\exp\left(-t^2\right)}{\mathring{\gamma}^3(t)} \frac{\left(1+2p^2/t^2\right)^2}{t^2 - z^2\left[\mathring{\gamma}(t)\right]^2}. \tag{11.143}$$

The anisotropy is obliged only to ensemble of librators, see Eq. 11.138. For $p \to 0$ Eq. 11.138 reduces to Eq. 11.132.

To find the spectral function $\mathring{L}(z)$ we double Eq. 11.143:

$$\overset{\circ}{L} = \frac{2\exp\left(2p^2\right)}{\sqrt{\pi}\, I_0\left(2p^2\right)} \int_{2p}^{\infty} dt\, t^2 \frac{\exp\left(-t^2\right)}{\left[\overset{\circ}{\gamma}(t)\right]^3} \frac{\left(1+2p^2/t^2\right)^2}{t^2 - z^2\left[\overset{\circ}{\gamma}(t)\right]^2}. \tag{11.144}$$

Further dielectric properties will be illustrated for the simple QHA, what is sufficient for a *qualitative* study.

11.6 Dielectric Behavior

11.6.1 Dielectric Spectra of an Isotropic Medium

The effective susceptibility will be calculated for the Gross collision model. We employ the same formulas, that were used previously (see Table 6.1):

$$\chi^* = Gz \frac{\overset{\vee}{L}(z) + \overset{\circ}{L}(z)}{x + \dfrac{iy}{\overline{\Delta}}\left[\overset{\vee}{L}(z) + \overset{\circ}{L}(z)\right]}. \tag{11.145}$$

The SFs $\overset{\vee}{L}$ and $\overset{\circ}{L}$ due to librators and hindered rotators are given respectively by Eq. 11.139 and 11.144 and the mean $\overline{\Delta}$ is given by Eq. 11.102a.

DIELECTRIC SPECTRA FOR $y = y_c$ (Fig. 11.6). For *small field* ($p <$ 0.1) loss / absorption curves practically coincide with those, calculated for the ED model (upper curves in Fig. a, b).

For the *moderate p-value* ($p = 0.5$) these curves become two-humped (lower curves in Fig. a, b), since $\overset{\vee}{L}$ and $\overset{\circ}{L}$ are commensurable. The low-frequency peak is stipulated by librators and the high-frequency one - by hindered rotators.

For *strong field* ($p > 1$, Fig. c, d) the contribution of hindered rotors becomes small, the spectral lines $\chi''(x)$ and $x\chi''(x)$ become narrow and their peaks are placed at frequency $x_L \approx p$. All plots in Fig. 11.6 are calculated for an "intermediate" value of normalized collision frequency ($y = y_c$). We remind the Reader, that rather frequent collisions ($y > y_c$) are typical for liquids, while rare collisions ($y << y_c$) are typical for the gas state.

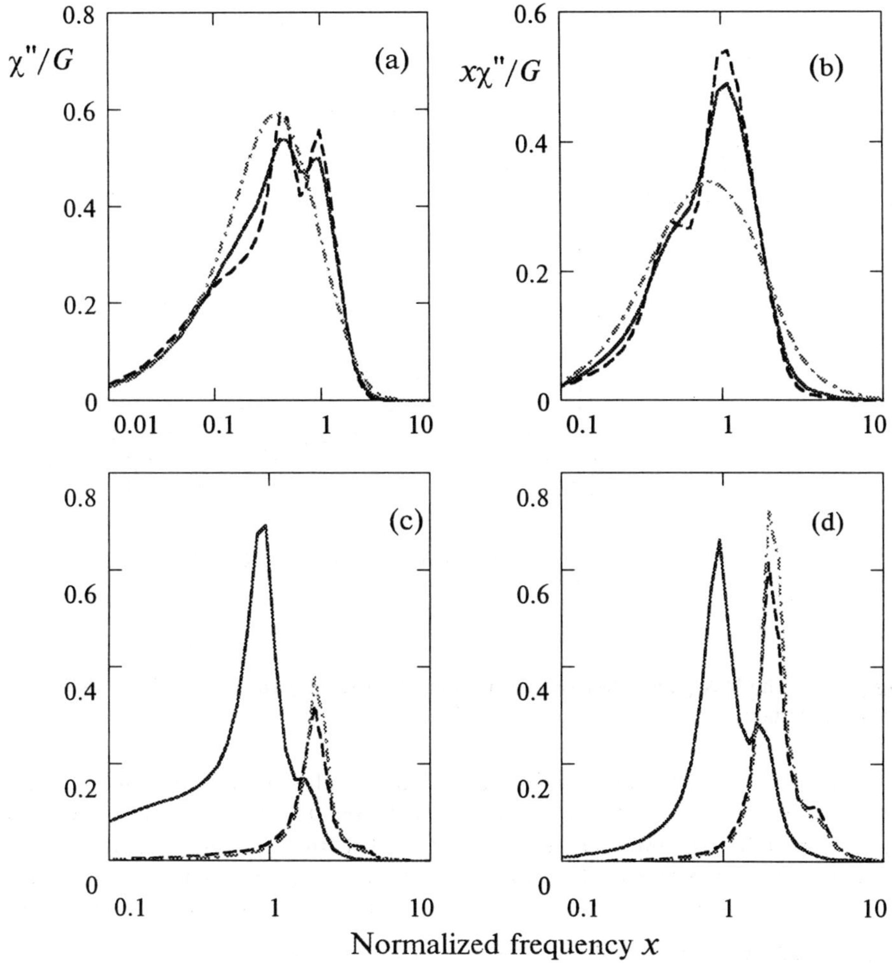

Fig. 11.6 Dielectric loss (a, c) and absorption (b, d) spectra, calculated in the quasi-harmonic approximation (solid lines) for an isotropic medium (the Gross collision model), $y = y_C$. a, b: $p = 0.1$, $y = 0.18$, ED model - dadots; $p = 0.5$, $y = 0.19$. c, d: $p = 1$, $y = 0.296$; $p = 2$, $y = 0.626$, EB model - dadots.

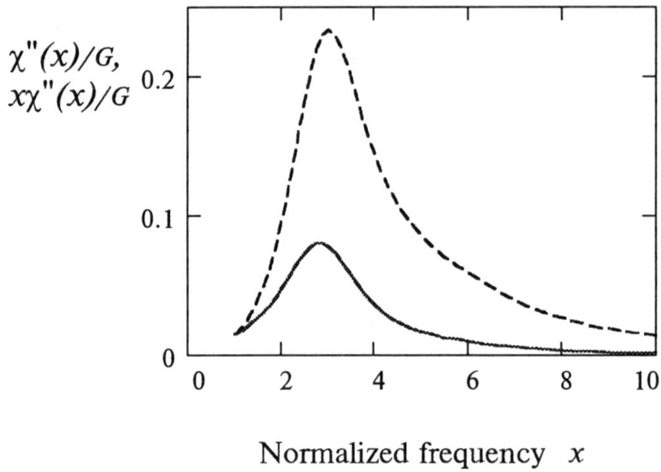

Fig. 11.7 The loss (solid line) and absorption (dashed line) vs. *x*; $p = 3$, $y = 2$. An isotropic medium.

Lorentz lines, described by the EB model (dadots in Fig. c, d for $p = 2$), are close to the loss / absorption lines, calculated in the QHA. For the field parameter p in the interval between 0.1 and 1 the loss curve $\chi''(x)$ has a shoulder (see Fig. a and c). One may consider the latter as some sign of Debye relaxation. However, in the case of an *isotropic medium* this relaxation is not pronounced because of multiplier $\left(\overline{\Delta}\right)^{-1}$ in Eq. 11.145. For $p = 2$ no signs of low-frequency relaxation are seen from Fig. 11.7, since the dispersion $\overline{\Delta}$ becomes small for large p:

$$\overline{\Delta} << 1 \text{ for } p > 1. \tag{11.146}$$

DIELECTRIC SPECTRA FOR STRONG FIELD AND FREQUENT COLLISIONS. These spectra are shown in Fig. 11.7 for $p = 3$ and $y = 2$. We shall show in Chap. 14, that the similar set of parameters is typical for *individual polar liquids.* Only high-frequency dielectric relaxation is pronounced now, the loss and absorption peaks being placed near the p-value, i.e. in our case for $x_m \approx x_L \approx 3$. *Since the low frequency Debye relaxation does not appear in the CF model, the latter is inapplicable for such liquids.*

11.6.2 Dielectric Behavior of an Anisotropic Medium

The complex dielectric susceptibility is given by

$$\chi^*_{\|}(x) = Gz \frac{\check{K}_{\|}(z) + \mathring{K}_{\|}(z)}{x + \left(iy/\sigma_{\|}\right)\left[\check{K}_{\|}(z) + \mathring{K}_{\|}(z)\right]}, \tag{11.147a}$$

$$\chi^*_{\perp}(x) = Gz \frac{\check{K}_{\perp}(z) + \mathring{K}_{\perp}(z)}{x + \left(iy/\sigma_{\perp}\right)\left[\check{K}_{\perp}(z) + \mathring{K}_{\perp}(z)\right]}. \tag{11.147b}$$

The dispersion $\sigma_{\|} = 3\Delta_{\|}(p)$, where $\Delta_{\|}(p)$, is given by Eq. 11.98c; while $\sigma_{\perp} = 3\Delta_{\perp};\ \Delta_{\perp}(p) = 3\left\langle q_{\perp}^2 \right\rangle$ - by Eq. 11.98b; the $\check{K}_{\|}(z)$ and $\check{K}_{\perp}(z)$ SFs - by Eq. 11.138 and $\mathring{K}_{\|}(z) = \mathring{K}_{\perp}(z)$ - by Eq. 11.143.

In Fig. 11.8 we present the loss and absorption curves, calculated for $y = y_c$ and $p = 0.5$; 1. For moderate p-values (e.g., $p = 0.5$) the transverse susceptibility is responsible for splitting of the loss / absorption curves (Fig. c, solid curve), while the longitudinal susceptibility stipulates the Debye loss peak near $x_D \approx 0.1$ (Fig. a). For a *strong field* ($p > 1$) the longitudinal Debye loss peak decreases (dashed curve in Fig. a); the absorption spectrum resembles a narrow resonance curve (dashed lines in Figs. b and d); the splitting of the $\chi^*_{\perp}(x)$ curve disappears.

11.6.3 The Limiting Loss / Absorption Lines

If the collision frequency y to tends to zero, the imaginary parts of integrands in Eq. 11.138 a, b and 11.143 transform into δ-functions:

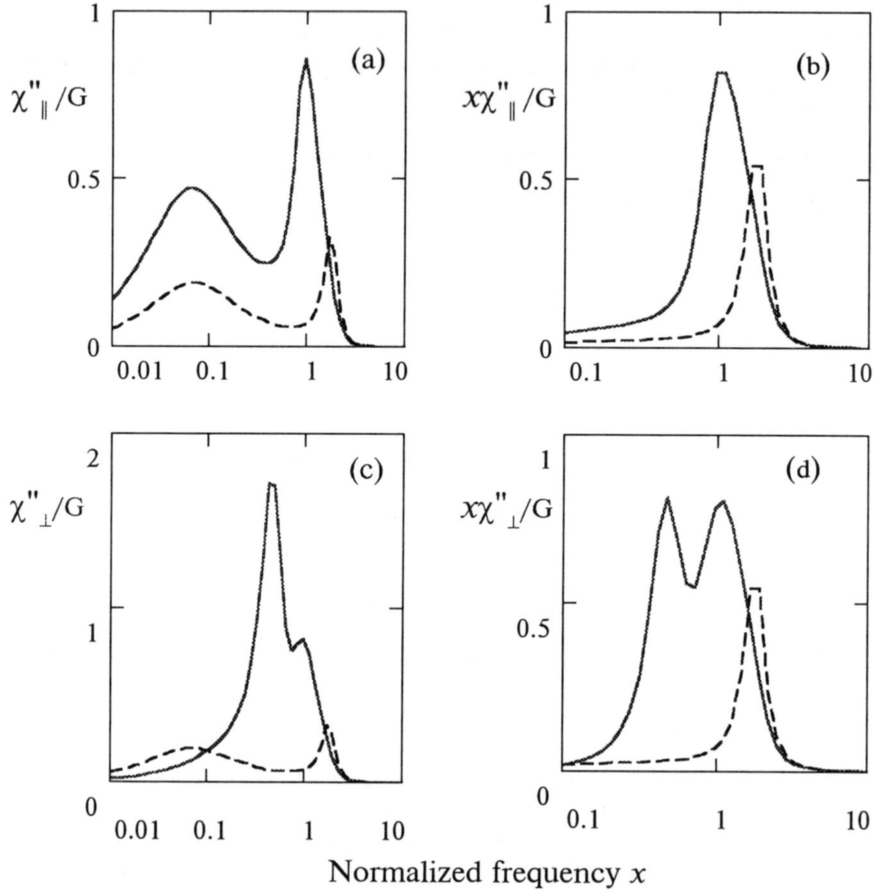

Fig. 11.8 Loss (a, c) and absorption (b, d) vs. frequency for the longitudinal (a, b) and transverse (c, d) orientations of alternating field. Solid lines for $p = 0.5$, dashed lines for $p = 1$. $y = y_c$.

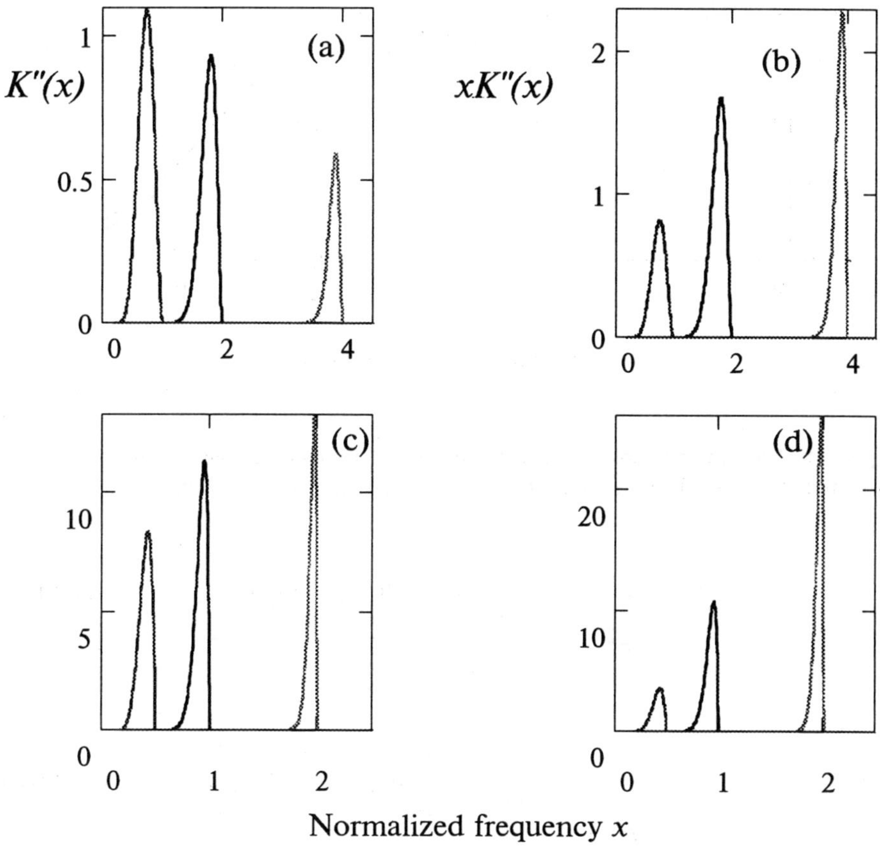

Fig. 11.9 Loss (a, c) and absorption (b, d) spectra for the longitudinal (a, b) and transverse (c, d) orientations of alternating field, calculated for the long lifetime limit. The left, middle and right curves for $p = 0.5$, 1, and 2, respectively.

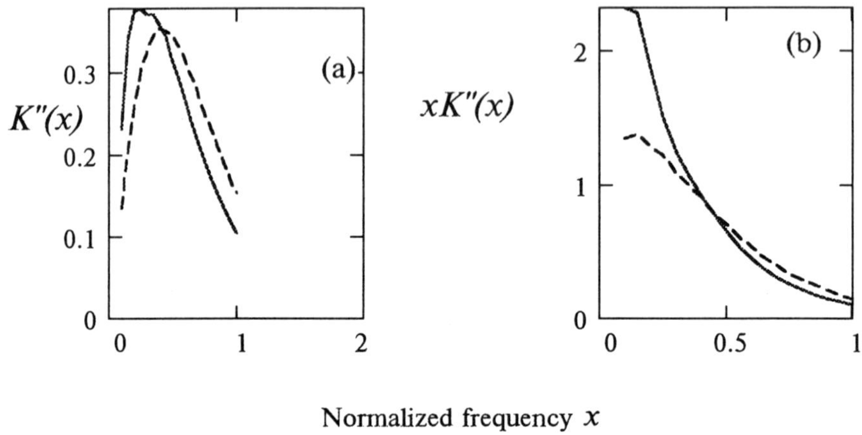

Fig. 11.10 Loss (a) and absorption (b) spectra due to subensemble of hindered rotors, calculated in the long lifetime limit for $p = 0.25$ (solid line) and 0.5 (dashed line).

$$\operatorname{Im}\frac{1}{\left(4p^2 - z^2 \overset{\vee}{\gamma}{}^2\right)} \to \frac{\pi}{4p}\,\delta\left[2p - x\overset{\vee}{\gamma}\left(\overset{\vee}{t}_{\parallel}\right)\right], \tag{11.148a}$$

$$\operatorname{Im}\frac{1}{\left(p^2 - z^2 \overset{\vee}{\gamma}{}^2\right)} \to \frac{\pi}{2p}\,\delta\left[p - x\overset{\vee}{\gamma}\left(\overset{\vee}{t}_{\perp}\right)\right], \tag{11.148b}$$

$$\operatorname{Im}\frac{1}{\left(t^2 - z^2 \overset{\circ}{\gamma}{}^2\right)} \to \frac{\pi}{2t}\,\delta\left[t - x\overset{\circ}{\gamma}\left(\overset{\circ}{t}\right)\right]. \tag{11.148c}$$

It is convenient to replace the variable of integration t by $\check{\gamma}\left(\check{t}_{\|}\right)$, $\check{\gamma}\left(\check{t}_{\perp}\right)$ and $\mathring{\gamma}\left(\mathring{t}\right)$. In view of Eq. 11.135 and 11.141 these values are determined as follows:

$$2p - x\check{\gamma}\left(\check{t}_{\|}\right) = 0, \text{ hence } \check{t}_{\|}(x) = 16p^2 \frac{2p-x}{x}, \ \check{\gamma} = \frac{2p}{x}; \tag{11.149a}$$

$$p - x\check{\gamma}\left(\check{t}_{\perp}\right) = 0, \text{ hence } \check{t}_{\perp}(x) = 16p^2 \frac{p-x}{x}, \ \check{\gamma} = \frac{p}{x}; \tag{11.149b}$$

$$\mathring{\gamma} = \mathring{t}/x;$$

$\mathring{t}(x)$ can be found as *real* root of the cubic equation

$$\mathring{t}^3 - x\mathring{t}^2 - xp^2 = 0. \tag{11.149c}$$

Let us denote

$$D(p) = \frac{\exp\left(p^2\right)}{4p\sqrt{\pi}\, I_0\left(2p^2\right)}. \tag{11.150}$$

In view of Eq. 11.135, 11.141, 11.148, 11.149, we have

$$\frac{d\check{\gamma}(t)}{dt} = \frac{1}{16p^2}, \left|\frac{d\mathring{\gamma}(t)}{dt}\right| = \frac{2p^2}{t^3},$$

$$\check{K}''_{\|}(x) = \frac{3\pi D(p)}{2p^2}\left(\frac{x}{2p}\right)^2\left(16p^2\frac{2p-x}{x}\right)^2 \exp\left(-16p^2\frac{2p-x}{x}\right), \ x < 2p;$$

$$\check{K}''_{\|}(x) = 0, \quad x \geq 2p; \tag{11.151a}$$

$$\check{K}''_{\perp}(x) = 24\pi D(p)\left(\frac{x}{p}\right)^2 \left(16p^2\frac{p-x}{x}\right)^2 \exp\left(-16p^2\frac{p-x}{x}\right),\ x < p;$$

$$\check{K}''_{\perp}(x) = 0,\ x \geq p; \tag{11.151b}$$

$$\mathring{K}''_{\|}(x) = \mathring{K}''_{\perp}(x) = 24\frac{3\pi D(p)}{p}\left(1+\frac{2p^2}{\mathring{t}^2}\right)^2 \mathring{t}x^2 \exp\left(-\mathring{t}^2\right), \tag{11.151c}$$

where $\mathring{t}(x, p)$ is found from Eq. 11.149c.

In an anisotropic medium the loss susceptibility is given for $\|$ and $\perp$ orientations respectively by

$$K''_{\|}(x) = \check{K}''_{\|}(x) + \mathring{K}''(x);\ K''_{\perp}(x) = \check{K}''_{\perp}(x) + \mathring{K}''(x). \tag{11.152}$$

In an *isotropic medium*

$$L''(x) = \frac{1}{3}\left(K''_{\|}(x) + K''_{\perp}(x)\right) = \frac{1}{3}\left(\check{K}''_{\|}(x) + \check{K}''_{\perp}(x) + 2\mathring{K}''(x)\right). \tag{11.153}$$

The loss / absorption spectra due to *librators* (Fig. 11.9) resemble frequency distributions $\check{f}\left(\check{p}\right)$, Fig.11.3 a, b, c, while spectra due to *hindered rotators* (see Fig. 11.10) resemble frequency distributions $\mathring{f}\left(\mathring{p}\right)$, Fig. 11.3 d, e.

The $K''_{\|}(x)$ and $xK''_{\|}(x)$ curves have far weaker peaks for $\|$-orientation, than corresponding dependencies $K''_{\perp}(x)$ and $xK''_{\perp}(x)$, calculated for $\perp$-orientation.

The absorption peak near x_L due to *librators* increases for $p > 0.5$ with the rise of conservative field, while for *hindered rotors* this frequency x_L depends weakly on p, cf. Fig.11.9 and 11.10.

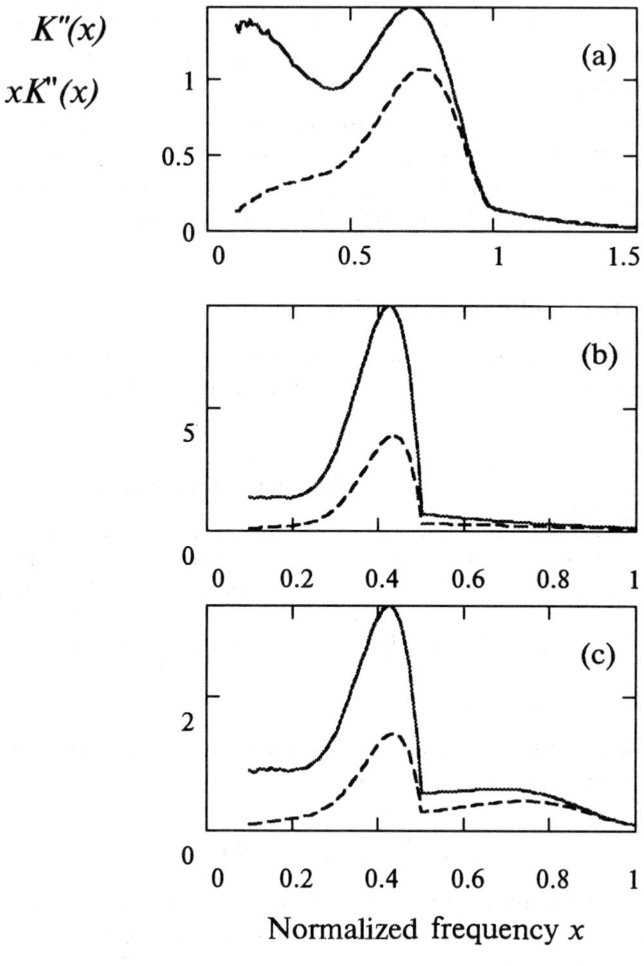

Fig. 11.11 Loss (solid lines) and absorption (dashed lines) spectra for the longitudinal (a) and transverse (b) orientations of an alternating field and for an isotropic medium (c). The long lifetime limit. $p = 0.5$.

For an intermediate p-value ($p = 0.5$) the absorption / loss spectra, calculated for anisotropic (Fig. 11.11 a, b) and isotropic (Fig. 11.11 c) media may become two-humped; these spectra resemble those, calculated for $y \neq 0$, cf. Fig.11.11 with Fig. 11.6, 11.8.

Resonance curves, pictured in this section for $y = 0$, constitute *limiting lines*; their bandwidths are minimal. We remind the Reader that in the elastic bond model, where the nonlinear dependence on the energy h of librational / hindered rotational frequency is neglected, the linewidth tends to zero for $y \to 0$.

11.7 The Stratified Approximation

We can improve the quasi-harmonic approximation by changing intervals of integration over energy h, as shown in the following tabulation:

The quasi-harmonic approximation		The stratified approximation		
Librators	Hindered rotators	Librators	Slow particles	Hindered rotators
$0 \le h \le 4p^2$	$4p^2 \le h$	$h \le 0 \le 4p^2$	$4p^2 \le h \le \frac{16}{3}p^2$	$\frac{16}{3}p^2 \le h$

For definiteness we consider here only the case of an isotropic medium; the SF $L(z)$ is represented as the sum

$$L(z) = \overset{\vee}{L}(z) + \tilde{L}(z) + \overset{\circ}{L}(z). \tag{11.154}$$

The first and last terms differ from those, calculated from Eq. 11.139 and 11.143 by the limits of integration over h. The middle term accounts for the response of particles, characterized by almost aperiodic law of motion. This term is calculated in GT, p. 310 in the assumption that the modulus $k = 1$. As a result, we have

$$\overset{\vee}{L}(z) = \frac{1}{3}\left(\overset{\vee}{K}_{\|}(z) + \overset{\vee}{K}_{\perp}(z)\right) = \frac{\exp(2p^2)}{4p\sqrt{\pi}\, I_0(2p^2)} \tag{11.155}$$

$$\times \int_0^{4p^2} \frac{h \exp(-h)}{\check{\gamma}^3(h)} \left\{ \frac{1}{p^2 - z^2 \check{\gamma}^2(h)} + \frac{h/(4p)}{4p^2 - z^2 \check{\gamma}^2(h)} \right\} dh,$$

$$\tilde{L}(z) = \frac{4\exp\left(-p^2\right)\left[1-\exp\left(-\frac{7p^2}{3}\right)\right]}{p\pi\sqrt{\pi}\, I_0\left(2p^2\right)} \left\{1 + \frac{iz}{p} - \frac{2iz^3}{p} \sum_{n=1}^{\infty} \frac{(-1)^n}{(pn - iz)^2}\right\}, \quad (11.156)$$

$$\mathring{L}(z) = \frac{\mathring{K}_{\parallel}(z) + \mathring{K}_{\perp}(z)}{3} = \frac{2\exp\left(2p^2\right)}{\sqrt{\pi}\, I_0\left(2p^2\right)} \int_{2p}^{\infty} \left(1 + \frac{2p^2}{t^2}\right)^2 \frac{t^2 \exp(-t^2)\, dt}{\mathring{\gamma}^3(t)\left[t^2 - z^2 \mathring{\gamma}^2(t)\right]}, \quad (11.157)$$

$$\check{\gamma}(h) = 1 + \frac{h}{16p^2}, \quad \mathring{\gamma}(t) = \frac{p^2}{t^2}. \quad (11.158)$$

For frequent collisions, viz. if

$$y > y_c, \quad (11.159)$$

expression (11.155), that describes contribution of librators to the SF, may be substantially simplified. Indeed, neglecting dependence on h in all denominators, we may eliminate all integrals:

$$\check{L}(z) = \frac{\exp\left(2p^2\right)}{4p\sqrt{\pi}\, I_0\left(2p^2\right)} \left\{ \frac{1 - \left(1 + 3p^2 + \frac{9p^4}{2}\right)\exp\left(-3p^2\right)}{2p^2\left(4p^2 - z^2\right)} + \frac{1 - \left(1 + 3p^2\right)\exp\left(-3p^2\right)}{p^2 - z^2} \right\}. \quad (11.160)$$

CHAPTER 12
THE COSINE SQUARED (CS) MODEL

In this chapter we follow calculations described in the previous one but for other intermolecular potential. The new features of dielectric relaxation and new understanding of relaxation-potential relation will transform in the next (13 th) chapter into a new class of molecular models.

12.1 The Law of Motion

We write intermolecular potential in the form

$$U(\vartheta) = U_0 (1 - \cos^2\vartheta), \qquad (12.1)$$

U_0 is the potential welldepth, i.e. the maximum U -value, since $U = 0$ is taken on the bottom of the well; ϑ is the angular shift of a dipole. Expression

$$E_s(\vartheta) = (U_0 / \mu) \cos \vartheta, \qquad (12.2)$$

for a static field, which corresponds to this $U(\vartheta)$, follows from Eq. 12.1. We see from Fig. 1.7b that:

a. Two bottoms of the potential are placed at $\vartheta = 0$ and $\vartheta = \pi$.

b. Static field, that orients dipoles, is directed near these bottoms in opposite directions. The resulting steady-state dipole moment is equal to zero, since $\langle q \rangle = 0$.

Variable ϑ presents an angular shift of a dipole relative to the bottom of one well, where $|\vartheta| < \pi/2$; the shift relative to the bottom of the other well is $|\pi/2 - \vartheta|$. Further we shall consider motion of a dipole and related quantities for $|\vartheta| < \pi/2$. (Noting the miror symmetry of the potential, the same results one can obtain also for $|\pi/2 - \vartheta| \le \pi/2$).

We employ the static-field parameter of the CS model

$$p = \sqrt{\frac{U_0}{k_B T}} \, . \tag{12.3}$$

The normalized Hamiltonian is given by

$$h = \frac{H}{k_B T} = \dot{\vartheta}^2 + p^2 \sin^2 \vartheta, \tag{12.4}$$

where $\dot{\vartheta} = d\vartheta / d\varphi$, , $\varphi = t / \eta$. The solution of Eq.12.4 is given by

$$\varphi + \varphi_0 = \int_0^{\vartheta} \frac{d\vartheta}{\sqrt{(h/p)^2 - \sin^2 \vartheta}} \, . \tag{12.5}$$

Putting in Eq. 12.4 $\dot{\vartheta} = 0$, we can relate the libration amplitude $\beta \equiv |\vartheta_{max}|$ to the energy h of a dipole:

$$\beta = a\sin\left(\frac{\sqrt{h}}{p}\right); \; h \le p^2; \tag{12.6}$$

$h_{th} = p^2$ being the threshold energy. In the subensemble of librators a dipole "reflects" (i.e. changes the sign of its angular velocity), when ϑ attains the β-value.

Dipoles (hindered rotators) of the second subensemble perform complete rotation in a plane. In this subensemble *the sign* of $\dot{\vartheta}$ does not alter in time for a dipole of a given energy *h*. As previously, we mark these subensembles by superscripts $\vee$ and $\circ$.

LIBRATORS. We introduce variable

$$w = \text{a}\sin\left(\frac{p}{\sqrt{h}\sin \vartheta}\right), \; h \le p^2, \tag{12.7}$$

so that

$$d\vartheta = \frac{\left(\sqrt{h}/p\right)\cos w\, dw}{\sqrt{1-\left(h/p^2\right)\sin^2 w}}\,; \quad \sqrt{\left(h/p^2\right) - \sin^2\vartheta} = \frac{\sqrt{h}}{p}\cos w.$$

Then we get from Eq. 12.5:

$$u = \left(\varphi + \varphi_0\right)p = \int_0^w \frac{dw'}{\sqrt{1-\left(h/p^2\right)\sin^2 w'}} = F(\lambda, \check{k})\,; \tag{12.8}$$

$$p\varphi = F(\lambda, \check{k}) - F(\lambda_0, \check{k})\,; \tag{12.9a}$$

$$\check{k} = \frac{\sqrt{h}}{p}, \ \sin\lambda = \mathrm{sn}(u, \check{k}) = \frac{p\sin\vartheta}{\sqrt{h}} \tag{12.9b}$$

Without loss of generality we may put $\varphi_0 = 0$; then we get from Eq. 12.9 the law of motion (in terms of Jacobi elliptic functions):

$$\check{q}_\perp(\varphi) = \sin\vartheta = \frac{\sqrt{h}}{p}\mathrm{sn}\left(p\varphi, \frac{\sqrt{h}}{p}\right), \qquad , \tag{12.10}$$

$$\check{q}_\parallel(\varphi) = \sqrt{1 - \check{q}_\perp^2(\varphi)} = = \sqrt{1 - \check{k}^2\ \mathrm{sn}^2\left(u, \check{k}\right)} = \mathrm{dn}\left(p\varphi, \frac{\sqrt{h}}{p}\right). \tag{12.11}$$

With account of Eq.12.8 we find the period $\check{\Phi}$of $\check{q}_\parallel(\varphi)$ function and of fundamental librational frequency:

$$p\check{\Phi} = 2\mathbf{K}(\check{k})\,; \tag{12.12}$$

$$\check{p} = \frac{\pi}{\check{\Phi}} = \frac{\pi p}{2\mathbf{K}(\check{k})}\ . \tag{12.13}$$

If $\check{k} << 1$, small-energy dipoles librate *harmonically* near the potential bottom with almost zero amplitude, since

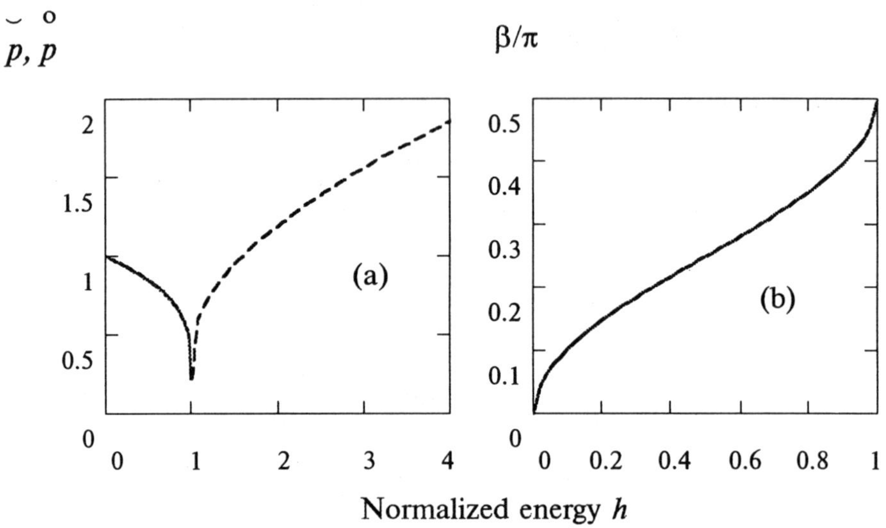

Fig. 12.1 **a:** The fundamental frequency of libration (solid line) and hindered rotation (dashed line) vs. energy h; b: the libration amplitude v s. energy h. In these examples the field parameter $p = 1$.

$$\mathrm{sn}\left(u, \check{k}\right) \approx \sin u; \; \check{k} = \frac{\sqrt{h}}{p} \approx \beta \tag{12.14}$$

$$\check{q}_{\perp}(\varphi) \approx \frac{\sqrt{h}}{p} \sin u \approx \beta \sin p\varphi; \; \check{q}_{\parallel}(\varphi) \approx 1 - \frac{\beta^2}{4} + \frac{\beta^2 \cos 2p\varphi}{4}. \tag{12.15}$$

Thus, the field parameter p coincides with the normalized librational frequency $\check{p}$, which in the limit $\check{k} \to 0$ characterizes harmonic oscillations, performed with infinitely small amplitude β.

In the opposite limit $\check{k} \to 1$ the energy h of a dipole is equal to its threshold energy $h_{\text{th}} = p^2$; now $\mathbf{K}\left(\check{k}\right) \to \infty$; the frequency $\check{p} \to 0$; rotational motion becomes aperiodic. For $\check{k} \to 1$ we obtain from Eq. 12.10, 12.11

$$\check{q}_{\perp}(\varphi) \approx \tan p\varphi;\ \check{q}_{\parallel}(\varphi) \approx \operatorname{sech} p\varphi. \tag{12.16}$$

If $\varphi \to \infty$, $\check{q}_{\perp}(\varphi) \to 1$, $\check{q}_{\parallel}(\varphi) \to 0$, the vector μ tends to direction, normal to symmetry axis Z.

HINDERED ROTATORS. If $h > p$, Eq. 12.5 yields:

$$u = \left(\varphi + \varphi_0\right)\frac{\sqrt{h}}{p} = \int_0^{\vartheta} \frac{d\vartheta}{\sqrt{1 - \left(p^2 / h\right)\sin^2 \vartheta}} = F(\vartheta, \mathring{k}), \tag{12.17}$$

$$\mathring{k} = \frac{p}{\sqrt{h}} = \frac{1}{\check{k}}. \tag{12.18}$$

Putting $\varphi_0 = 0$, we have

$$\begin{Bmatrix} \mathring{q}_{\perp}(\varphi), \\ \mathring{q}_{\parallel}(\varphi) \end{Bmatrix} = \begin{Bmatrix} \sin\vartheta, \\ \sqrt{1 - \sin^2\vartheta} \end{Bmatrix} = \begin{Bmatrix} \operatorname{sn}\left(\sqrt{h}\,\varphi, \mathring{k}\right), \\ \operatorname{cn}\left(\sqrt{h}\,\varphi, \mathring{k}\right) \end{Bmatrix} \tag{12.19}$$

It follows from Eq. 12.17 that the period $\mathring{\Phi}$ of a complete rotation and the fundamental frequency $\mathring{p} = 2\pi / \mathring{\Phi}$ of hindered rotation are given by

$$\overset{\circ}{\Phi} = \frac{4\,\mathbf{K}\left(\overset{\circ}{k}\right)}{\sqrt{h}}; \; \overset{\circ}{p} = \frac{\pi\, p}{2\overset{\circ}{k}\,\mathbf{K}\left(\overset{\circ}{k}\right)}. \tag{12.20}$$

For high-energy particles, i.e. in the limit $\overset{\circ}{k} \to 0$, we obtain from Eq. 12.19 the law of motion of a free rotor (which is encountered in the ED model):

$$\begin{Bmatrix} \overset{\circ}{q}_{\perp}(\varphi), \\ \overset{\circ}{q}_{\parallel}(\varphi) \end{Bmatrix} = \begin{Bmatrix} \sin\left(\sqrt{h}\,\varphi\right), \\ \cos\left(\sqrt{h}\,\varphi\right) \end{Bmatrix}. \tag{12.21}$$

In the other limit $\overset{\circ}{k} \to 1$, i.e. for rotation near the threshold energy, we get the previous result (12.16). Indeed, for $h = h_{\text{th}}$ we should not have any boundary between both subensembles.

We see from Fig. 12.1a, where we put $p = 1$, that the librational frequency $\overset{\vee}{p}$ (solid line) decreases with h, changing from maximum to zero. The frequency $\overset{\circ}{p}$ of a complete rotation (dashed line) increases with h from zero to ∞. When h varies from zero to h_{th}, the libration amplitude β increases from zero to $\pi/2$ (see Fig. 12.1b). We remark, that in the constant field model the maximum value of β is two times larger (max $\beta = \pi$, cf. Fig. 11.2 with Fig. 12.1b).

12.2 The Steady State

THE STEADY STATE DISTRIBUTION. To find the normalizing constant C we employ the same variables ϑ, $\dot{\vartheta}$, as in Eq. 11.40. Using Eq. 11.41 and 12.4, we have:

$$\frac{1}{C} = \int_{-\infty}^{\infty} \exp\left(-\dot{\vartheta}^2\right) d\dot{\vartheta} \int_{-\pi}^{\pi} \exp\left(-p^2 \sin^2 \vartheta\right) d\vartheta$$

$$4\int_{-\infty}^{\infty}\exp\left(-\dot{\vartheta}^{2}\right)d\dot{\vartheta}\int_{0}^{\pi}\exp\left(-p^{2}\sin^{2}\vartheta\right)d\vartheta=2\pi\sqrt{\pi}\exp\left(-\frac{p^{2}}{2}\right)I_{0}\left(\frac{p^{2}}{2}\right);\quad(12.22)$$

$$W=\frac{\exp\left(p^{2}/2\right)}{2\pi\sqrt{\pi}\,I_{0}\left(p^{2}/2\right)}\exp(-h).\qquad(12.23)$$

MEAN SQUARES OF DIRECTION COSINES. We write by analogy with Eq. 12.22:

$$\begin{Bmatrix}\langle q_{\parallel}^{2}\rangle\\ \langle q_{\perp}^{2}\rangle\end{Bmatrix}=\frac{\exp\left(p^{2}/2\right)}{2\pi\sqrt{\pi}\,I_{0}\left(p^{2}/2\right)}\int_{-\pi}^{\pi}d\vartheta\int_{-\infty}^{\infty}\exp\left[-\dot{\vartheta}^{2}-\frac{p^{2}}{2}(1-\cos 2\vartheta)\right]$$

$$\times\frac{1\pm\cos 2\vartheta}{2}\,d\dot{\vartheta}=\frac{I_{0}\left(p^{2}/2\right)\pm I_{1}\left(p^{2}/2\right)}{2I_{0}\left(p^{2}/2\right)}=\frac{1}{3}\begin{Bmatrix}\sigma_{\parallel}\\ \sigma_{\perp}\end{Bmatrix}.\qquad(12.24)$$

Here and in (12.22) we account for the contributions of dipoles rotating in *both* potential wells. We shall see just now that the situation changes, if one uses another pair of phase variables.

PROPORTIONS OF LIBRATORS AND HINDERED ROTATORS. Now we take energy h and initial phase φ_0 as phase variables.

For ensemble of librators, where $h \in [0, p^2]$ *and* $\varphi_0 \in [0, \frac{1}{2}\overset{\vee}{\Phi}]$, *we should double the result of calculations in order to take account of contributions of dipoles, oscillating in both potential wells. This multiplier 2 is not used for ensemble of rotators, where* $h \in [p^2, \infty]$ *and* $\varphi_0 \in [0, \overset{\circ}{\Phi}]$, *since all phases (angles) are included into the interval* $[0, 2\pi]$. (12.25)

This rule also holds for calculation of statistical parameters / spectral functions. With account of Eq. 12.12, 12.23 and property (12.25) we write by analogy with Eq. 11.66:

$$\check{r} = 2\int_0^{p^2} W(h)\frac{\check{\Phi}(h)}{2}dh \quad . \tag{12.26}$$

We introduce m instead of h as variable of integration:

$$m = \check{k}^2 = h/p^2;\ \text{we mean } \mathbf{K} = \mathbf{K}(\check{k}) = \mathbf{K}(\sqrt{m})\,; \tag{12.27}$$

$$\check{r} = \frac{2p\exp\left(p^2/2\right)}{\pi\sqrt{\pi}\,I_0\left(p^2/2\right)}\int_0^1 \exp\left(-p^2 m\right)\mathbf{K}\,dm \quad . \tag{12.28}$$

The proportion of hindered rotators is given by

$$\mathring{r} = \int_{p^2}^{\infty} W(h)\mathring{\Phi}(h)dh \quad . \tag{12.29}$$

We substitute $\mathring{\Phi}$ from Eq. 12.20 and use as a variable

$$m = \mathring{k}^2 = p^2/h\,;\ \text{we mean } \mathbf{K} = \mathbf{K}(\mathring{k}) = \mathbf{K}(\sqrt{m}) \tag{12.30}$$

Thus, $dh = -\left(p/m\right)^2 dm$, $h = p^2/m$. Finally, Eq. 12.28 yields

$$\mathring{r} = \frac{2p\exp\left(p^2/2\right)}{\pi\sqrt{\pi}\,I_0\left(p^2/2\right)}\int_0^1 \exp\left(-\frac{p^2}{m}\right)\mathbf{K}\,dm\,. \tag{12.31}$$

Numerical calculations shows, that the evident identity holds

$$\check{r}(p) + \mathring{r}(p) = 1. \tag{12.32}$$

For large and small p we have, respectively:

$$I_0\left(\frac{p^2}{2}\right) \approx \frac{\exp\left(p^2/2\right)}{p\sqrt{\pi}} \qquad \text{for } p >> 1 , \tag{12.33}$$

$$I_0\left(p^2/2\right) \approx 1 \qquad \text{for } p << 1. \tag{12.34}$$

If p is large, only small m-values contribute to the integral (12.28) due to $\exp(-p^2 m)$. Putting $\mathbf{K} = \mathbf{K}(0) = \pi/2$, we get:

$$\check{r} = \frac{2p^2}{\pi}\int_0^1 \exp\left(-p^2 m\right)\frac{\pi}{\mathbf{2}}\,dm = 1-\exp\left(-p^2\right); \ \mathring{r} \approx \exp\left(-p^2\right) \text{for } p >> 1. \tag{12.35}$$

Thus, $\check{r}(p)$ is about 1 for $p >> 1$, as it should be.

If p is small, the proportion $\check{r}$ tends to zero. Indeed, in view of Eq. 11.73 we have $\int_0^1 \mathbf{K}\,dm = 2\int_0^1 \mathbf{K}\,k\,dk = 2$. Then

$$\check{r} \approx \frac{2p}{\pi\sqrt{\pi}}\int_0^1 \mathbf{K}\,dm \approx \frac{4p}{\pi\sqrt{\pi}}; \ \mathring{r} \approx 1-\frac{4p}{\pi\sqrt{\pi}} \qquad \text{for } p << 1. \tag{12.36}$$

MEAN FREQUENCIES OF LIBRATION / HINDERED ROTATION. In view of Eq. 12.13, 12.23, 12.26 we find by analogy with Eq. 11.87a (the Reader should not miss to double the result for librators!):

$$\left\langle \check{p} \right\rangle = \frac{2\int_0^{p^2} W(h)\,\check{p}\,\frac{\check{\Phi}(h)}{2}\,dh}{2\int_0^{p^2} W(h)\,\frac{\check{\Phi}(h)}{2}\,dh} = \frac{2C\int_0^{p^2}\exp(-h)\frac{\pi}{\check{\Phi}}\check{\Phi}\,dh}{\check{r}}$$

$$= \frac{2\pi\exp\left(p^2/2\right)\left[1-\exp\left(-p^2\right)\right]}{2\pi\sqrt{\pi}\,I_0\left(p^2/2\right)\check{r}} = \frac{2\,\mathrm{sh}\left(p^2/2\right)}{\sqrt{\pi}\,I_0\left(p^2/2\right)\check{r}}. \tag{12.37}$$

Taking account of Eq. 12.33, 12.35, we write:

$$\left\langle \overset{\vee}{p} \right\rangle \approx \frac{2\,\mathrm{sh}\left(p^2/2\right)}{\sqrt{\pi}\,\exp\left(p^2/2\right)}\sqrt{\pi}\,p \approx p \quad \text{for} \quad p >> 1, \tag{12.38}$$

$$\left\langle \overset{\vee}{p} \right\rangle \approx \frac{2\pi\,\sqrt{\pi}\,\left(p^2/2\right)}{4\,p\sqrt{\pi}} \approx \frac{\pi p}{4} \ (\text{for } p << 1). \tag{12.39}$$

For ensemble of rotators we obtain by the same way:

$$\left\langle \overset{\circ}{p} \right\rangle = \frac{\int_{p^2}^{\infty} W(h)\,\overset{\circ}{p}\,\overset{\circ}{\Phi}(h)\,dh}{\int_{p^2}^{\infty} W(h)\,\overset{\circ}{\Phi}(h)\,dh} = \frac{2C}{\overset{\circ}{r}}\int_{p^2}^{\infty} \exp(-h)\,\pi\,\frac{\dot{\Phi}}{\overset{\circ}{\Phi}}\,dh \approx \frac{\exp\left(-p^2/2\right)}{\overset{\circ}{r}\sqrt{\pi}\,I_0\left(p^2/2\right)}\,. \tag{12.40}$$

For $p >> 1$ we have again the result (12.38):

$$\left\langle \overset{\circ}{p} \right\rangle \approx \frac{\sqrt{\pi}\,p\,\exp\left(-p^2/2\right)}{\exp\left(-p^2\right)\sqrt{\pi}\,\exp\left(p^2/2\right)} \approx p\,. \tag{12.41}$$

If $p << 1$, $I_0\left(p^2/2\right) \approx 1 + \frac{p^4}{16}$, and we find

$$\left\langle \overset{\circ}{p} \right\rangle \approx \frac{1-\left(p^2/2\right)}{\left(1-\frac{4p}{\pi\sqrt{\pi}}\right)\sqrt{\pi}\left(1+\frac{p^4}{16}\right)} \approx \frac{1}{\sqrt{\pi}}\left(1+\frac{4p}{\pi\sqrt{\pi}}\right). \tag{12.42}$$

Consequently,

Both mean frequencies, $\left\langle \overset{\vee}{p} \right\rangle$ *and* $\left\langle \overset{\circ}{p} \right\rangle$*, increase with the rise of the field parameter p.* (12.43)

Taking account of Eq. 12.37, 12.40, we find the mean (over two subensembles) frequency $\langle p \rangle$ and the effective collision-time parameter y_c:

$$\langle p \rangle = \overset{\vee}{r} \left\langle \overset{\vee}{p} \right\rangle + \overset{\circ}{r} \left\langle \overset{\circ}{p} \right\rangle = \frac{\exp\left(p^2/2\right)}{\sqrt{\pi}\, I_0\left(p^2/2\right)};\ y_c = \frac{\langle p \rangle}{\pi} \qquad (12.44)$$

THE MEAN LIBRATION AMPLITUDE. In view of Eq.12.6 we have

$$\bar{\beta} = \frac{2\int_0^{p^2} \beta \overset{\vee}{\Phi} W\, dh}{2\int_0^{p^2} \overset{\vee}{\Phi} W\, dh} = \frac{2C}{\overset{\vee}{r}} \int_0^{p^2} \operatorname{asin}(\overset{\vee}{k})\, \frac{\pi}{\overset{\vee}{\Phi}} \overset{\vee}{\Phi} \exp(-h)\, dh$$

$$= \frac{2p \exp\left(p^2/2\right)}{\overset{\vee}{r}\, \pi\sqrt{\pi}\, I_0\left(p^2/2\right)} \int_0^1 \operatorname{asin}(\sqrt{m})\, \mathbf{K} \exp\left(-p^2 m\right) dm. \qquad (12.45)$$

For $p \approx 1$ in view of the exponential factor in the integrand we may substitute $\mathbf{K}(m)$ by $\mathbf{K}(0)$ and $\operatorname{asin}\left(\sqrt{m}\right)$ by $\sqrt{m}$. Putting $\overset{\vee}{r} \approx 1$, we have:

$$\bar{\beta} \approx \frac{4p \exp\left(p^2/2\right) \sqrt{\pi}\ p}{\pi\sqrt{\pi} \exp\left(p^2/2\right)} \int_0^1 \frac{\pi}{2} \exp\left(-p^2 k^2\right) k^2\, dk \approx \frac{\sqrt{\pi}}{2p}. \qquad (12.46)$$

We have got the same result, as for the CF model, see Sec. 11.3.2. But for the shallow potential well ($p << 1$) we have only half of the corresponding result. Indeed, noting Eq. 11.79, we have: $\overset{\vee}{r} \approx \dfrac{4p}{\pi\sqrt{\pi}}$,

$$\bar{\beta} = \int_0^1 k\mathbf{K}\, a\sin(k)\, dk = \frac{\pi}{2}\left(1 - \frac{\pi}{2\mathbf{K}^2\left(1/\sqrt{2}\right)}\right). \qquad (12.47)$$

The distinction between CF and CS models is probably obliged to the number of wells.

DISTRIBUTIONS OVER FREQUENCIES OF LIBRATION / HINDERED ROTATION. First we write the normalizing condition:

$$\int W\,d\Gamma = C\int \exp(-h)\,d\Gamma = C\left[2\int_0^{p^2}\exp(-h)\overset{\vee}{\Phi}\,dh + \int_{p^2}^{\infty}\exp(-h)\overset{\circ}{\Phi}\,dh\right] = 1;$$

$$C\left[2\int_0^{p}\exp(-h)\frac{2\pi}{\overset{\vee}{p}}\left(-\frac{dh}{d\overset{\vee}{p}}\right)d\overset{\vee}{p} + \int_0^{\infty}\exp(-h)\frac{2\pi}{\overset{\circ}{p}}\left(\frac{dh}{d\overset{\circ}{p}}\right)d\overset{\circ}{p}\right] = 1. \quad (12.48)$$

The desired frequency distributions are given by

$$\overset{\vee}{f}\left(\overset{\vee}{p}\right) = -8\pi C\exp\left[-h\left(\overset{\vee}{p}\right)\right]\frac{dh}{d\overset{\vee}{p}{}^2} \text{ for } 0 \le \overset{\vee}{p} \le p\ ; \quad (12.49)$$

$$\overset{\circ}{f}\left(\overset{\circ}{p}\right) = 4\pi C\exp\left[-h\left(\overset{\circ}{p}\right)\right]\frac{dh}{d\overset{\circ}{p}{}^2} \text{ for } 0 \le \overset{\circ}{p} \le \infty. \quad (12.50)$$

By analogy with derivation, given in Sec. 11.2, we have:

$$\frac{d\mathbf{K}(k)}{dk} = \frac{\mathbf{E} - m\mathbf{K}}{m'\sqrt{m}}, \text{ where } m = k^2,\ m' = 1 - k^2 = 1 - m; \quad (12.51)$$

$$\frac{d\mathbf{K}}{dm} = \frac{\mathbf{E} - m\mathbf{K}}{m'\sqrt{m}\,2\sqrt{m}} = \frac{\mathbf{E} - m\mathbf{K}}{2m\,m'};$$

$$\frac{d}{dm}\frac{1}{m\mathbf{K}^2} = -\frac{1}{m\mathbf{K}^2}\left(\frac{1}{m} + \frac{2}{\mathbf{K}}\frac{d\mathbf{K}}{dm}\right) = -\frac{1}{m\mathbf{K}^2}\frac{\mathbf{E}}{m\,m'\mathbf{K}} = \frac{\mathbf{E}}{m^2\,m'\mathbf{K}^3} \quad (12.52)$$

For librators in view of Eq. 12.10, 12.13 $dm/dh = p^{-2}$,

$$-\frac{d\overset{\vee}{p}{}^2}{dh} = -\left(\frac{\pi p}{2}\right)^2\frac{d}{dh}\frac{1}{\mathbf{K}^2} = \left(\frac{\pi p}{2}\right)^2\frac{2}{\mathbf{K}^3}\frac{dm}{dh}\frac{d\mathbf{K}}{dm} = \frac{2}{\pi}\frac{\mathbf{E} - m\mathbf{K}}{m\,m'}\left(\frac{\overset{\vee}{p}}{p}\right)^3. \quad (12.53)$$

For hindered rotators, where $dm/dh = -(p/h)^2 = -(m/p)^2$, we get

$$\frac{d\mathring{p}^2}{dh} = \left(\frac{\pi p}{2}\frac{m}{p}\right)^2 \frac{\mathbf{E}}{m^2 m' \mathbf{K}^3} = \left(\frac{\pi p}{2}\right)^2 \frac{m\sqrt{m}}{p^2}\frac{\mathbf{E}}{m'}\left(\frac{2\mathring{p}}{\pi p}\right)^3 = \left(\frac{\mathring{p}}{p}\right)^3 \frac{2}{\pi}\frac{\mathbf{E}\, m\sqrt{m}}{m'}. \tag{12.54}$$

Substituting derivatives (12.53), (12.54) and normalizing constant C (see Eq. 12.23) into (12.49) - (12.50), we find:

$$\check{f}\left(\check{p}\right) = \frac{\sqrt{\pi}\,\exp\left(\frac{p^2}{2} - p^2 m\right)}{I_0\left(p^2/2\right)}\left(\frac{p}{\check{p}}\right)^3 \frac{m\,m'}{\mathbf{E} - m'\mathbf{K}}, \tag{12.55}$$

$$\mathring{f}\left(\mathring{p}\right) = \frac{\sqrt{\pi}\,\exp\left(\frac{p^2}{2} - \frac{p^2}{m}\right)}{I_0\left(p^2/2\right)}\left(\frac{p}{\mathring{p}}\right)^3 \frac{m'}{m\sqrt{m}\,\mathbf{E}}. \tag{12.56}$$

It is convenient to calculate Eq. 12.55 and 12.56 in a parametric form by taking m in the interval [0, 1], variables $\check{p}$ and $\mathring{p}$ being functions of m given by Eq. 12.13 and 12.20.

If conservative field is strong, the first function, Eq. 12.55, transforms into some sort of δ-function. Indeed, if $p >> 1$ and $\check{p}$ is near p, m is near zero, $\mathbf{K}$ is near $\pi/2$. Then we have from Eq. 12.13 and 11.62:

$$\frac{\check{p}}{p} \approx \frac{\pi/2}{(\pi/2)(1+m/4)} \approx \frac{1}{1+m/4};\; m \approx 4\left(\frac{p}{\check{p}} - 1\right).$$

With account of Eq. 11.62 we derive: $\mathbf{E} - m'\mathbf{K} \approx \pi m/4$,

$$\check{f}(\check{p}) \approx 4p\left[\left(\frac{p}{\check{p}}\right)^3\right]\exp\left[-4p^2\left(\frac{p}{\check{p}} - 1\right)\right]. \tag{12.57}$$

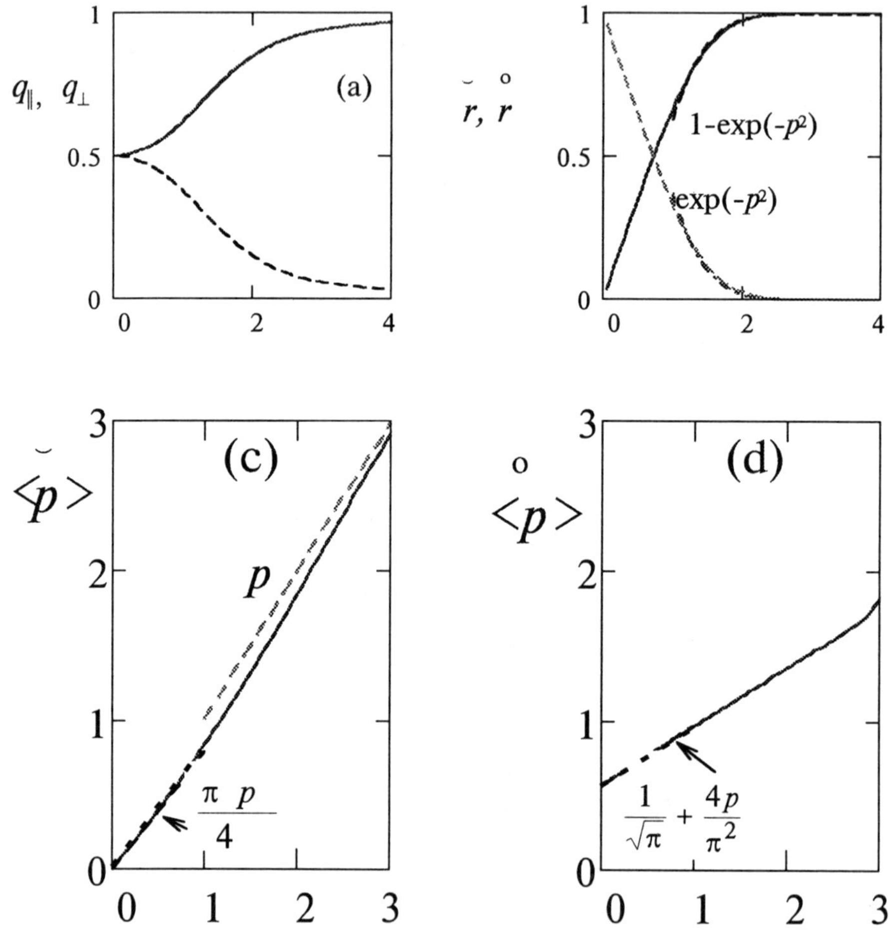

Fig. 12.2 a - d: See capture on the next page.

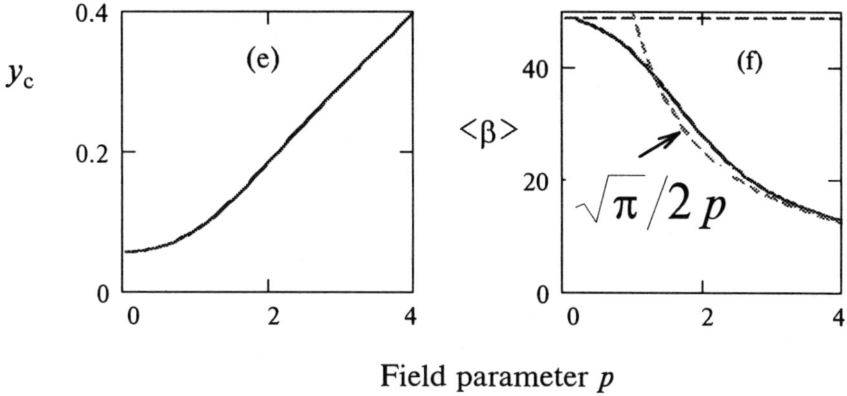

Fig.12.2 Dependence of the steady-state parameters vs. p: squares of direction cosines (a), proportions of librators and hindered rotors (b) and their mean frequencies (c, d), the effective collision-time parameter (e) and libration amplitude (f). The horizontal line on Fig. f refers to the value (12.47).

This function describes a narrow peak near $\check{p} \approx p$.

On the other side, Eq. 12.56 transforms into Maxwellian distribution $\frac{2}{\sqrt{\pi}}\exp\left(-\mathring{p}^2\right)$ for vanishing conservative field. Indeed, if

$p \to 0$, then $m \to 0$, $\mathbf{E} \to \pi/2$, $m' \to 1$, $\mathring{p} \approx p/k$,

$$\mathring{f}(\mathring{p}) \approx \sqrt{\pi}\exp\left[-\frac{p^2\mathring{p}^2}{p^2}\right]\left(\frac{p}{\mathring{p}}\right)^3\frac{2\cdot 1}{\pi}\left(\frac{\mathring{p}}{p}\right)^3 = \frac{2}{\sqrt{\pi}}\exp\left(-\mathring{p}^2\right). \quad (12.58)$$

INFLUENCE OF CONSERVATIVE FIELD ON MOLECULAR DYNAMICS. In Fig. 12.2 we compare corresponding dependencies with those, found in Chap. 11 (c.f. with Fig. 11.4, 11.5).

Parameter	CS model	CF model
Proportions $\check{r}$ and $\mathring{r}$	$\check{r}$ is near 1 and $\mathring{r}$ is near 0 at $p = 2$	$\check{r} \approx 1$ and $\mathring{r}$ is near 0 at $p = 1$
Frequencies $\check{p}$ and $\check{p}$	Rather slow increase with the rise of p	Rapid increase with the rise of p
Collision time y_c	In the CS model y_c is about 2 times lesser, than in the CF model	

(12.59)

If a dipole librates (in each model) near the bottom with the same frequency p, then influence of conservative potential on molecular dynamics is lesser in the model, characterized by *steeper* potential walls (in our example it is the CS model). Thus, for $p \geq 1$ the proportion $\mathring{r}$ of hindered rotors is appreciably higher in the CS model, than in the CF model.

The steady-state distributions are similar in both models (cf. Fig. 12.3 and 11.3). However, since the $\mathring{f}\left(\mathring{p}\right)$ distribution vanishes in the CF model for $p = 1$, we may suppose that

---

The dielectric response due to hindered rotors should be negligible in the CS model for the field parameter $p > 2$, while in the CF model - for $p > 1$. (12.60)

12.3 The Spectral Function: Rigorous Theory

Each term of the sum

$$K(z) = \check{K}(z) + \mathring{K}(z), \qquad (12.61)$$

corresponds to a specific subensemble of dipoles.

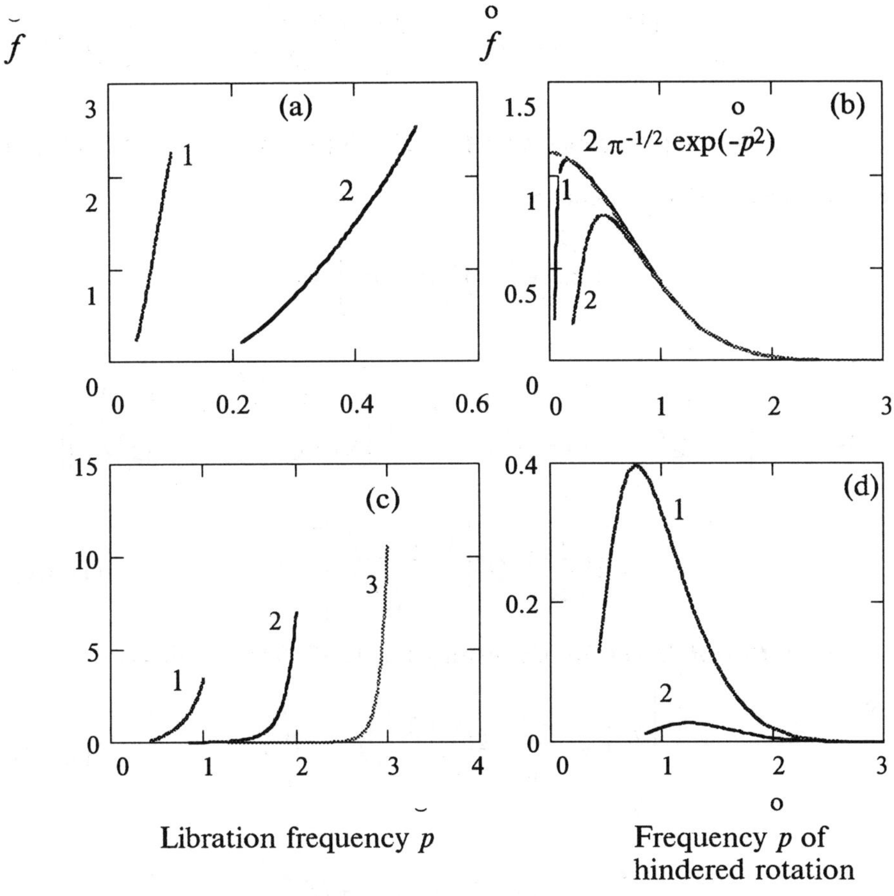

Fig. 12.3 Distribution functions vs. fundamental frequencies of librations (a, c) and hindered rotations (b, d). **a, b:** for the field parameter $p = 0.1$ (curves 1) and 0.5 (curves 2). **c, d:** for $p = 1, 2, 3$ (curves 1, 2, 3, respectively).

12.3.1 Fourier Amplitudes

LIBRATORS. Let us consider series (5.81). Comparing Eq. 11.104 and 11.106 respectively with Eq. 12.10 and 12.11, we derive:

$$\overset{\vee}{a}_{2n} = \frac{2\pi}{\mathbf{K}} \frac{Q^n}{1+Q^{2n}} \text{ and } \overset{\vee}{b}_{2n-1} = \frac{2\pi}{\mathbf{K}} \frac{Q^{n-1/2}}{1-Q^{2n-1}} \; ; \; k = \frac{\sqrt{h}}{p}; \quad (12.62a)$$

$$\mathbf{K} \equiv \mathbf{K}(k),\ Q = \exp\left(-\pi \mathbf{K}'/\mathbf{K}\right),\ \mathbf{K}' \equiv \mathbf{K}(k'),\ k' = \sqrt{1-k^2}\,. \quad (12.62b)$$

We write down useful combinations:

$$\overset{\vee}{a}_n^2 + \overset{\vee}{b}_n^2 + (-1)^n\left(\overset{\vee}{a}_n^2 - \overset{\vee}{b}_n^2\right) =$$

$$\left(\frac{2\pi}{\mathbf{K}}\right)^2 Q^n\left\{\left(\frac{1}{1+Q^n}\right)^2 + \left(\frac{1}{1-Q^n}\right)^2 + (-1)^n\left[\left(\frac{1}{1+Q^n}\right)^2 - \left(\frac{1}{1-Q^n}\right)^2\right]\right\}$$

$$= 2\left(\frac{2\pi}{\mathbf{K}}\right)^2 Q^n\left[\frac{1-(-Q)^n}{1-Q^{2n}}\right]^2 . \quad (12.62c)$$

HINDERED ROTATORS. Comparing Eq. 12.19, 5.89, we derive

$$\left\{\begin{matrix} \overset{\circ}{a}_n \\ \overset{\circ}{b}_n \end{matrix}\right\} = \frac{2\pi}{k\mathbf{K}} \frac{Q^{n-1/2}}{1 \pm Q^{2n-1}},\ k = \frac{p}{\sqrt{h}}; \quad (12.63)$$

$$\overset{\circ}{a}_n^2 + \overset{\circ}{b}_n^2 = \left(\frac{2\pi}{k\mathbf{K}}\right)^2 Q^{2n-1} \frac{1+Q^{4n-2}}{\left(1-Q^{4n-2}\right)^2} . \quad (12.64)$$

12.3.2 The SF of Librating Particles

Here we double the result of calculation.

AN ANISOTROPIC MEDIUM. We substitute into Eq. 5.86-5.87 expressions (12.23) for W, (12.13) for p, (12.62a) for $\overset{\vee}{a}_n$ and $\overset{\vee}{b}_n$. Replacing variable h by m, where $h = p^2 m$, we have:

$$\frac{\check{K}_{\|}(z)}{3} = \frac{8p^3 \exp\left(p^2/2\right)}{\sqrt{\pi}\, I_0\left(p^2/2\right)}$$

$$\times \int_0^1 dm \exp\left(-p^2 m\right)\left(\frac{\pi}{2\mathbf{K}}\right)^3 \sum_{n=1}^{\infty}\left(\frac{2n}{1+Q^{2n}}\right)^2 \frac{Q^{2n}}{\left(\frac{\pi p}{2\mathbf{K}}\right)^2 (2n)^2 - z^2}, \quad (12.65)$$

$$\frac{\check{K}_{\perp}(z)}{3} = \frac{8p^3 \exp\left(p^2/2\right)}{\sqrt{\pi}\, I_0\left(p^2/2\right)}$$

$$\times \int_0^1 dm \exp\left(-p^2 m\right)\left(\frac{\pi}{2\mathbf{K}}\right)^3 \sum_{n=1}^{\infty}\left(\frac{2n-1}{1-Q^{2n-1}}\right)^2 \frac{Q^{2n-1}}{\left(\frac{\pi p}{2\mathbf{K}}\right)^2 (2n-1)^2 - z^2}. \quad (12.66)$$

AN ISOTROPIC MEDIUM. With account of Eq.12.62c we find the SF:

$$\check{L}(z) = \frac{\check{K}_{\|}(z) + \check{K}_{\perp}(z)}{3} = \frac{8p^3 \exp\left(p^2/2\right)}{\sqrt{\pi}\, I_0\left(p^2/2\right)}$$

$$\times \int_0^1 dm \exp\left(-p^2 m\right)\left(\frac{\pi}{2\mathbf{K}}\right)^3 \sum_{n=1}^{\infty}\left(\frac{1-(-Q)^n}{1-Q^{2n}}\right)^2 \frac{n^2 Q^n}{\left(\frac{\pi p}{2\mathbf{K}}\right)^2 n^2 - z^2}. \quad (12.67)$$

12.3.3 The SF of Hindered Rotors

We find the l. h. s. of Eq. 5.90 by substituting W from (12.23), $\mathring{p}$ from (12.20), Fourier amplitudes from (12.63).

For an *anisotropic* medium we have

$$\frac{1}{3}\left\{\begin{matrix}\overset{\circ}{K}_{\parallel}(z)\\ \overset{\circ}{K}_{\perp}(z)\end{matrix}\right\} = \frac{8p\exp\left(p^2/2\right)}{\sqrt{\pi}\,I_0\left(p^2/2\right)}$$

$$\times\int_{p^2}^{\infty} dh\,\exp(-h)\left(\frac{\pi}{2k\mathbf{K}}\right)^3 \sum_{n=1}^{\infty}\left(\frac{2n-1}{1\pm Q^{2n-1}}\right)^2 \frac{Q^{2n-1}}{\left(\frac{\pi p}{2k\mathbf{K}}\right)^2(2n-1)^2 - z^2}.$$

We use $t=\sqrt{h}$ instead of h as a variable of integration. Since $k = p/\sqrt{h} = p/t$, $\mathbf{K} = \mathbf{K}(k) \equiv \mathbf{K}(p/t)$,

$$\frac{1}{3}\left\{\begin{matrix}\overset{\circ}{K}_{\parallel}(z)\\ \overset{\circ}{K}_{\perp}(z)\end{matrix}\right\} = \frac{16\exp\left(p^2/2\right)}{\sqrt{\pi}\,p^2\,I_0\left(p^2/2\right)}$$

$$\times\int_{p}^{\infty} dt\,t^4\exp\left(-t^2\right)\left(\frac{\pi}{2\mathbf{K}}\right)^3 \sum_{n=1}^{\infty}\left(\frac{2n-1}{1\pm Q^{2n-1}}\right)^2 \frac{Q^{2n-1}}{\left(\frac{\pi t}{2\mathbf{K}}\right)^2(2n-1)^2 - z^2}. \quad (12.68)$$

For an isotropic medium, taking into account Eq. 12.64, we have:

$$L(z) = \frac{1}{3}\left\{\overset{\circ}{K}_{\parallel}(z) + \overset{\circ}{K}_{\perp}(z)\right\} = \frac{32\exp\left(p^2/2\right)}{\sqrt{\pi}\,p^2\,I_0\left(p^2/2\right)}$$

$$\times\int_{p}^{\infty} dt\,t^4\exp\left(-t^2\right)\left(\frac{\pi}{2\mathbf{K}}\right)^3 \sum_{n=1}^{\infty}\left(\frac{2n-1}{1-Q^{4n-2}}\right)^2 \frac{Q^{2n-1}\left(1+Q^{4n-2}\right)}{\left(\frac{\pi t}{2\mathbf{K}}\right)^2(2n-1)^2 - z^2}. \quad (12.69)$$

12.4 The Spectral Function: the Quasi-Harmonic Approximation

As in Sec.11.5 we write:

$$\mathbf{K} \approx \frac{\pi}{2}\left(1+\frac{m}{4}\right);\, Q \approx \frac{m}{16}\left(1+\frac{m}{2}\right),\, m = k^2. \tag{12.70}$$

For ensemble of *librators,* where $m = h/p^2$, we have in view of Eq. 12.65 - 12.66:

$$\frac{\pi}{2\mathbf{K}} \approx \frac{1}{\check{\gamma}},\ \text{where}\ \check{\gamma} = 1+\frac{h}{4p^2};\ Q \approx \frac{h}{16p^2}\left(1+\frac{h}{2p^2}\right); \tag{12.71}$$

$$\frac{\check{K}_{\parallel}(z)}{3} = \frac{\exp\left(p^2/2\right)}{8\sqrt{\pi}\ p^3 I_0\left(p^2/2\right)} \int_0^{p^2} \frac{h^2\exp(-h)}{\check{\gamma}} \frac{\left(1+\dfrac{h}{2p^2}\right)^2 dh}{4p^2-\left(\check{\gamma}z\right)^2}, \tag{12.72}$$

$$\frac{\check{K}_{\perp}(z)}{3} = \frac{\exp\left(p^2/2\right)}{2\sqrt{\pi}\ p I_0\left(p^2/2\right)} \int_0^{p^2} \frac{h\exp(-h)}{\check{\gamma}\left(1-\dfrac{h}{16p^2}\right)^2} \frac{\left(1+\dfrac{h}{2p^2}\right) dh}{p^2-\left(\check{\gamma}z\right)^2}. \tag{12.73}$$

For an ensemble of *hindered rotors,* where $k = p/t$, we have from Eq. 12.70, 12.68:

$$Q(k) \approx \frac{p^2\left(1+\dfrac{p^2}{2t^2}\right)}{16t^2};\ \frac{\pi}{2\mathbf{K}} \approx \frac{1}{\mathring{\gamma}},\mathring{\gamma} \equiv 1+\frac{p^2}{4t^2}; \tag{12.74}$$

$$\frac{1}{3}\begin{Bmatrix} \overset{\circ}{K}_{\|} \\ \overset{\circ}{K}_{\perp} \end{Bmatrix} = \frac{\exp\left(p^2/2\right)}{\sqrt{\pi}\, I_0\left(p^2/2\right)} \int_p^{\infty} \frac{dt\, t^2 \exp\left(-t^2\right)}{t^2 - \overset{\circ}{\gamma}{}^2 z^2} \frac{1+\dfrac{p^2}{2t^2}}{\left(1+\dfrac{p^2}{4t^2}\right)\left(1\pm\dfrac{p^2}{16t^2}\right)} . \quad (12.75)$$

The QHA may be employed for a qualitative study of dielectric behavior. For the CS model this possibility was confirmed in the (Gaiduk, Gaiduk, Novskova, Tseitlin, and Novskova 1995) work, where the comparison with rigorous theory was made. For simplicity further we restrict calculations by the quasi-harmonic approximation.

TRANSITION TO THE ELASTIC BOND MODEL. For large field parameter ($p > 2$), only librators contribute to dielectric response of polar molecules, $\overset{\circ}{k} \to 0$. We put $\sqrt{\pi}\, p I\left(p^2/2\right) \approx \exp\left(p^2/2\right)$ and *neglect the anharmonicity of librators*. Then in Eq. 12.71 we replace $\overset{\vee}{\gamma}$ by 1 and Q by $h/\left(16p^2\right)$. After these simplifications Eq. 12.72, 12.73 reduce to Eq. 10.6 - 10.7 of the elastic bond model:

$$\overset{\vee}{K}_{\|} = \frac{3}{2}\frac{1}{4\ p^2\left(4p^2-z^2\right)}\int_0^{p^2} h^2 \exp(-h)dh \approx \frac{3}{4p^2\left(4p^2-z^2\right)} , \quad (12.76)$$

$$\overset{\vee}{K}_{\perp} = \frac{3}{2}\frac{1}{p^2-z^2}\int_0^{p^2} h \exp(-h)dh \approx \frac{3}{2}\frac{1}{p^2-z^2} . \quad (12.77)$$

We remark that the EB model may yield wrong results in the vicinity of the resonance-line center, since we operate with the resonance denominator of the type

$$\frac{1}{(ap-x)^2+y^2}, \quad (12.78)$$

where a = 1 or $a = 2$. In the QHA resonance denominator is given by

$$\frac{1}{\left(ap - x - \dfrac{hx}{16p^2}\right)^2 + y^2\left(1 + \dfrac{h}{16p^2}\right)}, \tag{12.79}$$

The linewidth, found in the quasi-harmonic approximation, is nonzero for $y \to 0$, while in the EB model the linewidth vanishes.

TRANSITION TO THE EXTENDED DIFFUSION MODEL. If $p \to 0$, $\check{k} \to 0$ and Eq. 12.75 yields the SF of the planar ED model (cf. with Eq. 5.98):

$$K_{\|}(z) = K_{\perp}(z) = \frac{3}{\sqrt{\pi}} \int_{p}^{\infty} \frac{t^2 \exp\left(-t^2\right) dt}{t^2 - z^2} = \frac{3}{2} L_F(z). \tag{12.80}$$

No distinction appears between expressions for the SFs, calculated for the CF and CS models under assumptions of weak or strong field. However, this distinction appears in the quasi-harmonic approximation and in the rigorous theory.

12.5 Limiting Lines

12.5.1 Librators

Resonance denominators $\left[ap^2 - \check{\gamma}^2 z^2\right]^{-1}$ in Eq. 12.72 and 12.73, where respectively $a = 2$ or $a = 1$ are transformed to

$$\lim_{y \to 0} \operatorname{Im} \frac{1}{2ap}\left[\frac{1}{ap - \check{\gamma}\, x - i\check{\gamma}\, y} + \frac{1}{ap + \check{\gamma}\, x + i\check{\gamma}\, y}\right],$$

$$\frac{\pi}{2ap}\, \delta\left(ap - \check{\gamma}\, x\right) = \frac{\pi}{2ap}\, \delta\left(ap - \left(1 + \frac{h}{4p^2}\right)x\right).$$

We substitute variable of integration h by $\check{\gamma}\, x$. Since $d\check{\gamma}/dh = \dfrac{1}{4p^2}$, we have:

$$\frac{\pi}{2ap}dh = \frac{\pi}{2apx}\frac{d\check{\gamma}\, x}{d\check{\gamma}(h)/dh} = \frac{2\pi p}{ax}d\check{\gamma}\, x.$$

Then imaginary parts of integrals (12.72), (12.73) transform to the following expressions:

(A). THE LONGITUDINAL ORIENTATION. Let $\check{\gamma}_{\|}$ be the root of equation $\check{\gamma}_{\|}\, x = 2p$, to which $h_{\|}(x) = 4p^2\left(\dfrac{2p}{x} - 1\right)$ corresponds; $h_{\|}(x)$ vanishes, when x attains $x_{\max}^{\|} = 2p$. For $x > x_{\max}^{\|}$ the loss $\check{K}''_{\|}$ vanishes. Hence,

$$\check{K}''_{\|} = \frac{3\exp\left(p^2/2\right)\sqrt{\pi}}{16 I_0\left(p^2/2\right)}\left\{h^2 \exp(-h)\left(1+\frac{h}{2p^2}\right)\right\}_{h=h_{\|}(x)} ; x < 2p;$$

$$\check{K}''_{\|}(x) = 0, \qquad x > 2p. \tag{12.81}$$

(B). THE TRANSVERSE ORIENTATION. If $\check{\gamma}_{\perp}\, x = p$, then

$$h_{\perp}(x) = 4p^2\left(\frac{p}{x} - 1\right)$$

is the root of this equation, so that $x_{\max}^{\perp} = p$ corresponds to the zero value of $h_{\perp}(x)$. Hence,

$$\check{K}''_{\perp} = \frac{3\exp\left(p^2/2\right)\sqrt{\pi}}{p I_0\left(p^2/2\right)}\left\{h \exp(-h)\left(1+\frac{h}{2p^2}\right)\left(1-\frac{h}{16p^2}\right)\right\}_{h=h_{\perp}(x)} ; x < p;$$

$$\overset{\vee}{K}{}''_{\perp}(x) = 0, \qquad x > p. \tag{12.82}$$

12.5.2 Hindered Rotors

In view of Eq. 12.74, 12.75,

$$\lim_{y\to 0} \operatorname{Im} \frac{1}{2t}\left[\frac{1}{t-\overset{\circ}{\gamma} z} + \frac{1}{t+\overset{\circ}{\gamma} z}\right] = \frac{\pi}{2t}\,\delta\left(t - \overset{\circ}{\gamma} x\right); \quad t \geq p.$$

The argument of δ-function is zero, if t is the root of the cubic equation

$$4t^3 - 4t^2x - p^2x = 0. \tag{12.83}$$

We remark that $t \approx x$. Then in the limit $y \to 0$:

$$\frac{1}{3}\begin{Bmatrix} \overset{\circ}{K}{}''_{\parallel} \\ \overset{\circ}{K}{}''_{\perp} \end{Bmatrix} = \frac{3\sqrt{\pi}}{2} \frac{t^2 \exp(-t^2)\left(1+\dfrac{p^2}{2t^2}\right)}{\left(1+\dfrac{p^2}{4t^2}\right)\left(1\pm\dfrac{p^2}{16t^2}\right)} \qquad \text{for} \quad \begin{Bmatrix} t \approx x \\ x \geq p \end{Bmatrix};$$

$$\overset{\circ}{K}{}''_{\parallel} = \overset{\circ}{K}{}''_{\perp} = 0 \quad \text{for } x < p. \tag{12.84}$$

Since usually $p < 4t$, anisotropy of dielectric response is obliged mostly to *librators.*

12.6 Influence of the Field parameter on Dielectric Spectra

AN ISOTROPIC MEDIUM. We employ the following expressions for the complex susceptibility:

$$\chi^*(x) = \frac{GzL(z)}{x + iyL(z)}; \;. \tag{12.85}$$

$$L = \frac{1}{3}\left(\overset{\vee}{K}_{\parallel} + \overset{\vee}{K}_{\perp} + 2\overset{\circ}{K}\right). \tag{12.86}$$

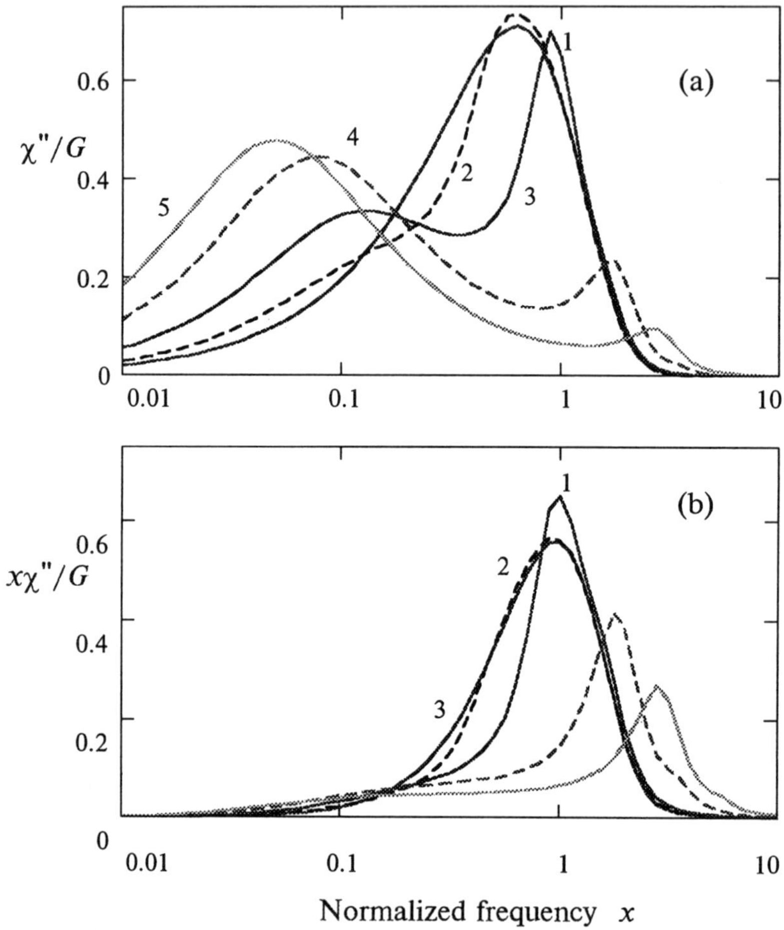

Fig. 12.4 The evolution of dielectric loss (a) and absorption (b) spectra due to the rise of the field parameter p. An isotropic medium. Curves 1, 2, 3, 4, and 5 for $p = 0$, 0.5, 1, 2, and 3. The normalized collision frequency $y = y_C$, viz. $y = 0.18$ (1), 0.203 (2), 0.278 (3), 0.582 (4), and 0.925 (5).

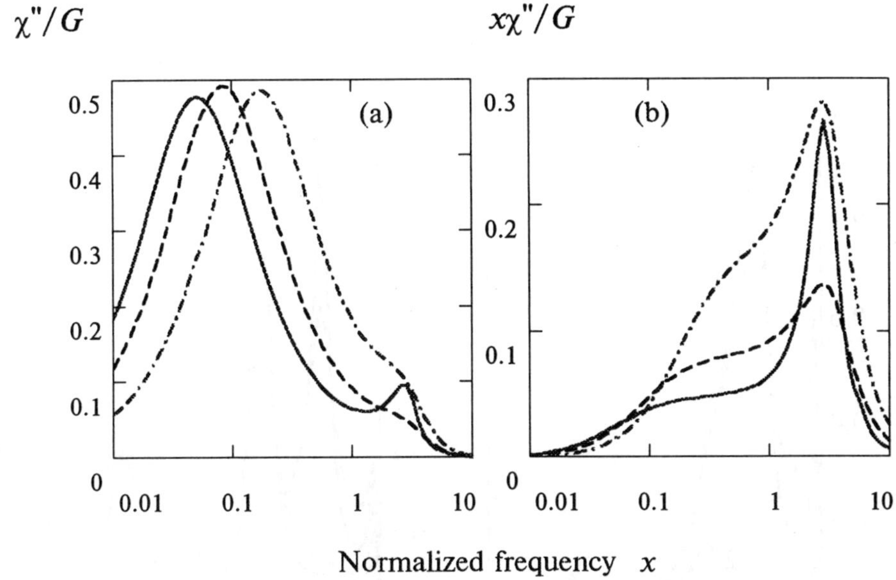

Fig. 12.5 Influence of collisions on loss (a) and absorption (b) spectra, calculated for an isotropic medium; the value $p = 3$ of the field parameter is typical for a strong absorbing liquid; $y = y_c = 0.925$ (solid curve), $y = 2$ (dashed curve). The dadotted curve corresponds to the doubled SF $L(z)$.

Putting $y = y_c$, we first consider the case of *rather rare collisions*. Evolution of dielectric spectra, stipulated by the rise of the intermolecular field parameter p, is illustrated by Fig. 12.4. The local anisotropy also increases with p. This leads to the appearance of a low-frequency (Debye) relaxation region. The frequency x_D of the Debye loss maximum χ''_D is much lower than the frequency x_L of the absorption maximum; x_L rises with the increase of p; the absorption curve $x\chi''(x)$ is localized near the peak frequency x_L.

The case of *frequent collisions* ($y > y_c$) is illustrated by Fig. 12.5 for $p = 3$ and $y = 2$ (dashed lines). We compare here the loss / absorption curves with those, calculated for $y_c = 0.925$ (solid curves).

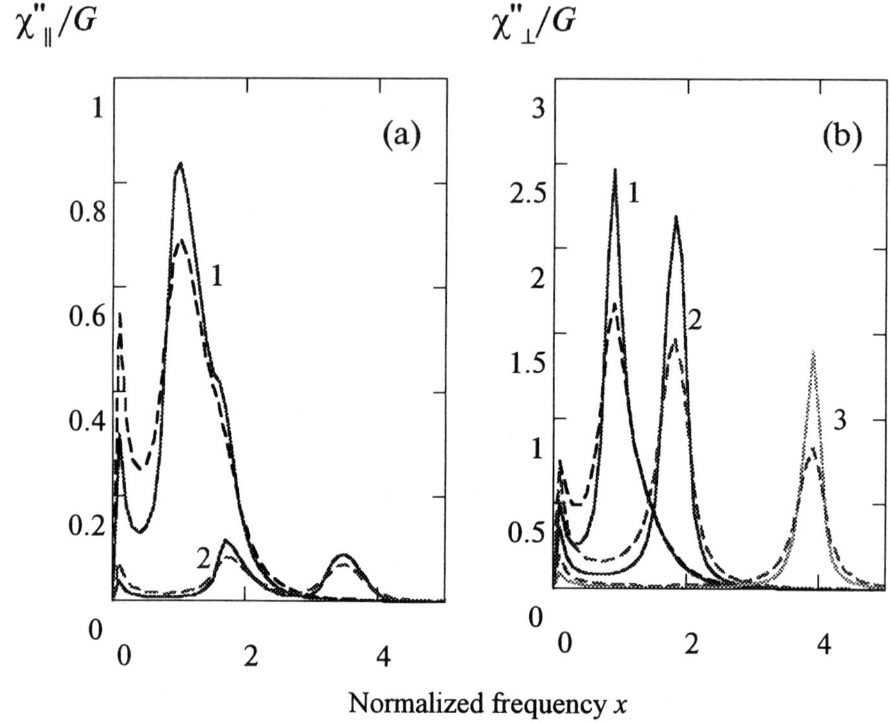

Fig. 12.6 The longitudinal (a) and transverse (b) loss spectra. The field parameter $p = 1$ (curves 1), 2 (curves 2) and 4 (curves 3). Solid lines for $y = 0.1$, dashed lines for $y = 0.2$.

The second loss maximum, seen for $y = y_c$, transforms into the shoulder for $y = 2$, while the absorption peak, that follows the quasi-Debye plateau, strongly decreases, since the absorption curve markedly widens. Dielectric spectra, calculated for $p = 3$ and $y = 2$, qualitatively resemble typical spectra of strongly absorbing liquids (see Chap. 14). It is especially true for the quasi-spatial calculation scheme with the *doubled* spectral function [in this case we replace $L(z)$ by $2L(z)$], see dadotted curves in Fig. 12.5.

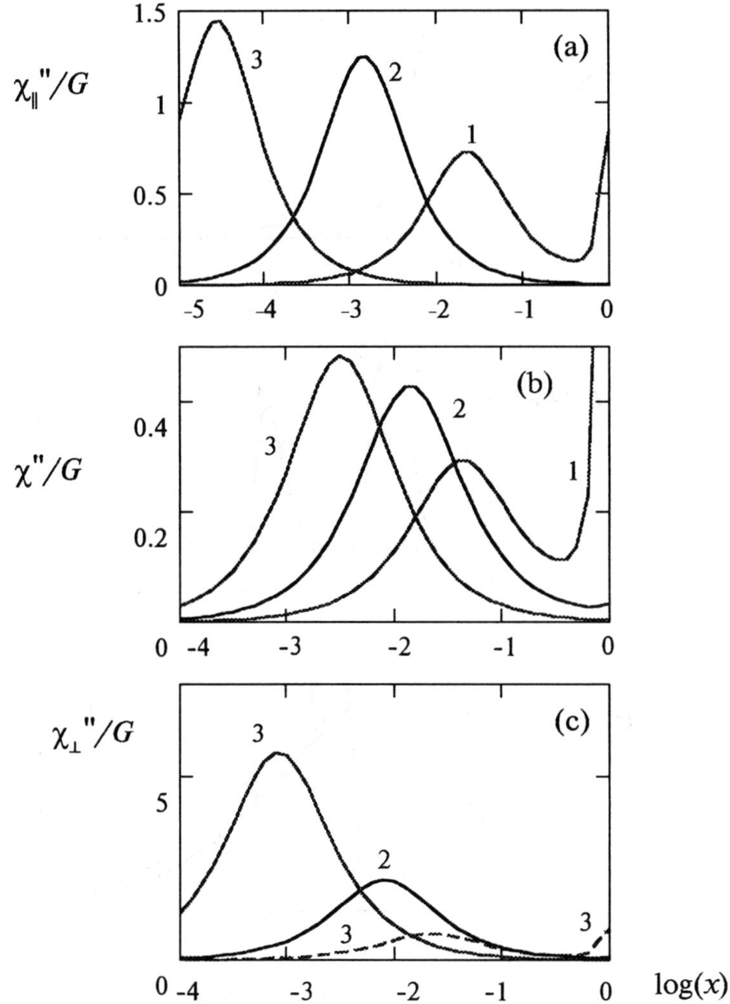

Fig. 12.7 The low-frequency loss vs. frequency dependence; curves 1, 2, 3 for $p = 1, 2, 3$. **a, c:** the longitudinal and transverse orientations of alternating field; **b :** an isotropic medium.

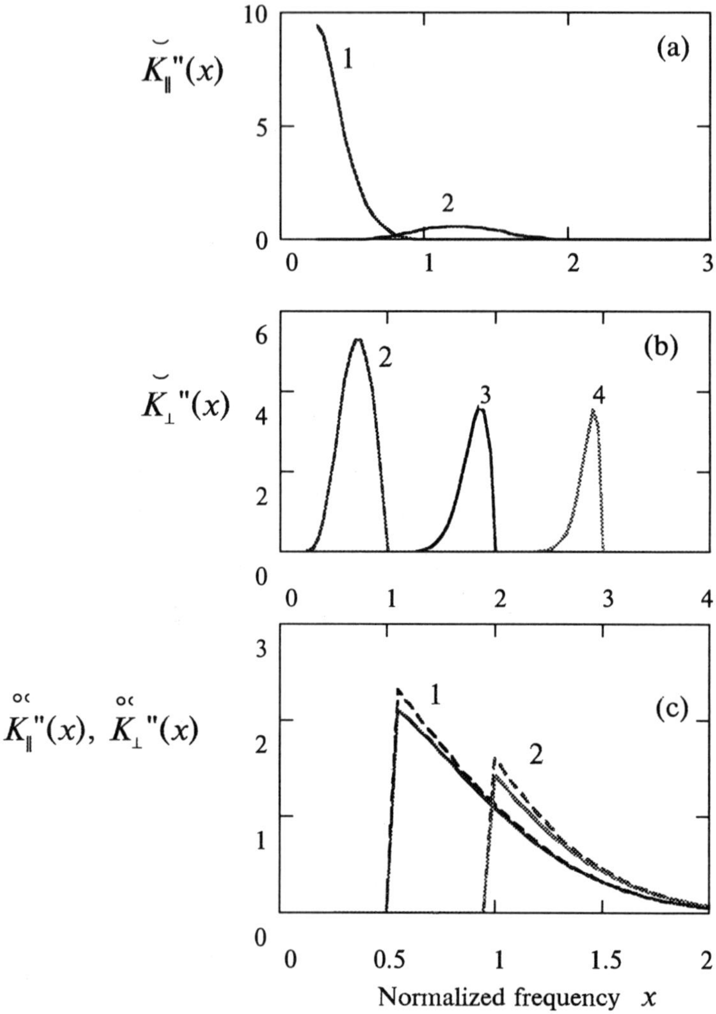

Fig. 12.8 The contributions to the longitudinal (a) and transverse (b) SFs due to librators and due to longitudinal (solid lines) and transverse (dashed lines) hindered rotators (c). The long lifetime limit for the field parameter $p = 0.5$ (curves 1), 1 (curves 2), 2 (curves 3), 3 (curve 4).

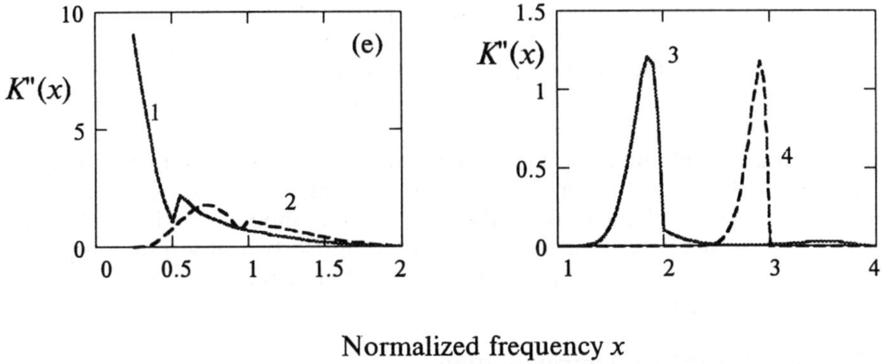

Fig. 12.9 Longitudinal (a, c) and transverse (b, d) SFs and the SF in the case of an isotropic medium (e, f). The long lifetime limit. Curves 1, 2, 3, and 4 for $p = 0.5$, 1, 2, and 3, respectively.

Note, that solid / dashed curves in Fig. 12.5 and Fig. 12.4 are calculated also for the quasi-spatial isotropic ensemble but without doubling of the planar SF $L(z)$. The comparison of Fig. 12.4 and 12.5 with Fig. 11.6 leads to the following conclusions:

a. For y about 1 the low-frequency maximum χ'' does not appear in the CF model (unlike in the CS model).

b. If $p > 1$, the loss curve is narrower in the CF model than in the CS model.

AN ANISOTROPIC MEDIUM. We employ Eq. 11.147. The loss spectra $\chi''(x)/G$ are drawn in Fig. 12.6 for the longitudinal (a) and transverse (b) orientations of alternating field. We have chosen here a rather small collision frequency ($y = 0.1$ and $y = 0.2$). When p rises, the frequency of the line center increases. If $p > 2$, max $\chi''_\perp >>$ max $\chi''_\parallel$, cf. curves 2, 3 in Fig. a and b; the peak intensity max $\chi''_\perp$ markedly exceeds that, calculated for the longitudinal orientation.

The low-frequency loss spectra are pictured in Fig. 12.7 for $\parallel$- and $\perp$-orientations of AC field and for an isotropic medium. The larger the field parameter p, then greater Debye loss maximum and lower its peak frequency $x_D = (2\pi c\tau_D)^{-1}$.

Dielectric behavior in the long lifetime limit is illustrated by Fig. 12.8. For $y = 0.1$ the intensity of limiting lines is rather close to those, calculated in the QHA (cf. Fig. 12.6 and 12.8). Dielectric spectra resemble those, found for the CF model (cf. Fig. 12.8a, 11.9 for librators and Fig. 12.8b, 11.10 for hindered rotors). It is seen from Fig. 12.8, that limiting lines, contributed by $\|$-librators, $\perp$-librators and hindered rotors, have "regular" form and are commensurable. But since the line centers are placed on different frequencies, the resultant limiting line may have rather complex form, see Fig. 12.9. Similar situation is characteristic also for the CF model (cf. Fig. 12.9 and 11.11. For instance, the limiting line may comprise two absorption maxima. This property seems to be important in view of spectra, observed in aqueous systems (see Chap. 15).

12.7 The Stratified Approximation

Analogously to the corresponding approximation, elaborated for the cosine potential profile (see Sec.11.7), we write down formula, applicable for an isotropic medium:

$$L(z) = \overset{\vee}{L}(z) + \tilde{L}(z) + \overset{\circ}{L}(z), \tag{12.87}$$

$$\overset{\vee}{L}(z) = \frac{\exp\left(p^2/2\right)}{2p\sqrt{\pi}\, I_0\left(p^2/2\right)} \int_0^{3p^2/4} \frac{h\exp(-h)}{\overset{\vee}{\gamma}} \left\{ \frac{1}{p^2 - z^2 \overset{\vee}{\gamma}^2} + \frac{h/(4p)}{4p^2 - z^2\overset{\vee}{\gamma}^2} \right\} dh, \tag{12.88}$$

$$\tilde{L}(z) = \frac{2i\exp\left(-\frac{p^2}{4}\right)\left[1 - \exp\left(-\frac{7p^2}{12}\right)\right]}{p\pi\sqrt{\pi}\, I_0\left(p^2/2\right)} \left\{ \frac{p}{z} - 2zp \sum_{n=1}^{\infty} \frac{(-1)^n}{(pn - iz)^2} \right\}, \tag{12.89}$$

$$\overset{\circ}{L}(z) = \frac{2\exp\left(p^2/2\right)}{\sqrt{\pi}\, I_0\left(p^2/2\right)} \int_p^{\infty} \frac{1 + \frac{p^2}{2t^2}}{1 - \left(\frac{p}{4t}\right)^4} \frac{t^2 \exp(-t^2)\, dt}{\overset{\circ}{\gamma}\left(t^2 - z^2 \overset{\circ}{\gamma}^2\right)}, \tag{12.90}$$

$$\check{\gamma}(h) = 1 + \frac{h}{4p^2}, \quad \mathring{\gamma}(t) = \frac{p^2}{4t^2}. \tag{12.91}$$

For frequent collisions, when condition (11.159) holds, Eq. 12.88 reduces to

$$\check{L}(z) = \frac{\exp\left(p^2/2\right)}{2p\sqrt{\pi}\, I_0\left(p^2/2\right)} \left\{ \frac{1 - \left(1 + \frac{3}{4}p^2 + \frac{9}{16}p^4\right) \exp\left(-\frac{3}{4}p^2\right)}{2p^2\left(4p^2 - z^2\right)} + \frac{1 - \left(1 + \frac{3}{4}p^2\right) \exp\left(-\frac{3}{4}p^2\right)}{p^2 - z^2} \right\}. \tag{12.92}$$

These expressions are derived in GT, p. 304.

Dielectric spectra, calculated rigorously, agree rather well with those based on this approximation. The simpler quasi-harmonic approximation yields for large p appreciably narrower absorption band. The stratified approximation will be applied to polar fluids in the last part of the book.

CHAPTER 13
THE HAT MODEL (HM)

Comparison of the bandwidths Δx_0 of limiting lines, calculated for the constant field, cosine squared, or hybrid models, shows that Δx_0 is minimal in the CF model and maximal in the HM. The purpose of this chapter is to demonstrate the possibility to control the Δx_0-value by choosing an appropriate potential profile. *We suggest to control the profile of the well from purely parabolic to rectangular.* Such a well should have the finite depth u_0. This form resembles a *hat.* First, we consider that such «hat» has flat brim. We consider the *planar* double well potential (DWP) configuration, where two wells have opposite directions.

The hat model is characterized by three geometrical parameters Θ, p, and u_0, controlling the form of the well, and by the lifetime τ (we recall the Reader that the last parameter is common for all models considered in the book):

$$u(\vartheta) = 0 \quad \text{for } |\vartheta| \leq \Theta,\ h \leq u_0, \tag{13.1}$$

$$u(\vartheta) = p^2 \left(|\vartheta| - \Theta\right)^2 \quad \text{for } \Theta \leq |\vartheta| \leq \beta_{max},\ h \leq u_0, \tag{13.2}$$

$$u(\vartheta) = 0 \quad \text{for } \ h > u_0, \tag{13.3}$$

p being the steepness of the well. Eq. 13.1, 13.2, and 13.3 refer respectively to flat bottom, parabolic potential and flat edge of the well. Small-energy dipoles ($h \leq u_0$), librating in ϑ-depending potential, are termed "librators", corresponding quantities are marked by superscript $^\vee$. High-energy particles, moving under effect of the constant potential (13.3), viz. in the homogeneous potential field, are free "rotators", we mark them by superscript $^\circ$. *We neglect the effect of the potential hole on the law of motion of particles, constituting this molecular group.*

The amplitude β of a librating particle attains its maximum value β_{max} near the edge of a well:

$$p^2(\beta_{max} - \Theta)^2 = u_0; \ \beta_{max} = \frac{\sqrt{u_0}}{p} + \Theta. \qquad (13.4)$$

Setting $2\beta_{max} = \pi/2$, we determine from Eq. 13.4 the minimal value p_{min} of the steepness p:

$$p_{min} = \frac{\sqrt{u_0}}{\frac{\pi}{2} - \Theta} \text{ for } \beta_{max} < \frac{\pi}{2}. \qquad (13.5)$$

If e.g. $\Theta_1 = \pi/8$ and $\Theta_2 = \pi/4$ and $u = 4$, we find that $p_{min1} = 1.84$ and $p_{min2} = 2.26$. For further calculations we chose $p = 2.5$, 5 and 20. In Fig.13.1 we present two examples of a hat well.

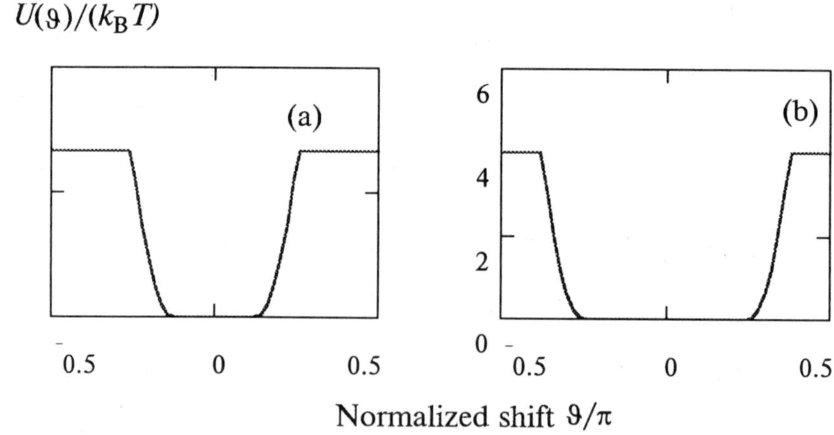

Fig. 13.1 The form of the potential well employed in the planar hat model. $u = 4$, $p = 5$, $\Theta = \pi/8$ (a) and $\pi/4$ (b).

13.1 The Law of Motion

We simplify notation of the dimensionless potential welldepth: $u_0 \to u$ and set $\vartheta \geq 0$. We put the normalized time $\varphi = 0$, when a dipole is oriented along the symmetry axis of the potential. In the potential region (13.1), where $h \leq u$, a dipole rotates by inertia. When the angular shift attains Θ (see Eq. 3.129, where we set: $p_m \equiv p$), the law of motion becomes harmonic:

$$\vartheta(\varphi) = \sqrt{h}\,\varphi,\ \vartheta < \Theta\,; \tag{13.6}$$

$$\vartheta(\varphi) = \Theta + \frac{\sqrt{h}}{p}\sin p\left(\varphi - \frac{\Theta}{\sqrt{h}}\right) \quad \text{for } \Theta \leq \vartheta \leq \beta\,. \tag{13.7}$$

The law of motion of high-energy dipoles ($h > u$) is the same, as in Eq. 13.6, if we substitute h by $h - u$:

$$\vartheta(\varphi) = \sqrt{h-u}\;\varphi, \qquad h > u\,. \tag{13.8}$$

13.2 The Steady-State Distribution $W = C\,exp\,(-h)$

We write:

$$\frac{1}{C} = \overset{\vee}{\mathrm{st}} + \overset{\circ}{st}\,. \tag{13.9}$$

The period of librations $\overset{\vee}{\Phi} = 2\Phi$, Φ being the period of function $q_{\|}(\varphi)$. By analogy with Eq. 3.134 we have:

$$\overset{\vee}{\mathrm{st}} = 2\int_0^u \exp(-h)\,dh \int_0^{\Phi} d\varphi_0 = \int_0^u \exp(-h)\,\overset{\vee}{\Phi}\,dh\ ,$$

$$\overset{\circ}{\mathrm{st}} = \int_u^{\infty} \exp(-h)\,\overset{\circ}{\Phi}\,dh = \int_u^{\infty} \exp(-h)\,\frac{2\pi\,dh}{\sqrt{h-u}}$$

$$= 4\pi\exp(-u)\int_0^{\infty}\exp\left(-t^2\right)dt = 2\pi\sqrt{\pi}\exp(-u),$$

$$\frac{1}{C} = 2\int_0^u \exp(-h)\frac{\check{\Phi}}{2}dh + 2\pi\sqrt{\pi}\exp(-u)\,. \tag{13.10}$$

To find the libration period $\check{\Phi}(h)$ we calculate angular velocity from Eq. 13.9 and set it equal to zero at the instant $\varphi = \check{\Phi}(h)/4$:

$$\dot{\vartheta}\left(\frac{\check{\Phi}}{4}\right) = \sqrt{h}\cos p\left(\frac{\check{\Phi}}{4} - \frac{\Theta}{\sqrt{h}}\right) = 0,$$

from where we get

$$\check{\Phi}(h) = \frac{2\pi}{p} + \frac{4\Theta}{\sqrt{h}} = \frac{2\pi f}{\sqrt{h}};\ f = \frac{\sqrt{h}}{p} + f_0,\ f_0 \equiv \frac{2\Theta}{\pi}. \tag{13.11}$$

Substitution of Eq. 13.11 into 13.10 yields:

$$\pi C = \left\{\frac{2}{p}\left[1-\exp(-u)\right] + \frac{4\Theta}{\sqrt{\pi}}\operatorname{erf}\left(\sqrt{u}\right) + 2\sqrt{\pi}\exp(-u)\right\}^{-1} = \frac{1}{f_0 D}, \tag{13.12}$$

$$D = 2\frac{1-\exp(-u)}{f_0 p} + 2\sqrt{\pi}\left[\operatorname{erf}\left(\sqrt{u}\right) + \frac{\exp(-u)}{f_0}\right]. \tag{13.13}$$

The energy depending libration amplitude $\beta(h)$ is found by substitution of Eq. 13.11 into Eq. 13.7:

$$\beta(h) = \Theta + \frac{\sqrt{h}}{\pi}\sin p\left(\frac{\check{\Phi}}{4} - \frac{\Theta}{\sqrt{h}}\right) = \Theta + \frac{\sqrt{h}}{p}. \tag{13.14}$$

We remark that for the transition to the planar CR model (Sec.8.1) we should set:

$$\Theta \to \beta = \text{const}(h);\ f_0 \to f \equiv \frac{2\beta}{\pi};\ p \to \infty,\ u \to \infty\,. \tag{13.15}$$

The averaged value $\langle X \rangle$ of any parameter $X(\Gamma)$ may be found from the integral

$$\langle X \rangle = C\left[2\int_0^u X(\overset{\vee}{\Gamma})\exp(-h)\,\frac{\overset{\vee}{\Phi}(h)}{2}\,dh + \int_u^\infty X(\overset{\circ}{\Gamma})\exp(-h)\,\overset{\circ}{\Phi}(h)\,dh\right]$$

$$= \frac{1}{\pi f_0 D}\left[2\int_0^u X(\overset{\vee}{\Gamma})\exp(-h)\left(\frac{\pi}{p}+\frac{2\Theta}{\sqrt{h}}\right)dh + \int_u^\infty X(\overset{\circ}{\Gamma})\exp(-h)\,\frac{2\pi\,dh}{\sqrt{h-u}}\right]$$

$$= \frac{1}{f_0 D}\left[2\int_0^u X(\overset{\vee}{\Gamma})\exp(-h)\left(\frac{1}{p}+\frac{f_0}{\sqrt{h}}\right)dh + \int_u^\infty X(\overset{\circ}{\Gamma})\exp(-h)\,\frac{2\,dh}{\sqrt{h-u}}\right]. \quad (13.16)$$

We shall return to this question in Sec. 13.6.

13.3 Fourier Amplitudes of Librating Molecules

We may find amplitudes $\overset{\vee}{a}_{2n}$ and $\overset{\vee}{b}_{2n-1}$ by integration in the time interval φ, equal to 1/4 of the librational period $\overset{\vee}{\Phi}$. With regard of series (5.81) and accounting for the symmetry relations we have:

$$\begin{Bmatrix}\overset{\vee}{a}_{2n}\\ \overset{\vee}{b}_{2n-1}\end{Bmatrix} = \frac{8}{\overset{\vee}{\Phi}}\int_0^{\overset{\vee}{\Phi}/4}\begin{Bmatrix}\cos\vartheta\,\cos 2n\overset{\vee}{p}\varphi\\ \sin\vartheta\,\sin(2n-1)\overset{\vee}{p}\varphi\end{Bmatrix} d\varphi\,, \quad (13.17)$$

where the libration frequency $\overset{\vee}{p}$ is found from Eq. 13.11:

$$\overset{\vee}{p} = \frac{2\pi}{\overset{\vee}{\Phi}} = \frac{\sqrt{h}}{f}. \quad (13.18)$$

We divide the interval of integration by the flat and parabolic regions:

$$\begin{Bmatrix} \overset{\vee}{a}_{2n} \\ \overset{\vee}{b}_{2n-1} \end{Bmatrix} = \Im_f + \Im_p;\ \Im_f \equiv \begin{Bmatrix} \overset{\vee}{a}_{2n} \\ \overset{\vee}{b}_{2n-1} \end{Bmatrix}_f;\ \Im_p \equiv \begin{Bmatrix} \overset{\vee}{a}_{2n} \\ \overset{\vee}{b}_{2n-1} \end{Bmatrix}_p. \tag{13.19}$$

For the flat region we write in view of Eq. 13.6, 13.11:

$$\Im_f = \frac{8}{\overset{\vee}{\Phi}} \int_0^{\Theta/\sqrt{h}} \begin{Bmatrix} \cos\sqrt{h}\,\varphi\ \ \cos 2n\overset{\vee}{p}\varphi \\ \sin\sqrt{h}\,\varphi\ \ \sin(2n-1)\overset{\vee}{p}\varphi \end{Bmatrix} d\varphi. \tag{13.20}$$

Changing φ by $t = \sqrt{h}\,\varphi$ and denoting $l = \begin{Bmatrix} 2n \\ 2n-1 \end{Bmatrix}$ for the longitudinal (upper line) and transverse (lower line) orientations of an alternating field, we find:

$$\Im_f = \frac{2}{\pi f} \int_0^{\Theta} \begin{Bmatrix} \cos(1-l/f)t\ \ \cos(1+l/f)t \\ \sin(1-l/f)t\ \ \sin(1+l/f)t \end{Bmatrix} dt;$$

$$\begin{Bmatrix} \overset{\vee}{a}_l \\ \overset{\vee}{b}_l \end{Bmatrix}_f = \frac{4/\pi}{l^2-f^2}\left[\begin{Bmatrix} l \\ f \end{Bmatrix}\cos\Theta\ \sin\left(\frac{l\Theta}{f}\right) - \begin{Bmatrix} f \\ l \end{Bmatrix}\sin\Theta\ \cos\left(\frac{l\Theta}{f}\right)\right]. \tag{13.21}$$

For the *parabolic region* in accord with Eq. 13.7

$$\Im_p = \frac{4}{\pi f} \int_{\Theta}^{\pi f/2} \left[\begin{Bmatrix} \cos \\ \sin \end{Bmatrix}\left(\Theta + \frac{\sqrt{h}}{p}\sin\frac{p(t-\Theta)}{\sqrt{h}}\right) \begin{Bmatrix} \cos \\ \sin \end{Bmatrix}\frac{lt}{f}\right] dt. \tag{13.22}$$

One can obtain simple analytic expressions, *assuming large steepness of the well* ($p > \sqrt{h}$):

$$\sin\frac{\sqrt{h}}{p}\sin\frac{p(t-\Theta)}{\sqrt{h}} \approx \frac{\sqrt{h}}{p}\sin\frac{p(t-\Theta)}{\sqrt{h}}; \tag{13.23a}$$

$$\cos\left(\frac{\sqrt{h}}{p}\sin\frac{p(t-\Theta)}{\sqrt{h}}\right) \approx 1 - \frac{h}{2p^2}\sin^2\frac{p(t-\Theta)}{\sqrt{h}}. \tag{13.23b}$$

After elementary but rather tedious transformations we obtain:

$$\Im_p = \mp\frac{4}{\pi l}\left(1+\frac{hf^2}{\pi l^2-4p^2f^2}\right)\begin{Bmatrix}\cos\\ \sin\end{Bmatrix}\Theta\ \begin{Bmatrix}\sin\\ \cos\end{Bmatrix}\frac{l\Theta}{f}$$

$$\pm\frac{4}{\pi}\frac{hf}{hl^2-p^2f^2}\begin{Bmatrix}\sin\\ \cos\end{Bmatrix}\Theta\ \begin{Bmatrix}\cos\\ \sin\end{Bmatrix}\frac{l\Theta}{f}\ . \tag{13.24}$$

Summing Eq. 13.24 and 13.26, we finally derive

$$\begin{Bmatrix}\overset{\vee}{a}_l\\ \overset{\vee}{b}_l\end{Bmatrix} = \pm\frac{4}{\pi l}\frac{f}{l^2-f^2}\begin{Bmatrix}C_l\\ S_l\end{Bmatrix};$$

$$\begin{Bmatrix}C_l\\ S_l\end{Bmatrix}=\frac{\left(\frac{s}{2p}\right)^2-1}{\left(\frac{sl}{2pf}\right)^2-1}f\begin{Bmatrix}\cos\\ \sin\end{Bmatrix}\Theta\ \begin{Bmatrix}\sin\\ \cos\end{Bmatrix}\frac{l\Theta}{f}-\frac{\left(\frac{s}{p}\right)^2-1}{\left(\frac{sl}{pf}\right)^2-1}l\begin{Bmatrix}\sin\\ \cos\end{Bmatrix}\Theta\ \begin{Bmatrix}\cos\\ \sin\end{Bmatrix}\frac{l\Theta}{f}, \tag{13.25}$$

$$s=\sqrt{h}=p\left(f-\frac{2\Theta}{\pi}\right). \tag{13.26}$$

13.4 The Spectral Function of Librators

We substitute into Eq. 5.86 - 5.87 the steady state distribution $W = C\exp(-h)$ from (13.12) - (13.13); $\overset{\vee}{p}$ from (13.18); a_l and b_l from (13.25). Noting that $l\equiv\begin{Bmatrix}2n\\ 2n-1\end{Bmatrix}$, we find:

$$\frac{1}{l^2\overset{\vee}{p}^2-z^2}\equiv\frac{f^2}{l^2s^2-z^2f^2}\quad\text{for}\quad\begin{Bmatrix}\overset{\vee}{K}_{\parallel}(z)\\ \overset{\vee}{K}_{\perp}(z)\end{Bmatrix}; \tag{13.27}$$

$$\left\{\begin{matrix}\check{K}_{\|}(z)\\ \check{K}_{\perp}(z)\end{matrix}\right\}=\frac{96}{\pi^2 D}\int_0^{\sqrt{u}}ds\,\frac{f^3 s^2\exp\left(-s^2\right)}{f_0}\sum_{n=1}^{\infty}\frac{\left\{\begin{matrix}C_l^2\\ S_l^2\end{matrix}\right\}}{\left(l^2-f^2\right)^2\left(l^2 s^2-z^2 f^2\right)}. \quad (13.28)$$

The SF of an isotropic medium is given by

$$\check{L}(z)=\frac{1}{3}\left[\check{K}_{\|}(z)+\check{K}_{\perp}(z)\right]. \quad (13.29)$$

TRANSITION TO THE ELASTIC BOND MODEL. We write:

$$\left\{\begin{matrix}C_l\\ S_l\end{matrix}\right\}=f^2\left[\frac{\left(f-\frac{2\Theta}{\pi}\right)^2-4}{\left(f-\frac{2\Theta}{\pi}\right)^2 l^2-4f^2}\, f\left\{\begin{matrix}\cos\\ \sin\end{matrix}\right\}\Theta\,\left\{\begin{matrix}\sin\\ \cos\end{matrix}\right\}\frac{l\Theta}{f}\right.$$

$$\left.-\frac{\left(f-\frac{2\Theta}{\pi}\right)^2-1}{\left(f-\frac{2\Theta}{\pi}\right)^2 l^2-f^2}\, l\left\{\begin{matrix}\sin\\ \cos\end{matrix}\right\}\Theta\,\left\{\begin{matrix}\cos\\ \sin\end{matrix}\right\}\frac{l\Theta}{f}\right]. \quad (13.30)$$

Let the flat region $\Theta\to 0$, then $f=\frac{2\Theta}{\pi}+\frac{s}{p}\to\frac{s}{p}$.

For the *longitudinal orientation* we have from Eq. 13.30:

$$C_{2n}\to f^2\left[\frac{\left(f^2-4\right)2n\Theta}{\left(n^2-1\right)4f^2-\frac{4n^2 4\Theta f}{\pi}}-\frac{\left(f^2-1\right)2n\Theta}{\left(4n^2-1\right)f^2-\frac{4n^2 4\Theta f}{\pi}}\right].$$

For $\Theta\to 0$ the only non-zero term corresponds to $n=1$, so that

$$C_2\to-\frac{\pi f}{8}\left(f^2-4\right),\quad \frac{1}{l^2 s^2-f^2 z^2}\to\frac{p^2}{s^2\left(4p^2-z^2\right)}, \quad (13.31)$$

$$\frac{1}{f_0 D} \to \frac{p}{B}, \text{ where } B \equiv 2+2\exp(-u)\left[\sqrt{\pi}\, p-1\right]. \tag{13.32}$$

The upper line in Eq. 13.28 reduces to

$$\check{K}_{\parallel}(z) = \frac{3\cdot 2}{8Bp^2\left(4p^2 - z^2\right)} \int_0^u h^2 \exp(-h)\, dh. \tag{13.33}$$

If $u \to \infty$, then $B \to 1$, integral $\to 2$, Eq. 13.33 transforms to Eq. 10.6 for the planar EB model.

For the *transverse orientation*, in accord with the lower line of (13.28),

$$S_{2n-1} \to S_1 \to -\frac{\pi\left(f^2 - 1\right)}{4},$$

$$\check{K}_{\perp}(z) = \frac{3\cdot 2}{2B\left(p^2 - z^2\right)} \int_0^u h \exp(-h)\, dh. \tag{13.34}$$

If $u \to \infty$, this expression yields Eq. 10.7.

TRANSITION TO THE CONFINED ROTATOR MODEL. If $p \to \infty$, then $f \to f_0 = 2\Theta/\pi$, the coefficients (13.25) reduce to

$$\begin{Bmatrix} C_l \\ S_l \end{Bmatrix} \to f_0 \begin{Bmatrix} \cos \\ \sin \end{Bmatrix} \frac{\pi f_0}{2} \begin{Bmatrix} \sin \\ \cos \end{Bmatrix} \frac{\pi}{2}\begin{pmatrix} 2n \\ 2n-1 \end{pmatrix}$$

$$-\begin{pmatrix} 2n \\ 2n-1 \end{pmatrix} \begin{Bmatrix} \sin \\ \cos \end{Bmatrix} \frac{\pi f_0}{2} \begin{Bmatrix} \cos \\ \sin \end{Bmatrix} \frac{\pi}{2}\begin{pmatrix} 2n \\ 2n-1 \end{pmatrix} = l \begin{Bmatrix} \sin \\ \cos \end{Bmatrix} \frac{\pi f_0}{2}. \tag{13.35}$$

$$D \to 2\sqrt{\pi}\left[\operatorname{erf}\left(\sqrt{u}\right) + \frac{1}{f_0}\exp(-u)\right]. \tag{13.36}$$

Then Eq. 13.28 for the SFs are simplified as follows:

$$\left\{\begin{matrix} \overset{\vee}{K}_{\|}(z) \\ \overset{\vee}{K}_{\perp}(z) \end{matrix}\right\} = \frac{6}{\sqrt{\pi}} \frac{\sum_{n=1}^{\infty} d_n \int_0^{\sqrt{u}} \frac{s^2 \exp\left(-s^2\right) ds}{s^2 - \left(z f_0 / n\right)^2}}{\operatorname{erf}\left(\sqrt{u}\right) + \frac{\exp(-u)}{f_0}}; \tag{13.37}$$

$$d_n \equiv \left(\frac{2 f_0}{\pi}\right)^2 \left[1 \pm (-1)^n\right] \frac{\sin^2 \frac{\pi}{2}(n - f_0)}{\left(n^2 - f_0^2\right)^2}, \quad \begin{pmatrix} n = 2, 4, 6, \ldots \\ n = 1, 3, 5, \ldots \end{pmatrix}. \tag{13.38}$$

This expression yields Eq. 8.15, 8.16a for the planar confined rotator model, where we must put $u \to \infty$, $f = f_0$, $s_n = d_n$.

THE NORMALIZED ABSORPTION $\overset{\vee}{A}(x) = x \overset{\vee}{K}''(x)$ (Fig. 13.2). This contribution of librators to FIR absorption is calculated for $\|$ and $\perp$-orientations. We take two values of the flat part of the potential bottom ($\Theta = \pi/8$ and $\pi/4$), rather high welldepth ($u = 5.5$) and rather low collision frequency ($y = 0.1$). The well's steepness p is altered in a wide range ($p = 5$, 10 and 20). When p rises, the absorption line $\overset{\vee}{A}(x)$ widens and shifts to higher frequencies, transforming for $p \to \infty$ to a line, specific for the confined rotator model. The $\overset{\vee}{A}_{\perp}$-peak is placed on lower frequency x_L, than the $\overset{\vee}{A}_{\|}$-peak. For large p (viz. for $p > 10$) the latter becomes much smaller than the $\perp$-peak; besides, the $\overset{\vee}{A}(x)$ lines become markedly non-symmetric, since their high frequency parts become narrower than low frequency parts.

Thus, we actually can control the width and the form of the FIR absorption line, obliged to librators, by varying the steepness parameter p. (13.39)

THE LONG LIFETIME LIMIT. For $y \to 0$

$$\lim_{y=0} \text{Im}\left(\int_0^{\sqrt{u}} \frac{F(s)\, ds}{s^2 l^2 - z^2 f^2}\right).$$

$$= \frac{\pi}{2l} \int_0^{\sqrt{u}} F(s)\, \delta(sl - xf)\, \frac{ds}{s} = \frac{\pi}{2lf}\left[\frac{F(s)}{s}\left|\frac{\partial s}{\partial(sl/f)}\right|\right]_{sl = fx} . \quad (13.40)$$

If $s \le \sqrt{u}$, $s(x)$ can be found from the equation $ls(x) = xf$. Since $xf = \left[f_0 + \frac{s(x)}{p}\right]x$, we have:

$$s(x) = \frac{x f_0 p}{pl - x};\ f = f(x) = \frac{f_0 pl}{pl - x};\ \left|\frac{\partial s}{\partial(sl/f)}\right| = \frac{ls^2}{x^2 f_0},\ s \le \sqrt{u} . \quad (13.41)$$

Therefore, imaginary parts of Eq. 13.28 are given by

$$\begin{Bmatrix} \check{K}''_{\parallel}(x) \\ \check{K}''_{\perp}(x) \end{Bmatrix} = \frac{48}{\pi D} \sum_{n=1}^{\infty} \begin{Bmatrix} C_l^2 \\ S_l^2 \end{Bmatrix} \left(\frac{f}{f_0 x}\right)^2 \frac{s^3 \exp\left(-s^2\right) ds}{\left(l^2 - f^2\right)^2} X_l , \quad (13.42)$$

$$X_l = 1 \text{ for } x \le x_{l\max} = \frac{\sqrt{u}\, pl}{pf_0 + \sqrt{u}};\ X_l = 0 \text{ for } x > x_{l\max} . \quad (13.43)$$

For infinitely steep walls ($p \to \infty$) we can employ Eq. 13.36, 13.37. Using transformations, analogous to (13.40), we reduce Eq. 13.43 to

$$\begin{Bmatrix} \check{K}''_{\parallel}(x) \\ \check{K}''_{\perp}(x) \end{Bmatrix} = \frac{3\sqrt{\pi}}{\text{erf}\left(\sqrt{u}\right) + \dfrac{2}{f_0}\exp(-u)} \sum_{n=1}^{\infty} s_n X_n d_n \exp\left(-s_n^2\right), \begin{Bmatrix} n = 2, 4, \ldots \\ n = 1, 3, \ldots \end{Bmatrix},$$

$$s_n(x) = \frac{xf_0}{n},\ X_n = \begin{cases} 1, \text{ if } x < n\sqrt{u}/f_0 \\ 0, \text{ if } x > n\sqrt{u}/f_0 \end{cases} . \quad (13.44)$$

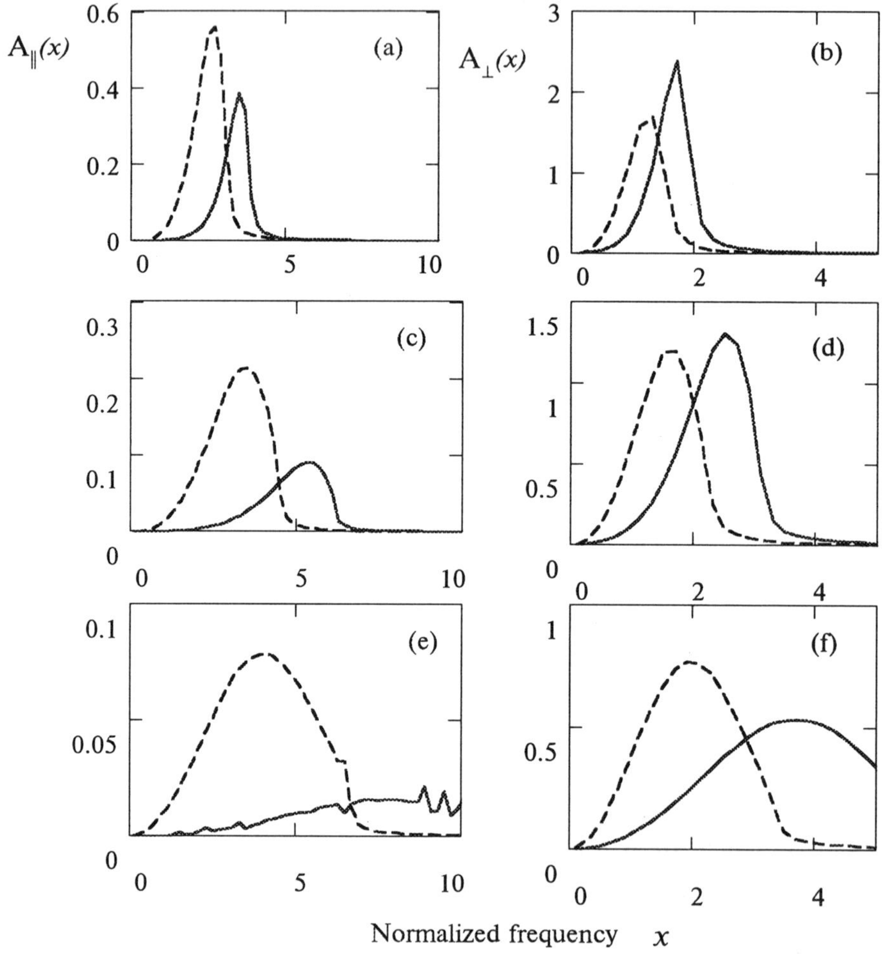

Fig. 13.2 Normalized absorption $A(x) = xK''(x)$ contributed in the FIR region by librating molecules. For $\parallel$ (a, c, d) and $\perp$ (b, d, e) orientations of an alternating field; the steepness of the well $p = 2.55$ (a, b), 5 (c, d) and 20 (e, f); the flat-bottom region of the well $\Theta = \pi/8$ (solid lines) and $\pi/4$ (dashed lines).

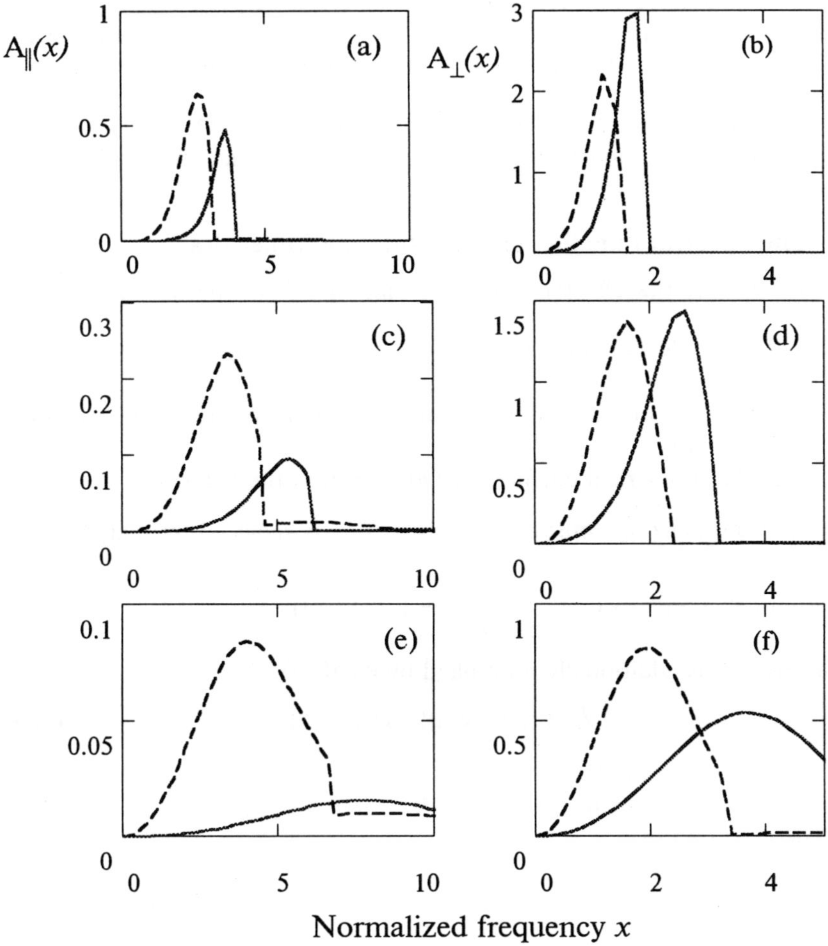

Fig. 13.3 The same, as in Fig. 13.1, but for $y = 0$.

We can compare Fig. 13.3 and 13.2, calculated respectively for $y = 0$ and $y = 0.1$. For $p = 5$ the main distinction between zero y-plots and those for $y = 0.1$ is related to the abrupt high frequency edge of the orientational polarization range. For $p > 20$ the difference between $\check{A}(x)$ plots, calculated for the HM and CR models, vanishes; the HF edge of the limiting lines is not pronounced in this case.

13.5 The Spectral Function of Free Rotating Molecules

Since we neglect the effect of the potential "hole" on the law of motion of rotators, the SFs $\mathring{K}_{\|}(z)$ and $\mathring{K}_{\perp}(z)$ have the same frequency dependence, as the SF $\mathring{K}(z)\Big|_{ED}$ of the extended diffusion model. However, we should employ another numerical coefficient. One can find the latter, if:

a. the normalizing constant $C_{ED} = \dfrac{1}{2\pi\sqrt{\pi}}$ is replaced by that, found for the hat model: $C_{ED} \to C = \dfrac{1}{\pi f_0 D}$ (see Eq. 13.13);

b. the SFs is additionally multiplied by $\exp(-u)$:

$$L_F(z) \to \exp(-u)\, L_F(z)\,. \tag{13.45}$$

This exponential multiplier appears, since the half-infinite interval $[0, \infty]$ of integration over h, used in the ED model, alters to $[u, \infty]$ (cf. Eq. 5.97 with Eq. 13.10). Noting that

$$\mathring{K}_{\|}(z)\Big|_{ED} = \mathring{K}_{\perp}(z)\Big|_{ED} = \frac{3}{2} L_F;\ L_F = \frac{2}{\sqrt{\pi}} \int_0^{\infty} \frac{t^2 \exp\left(-t^2\right) dt}{t^2 - z^2} \tag{13.46a}$$

and denoting $\mathring{K}(z) = \mathring{K}_{\|}(z) = \mathring{K}_{\perp}(z)$, we write

$$\mathring{K}(z) = \frac{C \exp(-u)}{C_{ED}} \frac{3}{2} L_F(z) = \frac{6 \exp(-u)}{f_0 D} \int_0^{\infty} \frac{t^2 \exp\left(-t^2\right) dt}{t^2 - z^2}. \tag{13.46b}$$

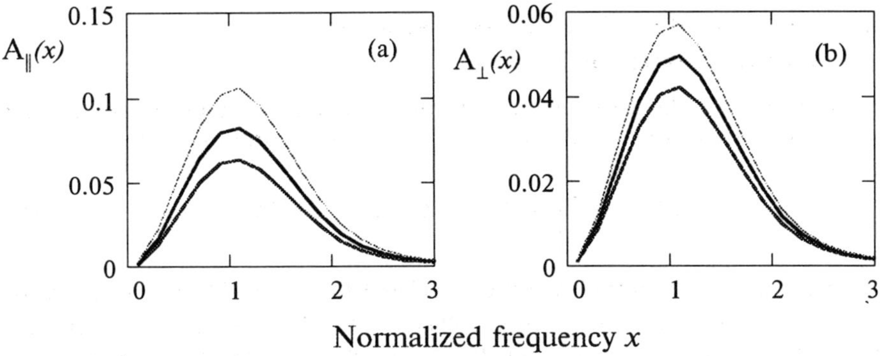

Fig. 13.4 Normalized absorption $A(x) = xK''(x)$ contributed in the FIR region by rotating molecules. Calculated for the flat region $\Theta = \pi/8$ (a) and $\pi/4$ (b); $y = 0.1$; $p =$ 2.5, 5 and 20 (from bottom to top).

NORMALIZED ABSORPTION $\overset{\circ}{A}(x) = x\overset{\circ}{K}''(x)$ (Fig. 13.4). When p rises, the $\overset{\circ}{A}(x)$ - peak increases, tending to some limit for $p \to \infty$. Since for $u = 4$ the well under consideration is rather deep, the absorption peak is much lower, than that found in the ED model, where $\max \overset{\circ}{A}(x) \approx 1$. This distinction is obliged to small (for $u = 4$) proportion $\overset{\circ}{r}$ of free librating particles, see later Sec. 13.7.

The absorption band of rotators and the absorption peak frequency do not depend, unlike those found for librators, on the steepness parameter p. (13.47)

13.6 The Effect of the Potential Well Steepness on Loss / Absorption Spectra

We consider here the case of an isotropic medium; we double the spectral function in order to get the true value of integrated absorption, so that

$$L(z) \to 2\left[\check{K}_{\parallel}(z) + \check{K}_{\perp}(z) + 2\mathring{K}(z)\right] \tag{13.48}$$

The complex susceptibility $\chi^*(x)$ is related to the SF $L(z)$ in accord with Table 6.1 (we consider the Gross collision model, $F = F_G$). Dielectric spectra are pictured in Fig. 13.5 for the same set of parameters, that was used in previous Figures 13.2 - 13.4. The features of these spectra can be "guessed" in view of the above discussed specific SFs $\check{K}_{\parallel}(z)$, $\check{K}_{\perp}(z)$, $\mathring{K}(z)$.

The FIR spectra are contributed mainly by ⊥-librators, while ∥-librators provide the HF tail of the loss / absorption curves. For $p > 5$ the spectra become wideband (especially for lower Θ-value), close to those, calculated for the CR model. These properties of the hat model confirm our suggestion that *one can control the absorption linewidth by varying the steepness p*.

The low-frequency loss spectrum (see Fig. 13.6) is calculated with use of the same formula for χ^* but with $L(x + iy)$ replaced by $L(iy)$. The p-dependence of the relaxation time τ_D , estimated from the expression

$$\tau_D/\eta = \frac{1}{y L(iy)}, \tag{13.49}$$

weakly depends on p and strongly - on Θ (see Fig. 13.7).

In this example we have chosen $y = 0.1$. For larger y the estimated relaxation time will be shorter.

13.7 Statistical Parameters

In Fig. 3.8 we present a number of plots: the proportion $\mathring{r}$ of free rotating molecules (a), the mean frequency of librators and rotators (b); the collision time parameter y_c (c) and the mean amplitude of librations (d).

PROPORTIONS OF ROTATORS AND LIBRATORS. $\overset{\circ}{r}$ and $\overset{\vee}{r}$ are equal to the ratio of corresponding statistical integrals (see Sec. 13.2) to their sum, i.e. to C^{-1}. Taking account, that $\overset{\circ}{L}(z) = 2\overset{\circ}{K}(z)$, we write:

$$\overset{\circ}{r} = \frac{\overset{\circ}{st}}{\overset{\circ}{st} + \overset{\vee}{st}} = C\,\overset{\circ}{st};\; \overset{\vee}{r} = \frac{\overset{\vee}{st}}{\overset{\circ}{st} + \overset{\vee}{st}} = C\,\overset{\vee}{st} = 1 - \overset{\circ}{r};; \qquad (13.50)$$

$$\overset{\circ}{r} = \frac{2\sqrt{\pi}\,\exp(-u)}{f_0\,D} = \frac{\overset{\circ}{L}(z)}{L_F(z)}. \qquad (13.51)$$

The proportion $\overset{\circ}{r}$ *of free rotators is exactly equal to the part of the SF* $L_F(z)$*, that determines their contribution to the total SF.* (13.52)

The comparison of $\overset{\circ}{r}$-values, calculated (see Table 13.1) for the hat model with large p (p = 100) and the hybrid model, where we put $\beta = \Theta$, leads to an important conclusion:

In the planar hat model with rectangular potential well the proportion of rotators is much lesser (for the same libration amplitude and potential u*), than* $\overset{\circ}{r}$ *in the spatial hybrid model. Consequently, models, based on consideration of planar rotation, may be used only for a qualitative study of molecular dynamics / dielectric response.*

We remark, that because of this property the *two-humped absorption curve is not pronounced in the hat model* (see Fig. 13.5, 13.6), unlike in the case of the hybrid model (cf. with Fig. 9.4b, curve 6). Consequently, for the study of absorption, that is observed in aqueous systems, one should employ the hybrid model (with rectangular well) or should elaborate the hat model, based on consideration of molecular rotation *in space*.

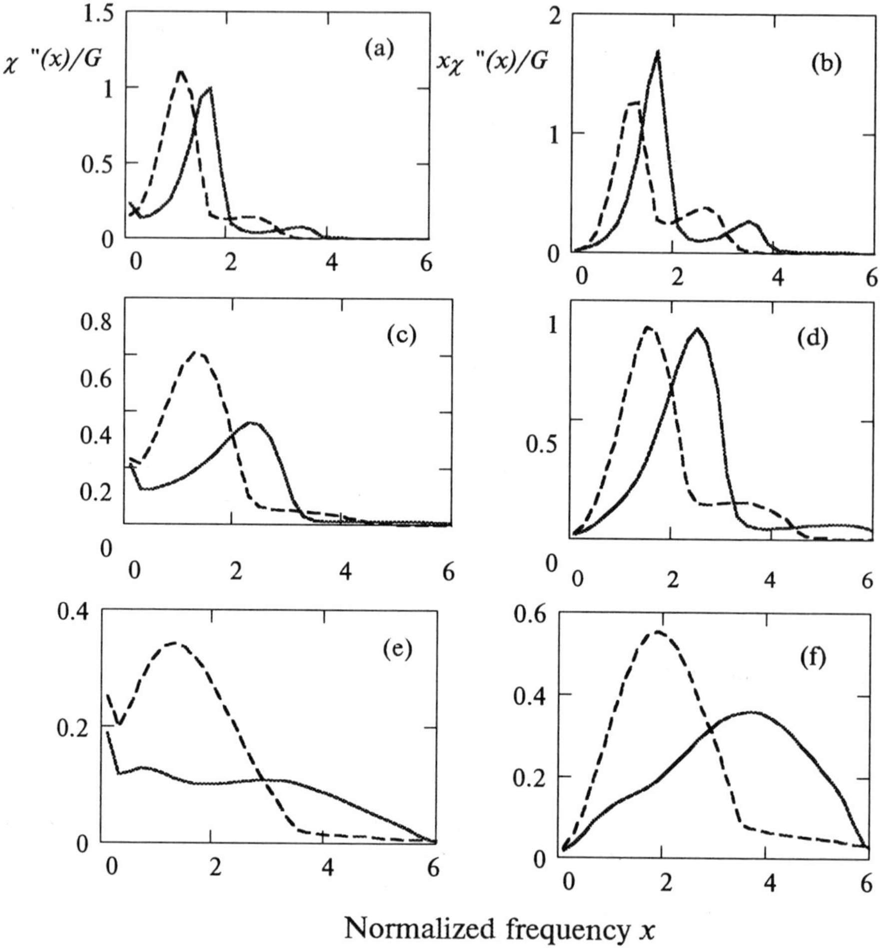

Fig. 13.5 The evolution of FIR loss (a, c, e) / absorption (b, d, f) spectra, calculated for an isotropic medium; the steepness of the well $p = 2.5$ (a, b), 5 (c, d) and 20 (e, f); the flat-bottom region of the well $\Theta = \pi/8$ (solid lines) and $\pi/4$ (dashed lines); the normalized welldepth $u = 4$.

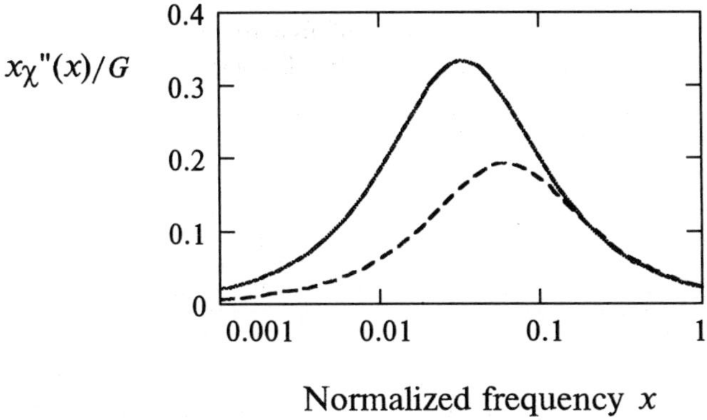

Fig. 13.6 The low-frequency plot of dielectric loss vs. normalized frequency x; $u = 4$, $y = 0.1$; $p = 5$; Θ = p/8 (solid line) and $\pi/4$ (dashed line).

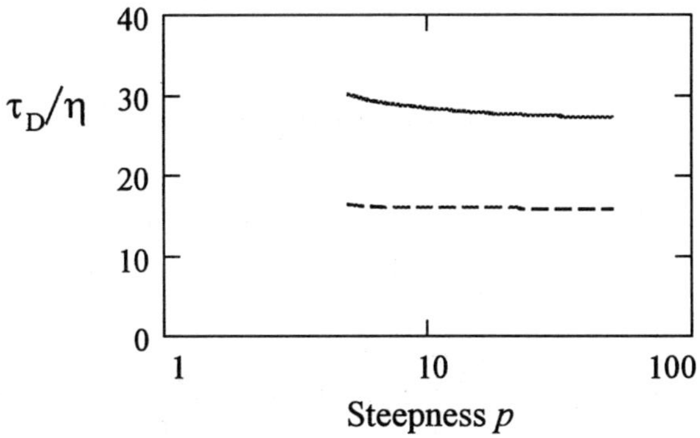

Fig. 13.7 The estimated normalized Debye relaxation time vs. steepness p of the well; $u = 4$, $y = 0.1$; $\Theta = \pi/8$ (solid line) and $\pi/4$ (dashed line).

Table 13.1 Proportion $\overset{\circ}{r}$ [per cent] of free rotators for a potential well with rectangular profile. A. For $\Theta = \beta = \pi/8$, B. For $\Theta = \beta = \pi/4$.

u	Case	4.0	4.5	5.0	5.5	6.0	6.5	7.0	7.5
Hat	A	6.7	4.2	2.6	1.6	1.0	0.6	0.4	0.2
	B	3.5	2.2	1.3	0.8	0.5	0.3	0.2	0.1
HM	A	19.7	12.9	8.2	5.1	3.2	1.9	1.2	0.7
	B	6.0	3.7	2.3	1.4	0.8	0.5	0.3	0.2

THE MEAN LIBRATION / ROTATION FREQUENCY. To find $\left\langle \overset{\vee}{p} \right\rangle$ or $\left\langle \overset{\circ}{p} \right\rangle$ we insert additionally multipliers $\left\langle \overset{\vee}{p} \right\rangle = \dfrac{2\pi}{\overset{\vee}{\Phi}}$ or $\left\langle \overset{\circ}{p} \right\rangle = \dfrac{2\pi}{\overset{\circ}{\Phi}}$ into the integrand of statistic integrals (see Eq. 13.10). Since

$$\overset{\vee}{st} = \frac{2\pi}{p}\left[1-\exp(-u)\right] + 2\pi\sqrt{\pi}\, f_0 \operatorname{erf}\left(\sqrt{u}\right), \tag{13.53}$$

$$\overset{\circ}{st} = 2\pi\sqrt{\pi}\, \exp\left(-u\right), \tag{13.54}$$

$$\left\langle \overset{\vee}{p} \right\rangle = \frac{2\pi}{\overset{\vee}{st}}\int_0^u \exp(-h)\, dh = \frac{p\left[1-\exp(-u)\right]}{1-\exp(-u) + f_0\sqrt{\pi}\, p \operatorname{erf}\left(\sqrt{u}\right)}, \tag{13.55}$$

$$\left\langle \overset{\circ}{p} \right\rangle = \frac{2\pi}{\overset{\circ}{st}}\int_u^\infty \exp(-h)\, dh = \frac{1}{\sqrt{\pi}}. \tag{13.56}$$

The mean frequency of a periodic motion and the effective collision-time parameter y_c are given by

$$\langle p \rangle = 2\pi C\left(\int_0^u \exp(-h)\, dh + \int_u^\infty \exp(-h)\, dh\right) = \frac{2}{f_0 D}, \tag{13.57}$$

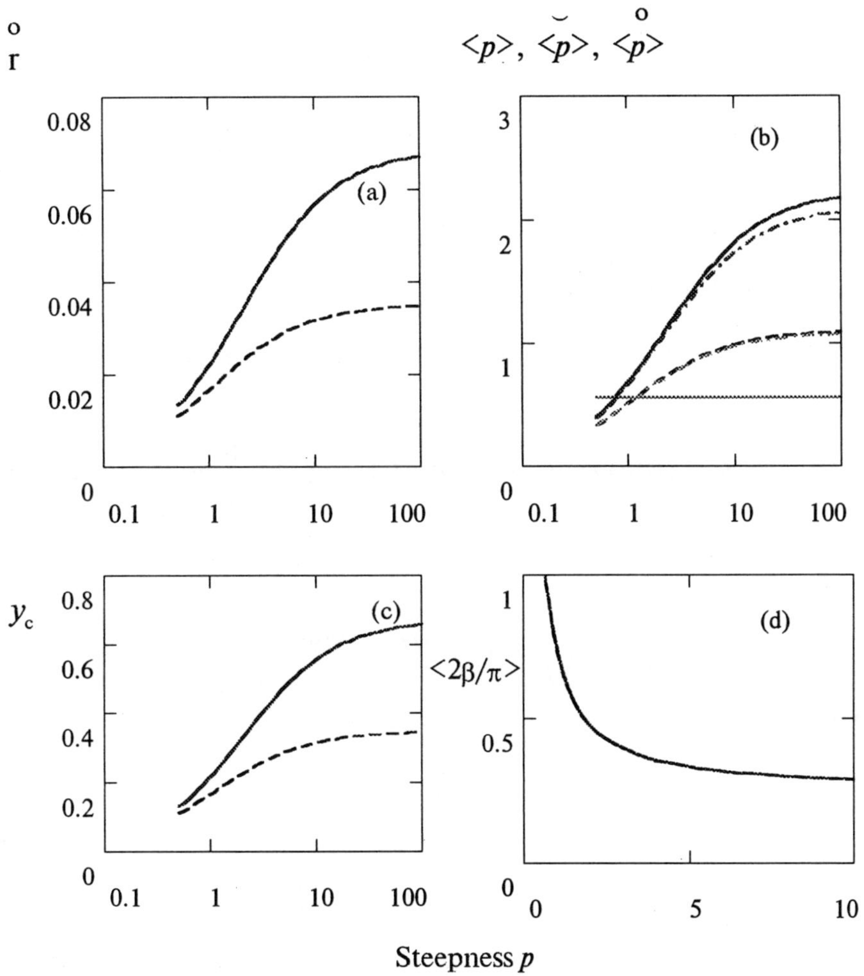

Fig. 13.8 Averaged parameters vs. steepness p; solid lines for $\Theta = \pi/8$, dashed lines for $\Theta = \pi/4$; $u = 4$. **(a):** the proportion of free rotors; **(b):** the mean frequency of librational motion (solid and dashed lines); dadots refer to the mean $\langle p \rangle$ and horizontal line refers to the subensemble of rotators; **(c):** the collision-time parameter; **(d):** the libration amplitude; in this Figure solid and dashed lines coincide.

$$y_c = \frac{\langle p \rangle}{\pi} = \frac{2}{\pi f_0 D} \, . \tag{13.58}$$

From Fig. 13.8 we see that dependencies of $\left\langle \overset{\vee}{p} \right\rangle$ and y_c on p resemble the $\overset{\circ}{r}(p)$ function. For $p \to \infty$ the mean librational frequencies are considerably higher, than $\left\langle \overset{\circ}{p} \right\rangle$.

THE MEAN LIBRATION AMPLITUDE. We write for the subensemble of librators:

$$\langle \beta \rangle = \int_0^u \beta(h) \exp(-h) \frac{\overset{\vee}{\Phi}}{2} \, dh \Bigg/ \int_0^u \exp(-h) \frac{\overset{\vee}{\Phi}}{2} \, dh \, . \tag{13.59}$$

Replacing $\sqrt{h}$ by t and noting, that

$$\frac{\overset{\vee}{\Phi}}{2} = \frac{\pi}{t}\left(\frac{t}{p} + f_0\right); \beta(h) = \frac{\pi f_0}{2} + \frac{t}{p},$$

we write

$$\langle \beta \rangle = \frac{\int_0^{\sqrt{u}} \left(\frac{\pi f_0}{2} + \frac{t}{p}\right) \exp\left(-t^2\right)\left(\frac{t}{p} + f_0\right) dt}{\int_0^{\sqrt{u}} \exp\left(-t^2\right)\left(\frac{t}{p} + f_0\right) dt} \, . \tag{13.60}$$

In accord with physical reasoning $\langle \beta \rangle$ decreases to the half-width Θ of the well at $p \to \infty$ (see Fig. 13.8d).

13.8 The Curved-Brim Hat Model

We have seen in Fig. 13.4 that the *form* of FIR absorption spectrum, contributed by rotators, does not depend on the parameters of the potential well. Here we shall change this result. Assuming again the DWP configuration, we replace u_0 in Eq. 13.3 by the parabolic potential:

$$u(\vartheta) = u_0 - S^2\left(\frac{\pi}{2} - \vartheta\right)^2 \quad \text{for} \left|\vartheta\right| \le \frac{\pi}{2} \text{ and } h > u_0, \quad , \qquad (13.61a)$$

$$u(\vartheta) = u_0 - S^2\left(\frac{\pi}{2} - \vartheta - \pi\right)^2 \quad \text{for } \frac{\pi}{2} \le \left|\vartheta\right| \le \frac{3\pi}{2} \text{ and } h > u_0. \qquad (13.61b)$$

Thus we introduce an additional parameter S, that is somehow related to intermolecular field, affecting rotating molecules. We assume that librators (that appear, when $h < u_0$) move in the same, as previously, potential (13.1)-(13.2).

GEOMETRICAL RELATIONS. Further we again put $\vartheta = |\vartheta|$ and return to simplified notation: $u_0 \to u$. The intersect point ϑ_c of two parabolic curves can be found as follows:

$$p^2(\vartheta - \Theta)^2 = u - S^2\left(\frac{\pi}{2} - \vartheta\right)^2 \quad \text{for } \vartheta = \vartheta_c; \qquad (13.62)$$

$$\vartheta_c = \frac{1}{p^2 + S^2}\left\{p^2\Theta + \frac{\pi}{2}S^2 + \sqrt{u\left(p^2 + S^2\right) - p^2S^2\left(\frac{\pi}{2} - \Theta\right)^2}\right\}. \qquad (13.63a)$$

Since the root here should be real, two conditions must be satisfied:

$$p \ge \frac{\sqrt{u}}{\frac{\pi}{2} - \Theta} \text{ and } S \le \frac{p\sqrt{u}}{\sqrt{p^2\left(\frac{\pi}{2} - \Theta\right)^2 - u}}. \qquad (13.63b)$$

Below we estimate $p_{\min}$ and $S_{\max}$ for $u = 4$:

for $\Theta = \frac{\pi}{8}$		for $\Theta = \frac{\pi}{4}$	
$p_{\min} = 1.7$	$S_{\max} = \begin{cases} 1.8, & p = 5 \\ 1.7, & p = 20 \end{cases}$	$p_{\min} = 2.6$	$S_{\max} = \begin{cases} 3.0, & p = 5 \\ 2.5, & p = 20 \end{cases}$

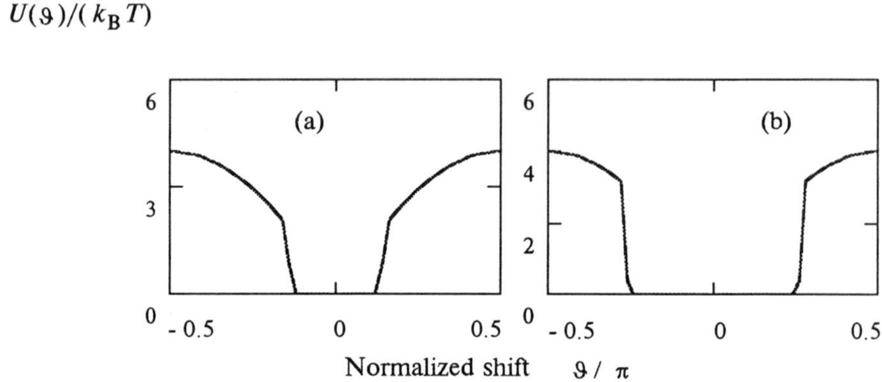

Fig. 13.9 The form of the potential well in a planar hat model with a parabolic brim. $u = 4$, $p = 20$, $S = 1.3$, $\Theta = \pi/8$ (a) and $\pi/4$ (b).

The S-depending form of the potential well is demonstrated by Fig. 13.9. For further calculation we put:

$$S = 1.3,\ p = 20,\ u = 4,\ \Theta = \frac{\pi}{8}. \tag{13.63c}$$

THE LAW OF HINDERED ROTATION. In view of Eq.13.61

$$h = u + \dot{\vartheta}^2 - S^2\left(\frac{\pi}{2} - \vartheta\right)^2;\ \dot{\vartheta} \equiv \frac{d\vartheta}{d\varphi};\ \varphi = \int_0^{\vartheta} \frac{d\varphi}{\sqrt{h - u + S^2\left(\frac{\pi}{2} - \vartheta\right)^2}}.$$

Substituting h by $h-u$ and denoting

$$a(h) \equiv \operatorname{asinh}\left(\frac{\pi S}{2\sqrt{h}}\right), \quad \text{where} \quad \operatorname{asinh}(X) \equiv \ln\left(X + \sqrt{X^2 + 1}\right),$$

we have:

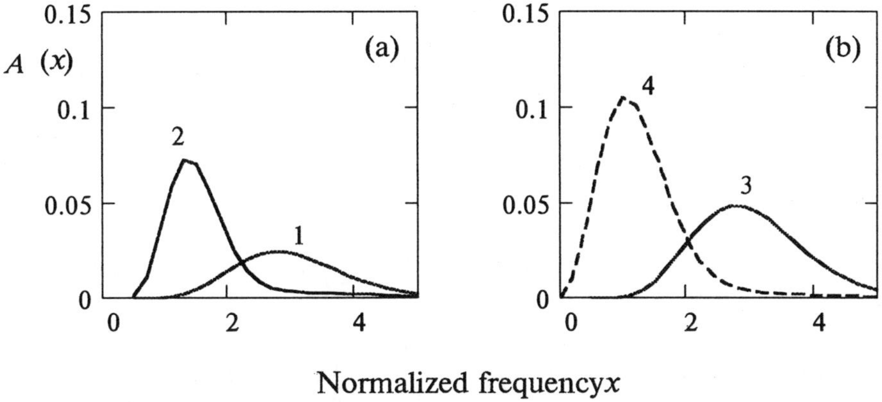

Fig. 13.10 Normalized absorption $A = xK''$ vs. x. The curved-brim (1, 2, 3) and flat-brim (4) hat model. p = 20, u = 4, $\Theta = \pi/8$. **a:** $A_{\|}$ (curve 1) and $A_{\perp}$ (curve 2); **b:** $A_{\|} + A_{\perp}$ (curve 3); $3xL''(x)$ for the ED model (curve 4). For curves 1, 2, 3: S = 1.3, y = 0; for curve 4: S = 0, y = 0.1.

$$S\varphi = \operatorname{asinh}\left[\frac{S}{\sqrt{h}}\left(\vartheta - \frac{\pi}{2}\right) + a(h)\right], \tag{13.64}$$

$$\mathring{\vartheta}(\varphi) = \frac{\pi}{2} + \frac{\sqrt{h}}{S}\sinh[S\varphi - a(h)]$$

$$= \pi \sinh\left(\frac{S\varphi}{2}\right)\left\{\sqrt{1+\frac{1}{a^2(h)}}\cosh\left(\frac{S\varphi}{2}\right) - \sinh\left(\frac{S\varphi}{2}\right)\right\}. \tag{13.65}$$

Taking $\varphi = \mathring{\Phi}/4$ and $\vartheta = \pi/2$, we find the period $\mathring{\Phi}$ of function $\mathring{q}(\varphi) = \cos\mathring{\vartheta}$ and the fundamental frequency $\mathring{p} = 2\pi/\mathring{\Phi}$:

$$\mathring{\Phi} = \frac{4a(h)}{S};\quad \mathring{p} = \frac{\pi S}{2a(h)}. \tag{13.66}$$

THE STEADY STATE. By analogy with Eq.13.10

$$\overset{\circ}{st} = \exp(-u) \int_0^\infty \exp(-h)\overset{\circ}{\Phi}\, dh = \frac{4\exp(-u)}{S} \int_0^\infty \exp(-h)\, a(h)\, dh \,. \quad (13.67)$$

Summing with $\overset{\vee}{st}$, we can find the *steady state distribution* and a number of averaged parameters:

$$W = C\exp(-h),\ C = \frac{p}{\pi E},, \quad (13.68a)$$

$$\frac{E}{2} = 1 - \exp(-u) + p\left[\sqrt{\pi}\, f_0 \operatorname{erf}\left(\sqrt{u}\right) + \frac{2\exp(-u)}{\pi S}\int_0^\infty \exp(-h)\, a(h)\, dh\right], \quad (13.68b)$$

$$\overset{\circ}{r} = C\,\overset{\circ}{st} = \frac{4p\exp(-u)}{\pi S E}\int_0^\infty \exp(-h)\, a(h)\, dh;\ \overset{\vee}{r} = 1 - \overset{\circ}{r}; \quad (13.69)$$

$$\left\langle \overset{\circ}{p} \right\rangle = \frac{2\pi}{\overset{\circ}{st}} \exp(-u) \int_0^\infty \exp(-h)\, dh = \frac{\pi S}{2}\, \frac{1}{\int_0^\infty \exp(-h)\, a(h)\, dh};; \quad (13.70)$$

$$y_c = \frac{\langle p \rangle}{\pi} = \frac{p}{\pi^2 E}\left(\int_0^u 2\pi\exp(-h)\, dh + \int_u^\infty 2\pi\exp(-h)\, dh\right) = \frac{2p}{\pi E}\,. \quad (13.71)$$

We estimate: $\overset{\circ}{r} \approx 3.1\%$, $\left\langle \overset{\circ}{p} \right\rangle \approx 1.17$. The distinction from the zero-S case, for which $\overset{\circ}{r} \approx 6.7\%$ and $\left\langle \overset{\circ}{p} \right\rangle = \frac{1}{\sqrt{\pi}} \approx 0.56$, is obliged to the decrease with S of the phase volume, occupied by hindered rotors (other mean parameters practically do not alter).

FOURIER AMPLITUDES AND SFs. By analogy with Eq. 13.17

$$\begin{Bmatrix} \overset{\circ}{a}_n \\ \overset{\circ}{b}_n \end{Bmatrix} = \frac{8}{\overset{\circ}{\Phi}} \int_0^{\overset{\circ}{\Phi}/4} \begin{Bmatrix} \cos\overset{\circ}{\vartheta}\ \cos 2\pi n\varphi/\overset{\circ}{\Phi} \\ \sin\overset{\circ}{\vartheta}\ \sin 2\pi n\varphi/\overset{\circ}{\Phi} \end{Bmatrix} d\varphi. \quad (13.72)$$

We take $t = \dfrac{4\varphi}{\overset{\circ}{\Phi}} = \dfrac{\varphi S}{a(h)}$ as new variable of integration, employ substitutions $\overset{\circ}{a}_n = 2\overset{\circ}{C}_n$, $\overset{\circ}{b}_n = 2\overset{\circ}{S}_n$ and express $\overset{\circ}{\vartheta}$ from Eq. 13.65:

$$\begin{Bmatrix} \overset{\circ}{C}_n \\ \overset{\circ}{S}_n \end{Bmatrix} = \int_0^1 \begin{Bmatrix} \cos \\ \sin \end{Bmatrix} \frac{\pi n t}{2} \begin{Bmatrix} \cos \\ \sin \end{Bmatrix} \Sigma(t,h)\, dt, \quad (13.73)$$

$$\Sigma(t,h) = \pi \sinh\frac{t\,a(h)}{2}\left[\sqrt{1+\frac{1}{a^2(h)}}\cosh\frac{t\,a(h)}{2} - \sinh\frac{t\,a(h)}{2}\right].$$

Denoting $\overset{\circ}{\gamma} = \dfrac{\sqrt{h}}{\overset{\circ}{p}} = \dfrac{2\sqrt{h}\,a(h)}{\pi S}$, we have:

$$\frac{\pi n^2 \overset{\circ}{p}}{\overset{\circ}{p}{}^2 n^2 - z^2} \begin{Bmatrix} \overset{\circ}{a}{}_n^2 \\ \overset{\circ}{b}{}_n^2 \end{Bmatrix} = \frac{4\pi\sqrt{h}\overset{\circ}{\gamma}}{h - \left(\overset{\circ}{\gamma} z/n\right)^2} \begin{Bmatrix} \overset{\circ}{C}{}_n^2 \\ \overset{\circ}{S}{}_n^2 \end{Bmatrix}; \quad (13.74)$$

$$\begin{Bmatrix} \overset{\circ}{K}_{\parallel}(z) \\ \overset{\circ}{K}_{\perp}(z) \end{Bmatrix} = \frac{12\,p\exp(-u)}{E} \int_0^\infty \sum_{n=1}^\infty \frac{\overset{\circ}{\gamma}\exp(-h)\sqrt{h}}{h - \left(\overset{\circ}{\gamma} z/n\right)^2} \begin{Bmatrix} \overset{\circ}{C}{}_n^2 \\ \overset{\circ}{S}{}_n^2 \end{Bmatrix} dh. \quad (13.75)$$

Thus, *anisotropy appears* along with the rise of S. If $S \to 0$, then

$$a(h) \to \frac{\pi S}{2\sqrt{h}},\ \sinh\frac{t\,a(h)}{2} \to \sinh\frac{t\pi S}{4\sqrt{h}} \to \frac{t\pi S}{4\sqrt{h}} \to 0,\ \cosh\frac{t\pi S}{4\sqrt{h}} \to 1,$$

$$\sqrt{1+h\left(\frac{2}{\pi S}\right)^2} \to \frac{2\sqrt{h}}{\pi S},\ \Sigma(t,h) = \frac{\pi t}{2},\ \overset{\circ}{\gamma} \to 1,$$

$$\begin{Bmatrix} \overset{\circ}{C}_n \\ \overset{\circ}{S}_n \end{Bmatrix} = \int_0^1 \begin{Bmatrix} \cos \\ \sin \end{Bmatrix} \frac{\pi n t}{2} \begin{Bmatrix} \cos \\ \sin \end{Bmatrix} \frac{\pi t}{2}\, dt = \begin{Bmatrix} 1/2 \\ 1/2 \end{Bmatrix} \text{ for } n=1,$$

$$\overset{\circ}{C}_n = \overset{\circ}{S}_n = 0 \text{ for } n>1\,,$$

$$\frac{E}{2} \to 1-\exp(-u) + p\left[\sqrt{\pi}\, f_0 \operatorname{erf}\left(\sqrt{u}\right) + \frac{2\exp(-u)}{\pi S}\int_0^\infty dh \exp(-h)\frac{\pi S}{2\sqrt{h}}\right]$$

$$\to 1-\exp(-u) + p\sqrt{\pi}\left[f_0 \operatorname{erf}\left(\sqrt{u}\right) + \exp(-u)\right].$$

Thus we have got Eq. 13.46b; an anisotropy disappears for $S \to 0$.

THE LONG LIFE TIME LIMIT. Taking $h = t^2$, $t(x) = x\overset{\circ}{\gamma}(x)/n$, we have for $y \to 0$:

$$\begin{Bmatrix} \overset{\circ}{K}''_{\parallel}(x) \\ \overset{\circ}{K}''_{\perp}(x) \end{Bmatrix} = \frac{12\, p\exp(-u)}{E}\int_0^\infty dt \sum_{n=1}^\infty t\overset{\circ}{\gamma}\, \exp\left(-t^2\right)\begin{Bmatrix} \overset{\circ}{C}{}_n^2 \\ S_n^2 \end{Bmatrix} \lim_{y\to 0}\frac{y}{\left(t-\left(\frac{x\overset{\circ}{\gamma}}{n}\right)\right)^2 + y^2}$$

$$= \frac{12\,\pi p\exp(-u)}{E}\sum_{n=1}^\infty t\overset{\circ}{\gamma}\, \exp\left(-t^2\right)\begin{Bmatrix} \overset{\circ}{C}{}_n^2 \\ S_n^2 \end{Bmatrix}; \qquad (13.76)$$

$$\overset{\circ}{\gamma} = \frac{nt}{x} = \frac{2t}{\pi S}\, \text{a} \sinh\left(\frac{\pi S}{2t}\right);\quad t(x) = \frac{\pi S/2}{\sinh\left(\frac{\pi n S}{2x}\right)},\ \overset{\circ}{\gamma}(x) = \frac{\pi n S/(2x)}{\sinh\left(\frac{\pi n S}{2x}\right)}$$

Let $X(t) = \frac{\pi n t S}{4x}$. In view of Eq. 13.74, noting that $\operatorname{asinh}\left(\frac{\pi S}{2t}\right) \equiv \frac{\pi n S}{2x}$, we finally obtain :

$$\sqrt{1+h\left(\frac{2}{\pi S}\right)^2} \to \sqrt{1+\frac{1}{\operatorname{sh}^2\left(\frac{\pi n S}{2x}\right)}} = \coth\left(\frac{\pi n S}{2x}\right), \frac{t\,a(h)}{2} \to \frac{\pi n t S}{4x},$$

$$\left\{\begin{matrix}\overset{\circ}{C}_n \\ \overset{\circ}{S}_n\end{matrix}\right\} = \int_0^1 \left\{\begin{matrix}\cos \\ \sin\end{matrix}\right\} \frac{\pi n t}{2} \left\{\begin{matrix}\cos \\ \sin\end{matrix}\right\} \left(\pi \sinh[X(t)]\, f(t)\right) dt \ , \qquad (13.77a)$$

$$f(t) = \coth \frac{\pi n S}{2x} \cosh X(t) - \sinh X(t) \ . \qquad (13.77b)$$

The FIR absorption spectrum is illustrated by Fig. 13.10. *The frequency of absorption maximum* (curves 1, 2, 3) *is appreciably larger, than for* $S = 0$ (curve 4). Absorption for $\perp$-orientation (curve 2) appreciably exceeds that, calculated for $\|$-orientation (curve 1), while for $S = 0$ no such distinction arises. These results correspond to the greater mean frequency $\left\langle \overset{\circ}{p} \right\rangle$ of hindered rotation.

Consequently, we have demonstrated that the *absorption maximum due to almost freely rotating molecules shifts to higher frequencies in the curved-brim hat model as compared with the flat-brim model.* This qualitative picture will be employed in Ch. 15, Sec. 15.1.6, wherein we shall discuss the fine «isotopic effect», observed in FIR spectra of liquid water.

PART FIVE

APPLICATIONS OF THE THEORY

CHAPTER 14
INDIVIDUAL NONASSOCIATED POLAR LIQUIDS

A *qualitative* consideration of dielectric relaxation in strongly absorbing liquids was given in Sec. 2.3 and 8.4. This chapter will be devoted to *quantitative* calculations / study of molecular dyna mics in such fluids. In Table 14.1 we point out main formulas used for description of dielectric spectra.

Table 14.1 Index of general relations and formulas

A. General relations	**Equations**
Absorption α	(4.46), (4.47)
Complex permittivity $\varepsilon^* = \varepsilon' + \varepsilon''$	(4.41)
For F_T (isothermal)	(6.21)
For F_T (correlation-orientation)	(6.17a)
Correlation factor g	(6.52)
B. Formulas for the SFs *L(z)*, the case of an isotropic model	**Equations**
The confined rotator model	(8.22)
The hybrid model	(9.41), (9.43), (9.47)
The cosine-squared model	(12.87)-(12.91)

The normalized concentration G, dipole moment μ in a liquid, the k_μ-coefficient and concentration N are related as follows:

$$G = \frac{\mu^2 N}{3k_B T},\ \mu = \mu_0 k_\mu \frac{n_\infty^2 + 2}{3},\ N = \frac{\rho N_A}{M}, \tag{14.1}$$

N_A being the Avogadro number, ρ density and M molecular mass;

$$G = 1454\ (\mu_0[D])^2 \bullet \rho\ [\mathrm{g{\bullet}cm^{-3}}]\ /\ (\ T[\mathrm{K}]\ M\ [\mathrm{g{\bullet}mole}]\ ). \tag{14.2}$$

14.1 Liquid Fluoromethane CH_3F

Experimental wideband ($0 < \nu/\mathrm{cm}^{-1} < 200$) loss $\varepsilon''(\nu)$ and absorption $\alpha(\nu)$ spectra, recorded in a wide temperature range, are known for a number of individual nonassociated liquids. These spectra can be rather well described on the basis of the simple confined rotator (CR) model with the correlation-orientation or isothermal induced distributions F_K or F_T being involved. The CR model has three fitting parameters, viz. libration angle β, lifetime τ and parameter k_μ, correcting the value of the dipole moment μ in a liquid:

(a) β actually determines the frequency ν_L of the FIR absorption peak α_L;

(b) τ mainly influences the position ν_D of the Debye loss-peak (the latter is usually placed at microwaves);

(c) the k_μ -correction permits to attain the coincidence of calculated and experimental integrated absorption $\int \alpha(\nu) d\nu$.

We start with application of the CR model to such typical strong absorbing nonassociated liquid, as fluoromethane CH_3F. Fitted parameters of the CR model and relevant molecular data are presented in Table 14.2. Here the normalized temperature is given by

$$T_R = (T - T_{tr}) / (T_{cr} - T_{tr}), \tag{14.3}$$

where T_{tr} and T_{cr} are triple and critical temperature.

Relations, used in these calculations, are pointed out in Table 14.1 and Eq. 14.1, 14.2. The CR model is described in Sec. 8.2; $L(z) \equiv L_{CR}$ is the SF, found for this model. Let $L_0 \equiv L(iy)$ be the "static" value of the SF (L_0 is real). The estimated value of the Debye relaxation time is given by

$$\tau_D^{est} = \frac{g\eta}{yL_0} \text{ for } F_K;\ \tau_D^{est} = \frac{g\eta}{yL_0}\left(1 - y^2 L_0\right) \text{ for } F_T; \tag{14.4}$$

Table 14.2 Molecular constants / fitted parameters calculated for the isothermal induced distribution F_T. Liquid CH_3F. ρ in g•cm^{-3}; τ_D and τ in ps; β in deg.

Temperature		Molecular constants[a)]				
T	T_R	ρ	ε_s	n_∞^2	τ_D	k_μ^{exp}
131	0.00	0.95				
133	0.01	0.95	49.0	1.79	6.6	1.17
153	0.12	0.93	36.0	1.77	4.2	1.16
173	0.23	0.90	31.4	1.75	2.8	1.14
193	0.33	0.86	25.4	1.72	1.9	1.11
233	0.55	0.78	17.2	1.62	1.0	1.05
270	0.74	0.66	12.0	1.52	0.5	1.06
293	0.87	0.57	9.0	1.43	0.4	1.05
318	1.00	0.28				

Dimensionless parameters			Fitted parameters of the CR model		
T	g	G	k_μ	τ	β
133	0.95	2.70	1.30	0.26	20.4
153	0.87	2.17	1.27	0.20	22.9
173	0.89	1.81	1.26	0.16	25.0
193	0.85	1.52	1.26	0.12	26.9
233	0.80	1.08	1.26	0.10	32.7
270	0.80	0.74	1.25	0.09	39.4
293	0.78	0.56	1.25	0.08	43.4

$I = 33 \bullet 10^{-40}$ g•cm^2, $M = 34$, $T_{tr} = 131$ K, $T_{cr} = 318$ K

μ = 1.81 D = 1.81•10^{-18} CGSE

[a)] From (Gerschel, Dimicoli, Jaffre, and Riou 1976)

$$z = x + iy,\ x = \omega\eta = 2\pi c\nu\eta,\ y = \frac{\eta}{\tau};\ \eta = \sqrt{\frac{I}{2k_B T}}. \tag{14.5}$$

THE EVOLUTION OF DIELECTRIC SPECTRA WITH TEMPERATURE was considered in Sec. 2.3.1. From Fig. 2.7 we see that theoretical and experimental spectra agree *quantitatively.* However, the calculated absorption $\alpha(\nu)$ decreases with ν in FIR region more slowly, than the observed $\alpha(\nu)$. Near critical temperature T_{cr} this discrepancy becomes more pronounced. For F_K this drawback becomes more evident than for F_T.

We can find dielectric spectra for *any* temperature. Starting from Table 14.2, we present below the empirical (given in the form of cubic polynomials) dependencies for a number of parameters *P*:

$$P(T_R) = \sum_{j=0}^{3} a_j T_R^j, \tag{14.6}$$

Parameter	a_0	a_1	a_2	a_3
G	2.74	-5.24	5.50	-2.74
β, deg	20.4	18.0	6.73	3.92
τ, ps	0.265	-0.690	0.985	-0.505
n_∞^2	1.79	-0.100	-0.378	0.01 8
g	0.948	-0.438	0.441	-0.179
ρ, g•cm^{-3}	0.960	-0.494	1.03	-1.20

(14.7)

Then we can apply the CR model with *P*(*T*) from (14.6), (14.7). The result of this procedure is illustrated by Fig. 14.1. The shoulder is seen at *T* near the triple point on the dependence $\varepsilon''(\nu)$ for $\nu \approx 70$ cm^{-1}; after a small increase of temperature this feature disappears (Fig. 14.1a). Fig. 14.1b demonstrates the low-frequency shift of the absorption α_L and the decrease of α_L with *T*, accompanied by the broadening of the $\alpha(\nu)$ spectrum. These phenomena are probably underlain by the destruction of near order in a liquid.

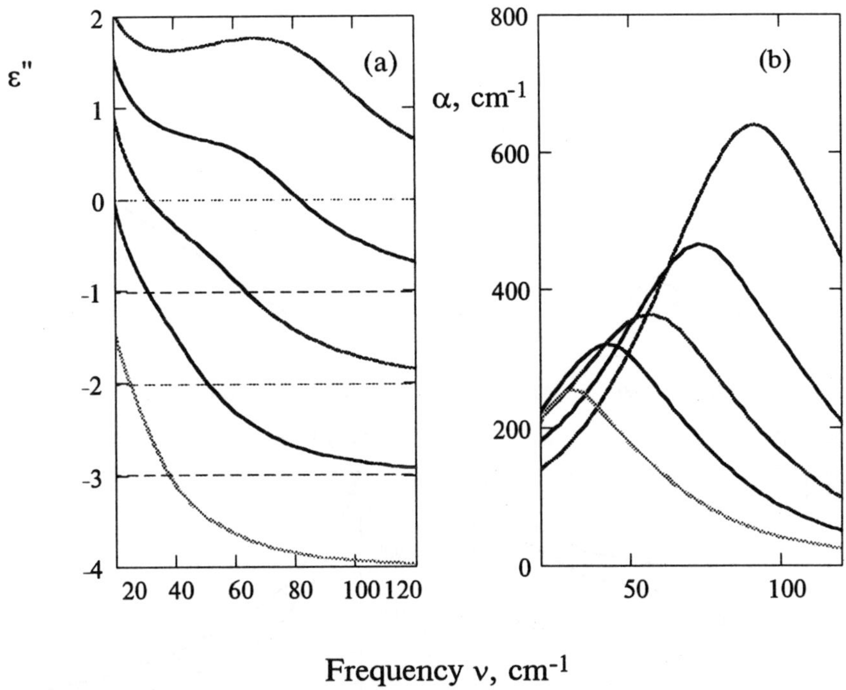

Fig. 14.1 Dielectric loss (a) and absorption coefficient (b) vs. frequency, predicted for liquid fluoromethane. T = 131, 171, 211, 251, 291 °C (from top to beneath). In Fig. a all curves, except the upper, are shifted downwards.

One may suggest that at $T \approx T_{\mathrm{tr}}$ orientational motion in a liquid resemble that in a solid state, while at $T \approx T_{\mathrm{cr}}$ it is like in a gas.

CRITICAL TEMPERATURE. We return to the idea, regarded in Sec. 2.3.2, 8.4.1. The calculated dependence τ_D/η on the normalized lifetime τ/η is plotted in Fig. 14.2c. In view of Eq. 14.4 for $F = F_K$

$$\frac{\tau_D^{\mathrm{est}}}{\eta} = f(T_R);\ f(T_R) \equiv \frac{g}{yL(y)} \text{ at } y \equiv y(T_R) \text{ and } g \equiv g(T_R)\ ; \quad (14.8)$$

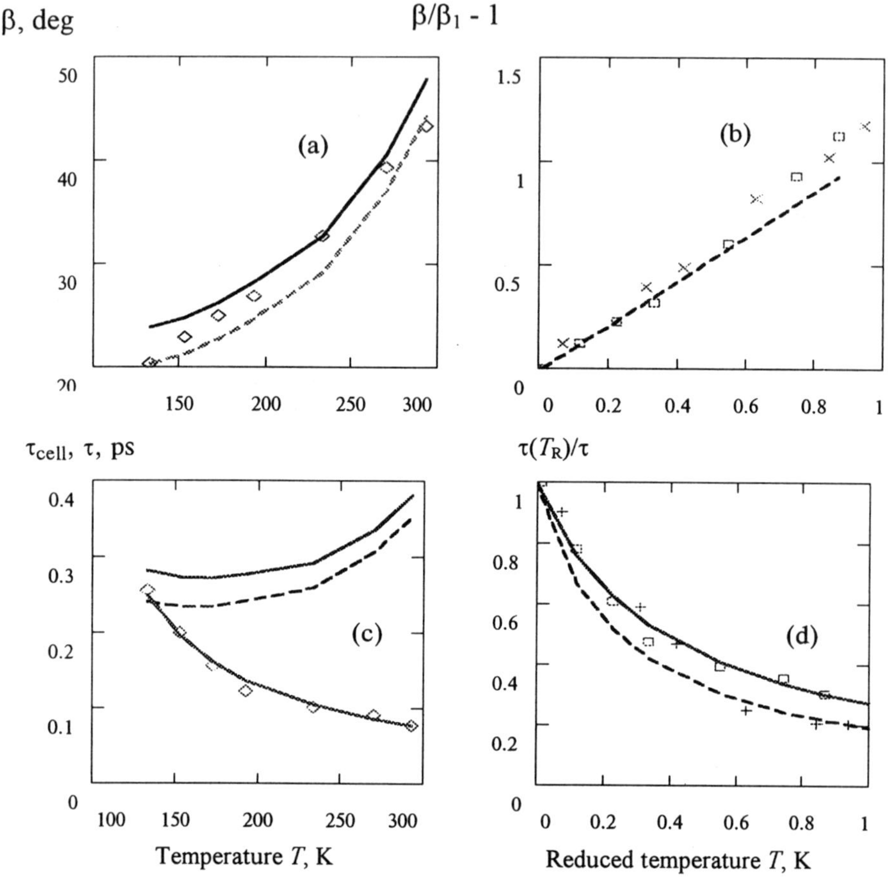

Fig. 14.2 **a:** fitted (dots) and analytical (solid line for $\sigma = 3.57$ Å and dashed line for $\sigma = 3.69$ Å) dependence of β vs. T for liquid CH_3F. **b:** fitted (squares for CH_3F and crosses for CHF_3) and empirical, Eq. 14.17, dependence of $\beta/\beta_1 - 1$ vs. T_R. **c:** fitted τ-values (dots); calculated time τ_{cell}, $\sigma = 3.57$ Å(1) and 3.69 Å (2); empirical dependence (14.33), $\sigma = 3.57$ Å. Liquid CH_3F. **d:** fitted τ for liquid CH_3F (squares) and CHF_3 (crosses); empirical dependence (14.6), (14.7) for $b = 2.65$ (solid line) and 4.15 (dashed line)

g and $y = \eta/\tau$ being functions of T, determined by Eq. 14.6, 14.7. In this point, where $f(T_R)$ goes trough a minimum:

$$T_R = \max T_R \approx 0.961,\ y = y_{cr} \equiv y(\max T_R) \approx 3.056,$$

$$\tau = \tau_{cr} \equiv \tau(\max T_R) \approx 0.0642 \text{ ps.}$$

Finally,

$$T_{cr}^{est} = \frac{1}{2k_B y_{cr}^2}. \tag{14.9}$$

This estimation yields $T_{cr} \approx 310.7$ K (about 2.3% discrepancy from the true value 318 K).

THE CONDITION OF RESONANCE INTERACTION. If y is not large, the frequency ν_L of the absorption maximum obeys an approximate relation

$$fx_L = f2\pi c\eta\nu_L = \sqrt{3/2}, \text{ where } f = 2\beta/\pi. \tag{14.9a}$$

Taking the experimental value of ν, one can estimate the $\beta(T)$ dependence; the latter agrees with the fitted $\beta(T)$. Since ν_L decreases with T, β should grow with T faster, than $\sqrt{T}$.

14.2 The Rotational Cell Model (RCM)

The free path of a molecule in a liquid can be estimated as the mean distance $\gamma N^{-1/3}$ between molecules minus the effective diameter σ of a molecule:

$$l_{cell}(T) = \frac{\gamma}{\sqrt[3]{N(T)}} - \sigma = \gamma \left(\frac{M}{N_A\rho}\right)^{1/3} - \sigma. \tag{14.10}$$

According to (Davis 1972), γ falls into the range [1, 2]. We take for γ the value $2^{1/6}$, corresponding to the face-centered lattice. The mean *translational* velocity $\langle v \rangle$ of a molecule and the mean lifetime τ_{cell} are defined as:

$$\langle v \rangle = \sqrt{\frac{8k_BT}{\pi m}}, \tag{14.11}$$

$$\tau_{\text{cell}} = \frac{l_{\text{cell}}}{\langle v \rangle} = \sqrt{\frac{\pi m}{8k_B T}} \left(\frac{\gamma}{\sqrt[3]{N}} - \sigma \right) . \tag{14.12}$$

Applying Maxwellian distribution W, we can estimate the mean rotational frequency $\langle \Omega \rangle$:

$$W = \exp\left(-p^2\right),\ p = \eta\Omega, \text{ so that } \int_0^\infty W\, dp^2 = 1; \tag{14.13}$$

$$\langle \Omega \rangle = \frac{1}{\eta} \int_0^\infty W\, p\, dp^2 = \frac{\sqrt{\pi}}{2\eta} . \tag{14.14}$$

If we assume that *during the lifetime* τ_{cell} *a dipole rotates just on the "sector of swinging"* 2β,

$$2\beta = \tau_{\text{cell}} \langle \Omega \rangle , \tag{14.15}$$

then

$$\beta(T) = \frac{\pi}{8} \sqrt{\frac{m}{I}} \left(\gamma \sqrt[3]{\frac{M}{N_A \rho(T)}} - \sigma \right) . \tag{14.16}$$

This is the principal relation of the "rotational" cell model. It is convenient to choose σ for some temperature $T = T_1$ near T_{cr}, for which $\beta = \beta_1$ is known or can be independently estimated:

$$\beta(T) = \beta_1 \text{ for } T = T_1. \tag{14.16a}$$

Analytical dependence (14.16) agrees with that, fitted for the amplitude β of the CR model (see Fig. 14.2a). It is seen from Fig. 14.2b that one may employ the linear approximation:

$$\beta / \beta_1 - 1 = f_\beta(T_R);\ f_\beta(T_R) = 1 + aT_R;\ a \approx 1.1 . \tag{14.17}$$

EMPIRIC RELATION FOR THE "ROTATIONAL" LIFETIME $\tau(T)$. Let us rewrite Eq. 14.12 in terms of molecular parameters:

$$\tau_{\text{cell}}(T) = \sqrt{\frac{\pi m_H M}{8k_B T}} \left(\gamma \sqrt[3]{\frac{M}{N_A \rho(T)}} - \sigma \right) . \tag{14.18}$$

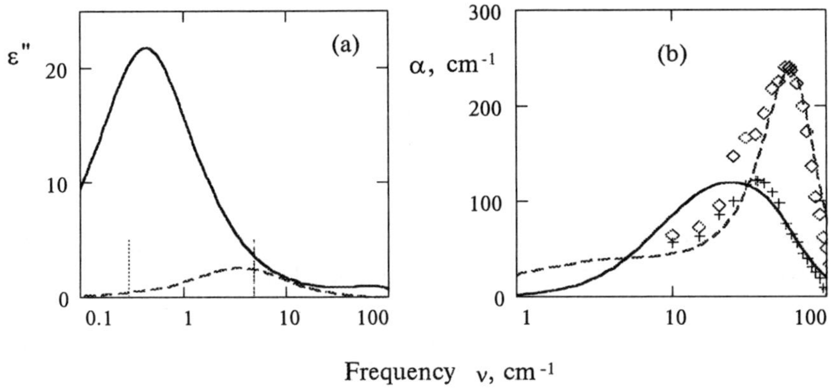

Fig. 14.3 The same as in Fig. 2.7 but for liquid trifluoromethane CHF_3. T = 130 K (solid lines) and 299 K (dashed lines).

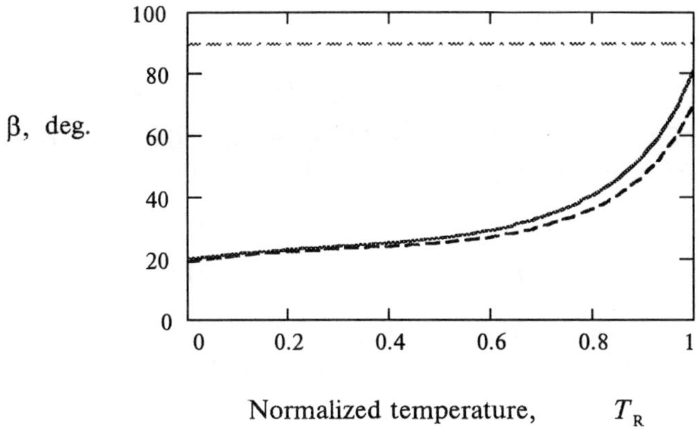

Fig. 14.4 The libration amplitude vs. normalized temperature, calculated for the rotational cell model: solid line - liquid CH_3F, dashed line - liquid CHF_3. The dashed horizontal line denotes the limiting β value ($\beta_{max} = \pi/2$).

Comparison with fitted $\tau(T)$ shows (see Fig. 14.2c) that τ_{cell} goes through a minimum and then *increases* with the rise of T, while τ decreases with T. Both times approximately coincide only near the triple point:

$$\tau \approx \tau_{cell} \text{ for } T \approx T_{tr}. \tag{14.19}$$

We assume that

$$\tau_{cell}\,(T_1) \approx \tau_{cell}\,(T_1) \equiv \tau_1 \tag{14.20}$$

and introduce into Eq. 14.18 the multiplier C_τ. In view of Eq. 14.17 the "rotational" lifetime τ, corresponding to the amplitude β_1, is given by

$$\tau_1 = C_\tau \sqrt{\frac{8I}{\pi k_B T_1}}\beta_1 \approx C_\tau \sqrt{\frac{8I}{\pi k_B T_1}} \frac{\beta(T_R)}{1+1.1T_R} \tag{14.21}$$

For liquid CH_3F, *where* T_1 = 133 K, τ_1 = 0.256 ps, we find from Eq. 14.21: $\sigma \approx 3.69\ \mathring{A}$, $C_\tau \approx 1.06$.

One can seek a *fundamental* basis for a "molecular" description of the lifetime vs. temperature dependence $\tau(T)$, see further Sec. 14.5. Here we restrict consideration by an empirical formula

$$\frac{\tau(T_R)}{\tau_1} \approx \frac{1}{1+bT_R}. \tag{14.22}$$

Values of τ, found from Eq. 14.22, agree with the fitted ones (see Fig. 14.2d; $b \approx 2.65$ in Eq. 14.22 for liquid CH_3F). Therefore,

$$\tau(T_R) = \frac{\tau(T_R=0)}{1+\beta T_R} \approx C_\tau \sqrt{\frac{8I}{\pi k_B T_1}} \frac{\beta(T_1)}{(1+\beta T_R)(1+1.1T_1)}. \tag{14.23}$$

For elevated temperatures the "rotational" lifetime τ is much lesser, than τ_{cell}, which corresponds to one turn of a dipole on the angle 2β. We conclude:

Orientational motion in a liquid is highly damped, since during lifetime τ a dipole turns, on the average, on the angle $\langle \delta\vartheta \rangle$, that is much lesser than "the sector of swinging" 2β. (14.24)

Table 14.3 Molecular constants / fitted parameters calculated for the isothermal induced distribution F_T. Liquid CHF_3.

Temperature		Molecular constants[a)]				
T, K	T_R	ρ, g•cm^{-3}	ε_s	n_∞^2	τ_D	k_μ^{exp}
117	0.00	1.67	45.8	1.82	18.0	1.05
130	0.07	1.63	39.2	1.81	14.2	1.04
173	0.31	1.59	24.0	1.74	5.8	1.03
193	0.42	1.46	19.2	1.69	3.8	1.03
231	0.63	1.30	12.6	1.60	1.9	1.02
270	0.84	1.10	8.1	1.47	1.2	1.01
288	0.94	0.88	6.0	1.38	1.1	1.01
299	1.00	0.52				

Dimensionless parameters			Fitted parameters of the CR model		
T, K	g	G	k_μ	τ, ps	β, deg.
117	1.26	1.88	1.20	0.52	19.3
130	1.28	1.59	1.18	0.47	21.7
173	1.11	1.10	1.17	0.31	26.9
193	1.10	0.89	1.17	0.24	28.8
231	1.01	0.62	1.16	0.13	35.2
270	0.93	0.41	1.16	0.11	39.0
288	0.87	0.32	1.20	0.11	42.0

$I = 81 \bullet 10^{-40}$ g•cm^2, $M = 70$, $T_{tr} = 117$ K,
$T_{cr} = 299$ K, $\mu = 1.65$ D $= 1.65 \bullet 10^{-18}$ CGSE

[a)]From (Gerschel, Dimicoli, Jaffre, and Riou 1976)

Note: coefficients a_j for $\rho(T)$ [g•cm^2] in Eq. 14.1 are given by:
$a_0 = 1.685$, $a_1 = 1.002$, $a_3 = 2.293$, $a_4 = 2.376$,

In view of this statement *molecular* interpretation of the lifetime τ may substantially differ from that, given above for the "translational" lifetime τ_{cell}.

14.3 Liquid Trifluoromethane CHF_3

Here we calculate also 0-THz spectra of liquid CHF_3 with a purpose to seek molecular-dynamics features, common for strong absorbing nonassociated liquids.

14.3.1 Application of the CR Model with Fitted Parameters

CHF_3 molecule presents an *oblate* symmetric top, while the considered above CH_3F molecule - a *prolate* symmetric top. The frequency dependencies of loss / absorption, calculated for the CR model with the *fitted* parameters β, τ, k_μ, are shown in Fig. 14.3 for two temperatures. Related data are presented in Table 14.3. Comparison with Fig. 2.7 shows, that if temperature is near T_{cr}, theory agrees worse with experiment (for both liquids), than for lower values of T.

14.3.2 Application of the CR Model / Rotational Cell Model

a. We find an analytic approximation for $\rho(T_R)$ with use of the experimental data, presented in Table 14.3, after that we fit the effective diameter σ (σ = 3.895 $\mathring{A}$) with use of Eq. 14.16. The latter is applied for the lowest temperature T_1 = 117 K, to which β^{fit} = 19.3° corresponds. The $\beta(T)$ values, predicted by Eq. 14.16, agree with the fitted β in all temperature range (see Fig. 14.2 a, b).

b. We determine multiplier C_τ in Eq. 14.23 with use of the following condition. Estimated (τ_1) and fitted (τ) lifetimes should coincide for temperature T_1. For liquid trifluoroform we find that $C_\tau \approx 1.34$.

c. We determine constant β in Eq. 14.23, demandind that the fitted τ–values should coincide for elevated temperatures with those, calculated for the rotational cell model. In our example (liquid CHF_3) we set T = 288 K and find that β = 4.15. The fitted and calculated $\tau(T_R)$ dependencies qualitatively agree (see Table 14.4).

We remark that the normalized theoretical dependencies $\beta(T_R)$, found for liquid CH_3F and CHF_3, practically coincide (see Fig. 14.2b). This remarkable result agrees with fitted $\beta(T)$.

Analytical dependencies $\beta(T_R)$ and $\tau(T_R)$, supplied by the rotational cell model, can be employed for the *prediction* of dielectric spectra in the range of temperatures.

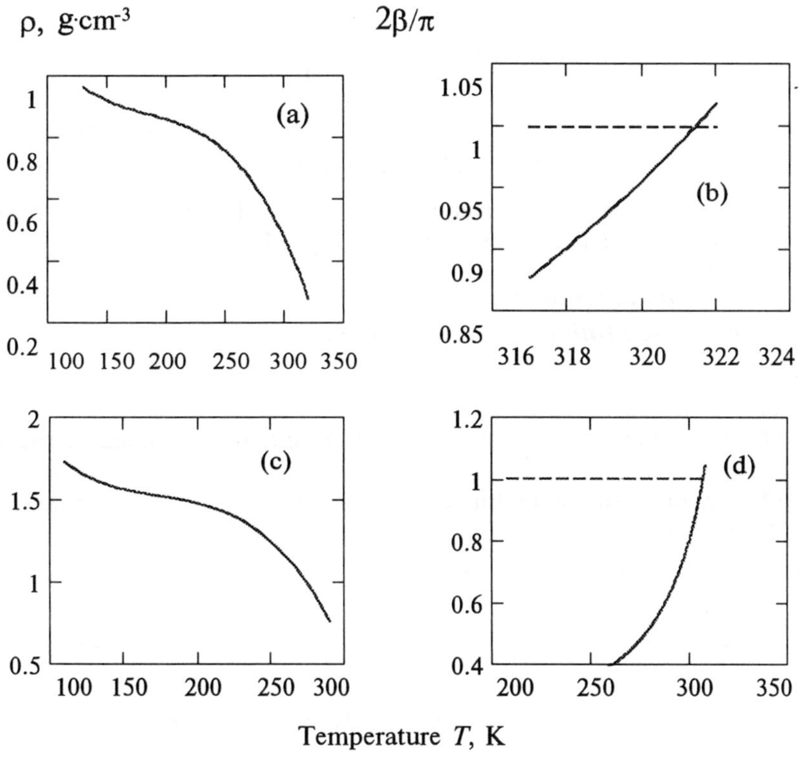

Fig. 14.5 The empirical $\rho(T)$ dependence and the graphs, pertaining to the prediction of the critical temperature with use of the rotational cell model. **a, b:** for liquid CH_3F; **c, d:** for liquid CHF_3.

14.4 Prediction of the Critical Temperature

14.4.1 Determination of T_{cr} from the Condition, Imposed on the Libration Amplitude

It follows from Table 14.4 and Fig. 14.4, that β approaches 90°, when T is near T_{cr}. Since still larger β-values have no physical sense, we suggest that the *critical* temperature obeys the relation

$$\beta\left(T = T_{cr}\right) = \frac{\pi}{2}, \tag{14.25}$$

where $\beta(T)$ is given by Eq.14.16. This criterion yields the following predicted values of T_{cr}^{est}: 321.5 K for CH_3F and 307 K for CHF_3 (the relative error about 1.1 and 2.7%, respectively). We remark, that the density ρ sharply decreases with T, when T tends to T_{cr} (see Fig. 14.5).

14.4.2 Determination of T_{cr} from the Minimum $\tau_D(\tau)$ Condition

We return to this approach (see Sec. 14.1), but now - on the basis of the rotational cell model (RCM). Namely:

(1) We employ theoretical $\beta(T_R)$, $\tau(T_R)$ functions, supplied by the RCM, Eq. 14.16, 14.23;

(2) we omit the g multiplier in Eq. 14.4, since the minimum under consideration is very sharp. For the isothermal induced distribution F_T we estimate T_{cr}^{est} from the condition

$$\frac{1 - y^2 L(iy)}{yL(iy)} = \min \quad \text{for } \tau = \tau(T);\ y = y(T);\ T = T_{cr}^{est}. \tag{14.26}$$

It can be shown that the last equation yields practically the same results, as the criterion (14.25).

14.4.3 On the Nature of Liquid → Solid Phase Transition

The mean distance $\langle s \rangle$ between molecules,

$$\langle s \rangle = \frac{1}{\sqrt[3]{N}} = \sqrt[3]{\frac{\rho M}{N_A}}, \tag{14.27}$$

has the same order of magnitude, as the effective diameter σ. Consequently, one can suppose that the *liquid → solid phase transition occurs because of the lack of free space required for librations of an individual polar molecule.*

Table 14.4 Parameters β and τ, fitted / estimated for the rotational cell model. Liquid CHF_3. $\sigma = 3.895$ Å, $C_\tau = 1.34$, $b = 3.14$

Parameter	Temperature T_R							
	.000	.071	.308	.418	.626	.841	.941	1.00
	β, deg							
Fitted for the CR model	19.3	21.7	26.9	28.8	35.2	39.0	42.0	
Calculated for the RCM	19.2	20.7	23.4	24.3	28.0	40.4	54.5	70.3
	τ, ps							
Fitted for the CR model	0.52	0.46	0.31	0.25	0.12	0.11	0.11	
Calculated for the RCM	0.52	0.40	0.23	0.19	0.15	0.12	0.11	0.10

We can check this idea by assuming, that the libration amplitude β attains its minimal value $\beta_{\min}$ just when $T \to T$cr. The minimal possible volume, occupied by a molecule, has an order of magnitude

$$V_{\min} = \frac{\pi\sigma^3}{6} = \frac{1}{N_{\min}(T_R)}. \tag{14.28}$$

We estimate $\beta_{\min}$, substituting $N_{\min}$ into Eq. 14.16:

$$\beta_{\min} = \frac{\pi\sigma}{8}\sqrt{\frac{m_H M}{I}}\left(\gamma\sqrt[3]{\frac{6}{\pi}} - 1\right). \tag{14.29}$$

One can see that, if γ is near 1, Eq. 14.29 indeed yields values of β_{min}, that are close to β_1, fitted / estimated for low T near T_{tr}. This qualitative reasoning confirms our idea concerning criteria (14.28), (14.29), to which T_{tr}. obey.

14.5 Molecular Interpretation of the "Rotational" Lifetime τ in the Light of the Debye Rotational Diffusion Theory

In accord with the previous arguments (see Sec. 14.2) it is difficult to identify the "rotational" lifetime τ with the lifetime τ_{cell}, during which a molecule in a liquid

(1) translates on the elementary cell-distance l_{cell} and

(2) rotates on the "sector of swinging" 2β.

Here we shall obtain an analytic expression, that describes a decreasing dependence $\tau(T)$. We shall relate parameters of the CR model to those, involved in the old Debye theory of rotational diffusion. Namely, we make use of the fact, that the kinetic equation (1.36) for the distribution F of electric moments μ over orientations involves the ratio $\overline{\Theta^2}/\delta t$. Indeed, according to (Debye 1929) certain probability function $W(\Theta)$, related to stochastic reorientation "must approach zero very rapidly (during interval ϑt), when Θ (the angle between two electric moments μ) exceeds few degrees".

We revise the meaning of the Debye terms $\overline{\Theta^2}$ *and* δt *by identifying* δt *with* $\tau(T)$ *and* $\overline{\Theta^2}$ *with* $[\beta(T)]^2 C_D$, (14.30)

---

where constant C_D is independent on T (one may expect that C_D is commensurable with 1). Using suggestion (14.30) and Debye relations (1.37), (1.39), we derive:

$$\tau(T) = (1/2)C_D\,[\beta(T)]^2\tau_D(T). \tag{14.31}$$

The applicability of Eq. 14.31 to nonassociated liquids is demonstrated, at least for a qualitative description, by Fig. 14.6; C_D was estimated for the lowest temperature T ($C_D = 0.62$ for liquid CH_3F and 0.51 for liquid CHF_3). Since calculated and fitted values of τ decrease with the rise of T, the decrease of τ_D should prevail over the increase of β^2.

We remark that the Debye relation (1.39) between relaxation time and "a constant ζ measuring the inner friction" may be expressed through free parameters of the confined rotator model:

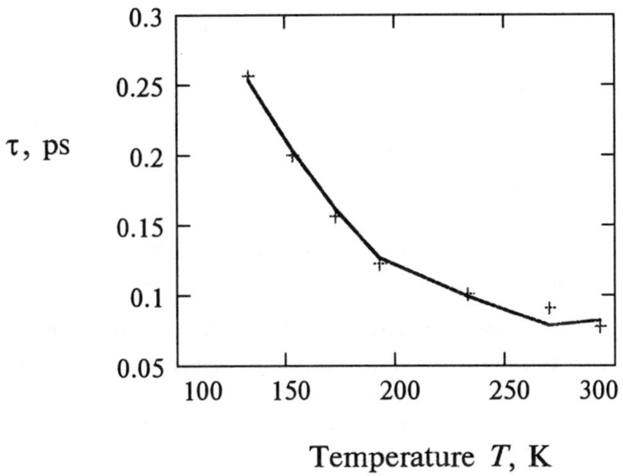

Fig. 14.6 Lifetime τ vs. temperature T, calculated from the Debye-like equation 14.31; points - fitted values. Liquid CH_3F.

$$\zeta(T) = 2k_B T \tau_D(T) = \frac{k_B T\, \tau(T)}{C_D\, \beta^2(T)}, \tag{14.32}$$

the friction ζ being a descending function of T.

Let us summarize results of our investigation:

(1). Librational and Brownian motions coexist in an individual nonassociated liquid.

(2). Stochastic reorientation determine the change of neighbors of a given polar molecule during its lifetime τ. Again we are obliged to conclude, that the interval τ "between strong collisions" means the change of a cage in a liquid, where a dipole is affected by a certain potential well.

(3). Librations with a characteristic frequency ν_L, present an immanent property of molecular dynamics in a typical polar liquid, having intense absorption spectrum $\alpha(\nu)$. The loss maximum ε'' reveals sooner the existence of a stochastic motion. One may regard, that the angular scale β presents a link between both extreme properties of a liquid. Indeed, β is the libration amplitude and also - the angular scale of Brownian reorientation during the lifetime.

(4). It must be admitted, that *two* time scales regulate molecular dynamics in "simple" liquids. The longer lifetime τ_{cell} is the characteristic of the translational motion. According to Eq. 14.12, 14.15, τ_{cell} determines the cage in a liquid, at least its angular scale 2β. The much *shorter lifetime* τ characterizes "rotational diffusion", viz. the process of collisions due to reorientation in a liquid, that leads to escape of a molecule from this cage. As we have seen previously, the mean rotation $\langle \delta\vartheta \rangle$, that occurs during lifetime τ, is close to 2β only near the triple point T_{tr}, being much smaller, than 2β, near the critical temperature T_{cr}. Although dielectric spectroscopy reflect mainly the response of a polar medium to molecular rotation, the interpretation of the observed data compels us to confess, that translations also play an important role.

In Sec. 14.6 we shall present another viewpoint concerning relation of molecular reorientation to translations.

14.6 Relation of the Potential Welldepth to Dielectric Spectra

The rectangular welldepth U_0 is *infinite* in the confined rotator model. Further we consider the same problem with use of the hybrid model (HM) and the cosine squared (CS) model. Both models are characterized by *finite* welldepth U_0. Again we assume the double well potential configuration, which is probably more specific for a liquid, then the single well potential. We also employ the normalized barrier $u = U_0/(k_B T)$. The basic relations, which we employ in calculations, are mentioned in Table 14.1. Our main purpose is to find the temperature dependence $u(T)$.

Two free-model parameters are employed in the HM: the libration amplitude β and the welldepth u, while in the CS model (that is also termed the field model) we have only one specific parameter

$$p = \sqrt{u} = \sqrt{\frac{U_0}{k_B T}}, \tag{14.33}$$

which characterizes the strength of intermolecular field.

In both models, HM and CS, we also employ two non-specific parameters: the lifetime τ and the k_μ coefficient. We remark that the SF $L = L_{CS}$ in Eq. 12.87 is doubled in order to take approximately into account the specificity of libration *in space*.

14.6.1 Dielectric Spectra and the Potential Welldepth

Dielectric spectra were calculated for liquid CH_3F and CHF_3, the fitted parameters and others, related to them, are presented in Tables 14.5, 14.6. The hybrid model is, in general, better than the field model, since in the former the positions ν_D and ν_L of the Debye loss ε'' and absorption α agree better with the experimental peaks (see Fig. 14.7 - 14.8). In particular, at low T the field model gives for liquid CHF *too narrow* absorption line. We note, that at high temperatures (near T_{cr}) the absorption spectrum, calculated with use of the HM or the CR model (see also Sec.14.1, 14.3), agrees worse with the experiment, than for lower T values.

14.6.2 Dependence of the Welldepth on Temperature

The field model yields at low T unreasonably high potential u (viz., $U_0 \approx 30$ $k_B T$), while the HM yields more reasonable values: about (6-9) $k_B T$.

Correspondingly, the proportion r of rotating molecules vanishes in the CS model (see Table 14.5). When T rises, u decreases, since intermolecular interactions weaken. At $T \to T_{cr}$ both potentials have nearly equal depths ($u \approx 3$), see Fig. 14.9. In view of Fig. 14.9b one may consider the $u(T_R)$ dependence, obtained in the HM model to be more reliable than that, found in the CS model, especially since the former model provides better description of dielectric behavior. Further in this chapter we shall apply only the hybrid model.

14.7 The Polarizability A''(ν) Spectra

We shall calculate also the imaginary part $A''(\nu)$ of the so called *complex polarizability* $A(\nu)$.

Table 14.5 The fitted / estimated parameters, found for the CS model. The stratified approach with the correlation-orientation induced distribution F_K.

Temperature		Fitted parameters (CS)			Estimated parameters (CS)		
T, K	T_R	k_μ	τ, ps	p	y	y_c	r
				Liquid CH_3F			
131	0.000						
133	0.011	1.28	.120	5.4	2.50	1.70	.000
153	0.118	1.28	.115	4.7	2.43	1.48	.000
173	0.225	1.21	.115	4.2	2.29	1.32	.000
193	0.332	1.20	.115	3.6	2.16	1.12	.000
233	0.545	1.37	.103	3.0	2.20	0.93	.000
270	0.743	1.40	.085	2.7	2.48	0.82	.001
293	0.866	1.80	.085	2.5	2.38	0.76	.003
318	1.000						
				Liquid CHF_3			
117	.000	1.20	0.30	5.65	1.67	1.78	.000
130	.071	1.13	0.30	5.00	1.58	1.58	.000
173	.308	1.23	0.25	3.95	1.65	1.24	.000
193	.418	1.23	0.20	3.50	1.95	1.09	.000
231	.626	1.34	0.17	3.15	2.10	0.97	.000
270	.841	1.44	0.13	2.75	2.54	0.84	.001
288	.940	1.44	0.105	2.60	3.04	0.79	.002
299	1.00						

Table 14.6 The same, as in Table 14.5, but for the hybrid model

T, K	T_R	k_μ	τ, ps	β, deg	u	y	y_c	r,%
Liquid CH_3F								
131	.000							
133	.011	1.35	0.25	20.4	9.0	1.20	1.6	0.2
153	.118	1.38	0.22	22.9	7.0	1.27	1.4	1.1
173	.225	1.39	0.20	25.0	6.0	1.31	1.2	2.6
193	.332	1.37	0.18	26.9	5.5	1.38	1.1	3.7
233	.545	1.50	0.13	32.7	4.5	1.74	.90	6.6
270	.743	1.65	0.10	39.4	3.7	2.10	.71	10
293	.866	1.80	0.08	43.4	3.2	2.52	.62	13.5
318	1.00							
Liquid CHF_3								
117	.000	1.20	0.90	20.4	6.0	0.56	1.51	3.8
130	.071	1.13	0.90	22.9	5.4	0.53	1.32	5.4
173	.308	1.23	0.70	25.0	4.6	0.59	1.16	9.8
193	.418	1.23	0.65	26.9	4.4	0.60	1.06	10.8
231	.626	1.34	0.50	32.7	3.7	0.71	0.84	13.5
270	.841	1.34	0.40	39.4	3.2	0.82	0.68	15.2
288	.940	1.44	0.30	43.4	3.0	1.06	0.60	16.1
299	1.00							

In accord with (Grochulski and Gerschel 1992) the *polarizability* loss $A''(\nu)$ "gives higher weight to the spectrum in the region of millimeter waves, than the more often used absorption or permittivity representation". It is known (see Scaife 1989, p.52) that for a spherical dielectric sample function $A''(\nu)$ is proportional to

$$\frac{\varepsilon''(\nu)}{[\varepsilon'(\nu)+2]^2+[\varepsilon''(\nu)]^2}. \tag{14.34}$$

We omit the proportionality coefficient, independent on T and ν, since only qualitative frequency dependencies are here considered.

Theoretical dependence of $A''(\nu)$ on T agrees (see Fig. 14.10) with that, observed for liquid CH_3F by (Grochulski and Gerschel 1992). In the last work a and n were measured. For these estimations we used formulas

$$\varepsilon''(\nu) = \frac{\alpha(\nu)n(\nu)}{2\pi\nu}, \quad \varepsilon'(\nu) = [n(\nu)]^2 - \left[\frac{\alpha(\nu)}{4\pi\nu}\right]^2,$$

that follow from Eq. 1.28 - 1.29.

When T rises, the two-humped dependence $A''(\nu)$ gradually transforms into the curve, having the only peak. The two-humped $A''(\nu)$ spectrum was declared in the aforecited article to be the evidence "of the observed new absorption band". In order to clarify situation it is useful to compare the $A''(\nu)$ curves, calculated for the HM, CR and ED models, in which u is respectively finite, infinite or zero. Since in the ED model, where $u = 0$, the two-humped curve disappears, it is evident that the latter is obliged to the subensemble of librators and can be described in terms of the "usual" mechanisms of orientational relaxation.

14.8 Distortion of the Cole-Cole Diagram due to Micro Heterogeneity

THE CLUE IDEA. The Cole-Cole diagram considerably differs in some experiments from the Debye semi-circle. A new approach (it was briefly touched in Sec. 6.9.2) will be described here.

a. We employ the SF $K(z, \Psi)$ and the dispersion $\sigma(\Psi)$, found for an *anisotropic* medium, characterized by an arbitrary angle Ψ between radiation-field amplitude $\boldsymbol{E}_m$ and the symmetry axis of the internal potential $U(\vartheta)$;

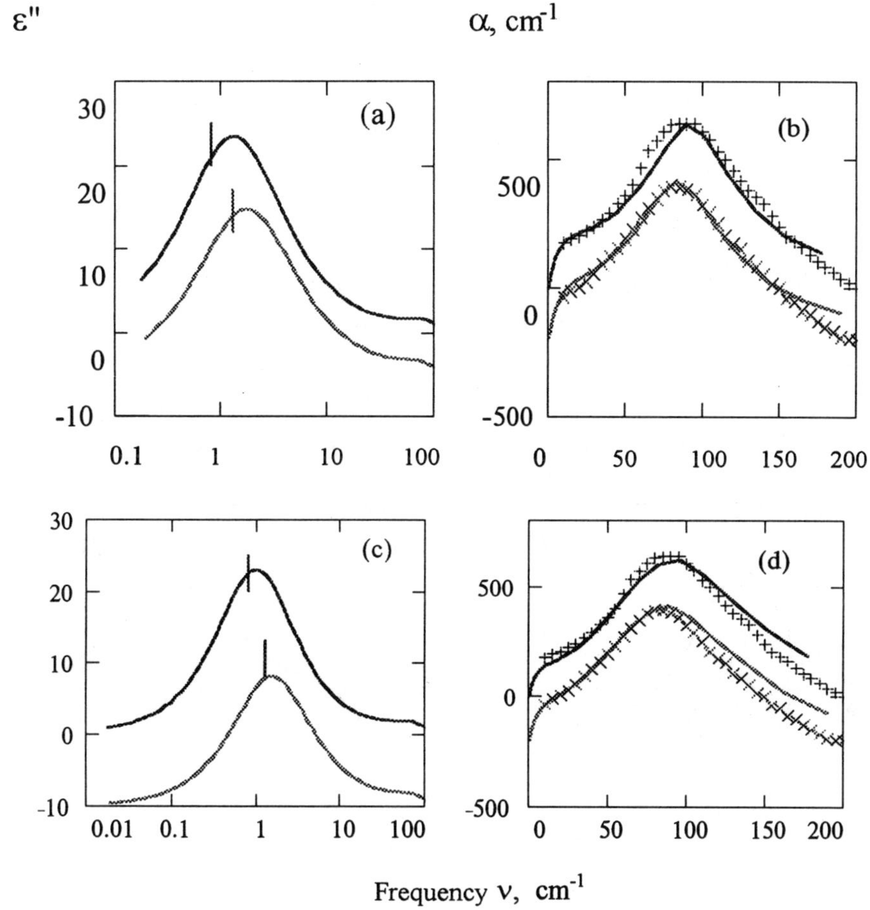

Fig. 14.7 Dielectric loss (a, c) and absorption (b, d) of liquid CH_3F, calculated for the stratified approach (a, b) and for the hybrid model (c, d). The range of low temperatures: $T = 133$ (upper curves) and 153 K (lower curves, shifted downwards). Experimental data (points) from (Gerschel, Dimicoli, Jafffre and Riou 1976): vertical lines refer to the positions of loss maxima and points - to the values of α_L.

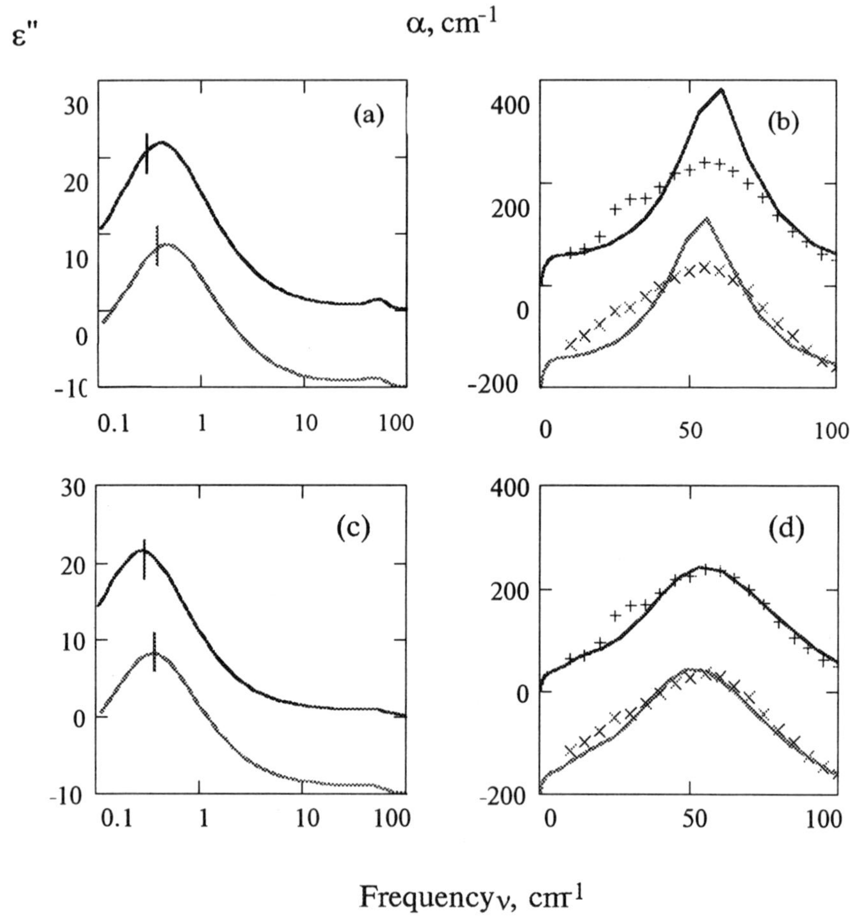

Fig. 14.8 The same, as in Fig. 14.1, but for liquid CHF_3. $T = 117$ and 130 K (from top to bottom).

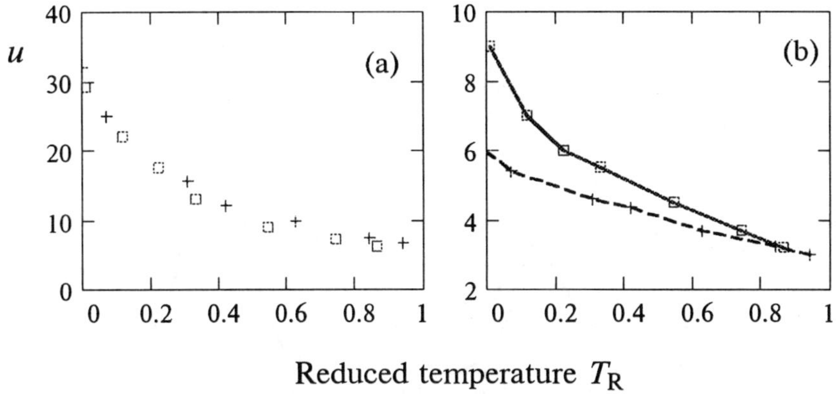

Fig. 14.9 Temperature dependence of the reduced potential welldepth *u*, fitted for the field (a) and hybrid (b) models. The boxes refer to liquid CH_3F and the crosses - to liquid CHF_3.

b. Transition to an isotropic medium is performed by averaging of the *susceptibility* χ^* over Ψ [not by averaging of $K(z, \Psi)$ and $\sigma(\Psi)$, as it was made in the previous sections].

THE CALCULATION SCHEME. We consider motion *in space*. For the correlation-orientation collision model:

$$K(z,\Psi) = K_\perp(z)\sin^2\Psi + K_\parallel(z)\cos^2\Psi; \tag{14.35a}$$

$$\frac{\sigma(\Psi)}{3} = \left\langle q_\perp^2 \right\rangle \sin^2\Psi + \left\langle q_\parallel^2 \right\rangle \cos^2\Psi; \tag{14.35b}$$

$$\chi^*(x) = gGz \int_0^{\pi/2} \frac{K(z,\Psi)}{gx + iK(z,\Psi)\sigma^{-1}(\Psi)} \sin\Psi \, d\Psi. \tag{14.36}$$

One can write for the HM:

$$K_\perp = \overset{\vee}{K}_\perp + \overset{\circ}{K};\; K_\parallel = \overset{\vee}{K}_\parallel + \overset{\circ}{K};\; \overset{\circ}{K} = \overset{\circ}{K}_\perp = \overset{\circ}{K}_\parallel; \tag{14.37}$$

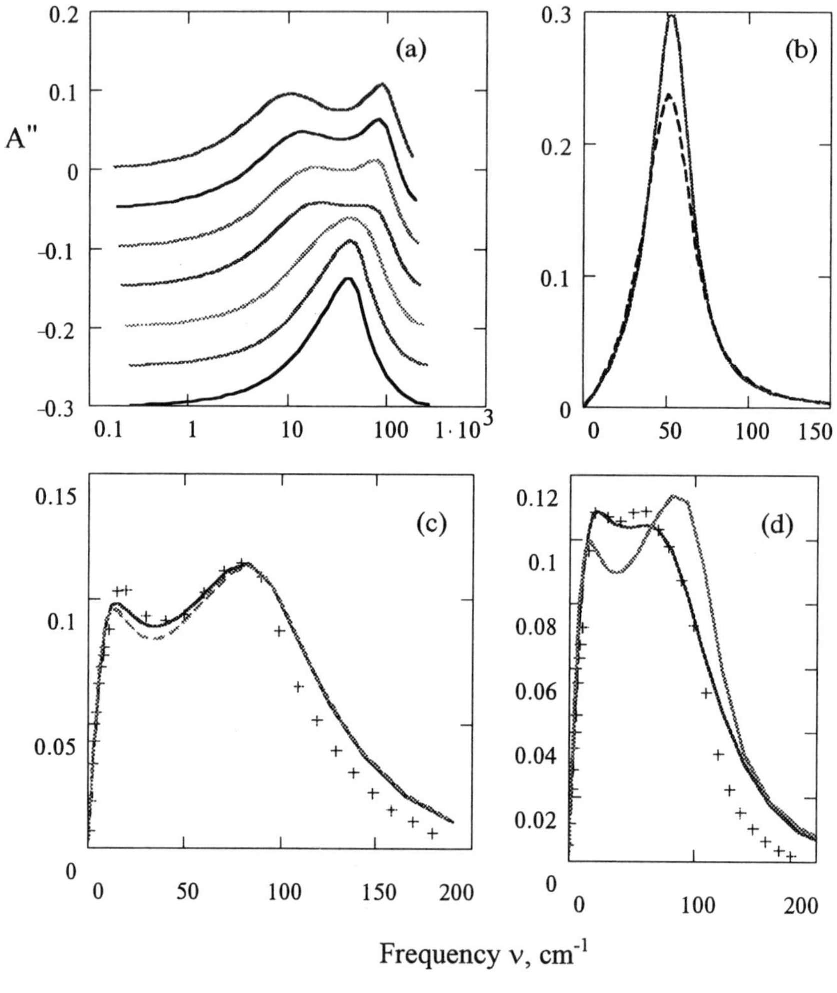

Fig. 14.10 The imaginary part of the polarizability vs. frequency. **a:** the hybrid model; T = 133, 153, 173, 193, 233, 270, and 293 K (from top to bottom, the curves are shifted downwards); **b:** the ED model, T = 153 K (solid curve) and 193 K (dashed curve); **c, d:** the hybrid model (solid curves) and the CR model (dashed curves), T = 153 K (c) and 193 K (d). Experimental data of (Grochulski and Gerschel 1992) are multiplied by an arbitrary coefficient (points).

$$K(z,\Psi) = \overset{\vee}{K}_{\perp}(z)\sin^2\Psi + \overset{\vee}{K}_{\parallel}(z)\cos^2\Psi + \overset{\circ}{K}; \tag{14.38a}$$

$$\sigma(\Psi) = \sigma_{\perp}\sin^2\Psi + \sigma_{\parallel}\cos^2\Psi + \sigma_0; \tag{14.38b}$$

$$\sigma_{\perp} = (1-\sigma_0)\left(1 - \frac{\cos\beta}{2} - \frac{\cos^2\beta}{2}\right); \ \sigma_{\parallel} = (1-\sigma_0)\left(1 + \cos\beta + \cos^2\beta\right). \tag{14.38c}$$

We remark that K in Eq. 14.36 depends also on z. Combining Eq. 14.36 - 14.38, one can see that the idea works. However, the distortion of the $\varepsilon''(\varepsilon')$ diagram seems to be *too pronounced*, and we have no possibility to control it!

One can improve this situation by introducing the *domain factor* γ, with which it is possible to *mix* two ways of anisotropy $\rightarrow$ isotropy transition. Indeed, if $\gamma = 0$, we substitute in the expression for χ^* *fully averaged* (over Ψ) values $\langle K\rangle$ and $\langle\sigma\rangle$. In the another limiting case, just considered ($\gamma = 1$), we apply the scheme (14.36)-(14.38). For an intermediate case, when $0 \le \gamma \le 1$, we write:

$$K(\Psi) = \overset{\vee}{K}_{\perp}\left[\frac{2}{3}(1-\gamma) + \gamma\sin^2\Psi\right] + \overset{\vee}{K}_{\parallel}\left[\frac{1}{3}(1-\gamma) + \gamma\cos^2\Psi\right] + \overset{\circ}{K}, \tag{14.39a}$$

$$\sigma(\Psi) = \sigma_{\perp}\left[\frac{2}{3}(1-\gamma) + \gamma\sin^2\Psi\right] + \sigma_{\parallel}\left[\frac{1}{3}(1-\gamma) + \gamma\cos^2\Psi\right] + \sigma_0. \tag{14.39b}$$

One can see, that the following integrals do not depend on γ:

$$\langle\sigma(\Psi)\rangle \equiv \int_0^{\pi/2}\sigma(\Psi)\sin\Psi\,d\Psi = 1; \ \chi^*(0) = \chi_s = gG,$$

$$\langle K(z,\Psi)\rangle \equiv \int_0^{\pi/2} K(z,\Psi)\sin\Psi\,d\Psi = \frac{2}{3}\overset{\vee}{K}_{\perp}(z) + \frac{1}{3}\overset{\vee}{K}_{\parallel}(z) + \overset{\circ}{K}(z). \tag{14.40}$$

APPLICATION TO LIQUID CH_3F. Taking $T = 153$ K, we demonstrate in Fig. 14.11, that when γ increases:

(a) the Cole-Cole diagram flattens,

(b) the loss $\chi''(x)$ curve widens, its maximum χ''_D decreases,

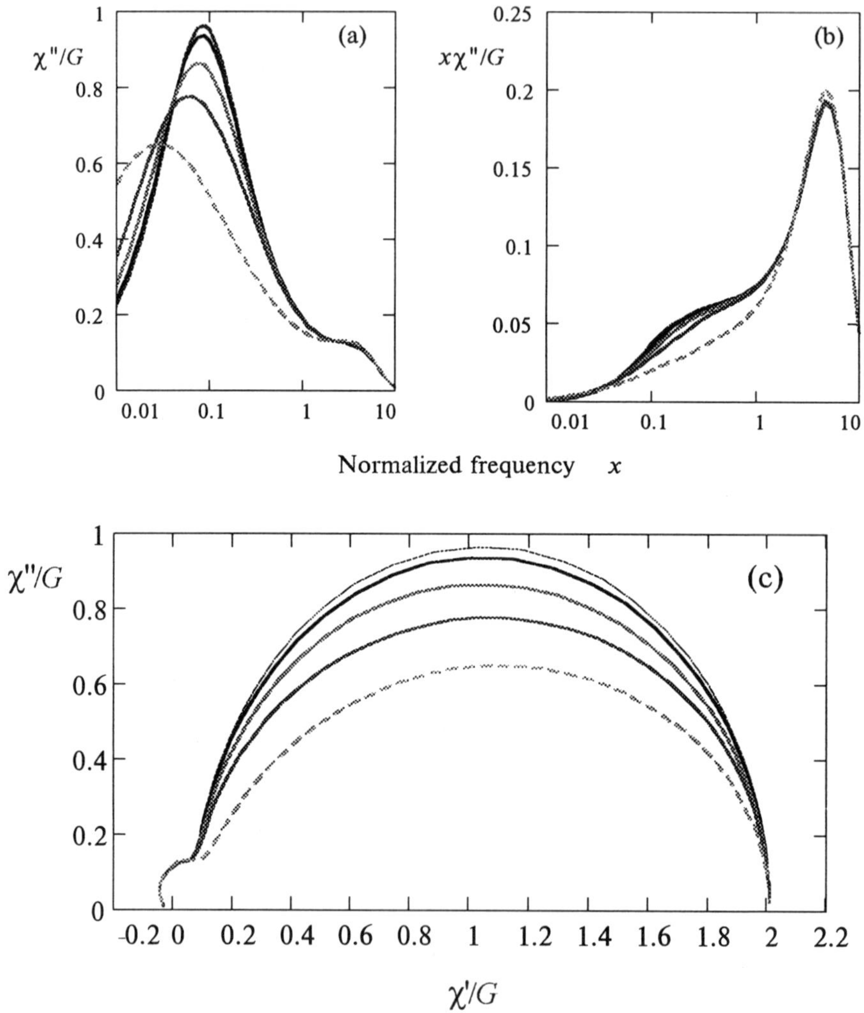

Fig. 14.11 Influence of the domain anisotropy on dielectric spectra; $\gamma = 0$, 0.2, 0.4, 0.6 (solid lines) and 1 (dashed line). The hybrid model; other parameters are taken as for liquid CH_3F at $T = 153\ K$.

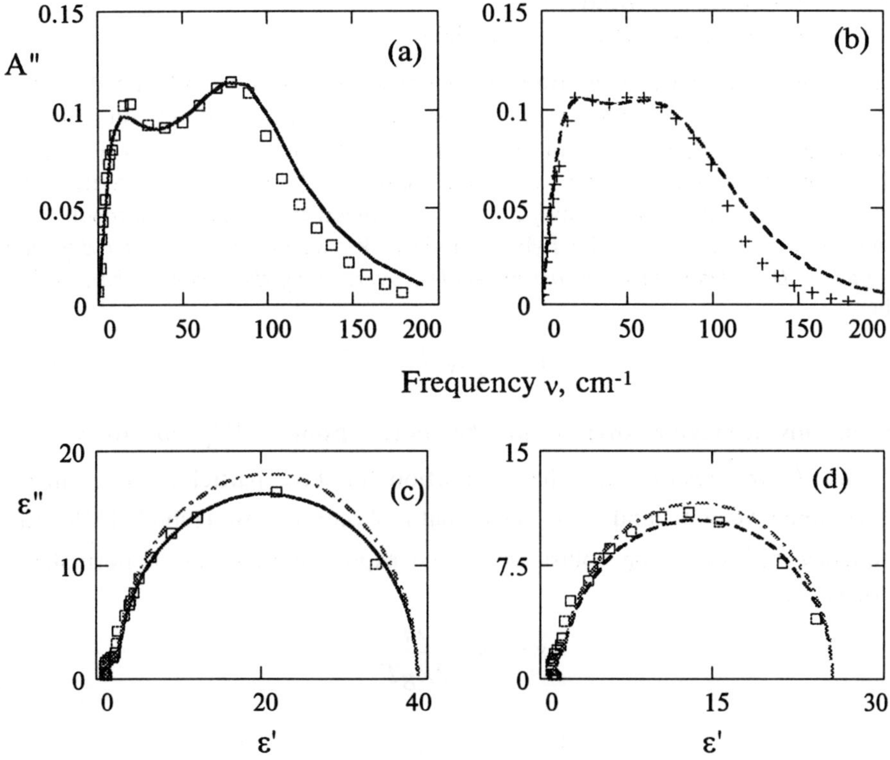

Fig. 14.12 Dielectric spectra of liquid CH_3F, calculated for T = 153 K (solid lines, boxes) and T = 153 K (dashed lines, crosses). The hybrid model, the domain factor γ = 0.4. Dadots in Fig. c, d refer to γ = 0. Experimental data (points) are taken from (Grochulski and Gerschel 1992) and from (Gerschel, Dimicoli, Jaffre, and Riou 1976).

(c) all spectra practically do not alter in the FIR region. From Fig. 14.12 we see that spectra, calculated for γ = 0.4, T = 153 and 193 K, agree with the observed data; a rather flat Cole-Cole diagram looks reasonable.

14.9 Isotropic Potential and the Self-Diffusion Coefficient

We consider translational motion in condensed medium as influenced by the potential U_v. We assume, that the latter does not depend on angular coordinates of a dipole moment μ, but does on the distance between particles. Choosing a simple model of hard spheres (HS), we separate fast ($mv^2/2 > U_v$) and slow ($mv^2/2 < U_v$) molecules (similar approach was elaborated in the case of the hybrid model). *We suggest that the self-diffusion coefficient D is determined by fast molecules, overcoming the potential barrier U_v.* Then:

$$D = \frac{1}{6} r_v \langle l \rangle \langle v \rangle, \tag{14.41}$$

$\langle \bullet \rangle$ means averaging over v in the range from $\sqrt{2U_v / m}$ to ∞; r_v, $\langle v \rangle$ and $\langle l \rangle$ are respectively the proportion of fast particles, their mean translational velocity and their mean path. According to (Liboff 1969) the frequency τ_v^{-1} of strong collisions can be expressed in terms of normalized velocity

$$s = v\sqrt{\frac{m}{2k_B T}}:$$

$$\tau_v^{-1} = \tau_{v0}^{-1}\, \varphi(s);\ \varphi(s) = \frac{1}{2}\left[\exp\left(-s^2\right) + \frac{\sqrt{\pi}}{2}\left(2s + \frac{1}{s}\right)\operatorname{erf}(s)\right]. \tag{14.42a}$$

Here

$$\tau_{v0}^{-1} = 2N\sigma^2\sqrt{\frac{2\pi k_B T}{m}}; \tag{14.42b}$$

σ is diameter of a molecule [not dispersion $\sigma(\Psi)$ as it was designated in the previous section!].

We employ Maxwellian distribution

$$W_M(s) = \frac{4}{\sqrt{\pi}} s^2 \exp\left(-s^2\right). \tag{14.42c}$$

Introducing the normalized potential barrier

$$V \equiv \frac{U_v}{k_B T}, \tag{14.42d}$$

we write integrals over s, exceeding the threshold quantity $\sqrt{V}$, for the proportion $r_v(V)$, averaged velocity $\langle v \rangle$ and averaged path:

$$\langle l \rangle = \langle v(s)\tau_v(s) \rangle. \tag{14.42e}$$

We have:

$$r_v(V) = \frac{4}{\sqrt{\pi}} \int_{\sqrt{V}}^{\infty} s^2 \exp\left(-s^2\right) ds; \langle l \rangle = \sqrt{\frac{2k_B T}{m}} \tau_{v0} \Phi(V); \tag{14.43a}$$

$$\Phi(v) = \frac{1}{\Im} \int_{\sqrt{V}}^{\infty} \frac{s^3 \exp(-s^2) ds}{\varphi(s)}, S(v) = \frac{1}{\Im} \int_{\sqrt{V}}^{\infty} \frac{s^2 \exp(-s^2) ds}{\varphi(s)}; \tag{14.43b}$$

$$\langle v \rangle = \sqrt{\frac{2k_B T}{m}} S(V); \quad \Im = \int_{\sqrt{V}}^{\infty} s^2 \exp(-s^2) ds. \tag{14.43c}$$

Note: averaging over s *in the absence of a barrier* (i.e. at $V = 0$) yields the well known expression for the lifetime τ_{HS} of hard spheres:

$$\tau_{HS} = \frac{1}{\left\langle \tau_v^{-1}(s) \right\rangle} = \frac{1}{N\sigma^2} \sqrt{\frac{m}{16\pi k_B T}}. \tag{14.44}$$

Consequently, the self-diffusion coefficient is given by

$$D(T) = \frac{1}{6} \sqrt{\frac{k_B T M}{2\pi m_H}} \frac{1}{N_A \rho \sigma^2} \sqrt{\frac{m}{16\pi k_B T}} r_v(V) \Phi(V) S(V). \tag{14.45a}$$

If D, ρ, σ, and T are expressed, respectively. in cm^2/s, $g \bullet cm^{-3}$, $\mathring{A}$, and K, then

$$D(T) \approx 10^{-5} \frac{\sqrt{TM}}{\rho\sigma^2} r_v(V) \Phi(V) S(V). \tag{14.45b}$$

THE PRESCRIBED $U_v(T)$ DEPENDENCE. Introducing constant T_0 as a free-model parameter, we assume, that U_v depends linearly on T:

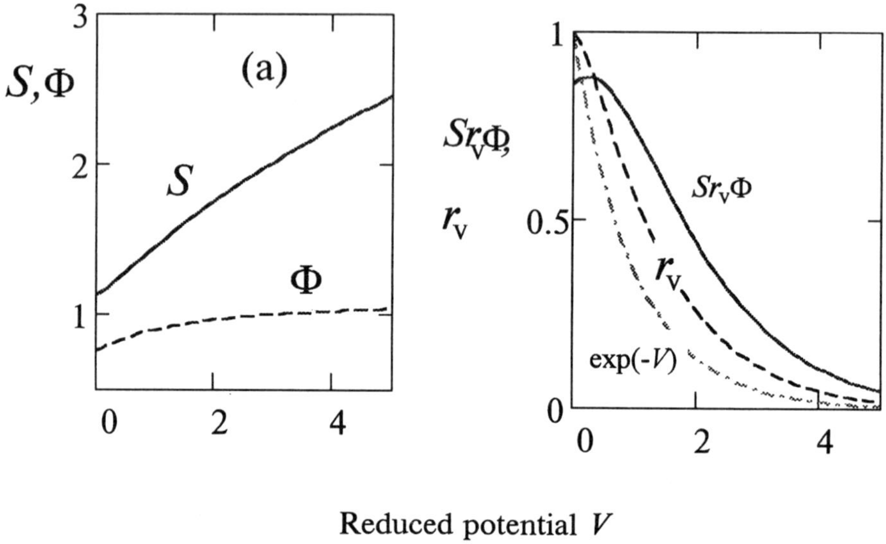

Fig. 14.13 Dependencies vs. normalized barrier height V of reduced parameters. (a): of the translational velocity S (solid line) and of the path Φ (dash); (b): of the self-diffusion coefficient $Sr_V\Phi$ (solid line), of the proportion r_V (dash) and of exp($-V$) (dadot).

$$U_V(T) = k_B T_0\left(1 - \frac{T}{T_{cr}}\right);\ V(T) = \frac{U_V(T)}{k_B T} = \frac{T_0}{T} - \frac{T_0}{T_c}. \qquad (14.46)$$

THE SELF-DIFFUSION IN NONASSOCIATED LIQUIDS. We calculate $D(T)$ for liquid CH_3F and CHF_3 from Eq. 14.45-14.46. Experimental values of $\rho(T)$, $D(T)$, fitted T_0 values and molecular constants σ and M are given in Tables 14.2 and 14.7. The dependence on V of mean normalized quantities (velocity S, path Φ and proportion r_V) is pictured in Fig. 14.13. S and Φ rise with the increase of T, while r_V decreases. The resulting dependence of the self-diffusion D on V is similar to the $r_V(V)$ function. We note that D and r_V decrease with the rise of V more slowly, than exp($-V$). The calculated $D(T)$ dependence (14.46) agrees with the observed one (see Fig. 2.10 a, b).

Table 14.7 Experimental values of $D(T)$ and fitted T_0-values

CH_3F	T, K	133	153	173	193	233	270	293
	$D \cdot 10^5$	1.1	2.1	3.4	5.1	9.3	13.6	16.3
CHF_3	T, K	117	130	173	193	231	270	288
	$D \cdot 10^5$	0.27	0.58	2.10	3.20	6.0	9.5	11.3
for CH_3F		$T_0 = 748.4$, $\sigma = 3.54$, $M = 34$						
for CHF_3		$T_0 = 849.7$, $\sigma = 3.95$, $M = 70$						

THE V(T) DEPENDENCE IN NONASSOCIATED LIQUIDS. The predicted reduced potentials V are pictured in Fig. 2.10 c, d. The following features are observed:

(i) V decreases with T similarly to $u(T)$ values, fitted on the basis of observed spectroscopic data;

(ii) V is far less / decreases faster, than u, with T;

(iii) the ration V/u varies ca. linearly with T.

We conclude: in such liquids it is "easier" for a dipole to perform translational motion than molecular rotation.

CHAPTER 15
LIQUID WATER AND AQUEOUS SYSTEMS

15.1 Liquid Water

In this section we employ the same formulas, that were mentioned in Table 14.1; in most cases we use the hybrid model with the isothermal induced distribution F_T .

15.1.1 The Experimental Data

THE LIEBE-HUFFORD-MANABE (1991) FORMULA FOR THE COMPLEX PERMITTIVITY. This formula contains two Debye and two Lorentz terms as functions of frequency $f = \omega/(2\pi)$ [Hz]:

$$\varepsilon^*(f) = \varepsilon_s - \left(\frac{\varepsilon_s - \varepsilon_{\infty 1}}{f + i\gamma_1} + \frac{\varepsilon_{\infty 1} - \varepsilon_{\infty 2}}{f + i\gamma_2}\right) f + \sum_{j=1,2} \left(\frac{A_j}{f_j^2 - f^2 - i\Gamma_j f} - \frac{A_j}{f_j^2}\right),$$

$$\theta = 1 - \frac{300}{T\,[\mathrm{K}]} = 1 - \frac{300}{273.15 + T^\circ C]};\ \varepsilon_s = 77.16 - 103.3\theta; + 7.52\theta$$

$$\varepsilon_{\infty 1} = 0.0671\varepsilon_s;\ \varepsilon_{\infty 2} = 3.52;$$

$$\gamma_1 = \left(20.2 + 146.4\theta + 316\theta^2\right) \bullet 10^9;\ \gamma_2 = 39.8\gamma_1$$

$$\begin{Bmatrix} f_1 \\ f_2 \end{Bmatrix} = \begin{Bmatrix} 5.11 \bullet 10^{12} \\ 18.2 \bullet 10^{12} \end{Bmatrix};\ \begin{Bmatrix} \Gamma_1 \\ \Gamma_2 \end{Bmatrix} = \begin{Bmatrix} 4.46 \bullet 10^{12} \\ 15.4 \bullet 10^{12} \end{Bmatrix};\ \begin{Bmatrix} A_1 \\ A_2 \end{Bmatrix} = \begin{Bmatrix} 25.03 \bullet 10^{24} \\ 282.4 \bullet 10^{24} \end{Bmatrix}; \qquad (15.1)$$

Table 15.1 Real (n) and imaginary (κ) parts of the refraction index n^*. Liquid water at T = 1 °C and 50 °C.

ν,	1°C		50°C		ν,	1°C		50°C	
cm^{-1}	n	κ	n	κ	cm^{-1}	n	κ	n	κ
400	1.54	0.35	1.52	0.39	600	1.40	0.42	1.32	0.42
410	1.54	0.36	1.51	0.40	610	1.39	0.42	1.31	0.42
420	1.54	0.36	1.51	0.40	620	1.37	0.42	1.29	0.41
430	1.54	0.37	1.50	0.41	630	1.36	0.42	1.28	0.41
440	1.53	0.37	1.49	0.41	640	1.35	0.42	1.27	0.40
450	1.53	0.38	1.48	0.41	650	1.34	0.41	1.26	0.39
460	1.53	0.38	1.48	0.42	660	1.33	0.41	1.25	0.38
470	1.52	0.39	1.47	0.42	670	1.31	0.41	1.24	0.38
480	1.51	0.39	1.46	0.42	680	1.30	0.40	1.23	0.37
490	1.51	0.39	1.45	0.42	690	1.28	0.40	1.21	0.36
500	1.50	0.40	1.44	0.43	700	1.27	0.39	1.20	0.35
510	1.49	0.40	1.43	0.43	710	1.26	0.39	1.19	0.34
520	1.48	0.41	1.42	0.43	720	1.24	0.38	1.18	0.33
530	1.47	0.41	1.41	0.43	730	1.23	0.38	1.17	0.32
540	1.47	0.41	1.40	0.43	740	1.21	0.37	1.16	0.31
550	1.46	0.41	1.39	0.43	750	1.20	0.36	1.15	0.30
560	1.45	0.42	1.37	0.43	760	1.19	0.35	1.14	0.29
570	1.43	0.42	1.36	0.43	770	1.17	0.34	1.13	0.27
580	1.42	0.42	1.35	0.43	780	1.16	0.32	1.12	0.26
590	1.41	0.42	1.34	0.43	790	1.15	0.31	1.12	0.24
					800	1.14	0.29	1.12	0.22
					810	1.14	0.27	1.12	0.21
					820	1.14	0.26	1.12	0.19

Table 15.2 **Real** (n), imaginary (κ) parts of the refraction index n^* and absorption α [cm^{-1}]. Liquid water. $T = 27°C$.

ν	n	κ	α	ν	n	κ	α
10	2.600	1.090	137	310	1.543	0.357	1401
20	2.225	0.718	192	320	1.546	0.353	1432
30	2.150	0.527	210	330	1.550	0.352	1472
40	2.110	0.460	240	340	1.552	0.356	1532
50	2.070	0.438	290	350	1.552	0.359	1593
60	2.040	0.444	360	360	1.552	0.363	1658
70	2.020	0.450	429	370	1.549	0.368	1724
80	2.010	0.466	509	380	1.545	0.372	1793
90	2.000	0.487	594	390	1.541	0.377	1862
100	1.997	0.507	678	400	1.537	0.382	1933
110	1.982	0.532	773	410	1.532	0.386	2004
120	1.960	0.557	872	420	1.527	0.390	2072
130	1.929	0.577	967	430	1.521	0.394	2143
140	1.890	0.593	1065	440	1.515	0.397	2210
150	1.848	0.608	1165	450	1.510	0.401	2280
160	1.801	0.622	1266	460	1.504	0.404	2347
170	1.746	0.629	1358	470	1.496	0.408	2423
180	1.689	0.618	1412	480	1.488	0.411	2494
190	1.640	0.597	1437	490	1.480	0.415	2565
200	1.600	0.571	1445	500	1.470	0.418	2638
210	1.657	0.539	1434	510	1.462	0.421	2709
220	1.542	0.505	1407	520	1.451	0.423	2779
230	1.528	0.469	1364	530	1.441	0.425	2842
240	1.525	0.436	1323	540	1.431	0.426	2903
250	1.529	0.414	1310	550	1.419	0.427	2964
260	1.532	0.398	1311	560	1.407	0.427	3022
270	1.534	0.385	1317	570	1.396	0.428	3077
280	1.537	0.375	1331	580	1.385	0.427	3126
290	1.539	0.368	1351	590	1.372	0.425	3167
300	1.541	0.361	1374	600	1.361	0.423	3203

Continuation of Table 15.2

ν	n	κ	α	ν	n	κ	α
610	1.348	0.420	3234	810	1.130	0.243	2467
620	1.335	0.417	3259	820	1.130	0.226	2309
630	1.324	0.412	3276	830	1.132	0.208	2143
640	1.313	0.408	3291	840	1.131	0.192	1987
650	1.303	0.403	3301	850	1.132	0.176	1833
660	1.289	0.397	3307	860	1.132	0.159	1692
670	1.277	0.392	3308	870	1.135	0.144	1533
680	1.264	0.386	3307	880	1.139	0.130	1396
690	1.249	0.379	3298	890	1.143	0.118	1270
700	1.236	0.373	3287	900	1.149	0.107	1165
710	1.223	0.365	3263	910	1.156	0.0973	1064
720	1.213	0.356	3231	920	1.162	0.0898	993
730	1.201	0.347	3192	930	1.168	0.0828	927
740	1.189	0.338	3150	940	1.174	0.0764	866
750	1.182	0.328	3100	950	1.181	0.0707	817
760	1.171	0.317	3040	960	1.189	0.0661	770
770	1.157	0.305	2969	970	1.194	0.0622	733
780	1.142	0.292	2883	980	1.202	0.0589	702
790	1.138	0.277	2760	990	1.208	0.0557	673
800	1.134	0.260	2618	1000	1.214	0.0534	651

Since the Lorentz-line parameters Γ_j, A_j are temperature independent, one may suppose that the FIR spectra are described only *qualitatively* in this approximation.

In Tables 15.1 and 15.2 we present respectively data of (Pincley, Setna, Williams, 1975) and (Downing and Williams, 1975), concerning FIR spectra. In Table 15.3 we summarize characteristic features of the observed spectra, that should be accounted for theoretically.

Table 15.3 Liquid water. Specific features of dielectric spectra and molecular constants.

Parameters	T°C		
	1	27	50
ε_s	87.4	77.6	70.3
$\tau_{D1} \equiv \tau_D$, ps	17.3	7.85	4.93
τ_{D2}, ps	0.494	0.197	0.124
$\varepsilon_{\infty 1} \equiv \varepsilon_\infty$	5.65	5.21	4.71
$\varepsilon_{\infty 2}$	2.81	3.52	4.06
$\nu_{D1} \equiv \nu_D$	0.307	0.676	1.077
ν_{D2}	12.2	26.9	42.9
ν_R	190	190-200	
α_R	849	1445	
α_{min}		1311	
$\nu(\alpha_{min})$		260	
$\Delta\nu_L$	710	670	630
ρ, g•cm^{-3}	1.0	0.9986	0.9881
Molecular constants	$I = 1.483 \bullet 10^{-40}$ g•cm^2, $n_\infty^2 = 1.7$, $M = 18$, $\mu_0 = 1.84$ D		

Note: here indexes 1 and 2 refer to the two Debye regions in water in terms of formula (15.1); $\Delta\nu_L$ means a rough estimate of the librational bandwidth near the ν_L peak.

15.1.2 Dielectric Spectra

In the hybrid model (HM) four fitting constants are used: halfwidth β of rectangular well; lifetime τ of reorienting molecules, normalized welldepth $u = U_0/(k_B T)$ and the correction factor k_μ. We shall try to fit:

(a) the libration absorption peak α_L and its frequency ν_L;

(b) the low-frequency (Debye) spectrum, that is characterized by the static permittivity ε_S, relaxation time τ_D, permittivity ε_∞ at "infinitely" high frequency;

(c) the two-humped absorption spectrum.

Consequently, we have only 4 parameters

$$\tau, \beta, u, \text{ and } k_\mu \tag{15.2}$$

for fitting ε_S, τ_D , ε_∞, ν_L, α_L, and two-humped absorption $\alpha(\nu)$ spectrum.

1st step. We estimate β from the resonance condition (14.9a).

2nd step. We estimate the potential u by supposing that the contributions of librators and rotators to the absorption $\alpha(\nu)$ should be commensurable near the frequency $\nu_R \approx 200$ cm^{-1}.

3d step. We calculate the lifetime τ from the Debye-like Eq. 14.31, where we put $C_D = 1$. Then

$$\tau(T) = \frac{1}{2}\beta^2(T)\tau_D(T). \tag{15.3}$$

4d step. We estimate the k_μ parameter from the condition

$$\max\{\alpha(\nu)\} = \alpha_L. \tag{15.4}$$

5th step. We correct u (if need) in order to achieve the agreement between the experimental and estimated relaxation time τ_D; Eq. 14.4 is applied.

After these five steps we repeat the fitting procedure in order to find an optimal set of parameters (15.2).

THE RESULTS OF CALCULATION. In Table 15.4 are presented the fitted HM-parameters and some related quantities. The calculated from Eq. 5.1 and observed spectra of water agree in general features (see also Sec. 2.4.2 and Fig. 2.11, 2.12). Consequently,

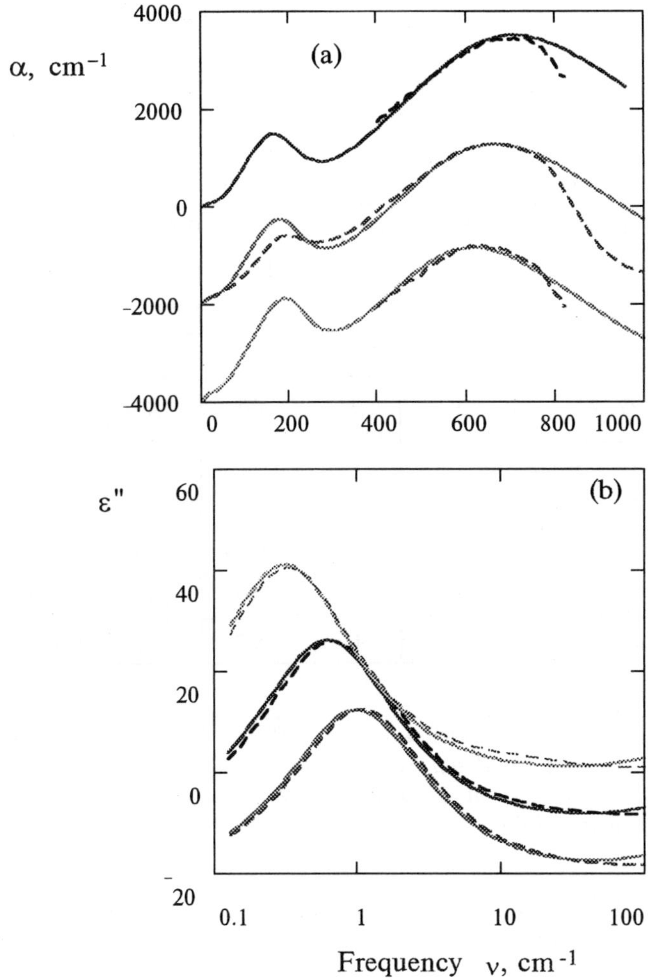

Fig. 15.1 Absorption (a) and loss (b) of liquid water. The hybrid model, $T = 1$, 27, and $50^{\circ}C$, solid lines, curves are shifted downwards (from top to bottom). Dashed lines - results of measurements, supplied by Tables 15.1, 15.2 (a) or calculated with use of the LHM approximation (b).

Table 15.4 Parameters of the hybrid model for liquid water (the isothermal distribution F_T). $\sigma = 2.9$ Å. The *l*-parameter will be introduced below/

T,°C	β, deg			u	k_μ	τ, ps
	from Eq. 14.9a	fitted		from Eq. 15.8		
1	18.6	18.0	5.6	20.8	0.85	1.16
27	20.6	19.9	5.7	21.1	0.48	1.19
50	22.7	22.1	4.6	22.1	0.37	1.10

T,°C	r, %		y	y_c	m_L	g	G	l, Å
	from (15.7)	from GNB						
1	7.1	3.8	0.05	1.6	16.0	2.24	2.1	0.21
27	9.3	7.3	0.09	1.5	8.2	2.35	1.7	0.21
50	12.2	11.5	0.11	1.3	5.3	2.38	1.6	0.22

The observed wideband [0 - 1000 cm^{-1}] dielectric spectra of liquid water are stipulated by reorientation of polar H 0 molecules and can be described in terms of the hat-like rectangular intermolecular potential $U(\vartheta)$.

Since the HM is applicable also to nonassociated liquids (see Chap. 14), the question arises, what is the *principal* difference between these two types of fluids.

We can find the answer, if estimate the number $m_L = \tau / \langle T_L \rangle$ of librations, performed by a dipole during its lifetime τ. We employ Eq. 9.30 for $j = 2$ and Eq. 9.31, from where we find:

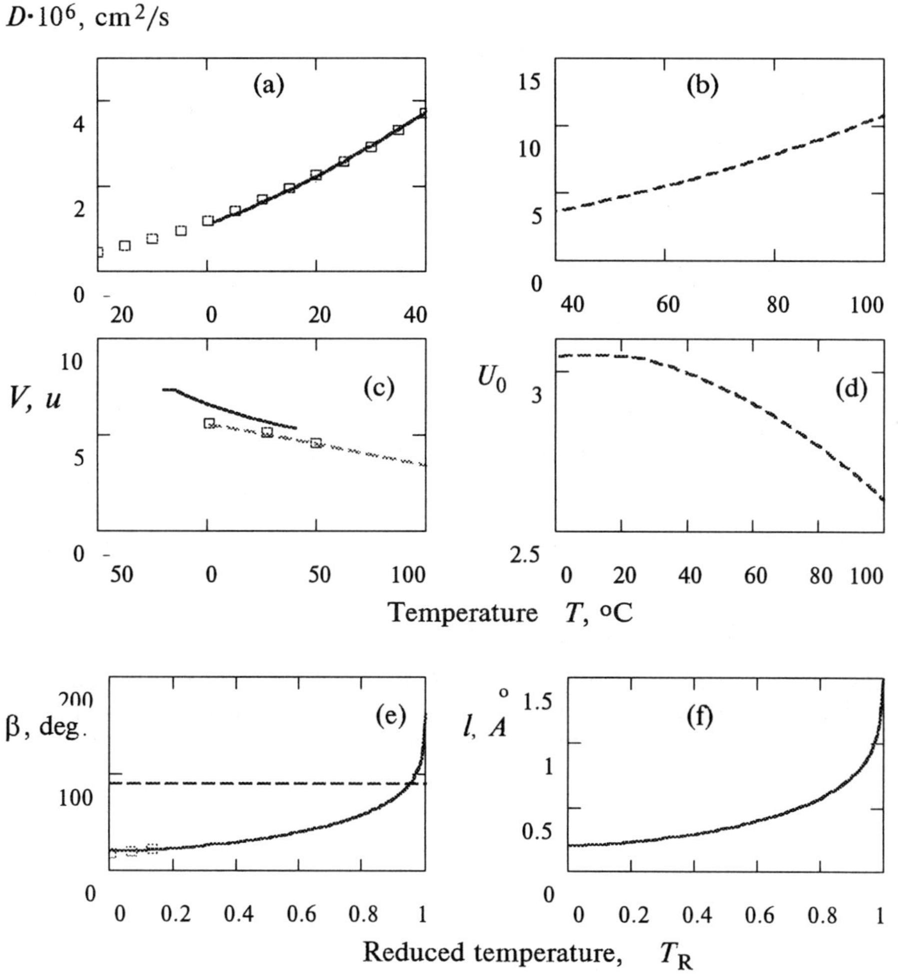

Fig. 15.2 Liquid water, temperature dependencies. **(a):** of the self-diffusion coefficient D, solid line - calculated values, dots - experimental data; **(b):** the predicted $D(T)$ dependence; **(c, d):** of isotropic (c, solid line) and of anisotropic (c, d, dashed lines) components of an intermolecular potential; dots and horizontal line correspond respectively to the fitted values and to the phase-transition condition.**(e, f):** of the libration amplitude β **(e)** and of the path (f).

$$m_L \approx \frac{\tau\langle p\rangle}{2\pi\eta} = \frac{\tau\, y_c}{2\eta} = \frac{y_c}{2y}. \tag{15.5}$$

It seen from Table 15.4 that m_L is about 10 in water, while m_L is about 1 or less in liquid CH_3F. This is an independent evidence, that *some structure (local order) exists in liquid water.* Its structure should be rather rigid since in the "usual" temperature range β weakly depends on T (see Table 15.4).

The proportion $\mathring{r}$ of rotators (for simplicity we denote it r) characterizes sooner the T-stipulated degradation of structure, than the latter itself. When T rises from 1 to 50°C, the "ice" component of water slightly "melts" and its "vapor" component increases from ca. 6.9 to 11.4%.

DRAWBACKS OF THE HM. **(1)** Too slow transition of absorption $\alpha(\nu)$ to transparency in the FIR range (cf. solid and dashed curves in Fig.15.1a). **(2)** The theoretical minimum of the $\alpha(\nu)$ curve, that appears between 200 and 300 cm^1, is evidently deeper than that, observed experimentally (cf. two middle curves in Fig. 15.3a, shown for T = 27°C). We could probably to improve theory, if would elaborate it for a more adequate potential $U(\vartheta)$. One such step is made in Sec. 15.1.6.

15.1.3 Translations as a Potential-Controlled Process

SELF-DIFFUSION AS A PROCESS OF RANDOM JUMPS. Analogously to the case of nonassociated liquids, Eq. 14.41, we suggest, that the self-diffusion coefficient D is contributed mainly by R-molecules, that have larger mobility than librators:

$$D = \frac{r}{6}\lim_{\Delta t\to 0}\left\langle\frac{\Delta l}{\Delta t}\right\rangle \approx \frac{r}{6\tau}\langle\Delta l\rangle, \tag{15.6}$$

$$r = \frac{\exp(-u)}{1-[1-\exp(-u)]\cos\beta};\ \tau = \frac{\beta^2\tau_D}{2};\ \tau_D = \frac{1}{2\pi\gamma_1}. \tag{15.7}$$

We regard the free constant $\langle\Delta l\rangle$ to be independent on T, since the water density ρ practically do not alter for [0, 50°C]. The relaxation time τ_D is calculated in the LHM approximation, Eq. 15.1. Using the data, given in Table 15.4, we apply the linearly interpolated $\beta(T)$ and $u(T)$ dependencies.

The observed and calculated $D(T)$ values agree well (see Fig. 15.2a and

Table 15.5). The fitted $\langle \Delta l \rangle$, being about 1 $\mathring{A}$, is commensurable with the radius of H_2O molecule. We may also *predict* $D(T)$ for elevated temperatures, say, in the interval [50, 100°C], see Fig. 15.2b and Table 15.5.

THE INTERMOLECULAR POTENTIAL V. Now we shall apply the theory, elaborated in Sec. 14.9 in order to find $V(T)$ from the condition, that the calculated $D[V(T)]$, Eq. 14.45, should coincide with the experimental or with $D(T)$, predicted above. The $V(T)$ function, calculated for water, is close (see Fig. 15.2c) to the $u(T)$ dependence, that was fitted for the HM. Moreover, we see, that $V > u$, while in nonassociated liquids $V << u$. Thus,

The isotropic component V of the intermolecular potential is probably not less in water, than the anisotropic component u.

We can interpret this result as follows. Translations / self-diffusion occur as the *sequence of small turns* under effect of strong H-bonds. Only a small part ($r < 12$ %) of thermally excited H_2O molecules participates in this motion. Consequently, the «isotropic» V-potential *as if originates from the anisotropic component u.* We shall prove in the next subsection that the rotational cell model is inapplicable to water for T < 50° C when, as we may suggest, the amplitude β has the «structural» origin. Small D and high V values support this viewpoint for water, since D is much larger and V is much lower in liquid CH_3F. Taking account that 1 Erg = $1.4396 \bullet 10^{13}$ kcal/mole, we find:

T, °C	1	27	50
u	5.6	5.1	4.6
U_0, kcal/mole	3.05	3.04	3.0

Table 15.5 Liquid water. The self-diffusion coefficient D in [cm^2/s].

T,°C	-20	-15	-10	-5
$D \bullet 10^6$	0.45	0.60	0.76	0.96

Observed or calculated values of *D(T)*										
T,°C	0	5	10	15	20	25	30	35	40	
$D\bullet10^6$	1.2	1.4	1.6	1.9	2.2	2.6	3.0	3.3	3.7	obs.
$D\bullet10^6$	1.2	1.4	1.7	2.0	2.3	2.6	3.0	3.4	3.8	cal.
Predicted values of *D(T)*										
T,°C	45	50	55	60	65	70	75	80	85	90
$D\bullet10^6$	4.2	4.7	5.1	5.7	6.2	6.8	6.4	8.0	8.7	9.4
Notes	The observed data from (Zatsepina 1997, p.155); $\langle\Delta l\rangle = 0.916$ Å									

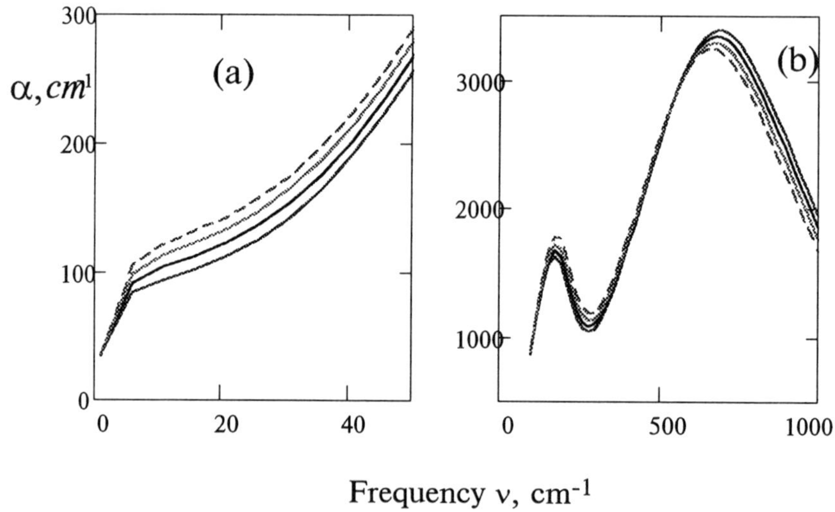

Fig. 15.3 Absorption spectra of liquid water, predicted in submillimeter (a) and FIR (b) regions for T °C = 10, 15, 20 (solid lines) and 25 (dash). Hybrid model with the isothermal induced distribution F_T involved.

The fitted U_0-values are commensurable with the H-bond-energy (ca. 5 kcal/mole); in a wide range [0, 50 C] U_0 is almost independent on T. Again this means, that the water structure is as if rigid. We remark, that for $T <$ 50°C, we may take approximately

$$\rho(T)\ [\mathrm{g{\bullet}cm^{-3}}] \approx 0.999585 + 0.465125{\bullet}10^{-4}\, T - 5.67572{\bullet}10^{-6}\, T^2,$$

while for $T >$ 50°C it is preferable to employ the tabulated experimental $\rho(T)$ dependence. Our estimates predict the faster decrease of V with T in this case (see Fig.15.2 d).

THE ROTATIONAL CELL MODEL OF LIQUID WATER. Contrary to Eq. 14.16 we replace γ by 1, since then σ approaches the diameter of H_2O molecule (ca. 3 $\mathring{A}$). Thus,

$$\beta(T) = \frac{\pi}{8}\sqrt{\frac{m_H M}{I}}\left(\sqrt[3]{\frac{M}{N_A\rho}} - \sigma\right);\ l(T) = \sqrt[3]{\frac{M}{N_A\rho}} - \sigma\,. \qquad (15.8)$$

We fit here σ by solving equation

$$\beta(T = 323\ \mathrm{K}) = 22.1°,$$

the last figure (22.1°) was fitted in the HM (see Table 15.4). Then we find that the rotational cell model is inapplicable for water at $T <$ 50°C, since Eq. 15.8 yields *lower* β values, than those, fitted in Table 15.4. Alteration of density ρ with T becomes more significant for higher temperatures, for which the cell model works! From Fig.15.2e we see that β increases rapidly, when T approaches T_{cr}. The criterion (14.25) permits to estimate the critical temperature: $T_{cr}^{est} \approx$ 628 K, the error Δ is ca. 3%, since for water $T_{cr} \approx 647.15$°C.

The path $l(T)$ varies with T similar to $\beta(T)$, see Fig.15.2 e, f. If $T \in$ [0, 50°C], $l \approx$ 0.2 $\mathring{A}$ (see Table 15.4); if $T \to T_{cr}$, the path l becomes commensurable with radius $\sigma/2$ of H_2O molecule.

15.1.4 Prediction of Wideband Spectra

We may estimate the free parameters of the HM:

(a) U_0 and k_μ may be set constant, $U_0 \approx$ 3.04 kcal/mole, $k_\mu \approx$ 1.18.

(b) β can be found by interpolation of values presented in Table 15.4;

(c) τ_D and τ can be calculated from Eq. 15.7.

Taking $T = 15, 20, 25, 30$°C we show in Fig. 15.3a the transition from microwave to submillimeter spectra, where α increases with the rise of T. Simple Debye formula (6.35) (see also Table 6.2) becomes invalid in this frequency range, but the quasi-resonance absorption is still not pronounced. Some sort of the Debye plateau is followed by markedly "over-Debye" region.

In Fig. 15.3b we show the transition from the translational to the librational absorption peak. The frequency ν_R of the first peak is almost independent on T, while the frequency ν_L of the second one decreases with T. The peak intensities $\alpha_R(T)$ and $\alpha_L(T)$ alter in opposite directions: α_R substantially increases and α_L slightly decreases with the rise of T.

15.1.5 The Autocorrelation Function

We shall apply Eq. 6.79a for the orientational ACF $\Phi(t)$ and for its normalized second derivative $\Phi_{rv}(t)$ by integrating the imaginary part $\chi''(x)$ of the susceptibility χ^* over normalized frequency x. We follow (Gaiduk, Novskova, and Brekhovskikh 1993) work, that is referred further as GNB. The two fractional confined rotator / extended diffusion model was applied in GNB for water at $T = 27$°C, where χ^* was calculated from the sum

$$\chi^*(x) = gGz\left(\frac{(1-r)L_L(z)}{gx + iyL_L(z)} + \frac{rL_R(z)}{gx + iyL_R(z)}\right), \qquad (15.9)$$

where the first term in (•) is obliged to librators and the second one - to free rotators. CR and ED models were employed respectively for calculation of $L_L(z)$ and $L_R(z)$. The proportion r is the free parameter of this CR/ED model; the lifetime τ was taken the same in each fraction. In GNB the following set of parameters was employed:

$$r = 7.3\%,\ \beta = 19.9°,\ \tau = 0.15 \text{ ps},\ k_\mu = 1.24. \qquad (15.10)$$

The comparison with Table 15.4 shows, that these parameters and calculated spectra are close to those, found for the HM.

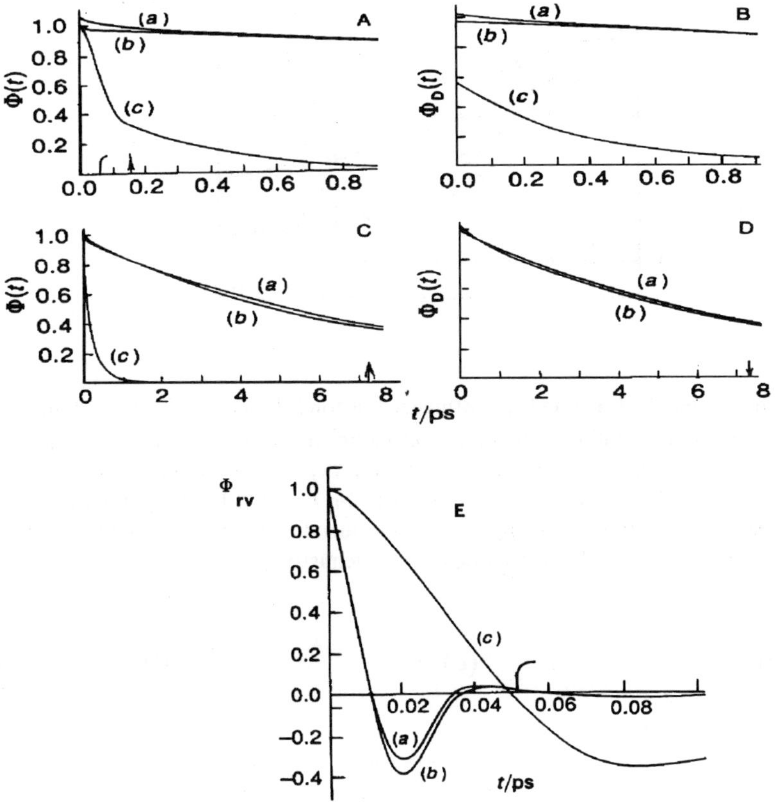

Fig. 15.4 Orientational ACF (A, C), simplified ACF (B, D), corresponding to the Debye process; second derivative of orientational ACF (E). Other symbols are explained in the text. Reproduced from GNB by permission of the Royal Society of Chemistry.

(a)

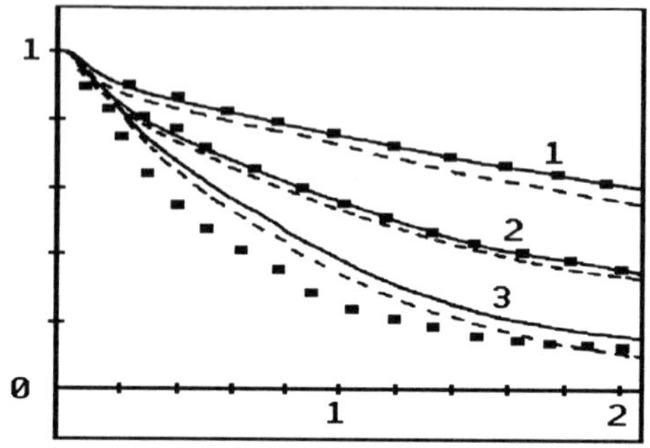

Fig. 15.5 Orientational ACF (a) and its normalized second derivative (b, c, d). Liquid CH_3F. Calculation for the confined rotator model (solid lines) and experimental values from (Gerschel, Dimicoli, Jaffre, and Riou, 1976) (dots). **(a):** at T = 133 (1), 193 (2) and 270 K (3); **(b):** for T = 133 K, **(c):** for T = 193 K, **(d):** for T = 270 K. Dashed lines - the single particle ACF, that corresponds to g = 1. Reproduced from (Brekhovskikh, 1993) and from GNB by permission of the Royal Society of Chemistry.

(b) (c) (d)

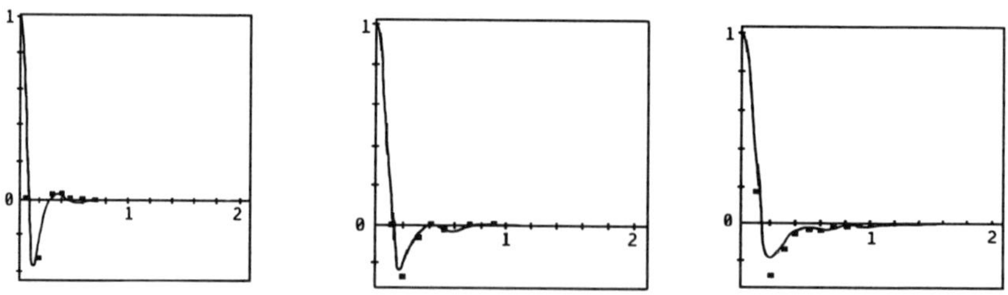

Fig. 15.5 **(b, c, d):** The capture is given above.

The ACFs spectra, calculated in GNB, are reproduced in Fig. 15.4 for short (A) and longer (C) time t intervals. In this Figure three marks are employed which are related to dynamics of polar molecules: the Debye time τ_D (shown by arrow ↓), the lifetime τ (arrow ↑), the mean half-period $\langle T_L \rangle$ (the symbol ⌠). The last quantity is estimated as

$$\langle T_L/2 \rangle \approx \int_0^\infty \frac{2\beta\eta}{p} \exp\left(-p^2\right) dp^2 = 2\sqrt{\pi}\,\beta\eta.$$

Orientational ACF $\Phi(t)$ is *monotonically decreasing,* while $\Phi_{IV}(t)$ tends to zero after *few oscillations,* see curves (a). By curves (b) we show some conditional ACF, obliged to librators. This ACF is obtained, if we put $r = 1$ in Eq. 15.9; in this case all molecules rotate in the rectangular well, so that the translational absorption peak disappears. Analogously, by curves (c) we show ACF of *rotators,* that is obtained from Eq. 15.9 for $r = 0$. All molecules perform in this case free rotation between strong collisions; the librational absorption peak α_L disappears, while the peak α_R , placed near $\nu_R \approx 200$ cm^{-1}, greatly increases. In Fig.15.4 B, D the simplified orientational ACF $\Phi_D(t)$ is shown. It is calculated with use of Eq. 15.9 by setting $x = 0$ in the arguments z of the SFs $L(z)$.

As seen, $\Phi(t)$ and $\Phi_{IV}(t)$ decay markedly at different intervals. The ACF $\Phi_D(t)$ corresponds to the low-frequency (Debye) spectrum, it is close to the exponent and decays e times for the time $t = \tau_D$. For short time interval ($t < \eta$) functions $\Phi(t)$ and $\Phi_{IV}(t)$ are determined mainly by the Fourier image of the SFs $L_L(z)$, $L_R(z)$.

Curve (c), plotted for the R-fraction, has two well pronounced portions. As seen from Fig. 15.4A,

(i) the initially rapidly decaying portion, close to the lifetime τ and

(ii) the slowly decaying (exponential) portion, in which $t > \tau$.

Similar portions can be found in the resultant ACF and in that, obliged to librators, curves (a) and (b), respectively. As seen from Fig. 15.4C, in the second time portion ($t > \tau$) the (a)- and (b)- curves decay in time τ_D (the picosecond time scale). The ACF, obliged to rotators, evidently decays in shorter time. These pictures prompt the idea that

Lifetimes of librators and rotators probably should differ in such manner, that $\tau_L > \tau_R$, since lifetimes should be commensurable with the decay times of the corresponding ACFs. (15.11)

Consequently, although lifetimes in Eq. 15.9 were taken equal for both fractions, $\tau_R = \tau_L = \tau$, the above picture suggests an idea that *these lifetimes should differ*. This property can be interpreted as follows. The "outer" shell of fast dipoles, rotating over the potential barrier, live shorter time than low-energy librators live.

The second derivative Φ_{rv} of the resultant process and of librators [see curves (a), (b) in Fig. 15.4E] decays over short time (time scale of hundredths of ps), commensurable with the half-period $\langle T_L/2 \rangle$ of librators and with time $\eta = \sqrt{\frac{I}{2k_BT}}$. Function $\Phi_{rv}(t)$, obliged to rotators, curve (c) in Fig.15.4E, decays in *longer time*, commensurable with τ. This property can be explained as follows. Rotators have lesser angular frequency / greater energy.

In Fig. 15.5 we show ACFs $\Phi(t)$ and $\Phi_{rv}(t)$, calculated for a nonassociated liquid CH_3F. Again we use the GNB results, see also GT, p. 362. In this case the CR model was employed, viz. Eq. 15.9 with $r = 1$; the fitted parameters are close to those, used in Sec. 14.1. Time intervals, in which $\Phi(t)$ and $\Phi_{rv}(t)$ decay, turn out to be approximately close one to another. This supports our conclusion that librations are strongly damped in nonassociated liquids. Indeed, $\langle T_L/2 \rangle < \tau$, while for water $\langle T_L/2 \rangle >> \tau$. Water molecule performs *many* librations in time τ because of structural stiffness of H-bond network; during time τ the ACF $\Phi(t)$ decreases sharply. In contrary to this CH_3F molecule could hardly perform a single libration during τ.

The ACFs-supplied picture supports our understanding of molecular dynamics, gained after the study of orientational relaxation in liquids. This picture may provide some improvement of the calculation scheme: we mean the ratio of lifetimes for librators and rotators.

15.1.6 The Theory of an Isotopic Effect (IE)

IE in water reveals as approximate constancy of the translational-peak frequency ($\nu_R \approx 200$ cm^{-1}) both in water H_2O and in water D_2O in spite of large distinction of moments of inertia I of these molecules: I(water H_2O) $/I$ (water D_2O) ≈ 2.

It is *impossible* to explain this phenomenon in terms of *free rotation* of water molecules. The IE can be qualitatively described in frames of the hybrid model, *if free rotors are replaced by hindered rotors.* This idea is compatible with another one, that in associated liquid H_2O molecules are incorporated into the H-bond network. In accordance with the result (15.11) we take different lifetimes τ_L and τ_R .

THE SF $L_{HM}(z_L)$ OF LIBRATORS. For this ensemble we apply the HM, Sec. 9.2, characterized by β, τ_L, u. The proportion $\overset{\circ}{r}(u,\beta)$ of librators is calculated from Eq. 9.22. The SFs $\check{K}_{\parallel}$ and $\check{K}_{\perp}$ are given by Eq. 9.41, 9.42, where we put

$$z \to z_L = x + iy_L,\ y_L = \frac{\eta}{\tau_L}, \quad \text{so that } L_{\text{HM}} = \frac{\check{K}_{\parallel}(z_L) + 2\check{K}_{\perp}(z_L)}{3}.$$

THE SF $L_{LH}(z_L)$ OF HINDERED ROTATORS. Their proportion $\overset{\circ}{r}$ is evidently equal to $1 - \check{r}$. For this fraction we apply the cosine squared (CS) model, characterized by the field parameter p and the lifetime τ_R. The normalized complex frequency $z_R = x + iy_R,\ y_R = \dfrac{\eta}{\tau_R}$. We employ the part (12.75) of the SF, related to rotors. Replacing the coefficient before integral (12.75) by constant a, we write:

$$L_{\text{CS}}(z_R) = \frac{\overset{\circ}{K}_{\parallel}(z_R) + 2\overset{\circ}{K}_{\perp}(z_R)}{3} = a \int_p^{\infty} \frac{f(p,t)\,dt}{t^2 - z_R^2 \gamma^2(p,t)},$$

$$f(p,t) = \frac{\left(1+\frac{p^2}{2t^2}\right) t^2 \exp\left(-t^2\right)}{\left[1-\left(\frac{p}{4t}\right)^4\right]\gamma(p,t)}, \quad \gamma(p,t) = 1+\frac{p^2}{4t^2}. \tag{15.12}$$

Let the integrated absorption Π_{CS} of an isotropic medium, obliged to hindered rotors, satisfy the Gordon sum rule for a spatial ensemble of dipoles, multiplied by the proportion r:

$$\Pi_{CS} = \int_{-\infty}^{\infty} x L''_{CS}\, dx = \pi\, r(u,\beta).$$

We remark that in Eq. 15.12 we take $2\overset{\circ}{K}_\perp$ instead of $\overset{\circ}{K}_\perp$ in order to account for approximately more realistic spatial rotation instead of planar one (we use here the theory, elaborated in Chap. 12 for planar motion).

Combining this equation with (15.12), we evaluate a:

$$a \equiv a(p,\beta,u) = r(u,\beta)\left(\int_p^{\infty} \frac{f(p,t)\, dt}{\gamma^2(p,t)}\right)^{-1}. \tag{15.13}$$

The complex susceptibility of the CR/CS model (with the isothermal collision model F_T involved) in view of Eq. 6.21, 6.80 is given by

$$\frac{\chi^*(x)}{gG} = \frac{z_L L_{CR}(z_L) + z_R L_{CS}(z_R)}{gx + iy_L\left(1+gxz_L\right)L_{CR}(z_L) + iy_R\left(1+gxz_R\right)L_{CS}(z_R)}. \tag{15.14}$$

The relaxation time is estimated by setting $x = 0$ in the SFs $L(z)$:

$$\frac{\tau_D^{est}}{g\eta} \approx \frac{1 - y_L^2 L_{CR}(iy_L) - y_R^2 L_{CS}(iy_R)}{y_L L_{CR}(iy_L) + y_R L_{CS}(iy_R)} \tag{15.15}$$

APPLICATION OF THE THEORY. FIR absorption / Debye spectra are shown in Fig. 15.6 for $T = 27°C$, the fitted parameters are presented in Table 15.6. For water H_2O this model yields better description of absorption $\alpha(\nu)$, than was obtained with use of the "pure" HM, cf. with Fig. 15.1a. In particular, FIR spectra near absorption minimum agree better with those experimentally recorded.

The wideband spectra of water D_2O are investigated rather poor. We find that $p_{D_2O} > p_{H_2O}$ and that β_{D_2O} occurs to be slightly less, than β_{H_2O}. Both results may serve as the evidence, that water D_2O is characterized by stronger structure, than water H_2O.

Table 15.6 Fitted and related parameters of the CR / CS model. Liquid water. $T = 27$°C. ν_L and ν_L in [cm^{-1}].

Parameter	Water		Parameter	Spectral feature		Band
	H_2O	D_2O	observed	H_2O	D_2O	
β, deg	19.93	19.28	ν_L	670	505	
τ_L, ps	0.4	0.5	α_L	3308		$\alpha(\nu)$
u	5.45	5.0	ν_R	≈200	≈200	
			α_R	1445		
p	1.5	2.0	ν_L	570	482	
k_μ	1.106	1.0	α_L	0.482	0.420	
r, %	6.7	10.8	ν_R	170	167	$\kappa(\nu)$
			κ_R	0.629	0.630	
Experimental data			Estimated	H_2O	D_2O	
τ_D, ps	7.85	8.9	ν_L	665	499	
ε_s	77.6	76.2	α_L	3294	1697	$\alpha(\nu)$
$I \bullet 10^{-40}$, g•cm^2	1.483	2.765	ν_R	200.1	180.5	
M	18	20	α_R	1427	715	

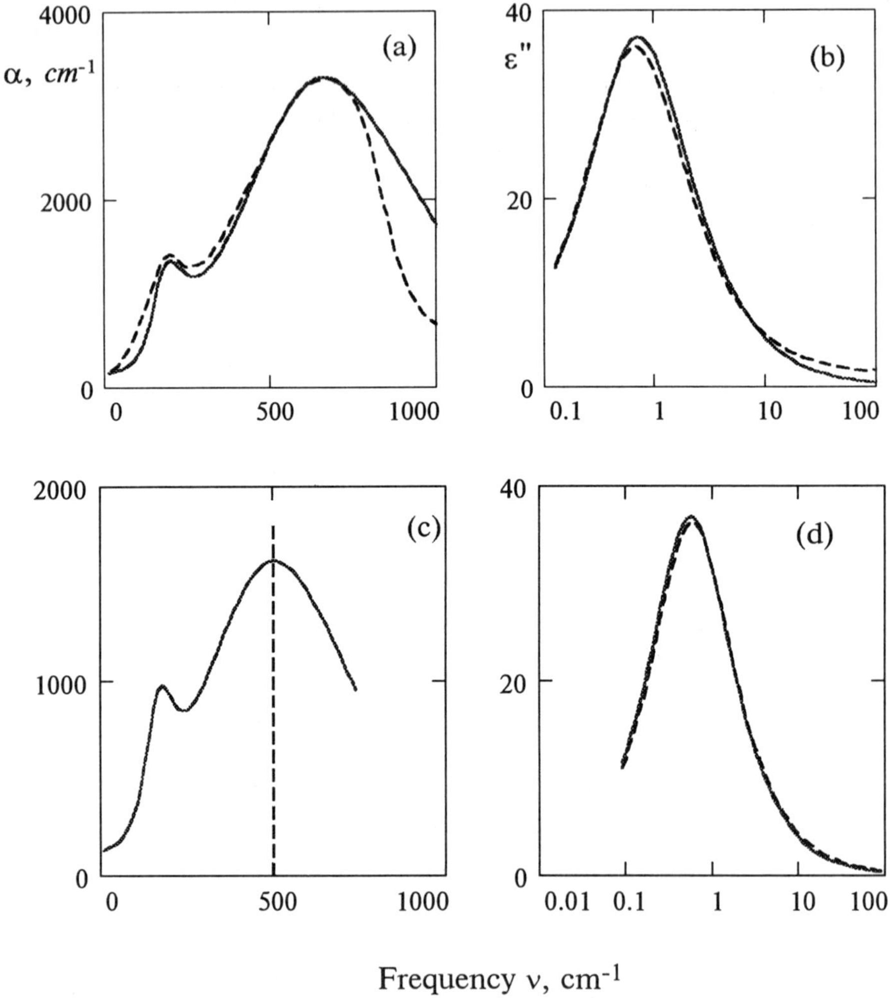

Fig. 15.6 Absorption (a, c) and loss (b, d) wideband spectra of water H_2O (a, b) and water D_2O (c, d), $T = 27\ ^{\circ}C$. Calculation with use of the CR/CS model (solid lines). Experimental data are shown by dashed lines; vertical line in Fig. c refers to the observed position of the librational absorption peak.

Absorption peak near 200 cm^{-1} is obliged to hindered rotation, specific for a condensed phase. The mean angular frequency of hindered rotation is in water D_2O about 1.33 times higher, than in water H_2O, since $p_{D_2O} \;/ p_{H_2O} \approx 2/1.5 \approx 1.33$.

Experiments show that in a liquid the amount of vapour H_2O molecules is negligible. This fact proves that H-bond network is *continuous* in water, see e.g. (Zatsepina 1987, p.122). The latter property does not contradict to our conception, that two types of molecular rotation coexist in a liquid, viz. libration and retarded rotation. We also remark that the fitted lifetime τ is commensurable in water with that of volume viscosity, for the latter see (Zatsepina 1987, p.148).

Experimental data and spectral features are taken above in this section from (Sethna, Palmer and Williams 1978), (Downing and Williams 1975), (Water 1972, p.162, 170).

15.2 Electrolytes in Aqueous Solution

This section presents as if an extensive comment on the brief review, given in the introductory Chap. 2, Sec. 2.6.1. We employ here again the hybrid / rotational cell models, developed for water in Sec. 5.1.1 -15.1.3, however with parameters u, β, τ, modified by ions due to their effect on H-bond network, specific for liquid water.

For simplicity the k_μ parameter is taken equal to that, fitted for liquid water. Besides, we employ expression for the complex permittivity as a sum:

$$\varepsilon^*(\nu) = \varepsilon' + i\varepsilon'' = \varepsilon^*(\nu)\big|_{\text{orientational relaxation}} + \frac{2i\sigma_s}{c\nu}. \qquad (15.16)$$

The first term is found for the HM and the second corresponds to the usual, see (Debye and Zakk, 1936), phenomenological description of dielectric loss due to ions, σ_s being the electric conductivity of the solution.

15.2.1 NaCl - H_2O Solution as a Medium, Characterized by the Positive Hydration

EXPERIMENTAL DATA. In Table 15.7 we present the experimental (Lane and Saxton, 1952) parameters ε_s, τ_D, σ_s of Debye relaxation for NaCl-H_2O solution and in Table 15.8 - the solution density vs. concentration dependence (the latter is taken from the Chemist's Reference Book). The low-frequency experimental permittivity is described by Debye formula:

$$\varepsilon^*(\nu) = \varepsilon' + i\varepsilon'' = \varepsilon_\infty + \frac{\varepsilon_S - \varepsilon_\infty}{1 - i2\pi c\nu\tau_D} + \frac{2i\sigma_S}{c\nu}, \tag{15.17}$$

in which $\varepsilon_\infty \equiv \varepsilon_{\infty 1}$ is calculated from the LHM formula (15.1).

APPLICATION OF THE HYBRID MODEL. In Table 15.9 we present the fitted parameters u, β, τ. For T = 30°C the low-frequency loss dependence agrees with Debye curve (15.17), see Fig. 15.7a. Calculated FIR absorption curves $\alpha(\nu)$ are shown in Fig. 2.19. When molar concentration C_M rises from 0 to 5 g•mole/l, then:

(i) the lifetime τ significantly increases;

(ii) the potential u slightly increases and

(iii) the libration angle β slightly decreases, so that NaCl solvent as if produces *structural cooling* of water.

INFLUENCE OF TEMPERATURE ON DIELECTRIC BEHAVIOUR. For T = 20°C we have no experimental data for FIR spectral range. Starting from the *predicted spectra* of liquid water (see Sec.15.1.4), we can estimate parameters of the HM for solution by assuming that the concentration dependencies of these parameters are similar to those, calculated for T = 30°C. Further calculations show, that the given temperature shift ΔT = -10°C produces small change of low-frequency and FIR spectra in comparison with the effect of the solvent.

We conclude that electrostatic interactions of H_2O molecules with Na^+ and Cl^- ions as if strengthen water structure, since the potential well becomes narrower (β decreases), the welldepth u increases and the proportion r of rotors decreases. We may suppose that the self-diffusion coefficient D should decrease because of these alterations. Librations live longer since τ increases.

THE SELF-DIFFUSION AND STRUCTURAL COOLING OF WATER. Let us calculate the $D(C_M)$ dependence. We use Eq. 14.45, where in view of Sec.15.1.3 we identify V and u potentials. At first, we put T = 30°C and fit diameter σ from the condition for liquid water:

Table 15.7 The observed parameters of Debye relaxation for NaCl-H_2O solution. From (Lane and Saxton 1952)

T°C	C_M, g•mole/l	ε_s	τ_D, ps	$\sigma_s \bullet 10^{11}$, s^{-1}
10	0.0	84	13.6	0.0
	0.5	74	12.2	0.31
	1.0	65	11.8	0.56
	1.5	60	11.3	0.72
	2.0	54	10.9	0.98
	2.5	48	10.6	1.15
	3.0	43	10.3	1.28
20	0.0	80	10.1	0.0
	0.5	71	9.2	0.40
	1.0	63	9.0	0.69
	1.5	57	8.7	0.97
	2.0	51	8.3	1.24
	2.5	46	7.9	1.46
	3.0	41	7.6	1.60
30	0.0	77	7.5	0.0
	0.5	68	7.2	0.47
	1.0	60	7.1	0.83
	1.5	55	6.9	1.17
	2.0	49	6.7	1.48
	2.5	44	6.4	1.74
	3.0	39	6.2	1.93

Table 15.8 NaCl-H_2O solution: experimental density-concentration dependencies.

Temperature T, °C			20		32	
Concentration			Density, g•cm^{-3}			
C, weight %	C_M, g·mole/l	C_w, weight g/l	ρ	ρ_{eff}	ρ	ρ_{eff}
1	0.1720	10.5	1.0053	0.9948	1.0075	0.9974
2	0.3464	20.25	1.0125	0.9923	1.0151	0.9948
4	0.7026	41.07	1.0268	0.9857	1.0304	0.9892
6	1.0688	62.47	1.0413	0.9788	1.0457	0.9830
8	1.445	84.47	1.0559	0.9714	1.0612	0.9763
10	1.831	107.07	1.0707	0.9636	1.0768	0.9691
12	2.228	130.2	1.0857	0.9555	1.0924	0.9613
14	2.636	154.1	1.1009	0.9468	1.1082	0.9531
16	3.055	178.5	1.1162	0.9377	1.1242	0.9443
18	3.485	203.7	1.1319	0.9228	1.1403	0.9350
20	3.927	229.5	1.1478	0.9183	1.1566	0.9253
22	4.380	256	1.1640	0.9080	1.1732	0.9151
24	4.846	283.2	1.1804	0.8972	1.1899	0.9043
26	5.325	311.2	1.1972	0.8860	1.2071	0.8933
Note	$\rho = \frac{\text{weight of NaCl + weight of water}}{\text{volume}} = \frac{C_M M}{1000} + \rho_{eff}$. $M_{NaCl} = 22.99 + 35.453$					

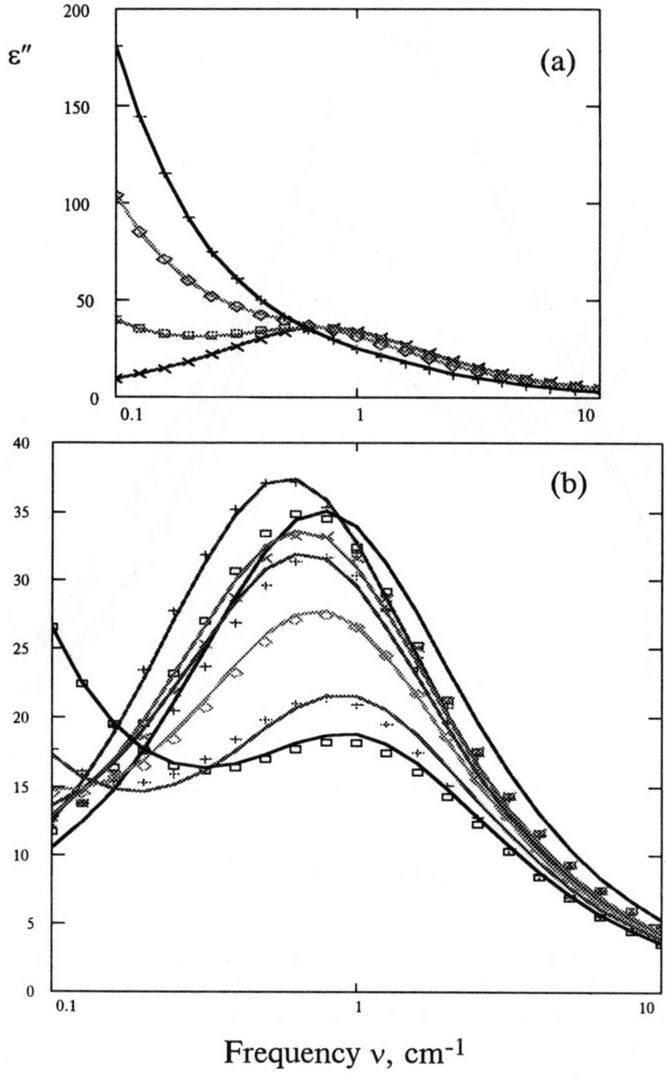

Fig. 15.7 **(a, b):** See capture to this Figure on the next page.

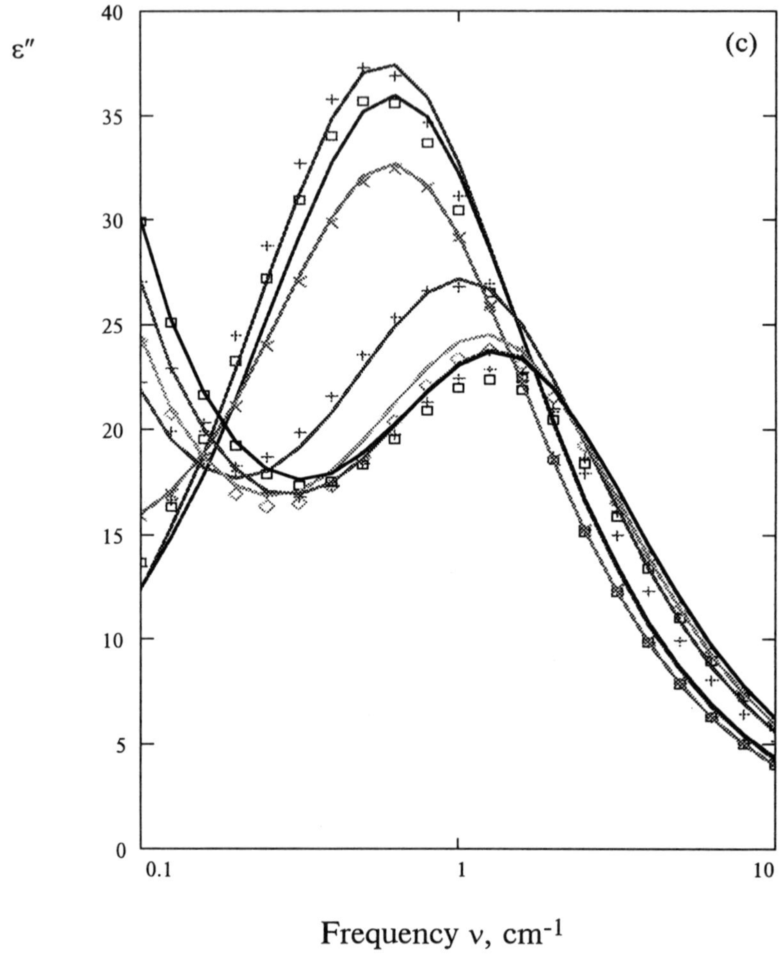

Fig. 15.7 Dielectric loss vs. frequency for NaCl-water (a), NaI-water (b) and KI-water (c) solutions, calculated for the hybrid model (solid lines) and from Debye formula (15.17) (dots). Concentrations [mole/l] from top to bottom: 0, 0.5, 1, 1.5 (a); 0, 1/3, 2/3, 1, 2, 4, 8 (b) and 0, 0.2, 1, 3, 4, 5, 6 (c).

$$D(C_M = 0) = D\big|_{\text{liquid water}}, \tag{15.18}$$

where the r.h.s. is known from the experiment (the Reader should not mix this σ with the electrical conductivity σ_S !). Substituting $\rho = \rho_{eff}$ into Eq.14.45 we find that $D(C_M)$ is *decreasing function* (Fig. 15.8a). This $D(C_M)$ agrees qualitatively with the recorded curves (Samoilov 1957, p.101).

We may also confirm the idea, that some kinetic properties of the solution may be described in terms of the effective temperature T_{eff} of water, see (Bernal and Fowler 1933); (Samoilov 1957, p.86). Structural temperature decreases for NaCl-H_2O due to interaction of H_2O molecules with the solute. The latter effect, characterized by the parameter ΔT_{eff}, can be estimated as follows. We find $T_{eff}(C_M)$ from the following condition:

$$D(T, C_M)\big|_{\text{solution}} = D\big|_{\text{liquid water}}(T_{eff}), \tag{15.19}$$

the r.h.s. is calculated with r, τ fitted for the solution:

$$D\big|_{\text{liquid water}}(T_{eff}) = \frac{1}{6}\left(\frac{r}{\tau}\right)\langle\Delta l\rangle^2\Big|_T . \tag{15.20}$$

First. We put $C_M = 0$, $T_{eff} = T$ and fit $\langle\Delta l\rangle$ from Eq. 15.20 with account of the experimental D-value, measured for liquid water.

Second. Let $C_M > 0$. Since the translational path $\langle\Delta l\rangle$ characterizes the water structure and weakly depends on temperature and on concentration [see e.g. (Samoilov 1957, p.89)], we estimate T_{eff} and $\Delta T_{eff} = T_{eff} - T$ by solving Eq. 15.19, 15.20 under assumption, that $\langle\Delta l\rangle$ does not depend on T.

$D(T)$, calculated with use of this procedure, is illustrated by Fig. 15.8b. We see that indeed $\Delta T_{eff} < 0$, so we actually have the case of the structural cooling. For a strong concentration (C_M = 5 mole/l), T = 20°C and 30°C we obtain respectively: $\Delta T_{eff} \approx$ -8°C and -5°C.

Our study of molecular dynamics / dielectric behaviour of NaCl - water solution confirms the Samoilov's picture of positive hydration in the vicinity of hydrated ions, viz.:

(i) the increase of the residence time τ of water molecule in its state of temporary equilibrium;

(ii) the decrease of the self-diffusion coefficient and of structural temperature.

We may estimate these parameters from our model of orientational relaxation. We have found that translations of H_2O molecules are closely related in aqueous solutions to their reorientation. The Samoilov's parameters τ and τ_i can be calculated from the prescribed molecular model of the solution, viz. the lifetime τ of H_2O molecule in water surroundings may be identified with our lifetime $\tau(C_M = 0)$ and the lifetime τ_i of H_2O molecule in the nearest (to the ion) equilibrium state - with the fitted $\tau(C_M \neq 0)$. We have found that $\tau > \tau_i$ for NaCL-H_2O solution.

15.2.2 Media with Negative Hydration

If at least one ion is large, the FIR absorption curve $\alpha(\nu)$ drastically changes.

DIELECTRIC BEHAVIOUR OF NaI-WATER SOLUTION. According to (Bennouna, Cachet, Lestrade, and Birch 1981) only a shoulder appears near frequency $\nu_L \approx 200$ cm^{-1}, see Fig. 2.20. This feature is evidently attributed by the influence of large I^- anion. We shall see just below that in this case we must take *larger libration amplitude* β *and lower potential welldepth* u. We fit u and β for T = 20°C (see Table 15.10) with account of the experimental parameters, that are found by approximation of various data, presented in the aforecited article, in the Reference Chemist's book and in the (Akhadov 1977, p. 314 - 315) book.

The low-frequency spectra are shown in Fig. 15.7b and FIR absorption spectrum - in Fig. 2.20. The distinction of the loss curves $\varepsilon''(\nu)$, that correspond to different concentrations, is larger than for NaCl - H_2O solution (cf. Fig. 15.7a with 15.7b), because of *lesser* conductivity σ_S.

The FIR absorption spectrum also differ. In Fig. 15.8 c, d we demonstrate, that the self-diffusion coefficient D of NaI-H_2O electrolyte increases with the rise of concentration, the shift ΔT_{str} being now positive, e.g. we have the case of structural heating.

All calculations are made by the same fashion, as previously for NaCl-water solution. Consequently, a negative-hydration system, viz. water-electrolyte network, as if loosens, since intermolecular barrier decreases with the rise of C_M; rotational motion becomes more pronounced and chaotic; the number m_L of librations during lifetime τ decreases, while for positive hydration $m_L(C_M)$ is the increasing function.

We may summarize the results as follows:

Positive hydration		Negative hydration	
Observed property	Fitted parameters (HM)	Observed property	Fitted parameters (HM)
$D\downarrow$	$u\uparrow, \beta\downarrow, \tau\uparrow$	$D\uparrow$	$u\downarrow, \beta\uparrow, \tau\downarrow$
Note: symbols $\uparrow$ and $\downarrow$ denote respectively the increase and decrease of any quantity due to rise of concentration of electrolyte			

DIELECTRIC BEHAVIOUR OF KI-WATER SOLUTION. Both ions of this electrolyte are comparatively large. One may await that the effect of negative hydration should be still more pronounced. This idea is actually confirmed by the observed FIR absorption spectrum (see Fig. 2.21), in which we however see two new features:

(1) The centers ν_R and ν_L of the $\alpha(\nu)$ curve are shifted to *lower frequencies* in comparison with the case of NaCl-H_2O solution.

(2) The square under absorption curve $\alpha(\nu)$ becomes appreciably lesser, than that, found for water with the same number density N_{eff}. One may suggest that some water molecules, let their part be $(1 - r_z)$, lose rotational mobility and become as if frozen-in. In (Lyashchenko, Lileev, Zasetsky, Novskova, and Gaiduk 1993) work the so called Z-fraction of water molecules was introduced, which describes "the presence of «frozen-in» molecules among the molecules from the first ionic sphere [the hydrated complexes, the separate ionic pairs, the joint hydration spheres for the cations and anions in a concentrated solution and so on]".

Thus, the number of water molecules that participate in the orientational relaxation, is now given by

$$r_Z N_{eff} = r_Z N_A M / \rho_{eff} , \tag{15.21}$$

where $\rho_{eff}(C_M)$ can be estimated from the standard experimental data and r_Z - from the comparison of calculated and observed spectral dependencies $\alpha(\nu)$ for 6M solution, to which $\rho_{eff} \approx 0.72$ corresponds. For lower concentrations we take the linear dependence $r_Z(C_M)$.

Table 15.9 NaCl-water solutions: the fitted and other parameters of the hybrid model.

T, °C	C_M, $\frac{\text{mole}}{\text{l}}$	u	β, deg	τ, ps	r, %	y	m_L	k_μ
20	0.0	5.51	19.4	0.48	6.7	0.090	8.6	1.13
	0.5	5.54	19.1	0.53	6.7	0.081	9.7	
	1.0	5.56	18.8	0.58	6.7	0.074	10.1	
	1.5	5.59	18.4	0.64	6.8	0.068	12.0	
	2.0	5.62	18.2	0.69	6.8	0.062	13.3	
	2.5	5.65	17.8	0.75	6.9	0.057	14.7	
	3.0	5.68	17.5	0.81	6.9	0.053	16.2	
	5.0	5.80	16.0	1.07	7.3	0.040	23.3	
30	0.0	5.30	19.9	0.36	7.7	0.117	6.4	1.12
	0.5	5.31	19.8	0.39	7.7	0.111	6.7	
	1.0	5.35	19.5	0.42	7.7	0.100	7.6	
	1.5	5.38	19.2	0.47	7.6	0.090	8.6	
	2.0	5.42	18.9	0.48	7.6	0.081	9.7	
	2.5	5.46	18.6	0.51	7.6	0.074	10.8	
	3.0	5.50	18.2	0.56	7.6	0.068	12.1	
	5.0	5.68	16.7	0.85	7.5	0.050	18.0	

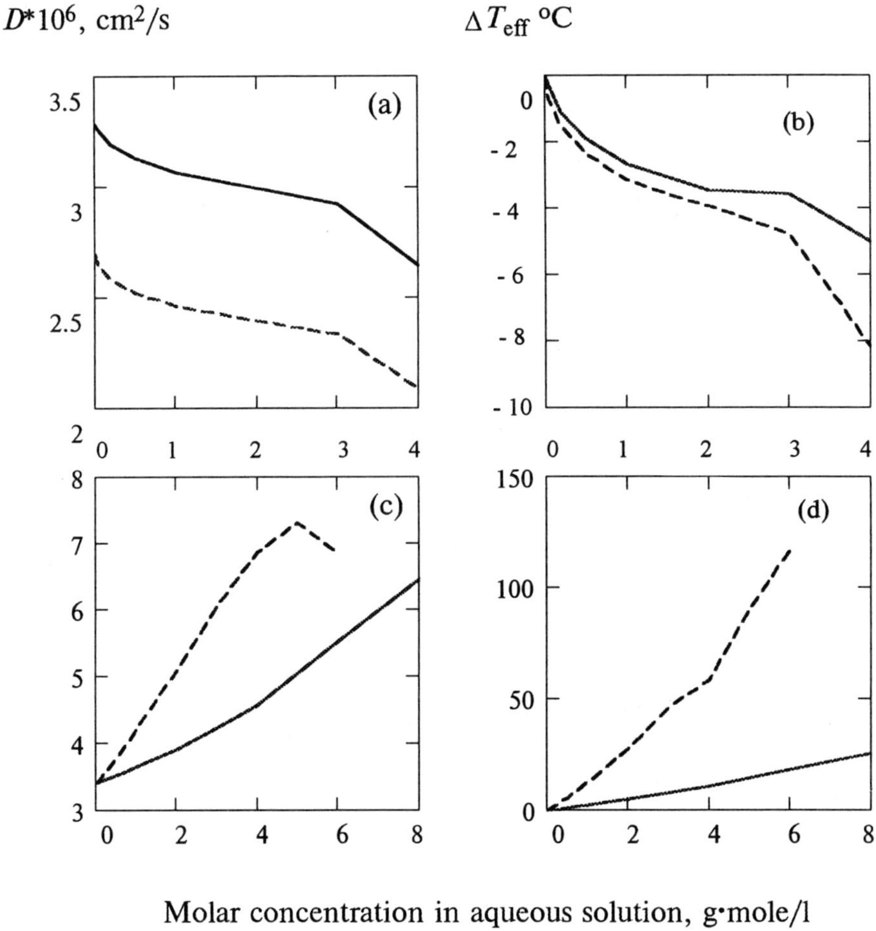

Fig. 15.8 Predicted concentration dependence of the self-duffusion coefficient (a, c) and the effective temperature cooling (b) or heating (d). $\langle \Delta l \rangle = 1.13 \cdot 10^{-8}$ cm^{-1}. **a, b:** NaCl-H_2O for T = 30 (solid curves) and 20°C (dashed curves). **c, d:** NaI-H_2O (solid lines) and KI-H_2O (dashed lines), T = 20°C.

Table 15.10 NaI-water solution: the experimental data, fitted and other parameters of the reorientation model. $T = 20°C$.

C_M, mole / l	Experimental data			
	ρ_{eff}, g•cm^{-3}	ε_s	τ_D, ps	σ•10^{11}, s^{-1}
0.0	0.998	80.1	9.4	0.0
0.33	0.987	74.7	7.8	0.026
0.66	0.975	71.3	7.4	0.048
1	0.963	67.5	7.1	0.069
2	0.926	58.3	6.7	0.120
4	0.849	45.0	6.2	0.198
8	0.691	37.0	5.0	0.353

C_M, mole / l	Fitted parameters			Other parameters		
	u	β, deg	τ, ps	r, %	y	m_L
0.0	5.24	19.4	0.48	8.6	0.090	8.5
0.33	5.22	19.7	0.48	8.5	0.090	8.3
0.66	5.21	19.9	0.47	8.4	0.090	8.2
1	5.20	20.2	0.47	8.3	0.091	8.0
2	5.16	21.1	0.47	8.0	0.092	7.6
4	5.07	22.8	0.45	7.5	0.094	6.9
8	4.90	26.4	0.43	6.7	0.100	5.7

Table 15.11 KI-water solution: the experimental data, fitted and other parameters of the reorientation model. $T = 20°C$.

C_M, mole / l	Experimental data			
	ρ_{eff}, g•cm^{-3}	ε_s	τ_D, ps	σ•10^{11}, s^{-1}
0.0	0.998	80.1	9.9	0.0
0.05	0.996	79.8	9.8	0.005
0.2	0.989	76.6	9.5	0.020
0.5	0.975	73.9	9.2	0.048
1	0.952	68.3	8.3	0.094
2	0.904	60.5	6.7	0.181
3	0.854	55.5	5.3	0.258
4	0.806	49.6	3.	0.314
5	0.756	47.4	4.0	0.360
6	0.705	46.0	3.8	0.405

C_M, mole / l	Fitted parameters			Other parameters			
	u	β, deg	τ, ps	r, %	r_Z	y	m_L
0.0	5.2	19.4	0.48	8.6	1.0	0.09	8.5
0.05	5.2	19.5	0.47	8.7	0.994	0.09	8.4
0.2	5.2	20.6	0.46	8.8	0.997	0.09	8.1
0.5	5.1	20.2	0.45	9.1	0.942	0.10	7.5
1	4.9	21.8	0.50	9.6	0.882	0.10	6.6
2	4.5	24.3	0.39	11.0	0.759	0.12	5.0
3	4.1	27.0	0.31	13.0	0.630	0.14	3.7
4	3.8	29.5	0.25	15.3	0.507	0.17	2.7
5	3.4	32.2	0.19	18.5	0.379	0.23	1.8
6	3.0	34.8	0.13	22.5	0.250	0.33	1.1

Experimental, fitted and other parameters are presented in Table 15.11. The low-frequency spectra agree (see Fig. 15.7c) with those, calculated from Eq. 15.17 with use of the observed Debye parameters. On the other side, from Fig. 2.21 we see that a rather large distinction exists between absorption of liquid water (dashed curve) an d that of concentrated 6M solution, since the number of frozen-in molecules is ca. 75%. The hybrid model (solid curve) yields only rough qualitative description of the observed (dots) spectrum. The theoretical frequency ν_R of the first peak is ca. 200 cm^{-1}, while the observed $\nu_R \approx 100$ cm^{-1}. The calculated position of the second peak (near 350 cm^{-1}) approximately could agree with the recorded ν_L, if we choose a large libration amplitude β (about 35°) and a low potential barrier ($U_0 \approx 3\ k_B T$).

We remark, that the simple Debye-like approximation (15.17) does not fit experimental data for concentrated NaI-water and KI-water solutions, see e.g. (Akhadov 1977); (Lyashchenko, Lileev, Zasetsky, Novskova and Gaiduk 1993). One can try to relate this property (that is not yet described) with the effect of each hydrated sheath of an ion on dielectric properties of concentrated electrolyte solutions.

The predicted coefficients $D(C_M)$ of NaI-H_2O and KI-H_2O electrolytes, as well as their structural heating ΔT_{eff}, are compared in Fig. 15.8 c, d, the r.h.s of Eq. 15.20 being multiplied for KI-H_2O by the factor r_Z, since only dipoles with substantial rotational mobility may contribute the self-diffusion in the solution.

For low and moderate concentrations ($C_M < 5$ mole/l) the self-diffusion coefficient of KI-H_2O electrolyte, its steepness and $|\Delta T_{eff}|$ are larger, than these parameters, calculated for NaI-H_2O system. For large concentrations the decrease of $D(C_M)$ function occurs due to the appearance of many frozen-in molecules. This property qualitatively corresponds to the observed $D(C_M)$ curve, see (Samoilov 1957, p.101). The concentration dependencies of fitted u, τ, β and of calculated number m_L confirm the aforecited conclusion, that KI-H_2O solution is more "negatively hydrated" than NaI-H_2O one.

For large concentrations the former system has rather weak H-bond network. However, it is still so rigid to make the structure of a liquid more pronounced in comparison with that which characterizes nonassociated liquids. Indeed, we have two absorption maxima α_L and α_R in KI-H_2O electrolyte, while e.g. in liquid CH_3F only one maximum α_R is observed.

15.3 Water in a Bound State

For calculation of absorption and of complex permittivity we apply formulas, mentioned in Table 14.1. The complex susceptibility χ^* is now found as the sum $\Sigma\chi_j^*$ over a number of fractions, each χ_j^* being found for the hybrid model with use of formulas, mentioned in Table 14.1. This calculation scheme is chosen, because the observed spectra usually exhibit a *number of low-frequency regions,* characterized by the corresponding relaxation times τ_{Dj}. The brief review of observed phenomena was given in Sec. 2.5.1 - 2.5.3.

15.3.1 A Protein Molecule in Aqueous Solution

EXPERIMENTAL DATA. We shall consider concentrated (20%) aqueous solution of hemoglobin (Hb). In Sec. 2.5.3 we have shown that the observed, see (Grant, Sheppard, and South 1978), low-frequency spectrum of this solution, that comprises β, γ, δ relaxational regions, originates due to reorientation: of macromolecules in the solution (β), of bound H_2O molecules (δ), and of H_2O molecules in bulk water (γ). Appropriate experimental data are presented in Table. 15.12.

PARAMETERS OF A MONOLAYER-HYDRATED Hb MOLECULE. For the moment of inertia I_{hyd} of hydrated Hb molecule we use formula

$$I_{\text{hyd}} = \frac{2}{5} M_{\text{Hb}}\, m_H \left(R_{\text{Hb}} + 2l\, R_{\text{H}_2\text{O}}\right)^2 \qquad (15.22)$$

for the spherical top. Its radius is taken equal to that of a protein (R_{Hb}) plus the width of a hydrated ring, containing l water layers, the width of each being equal to the diameter $2R_{\text{H}_2\text{O}}$ of H_2O molecule; M_{Hb} is molecular mass of the protein (viz. of Hb). Taking one layer ($l = 1$), we have:

$$M_{\text{Hb}} = 68000,\ R_{\text{Hb}} = 32.9\ \mathring{A},\ R_{\text{H}_2\text{O}} = 1.5\ \mathring{A}, \qquad (15.23)$$

$$I_{\text{hyd}}(l = 1) \approx 7.4 \bullet 10^{-33}\ \text{g} \bullet \text{cm}^2;\ I_{\text{Hb}}(l = 0) \approx 4.9 \bullet 10^{-33}\ \text{g} \bullet \text{cm}^2.$$

We also assume that one layer contains $N_l = 1000$ water molecules (for maximal possible number N_l we estimate the greater value, viz.

$$\max N_l \approx \frac{\text{surface of a protein molecule}}{\text{cross section of a water molecule}} = \frac{4\pi R_{\text{Hb}}^2}{\pi R_{\text{H}_2\text{O}}^2} \approx 1900\,).$$

If $l = 1$, molecular mass M_{hyd} and concentration N_{Hb} are approximately

$$M_{\text{hyd}}\,(l = 1) = M_{\text{Hb}} + 1000\,M_{\text{w}} = 68000 + 18000 = 86000; \quad (15.24)$$

$$N_{\text{Hb}}(l=1) = \frac{r_{\text{Hb}}\,N_A\,\rho_{\text{Hb}}}{M_{\text{Hb}}}, \qquad (15.25)$$

where $M_{\text{w}} = 18$ is molecular mass of water molecule, $r_{\text{Hb}} = 0.2$ is the proportion of water in 20% Hb-water solution, ρ_{Hb} is given in Table 15.10. We put $T = 300$ K and take $n_\infty^2 = 1.7$ (as in liquid water).

Table 15.12 Experimental data pertaining to 20% aqueous solution of hemoglobin. $T = 27°$C. From (Grant, Sheppard and South 1978, p. 3), (Gaiduk and Novskova 1996)

Parameter	Hb (β–band)	Bound water (δ–band)	Bulk water (γ–band)
τ_D, s	$1.6 \bullet 10^{-7}$	$5.9 \bullet 10^{-10}$	$12 \bullet 10^{-12}$
max $[\varepsilon''(\nu)]$	58.7	10.9	30.4
M	68000 (isolated) 12200 (hydrated)$^{a)}$	18	18
I, g•cm^2	$4.9 \bullet 10^{-33}$ $14.3 \bullet 10^{-33}$		
μ_0, D	531.5 $^{a)}$	1.84	1.84
R_{Hb}, Å	32.9	3	3
ρ, g•cm^{-3}	1.34	1	1
Note	$^{a)}$ Fitted		

DIELECTRIC RESPONSE OF A MONOLAYER-HYDRATED Hb MOLECULE. Rotational mobility of a macromolecule is determined by its water surroundings. Dielectric response of Hb molecules, reorienting as a whole in water solution, may be described in terms of the *extended diffusion model.* We use Eq. 7.58 for the SF $L_{spf}(z)$ of the spherical top, that is expressed through the SF $L_F^{pl}(z)$ of planar rotators, Eq. 5.99.

For very frequent collisions, when $y > 1000$, it is preferable to replace Eq. (7.58) by the SF $L_F^{sp}(z)$ of linear molecules, Eq. 7.11, 7.12. For rough estimation, taking $g = 1$, we may use the Gross collision model. Introducing

$$G_{Hb} = \frac{\mu_{Hb}^2 N_{Hb}}{3k_B T},\ z_{Hb} = x + iy_{Hb},\ x = \eta_{Hb}\omega,\ y_{Hb} = \frac{\eta_{Hb}}{\tau},\ \eta_{Hb} = \sqrt{\frac{I_{Hb}}{2k_B T}},$$

we find the complex susceptibility

$$\overset{*}{\chi}_{Hb} = \frac{G_{Hb}\, z_{Hb}\, L_{Hb}(z_{Hb})}{x + iy_{Hb}\, L_{Hb}(z_{Hb})};\ \overset{*}{\chi}_{Hb}(y_{Hb} >> 1) \approx \frac{G_{Hb}\, z_{Hb}\, L_F^{sp}(z_{Hb})}{x + iy_{Hb}\, L_F^{sp}(z_{Hb})}. \quad (15.26)$$

and the contribution of macromolecules to permittivity

$$\Delta\overset{*}{\varepsilon}_{Hb} \approx 4\pi\overset{*}{\chi}_{Hb}. \quad (15.27)$$

The loss curve $\varepsilon''(f)$, where f [Hz] $= \omega/(2\pi) = c\nu$, has the only maximum (see Fig. 15.9). It agrees with the observed maximum (ca. 59), if we take $\mu_{Hb} \approx 500$ D. Frequency ν_D of maximum loss corresponds to large y_{Hb} (y_{Hb} between 100 and 1000).

DIELECTRIC RESPONSE OF HYDRATED MONOLAYER. We find concentration

$$N_{bd} = 1000\ N_{Hb} \approx 3.3 \bullet 10^{21}\ \text{cm}^{-3}. \quad (15.28)$$

Let ρ_{bd} be density of bound water. Then

$$N_{bd} = \frac{N_A \rho_{bd}}{M_w} \text{ and } h_{bd} = \frac{\text{Weight of water}}{\text{Weight of protein}} = \frac{\rho_{bd}}{\rho_{Hb}} = \frac{N_{bd} M_w}{N_{Hb} M_{Hb}}. \quad (15.29)$$

The hydration

$$h_{bd} = \frac{1000 \bullet 18}{68000} \approx 0.265$$

falls in our case into the standard range [0.2, 0.3] g/g.

Further we employ the hybrid model (like for ordinary water, Sec.15.2). But now we have:

(a) lower concentration (15.28);

(b) less max $\varepsilon''(f)$ (viz. $\approx$ 11);

(c) longer relaxation time: $\tau_D^{\mathrm{bd}} \approx 5.9 \cdot 10^{-10}$ s.

One may await that the potential welldepth u_{bd} should be greater and lifetime τ_{bd} longer, than in the case of liquid water. We shall take the same Kirkwood factor g, as for ordinary water ($g \approx 2.35$) and estimate the low-frequency permittivity for the isothermal collision model [see Table 6.1 with $F = F_{\mathrm{T}}$, where $L(z)$ is replaced by $L(iy)$].

In Fig. 15.10 we plot theoretical frequency dependencies of real and imaginary parts of permittivity; fitted τ = 23.5 ps and u = 6.2; the reduced collision frequency y is very small, $y \approx 0.002$. These estimations show, that the absorption coefficient $\alpha_{\mathrm{bd}}(f)$ goes through maximum at microwaves ($f_L^{\mathrm{bd}} \approx 1$ GHz), the maximum absorption being very low (only few cm^{-1}).

The loss peak max $\varepsilon''_{\mathrm{bd}}(f)$, *estimated for the hybrid model, is about tree times lesser, than the recorded value. One may remove this drawback by taking three hydrated shells (l = 3) instead of one monolayer (l = 1), used in the above calculation.*

APPLICATION OF THE COSINE SQUARED MODEL. Taking the quasi-harmonic approximation (Sec. 12.4) and doubling the corresponding SF in order to get true value of integrated absorption, we find, that fitted lifetime τ_{bd} is long (about 10 ps). The position of low-frequency loss maximum agrees with the observed ν_D^{bd}, if we take $p_{\mathrm{bd}} \approx 5.5$ and $\tau_{\mathrm{bd}} \approx 8.1$ ps. The predicted absorption curve $\alpha(\nu)$ looks like a narrow peak, falling into FIR region; its value is commensurable with maximum absorption α_{L} of liquid water. Since such picture contradicts recorded data, it seems that the previous (hybrid) model is more preferable for description of dielectric response of bound water.

3-FRACTIONAL MODEL OF HB-WATER SOLUTION. We write the sum

$$\chi^*(\omega) = \chi^*_{\mathrm{Hb}}(\omega) + \chi^*_{\mathrm{bd}}(\omega) + \chi^*_{\mathrm{bl}}(\omega) \tag{15.30}$$

and apply the hybrid model for each term. In Eq. 15.30 subscripts bd and bl refer to bound and bulk water. The fitted parameters are obtained (see Table 15.13) with use of data, given in Table 15.12, on the basis of the preliminary information, gained above for hydrated protein and bound water. In particular, we account for:

(a) the values of maximum loss in bands β, δ, γ respectively on frequencies $\omega = \frac{1}{\tau_D^{\mathrm{Hb}}}, \frac{1}{\tau_D^{\mathrm{bd}}}, \frac{1}{\tau_D^{\mathrm{bl}}}$;

(b) the value $\varepsilon'(\omega) \approx 180$ on the low-frequency edge of this region, i.e. at $f = \omega/(2\pi) = 0.1$ MHz. The result of these calculations is illustrated by Fig. 2.18.

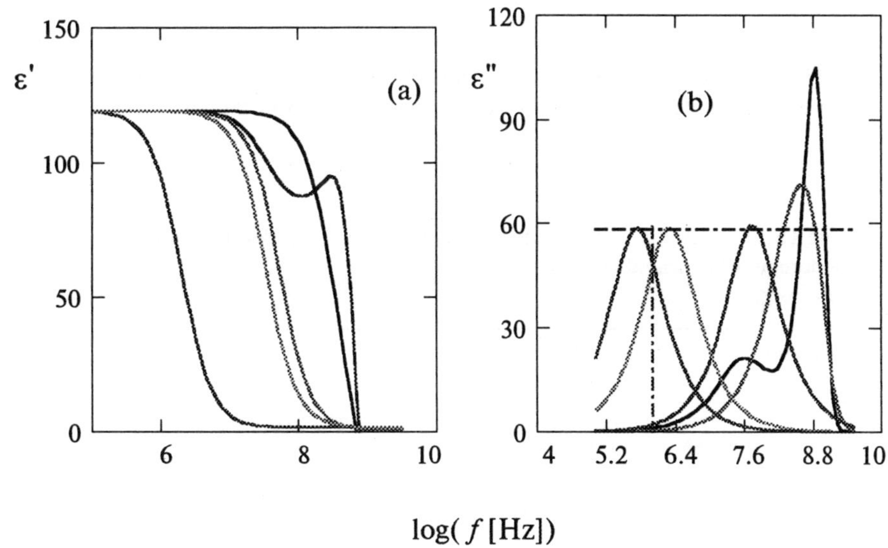

Fig. 15.9 Real (a) and imaginary (b) parts of ε^* vs. $\log f$. ED model for hydrated Hb. The normalized collision frequency $y_{\mathrm{Hb}} = 0.1, 1, 10, 100, 1000$ (from right to left). $T = 300\ K$. Horizontal / vertical lines correspond to the recorded data.

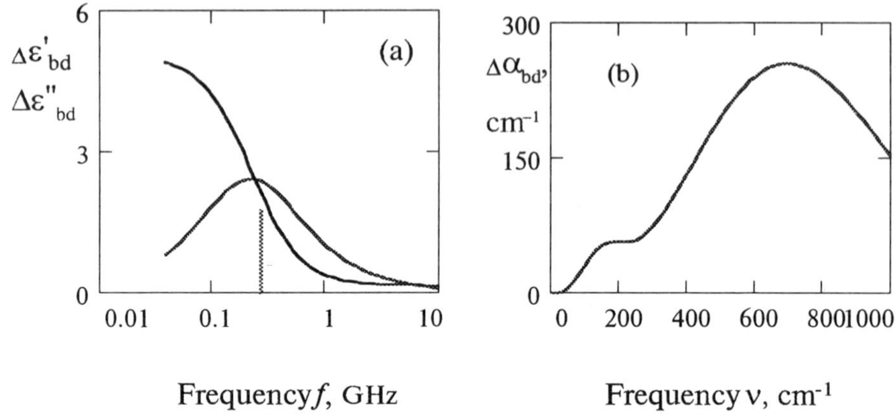

Fig. 15.10 Contribution to dielectric spectra of bound water, calculated for the hybrid model. Vertical line denotes the observed frequency position of the loss maximum in the δ-band. $T = 300$ K.

Table 15.13 Fitted and other parameters. 20%-hemoglobin solution. The number of hydrated layers $l = 3$.

Parameter	Hydrated Hb molecule	Bound water	Bulk water
Model	The ED model	The HM	The HM
N, cm^{-1}	$2.4 \bullet 10^{11}$	$7 \bullet 10^{21}$	$1.96 \bullet 10^{22}$
τ, ps	1.08	23.5	0.5
β, deg	19.3		
k_μ	1.08		1.195
g	1	2.8	2.35
u	6.2		5.5

Hb molecule, rotating as a whole. Taking $l = 3$, we obtain (see Table 15.12) dipole moment μ_{Hb}, moment of inertia I_{hyd} and molecular mass M_{hyd}. We regard the static permittivity ε_s^{Hb} to be approximately equal to the drop $\Delta\varepsilon'_{Hb} \approx 108$, observed in the β–band. We estimate the Kirkwood correlation factor g_{Hb} (g_{Hb} is about 1) from formula

$$g_{Hb} \approx \frac{\varepsilon_s^{Hb} - n_{inf}^2}{4\pi G_{Hb}} \frac{2\varepsilon_s^{Hb} - n_{inf}^2}{3\varepsilon_s^{Hb}}, \; G_{Hb} = \frac{\mu_{Hb}^2 N_{Hb}}{3k_B T} \approx 5.4. \quad (15.31)$$

The theoretical $\max \varepsilon''_{Hb}$ is slightly lesser, than the observed value; the distinction probably may be assigned to the contribution of ions into ε'' (their effect was neglected in our calculations).

Bound-water molecules. The estimated concentration and hydration of bound water at $l = 3$ are given by:

$$N_{bd} \approx 1000 \bullet 3 \bullet N_{Hb} \approx 7.1 \bullet 10^{21} \text{ cm}^{-3}, \; h_{hyd} \approx 0.8 \text{ g/g} \quad (15.32)$$

Dielectric measurements yield larger hydration than some other methods. The influence of protein on reorientation of water molecules may be detectable for distances up to ca. 3•2•3 ≈ 18 Å. For parameters, presented in Table 15.13, theoretical loss peaks agree with the observed one in δ- and γ-bands.

Concentration N_{bl} of bulk water is estimated as the difference N_w - N_{bd}. Since density ρ_s of solution is near density ρ_w of liquid water,

$$N_{bl} \approx \frac{0.8 N_A \rho_w}{M_w} \approx 2.67 \bullet 10^{22} \text{ cm}^{-3}.$$

Thus, in the solution about 79% of water exists in a form of bulk water. Contribution of this fraction to the total susceptibility is described just as for bound water but with different parameters of the hybrid model.

ABOUT RESONANCE ABSORPTION. We estimate the absorption $\alpha_{Hb}(\nu)$ due to Hb molecule by taking into account the refraction index $\langle n \rangle \approx \sqrt{73}$ of the solution and the collision frequency y_{HB} ($y_{HB} \approx 385$). The calculated dependence

$$\alpha_{Hb}(\nu) \approx 2\pi\nu \bullet \frac{4\pi\chi_{Hb}(\nu)}{\langle n \rangle}$$

spreads from 10^{-5} to 10 cm^{-1}, its absorption peak being extremely small (about 0.01 cm^{-1}). The absorption curve $\alpha_{bd}(\nu) + \alpha_{bl}(\nu)$ is similar to that, found for liquid water.

PHYSICAL PICTURE OF MOLECULAR ROTATION.

(a). Let us estimate the angular scale, that characterizes rotation of a macromolecule during its lifetime τ_{Hb}. By applying Eq. 15.7, where β^2 is substituted for $\langle \delta\Theta^2 \rangle$, we find:

$$\sqrt{\langle \delta\Theta^2 \rangle} \approx \sqrt{\frac{2\tau^{Hb}}{\tau_D^{Hb}}} \approx 0.2^\circ .$$

Actually we have a typical picture of Brownian motion, viz. small angular jumps of a rather large particle (a molecule with its hydration layer), surrounded by many small water molecules.

(b). Interesting feature of bound water is related to its long lifetime. Large ratio τ_{bd}/τ_{bl} corresponds to large ratio of relaxation times:

$$\frac{\tau_{bd}}{\tau_{bl}} \approx \frac{23.5}{1.06} = 21.8, \text{ while } \frac{\tau_D^{bd}}{\tau_D^{bl}} \approx \frac{510}{12} = 49 .$$

Thus, the low-frequency dielectric spectroscopy characterizes the change in the state of water under influence of specific surroundings (of protein molecules). The higher potential barrier ($u_{bd} \approx 6.2 > u_{bl} \approx 5.5$) correlates with the increase of the relaxation time. Greater values of u_{bd}, τ_{bd} comparing with u_{bl}, τ_{bl} testify that rather thick (ca. 18 $\mathring{A}$) hydrate layer has as if lower structural temperature.

(c). Greater, than in liquid water, bulk-water relaxation time τ_D^{bl} (cf. 12 ps with 7.9 ps) could be an indicative of more rigid structure that the bulk-water fraction possesses.

15.3.2 Humidified Protein Molecule

CALCULATING SCHEME. In accordance with known data (Mashimo, Kuwabara, Yagihara, and Higasi 1987), see also Sec. 2.5.2, only two low-frequency bands (δ and γ) are observed in this object. We consider dielectric behavior of collagen (Cg) as a "structural" protein of skin. Since Cg practically does not solves in water, it has no rotational mobility as a separate molecule.

We regard water, contained in Cg, to be distributed in four fractions (see also Table 15.14), viz. bound water (fractions 0, 1, 2) and bulk water (fraction 3). We put hydration h_0 = 0.074 g/g (g of water per one g of protein) of *structurally bound water* (fraction 0) and consider it to be non - absorbing.

The hydration h_1 of the *first hydrate layer (fraction 1)* is taken about 0.2 g/g. We account for experimental relaxation time / loss maximum ($\tau_{D1} \approx$ 1180 ps and $\varepsilon'' \approx 8.12$).

For the *second hydrate layer* [see (Water in polymers 1980, p. 114] *(fraction 2)* we introduce parameters h_2 and τ_{D2} ($\tau_{D2} \approx 20$ ps).

For the *bulk water (fraction 3)* we take the parameters h_3 and τ_{D3} ($\tau_{D3} \approx$ 7.1 ps, e.g. about the τ_D -value, known for liquid water). We remark that τ_{D2} and τ_{D3} are rather close; for simplicity we assume that hydration h_2 is equal to h_3. We fit h_3 by taking into account, that the only loss maximum $\varepsilon''_{2,3} \approx$ 8.13 corresponds to both fractions 2 and 3.

We take ρ_{Cg} = 1.17 g•cm^{-3} (e.g. ≈ density of skin in terms of humidified Cg). Thus, concentration of water in j th fraction is given by

$$N_j \approx \frac{N_A \rho_{Cg} h_j}{M_w}; \; j = 0,1,2,3; \; M_w = 18. \tag{15.33}$$

Applying the hybrid model for fractions 1, 2, 3, we put amplitudes β_j equal to the β -value, found for liquid water at T = 27 °C, viz. = 19.9°. Potential barriers u_j, Kirkwood correlation factors g_j, and dipole-moment correction k_μ are taken for the first hydrate layer (j = 1) and for bulk water (j = 3) approximately equal to those, chosen in Sec. 15.3.1. For the second hydrate layer these parameters are found with use of linear approximation of values corresponding to j = 1 and 3; lifetimes τ_j are fitted (see Table 15.14).

The complex susceptibility χ^* is found as the sum

$$\chi^* = \chi_1^* + \chi_2^* + \chi_3^*, \text{ with } F = F_T. \tag{15.34}$$

For χ_j^*, ε^* and α we use formulas, mentioned in Table 14.1.

THE RESULTS OF CALCULATIONS. Real and imaginary parts of permittivity $\varepsilon^*(f)$ are plotted vs. frequency f in Fig. 2.16a. For the first hydrate layer we have the fitted lifetime $\tau_1 = 35$ ps, this value is near the lifetime, estimated previously for water, bound by Hb molecule ($\tau_{bd} \approx 23.5$ ps). For the second hydrate layer the fitted $\tau_2 \approx 0.75$ ps, e.g. it has an intermediate value between lifetimes τ_1 and τ_3; $\tau_3 \approx 0.25$ ps being typical for liquid water.

The fitted hydration $h_1 = 0.16$ and $h_2 = h_3 = 0.11$ g/g. Thus, the total hydration

$$h_{Cg} = h_0 + h_1 + h_2 + h_3 = h_{bd} + h_{bl} = 0.454 \text{ g/g}; \quad (15.35)$$

$$h_{Cg} = 0.074 + 0.16 + 0.11 \approx 0.34; \; h_{bl} \approx 0.11 \text{ g/g}. \quad (5.36)$$

We see, that the contents of bulk water is in Cg lesser, than in Hb-H_2O solution. The total water concentration

$$N_{\text{water in Cg}} = N_0 + N_1 + N_2 + N_3 \approx 1.78 \bullet 10^{22} \text{ cm}^{-3} \quad (15.37)$$

comprises about half of that (N_w) in liquid H_2O, since

$$N_w = N_A \rho_w / M_w \approx 3.34 \bullet 10^{23} \text{ cm}^{-3}; \; \rho_w \approx 1 \text{ g} \bullet \text{cm}^{-3}. \quad (15.38)$$

Absorption coefficient of Cg is lesser, than that of liquid water, cf. lower and upper curves in Fig. 2.16b. The other feature concerns the low contribution of free rotors to absorption near 200 cm^{-1} because of lower proportion r of rotors, since the welldepth u is deeper in Cg, than in Hb-H_2O solution.

The loss contribution

$$\varepsilon_j'' \approx 4\pi \operatorname{Im}\left(\chi_j^*\right)$$

at microwaves of j th subensemble to the total loss ε'' is illustrated by Fig. 2.17a. The position ν_{Dj} of each loss maximum is determined mainly by the value τ_j of the corresponding lifetime.

THE ROUGH MODEL OF SKIN. We assume that *Cg bounds about 64% of water, contained in skin.* The concentration N_{add} of additional water in skin is thus

$$N_{add} \approx 0.36 N_{Cg} \approx 0.64 \bullet 10^{22} \text{ cm}^{-3}.$$

We suppose that dielectric behavior of this additional fraction is analogous to that of liquid water, the only distinction being the lower value (comparing with G in water) of the parameter $G_{add} = \mu^2 N_{add} / 3k_B T$. Introducing the corresponding susceptibility χ^*_{ad}, we take into account that the complex susceptibility of skin χ^*_{sk} differs from Eq. 15.34 just by term χ^*_{ad}:

$$\chi^*_{sk} = \chi^*_1 + \chi^*_2 + \chi^*_3 + \chi^*_{ad}, \text{ with } F = F_T . \tag{15.39}$$

In Fig. 2.17b we compare calculated absorption α of liquid water, of skin and of hydrated Cg. We see, that all frequency dependencies are similar. Here α is expressed in terms [dB/mm], where

$$\alpha\left[\frac{\text{dB}}{\text{mm}}\right] = 10 \log \left\{\exp\left(\alpha\left[\text{cm}^{-1}\right] \bullet 0.1 \text{ cm}\right)\right\} = \alpha\left[\text{cm}^{-1}\right] \bullet \log e . \tag{15.40}$$

Note: 10 or 20 dB/mm means attenuation of radiation is equal to 10 or 100 times on the 1 mm distance.

The estimated absorption coefficient of skin, being greater than α_{Cg}, is considerably lesser, than that of liquid water. Microwave radiation penetrates into skin on greater depth, than into water.

For more detailed comparison see Table 15.15.

THE REFLECTION COEFFICIENT $|R^2|$. In Fig. 2.17 c, d we present the spectrum of $|R^2|$, calculated for our simple model from expression

$$|R^2| = \frac{1 + |\varepsilon| - \sqrt{2}\sqrt{|\varepsilon| + \varepsilon'}}{1 + |\varepsilon| + \sqrt{2}\sqrt{|\varepsilon| + \varepsilon'}}, \quad |\varepsilon| \equiv \sqrt{\varepsilon'^2 + \varepsilon''^2}, \tag{15.41}$$

valid for normal incidence of radiation onto a smooth surface. We see that $|R^2|$ smoothly decreases with the rise of frequency f. It is important, that $|R^2|$ is in the case of skin considerably lesser, than in water. This result can be interpreted as follows: bound water molecules have restricted rotational mobility. The observed frequency dependence on f of $|R^2|$ is also smooth due to high absorption of bulk water, contained in skin.

Table 15.14 Experimental data and fitted parameters pertaining to strongly humidified collagen and to the model of skin.

Parameters	Bound water		Bulk water	
	Fraction 1	Fraction 2	Fraction 3	Addition (skin)
A. Experimental data				
τ_D, ps	1180	20	7.1	7.85
max [$\varepsilon''(\nu)$]	8.12		8.13	
ν_D, cm^{-1}	0.004	0.27	0.75	0.68
B. 3-fractional model				
h, g/g	0.16	0.11	0.11	
$N \bullet 10^{-22}$, cm^{-3}	0.63	0.43	0.43	0.64
τ, ps	35	0.75	0.25	0.47
u	6.2	5.77	5.0	5.1
k_μ	1.08	1.12	1.20	1.20
g	2.8	2.6	2.35	2.35
$\beta = 19.9°$	$\rho = 1$ g•cm^{-3}	$M = 18$	$\mu_0 = 1.84$ D	
$I = 1.483 \bullet 10^{-40}$ g•cm^2			$n_\infty^2 = 1.7$	$T = 27°$

15.3.3 Weakly Humidified Collagen

Here we consider the dielectric behavior of the layer of bound water (i.e. of fraction 1) by assuming that hydration h of this layer varies from zero to $h_1 = 0.16$ g/g (this h value was employed in the previous subsection). Consequently, we omit the second and last terms of the sum (15.34). We take the following parameters:

$$n_\infty^2 = 1.7, \ \rho_{Cg} = 1.2, \ u_1 = 6.2, \ g = 2.8, \ \tau_1 = 0.35 \text{ps}, \ k_\mu = 1.08 \ .$$

We also alter the normalized concentration G:

$$G(h) = \mu_0^2 \left(\frac{n_\infty^2 + 2}{3} \right)^2 \frac{k_\mu^2 N_A \rho_{Cg} h}{3 M_w k_B T} \qquad (5.42)$$

The calculated frequency dependencies of two components of permittivity are shown in Fig. 2.14 a, b. When h increases, the maximum loss ε''_D, the value $\varepsilon'(\omega_D)$, found for frequency $\omega_D = \frac{1}{\tau_D}$, and the static permittivity ε'_s rise.

Table 15.15 Microwave absorption, calculated for strongly humidified collagen, model of skin and liquid water. $T = 27°C$

f, GHz	ν, cm^{-1}	α, dB/mm		
		Cg	Skin	H_2O
2.8	0.09	0.27	0.29	0.30
3.6	0.12	0.40	0.45	0.49
4.6	0.15	0.59	0.68	0.80
5.8	0.20	0.86	1.02	1.29
7.5	0.25	1.22	1.51	2.05
9.6	0.32	1.69	2.20	3.19
12.4	0.41	2.29	3.13	4.84
15.9	0.53	3.08	4.37	7.08
20.3	0.68	4.13	5.98	9.93
26.1	0.87	5.50	7.99	13.3

Continuation of Table 15.15.

f, GHz	ν, cm^{-1}	α, dB/mm		
		Cg	Skin	H_2O
33.5	1.12	7.26	10.41	17.44
43.0	1.43	9.37	13.18	21.22
55.2	1.84	11.71	16.21	25.44
70.8	2.36	14.11	19.33	29.70
90.9	3.03	16.39	22.39	33.88
116.6	3.89	18.35	25.20	37.88
149.6	4.99	19.86	27.62	41.57
192.0	6.40	20.89	29.53	44.84
246.4	8.21	21.59	30.99	47.68
316.2	10.54	22.15	32.17	50.23

We remark that if $h = 0$, total hydration $h_{tot} \neq 0$, since we assume that structural water also exists (see Fig. 2.14a).

In Fig. 2.14b we picture the h-dependence of the quantity

$$f_K(h) = \frac{\varepsilon_s(h) - \varepsilon_\infty}{(2+\varepsilon_\infty)^2} \frac{2\varepsilon_s(h) + \varepsilon_\infty}{\varepsilon_s(h)}, \tag{5.43}$$

where we put $\varepsilon_\infty \equiv n_\infty^2 = 1.7$. If $h_{tot} > h_0$, $f_K(h)$ rises, when h and ε_s increase. This property qualitatively corresponds to the observed dependence (Bone and Pethig 1985). The rise with h of components ε', ε'' and of absorption α is a typical property of proteins and other biological objects, recorded in a wide frequency range. On the example of collagen such dependence is illustrated by Fig. 2.15 for millimeter waves.

15.4 Nonelectrolyte in Aqueous Solution

We regard dielectric response of aqueous-organic solution, in which the specific solvation takes place, viz. formation of water-organic complexes. It is known, see e.g. (Khurgin, Kudryashova, Zavizion, and Betski 1994, p. 507), that dimethyl sulfoxide (DMSO) molecules $(CH_3)_2SO$ are stable chemically in individual liquid as well as in aqueous surroundings. Due to polar S-O group DMSO is a potent H-bond acceptor and is characterized by a high static dielectric constant / dipole moment ($\varepsilon_s = 46$, $\mu_0 = 4.1$ D). Liquid DMSO is a strong absorbing medium. Two nonpolar CH_3-groups of DMSO also interact with water (such interaction is termed *hydrophobic*) and influence on its structure / molecular dynamics. DMSO is widely used in chemical technology and in biology as an aproton solvent. In *dilute* water solution DMSO molecule is fully hydrated by at least two or three H-bonded H_2O molecules. However such hydration becomes impossible in the concentrated DMSO-H_2O solution because of shortage of water content. Unlimited solubility of DMSO in water allows one to obtain complete concentration-loss and concentration-absorption curves (see Sec. 2.6.2).

Here we try to elaborate a simplified two-fractional molecular model of DMSO-H_2O solution. We intend to describe:

(a). *considerable (about 6 times) increase of the Debye relaxation time* τ_D, that is occurs, when the volume-part concentration C_V falls into [0.3, 0.7] range.

(b). *considerable distinction* between the observed microwave absorption α and the "additive" one,

$$\alpha_{\text{add}}(\nu, C_V) = \alpha_{\text{w}}(\nu, C_V) + \alpha_{\text{d}}(\nu, C_V), \tag{15.44}$$

calculated with account of the absorption coefficients α_w and α_d of water and of DMSO, *estimated for individual liquids* but with account of corresponding concentrations C_w and C_d in the solution.

Further calculations are made for $T = 20°C$.

EXPERIMENTAL DATA. Number densities N_w and N_d are determined by the observed molar concentrations C_w and C_d, expressed in mole/l (see Table 15.16):

$$N_w = 0.001\, N_A\, C_w \text{ and } N_d = 0.001\, N_A\, C_d. \tag{15.45}$$

It is also useful to employ the molar-part concentration

$$C_p = \frac{C_{\mathrm{d}}}{C_{\mathrm{w}} + C_{\mathrm{d}}} \tag{15.46}$$

Few molecular constants of individual liquids and concentration dependencies $C_{\mathrm{p}}(C_{\mathrm{V}})$, $\varepsilon_{\mathrm{s}}(C_{\mathrm{V}})$, $\tau_{\mathrm{D}}(C_{\mathrm{V}})$, $n_\infty(C_{\mathrm{V}})$ are also given in Table 15.16 for a full [0, 1] range of volume-part concentrations C_{V} of DMSO in water. We remark that n_∞ (refraction on the orientational-range edge) is identified with the recorded optical index n_{D}. The observed microwave data are given in Table 15.17; these data were kindly presented to the author by Dr. V. N. Apletalin [private communication, see also (Gaiduk, Liberman, Apletalin, and Meriacri in press]. All recorded dependencies were smoothed by the author.

TWO-FRACTIONAL MODEL. In this model the solvent-fraction is characterized by the parameters N_{w}, I_{w}, $\mu_{0\mathrm{w}}$ and the solute-fraction - by the parameters N_{d}, I_{d}, $\mu_{0\mathrm{d}}$; moments of inertia I_{w} and I_{d} and isolated-dipole moments $\mu_{0\mathrm{w}}$ and $\mu_{0\mathrm{d}}$ are taken for pure water and pure DMSO. The SF L(z) is represented as the sum

$$L(z) = \overset{\vee}{L}_{\mathrm{w}}(z_{\mathrm{w}}) + \overset{\circ}{L}_{\mathrm{w}}(z_{\mathrm{w}}) + L_{\mathrm{d}}(z_d) \tag{15.47}$$

of terms, obliged to DMSO (subscript d) and to water (subscript w). As for water, we consider contributions of librators (subscript $\vee$) and rotators (subscript $\circ$). We also introduce normalized concentrations G_{d} and G_{w}:

$$\overset{\vee}{z} = \eta_{\mathrm{w}}\left(\omega + \frac{1}{\overset{\vee}{\tau}}\right),\ \overset{\circ}{z} = \eta_{\mathrm{w}}\left(\omega + \frac{1}{\overset{\circ}{\tau}}\right);\ \eta_{\mathrm{w}} = \sqrt{\frac{I_{\mathrm{w}}}{2k_B T}};, \tag{15.48}$$

$$z_{\mathrm{d}} = \eta_{\mathrm{d}}\left(\omega + \frac{1}{\tau_{\mathrm{d}}}\right),\ ;\ \eta_{\mathrm{d}} = \sqrt{\frac{I_{\mathrm{d}}}{2k_B T}};\ G_{\mathrm{w}} = \frac{\mu_{\mathrm{w}}^2 N_{\mathrm{w}}}{3k_B T};\ G_{\mathrm{d}} = \frac{\mu_{\mathrm{d}}^2 N_{\mathrm{d}}}{3k_B T}; \tag{15.49}$$

$$\mu_{\mathrm{w}} = \mu_{0\mathrm{w}}\, k_{\mu\mathrm{w}}\left(n_\infty^2 + 2\right)/3;\ \mu_{\mathrm{d}} = \mu_{0\mathrm{d}}\, k_{\mu\mathrm{d}}\left(n_\infty^2 + 2\right)/3. \tag{15.50}$$

Table 15.16 Experimental data / fitted parameters for DMSO-water solution. C_V, C_d in mole/l, τ_d, $\check{\tau}$, $\mathring{\tau}$ in ps

Solute (DMSO)					Solvent (water)			
C_V	C_d	C_p	τ_d	u_d	C_w	τ_L	τ_R	u_w
0.0	0.0	0.0	2.6	8.9	55.4	0.55	0.27	6.05
0.05	0.71	0.01		8.9	52.7			
0.1	1.42	0.03	2.7	9.0	50.0	0.69	0.35	6.12
0.2	2.84	0.06	2.7	9.0	44.7	0.80	0.40	6.17
0.3	4.29	0.10	2.7	9.0	39.3	1.11	0.55	6.31
0.4	5.75	0.15	2.7	9.0	33.9	1.41	0.71	6.45
0.5	7.22	0.20	2.7	9.0	28.4	1.72	0.86	6.59
0.6	8.69	0.28	2.7	9.4	22.8	2.51	0.63	6.91
0.7	10.1	0.37	3.7	9.4	17.1	2.50	0.63	6.91
0.8	11.5	0.50	2.9	9.3	11.3	2.63	0.86	6.91
0.9	12.8	0.70	1.8	9.2	5.6	2.50	0.63	6.91
0.95	13.5	0.83	1.6	9.2	2.5	2.50	0.63	6.91
1.0	14.1	1.00	1.3	9.1	0.0	2.51	0.63	6.91

	Solution		
C_V	n_D^2	τ_D, ps	ε_s
0.0	1.77	9.4	80.1
0.05	1.79		
0.1	1.82	13.2	79.7
0.2	1.85	17.0	79.4
0.3	1.90	26.3	78.4
0.4	1.95	36.7	76.9
0.5	1.99	47.1	75.4
0.6	2.04	57.9	73.2
0.7	2.08	60.7	69.2
0.8	2.12	52.5	65.0
0.9	2.16	34.4	56.6
0.95	2.17	28.4	51.8
1.0	2.18	20.5	46.0
$I_w = 1.483 \bullet 10^{-40}$ g•cm^2 $\mu_{0w} = 1.84$ D $\beta_w = 18.8°$		$I_w = 1.49 \bullet 10^{-38}$ g•cm^2 $\mu_{0d} = 3.96$ D, $k_{\mu d} = 1.06$ $\beta_d = 24.0°$	

Table 15.17 Microwave absorption α and components ε', ε'', observed in water-DMSO solution at $T = 20°C$. Data belong to V.N. Apletalin (private communication).

ν,	Volume-part concentration C_V										
cm^{-1}	0	0.1	0.2	0.3	0.4	0.5	0.6	0.7	0.8	0.9	1.0
Absorption α [dB/mm]											
9.10	51.5	43.5	37.0	30.8	26.0	22.5	21.3	21.3	22.5	24.5	27.3
10.6	55.5	47.5	39.5	33.0	29.0	25.5	23.8	24.0	25.6	27.5	30.2
12.5	60.0	50.5	43.0	37.0	32.0	28.5	26.8	27.5	28.8	30.5	33.4
14.5	64.3	55.6	47.5	41.0	36.0	32.0	31.5	32.5	34.4	36.3	39.0
24.4	86.5	77.5	70.0	62.5	56.0	52.0	49.6	50.0	52.5	54.6	56.5
Real part ε' of permittivity											
9.10	4.33	4.27	4.22	4.14	4.04	3.97	3.88	3.81	3.70	3.46	3.36
Imaginary part ε'' of permittivity											
9.10	2.80	2.47	2.22	1.95	1.70	1.56	1.49	1.50	1.52	1.56	1.65

Dielectric response of *each* fraction is described on the basis of the HM.

For water: $\check{L} = \dfrac{\check{K}_{\|} + 2\check{K}_{\perp}}{3}$, Eq. 9.41, 9.42, and $\mathring{L} = \mathring{K}$, Eq. 9.43.

For DMSO: $L_d = L$, Eq. 9.47.

We suppose that low-frequency spectra are characterized by the *single* relaxation time τ_D / *single* Kirkwood factor g.

Let m denote any of symbols $\check{\ }$ or $\mathring{\ }$ or of subscript «d» (summation is made over m). We write down, on the basis of general theory, expressions for the complex susceptibility χ^* (with $F = F_T$) and for estimated relaxation time τ_D^{est} (see Sec. 6.8.2), elaborated for solutions:

$$\chi^*(\omega) = \frac{g\sum L_m G_m\left(\omega + \frac{1}{\tau_m}\right)}{g\omega + i\sum \frac{L_m}{\tau_m} R_m\left(1 + g x_m z_m\right)}; \; R_m = \frac{G_m}{\sum G_m}; \tag{15.51}$$

$$\tau_D^{\text{est}} = \frac{\sum G_m\left[1 - y_m^2\, L_m(iy_m)\right]}{\sum \frac{G_m\, L_m(iy_m)}{\tau_m}} \tag{15.52}$$

$$g = \frac{\left(\varepsilon_s - n_\infty^2\right)\left(2\varepsilon_s + n_\infty^2\right)}{12\pi\,\varepsilon_s \sum G_m}. \tag{15.53}$$

For loss ε" and absorption α we use standard expressions, mentioned in Table 14.1.

FREQUENCY DEPENDENCIES OF LOSS / ABSORPTION. We fit potentials / lifetimes $\left(u_{\text{w}}, \overset{\vee}{\tau}, \overset{\circ}{\tau}\right)$ and $\left(u_{\text{d}}, \tau_{\text{d}}\right)$ for a number of C_{V}-values with use of the following equation for τ_{D}:

$$\tau_D^{\text{est}} = \tau_D \quad \text{for} \quad C_V \in [0, 1], \tag{15.54}$$

other free parameters being kept independent on C_{V}. For low and moderate C_{V} (say, for $C_{\text{V}} \in [0, 0.5]$) *all DMSO molecules are bounded by water*; we regard that τ_{d} and u_{d} do not depend on C_{V}.

In concentrated solution all water molecules are probably bounded; for large C_{V} we assume that u_{w} does not depend on C_{V}.

This fitting procedure still remains ambiguous, since we use only one equation (15.54) for a number of free parameters. One set of fitted parameters, for which the observed τ_{D} agrees with τ_D^{est} (see Fig. 2.22), is given in Table 15.16.

Calculated and observed spectra are plotted in Fig. 15.11 for pure individual liquids and in Fig. 15.12 a, b, c - for DMSO-water solution.

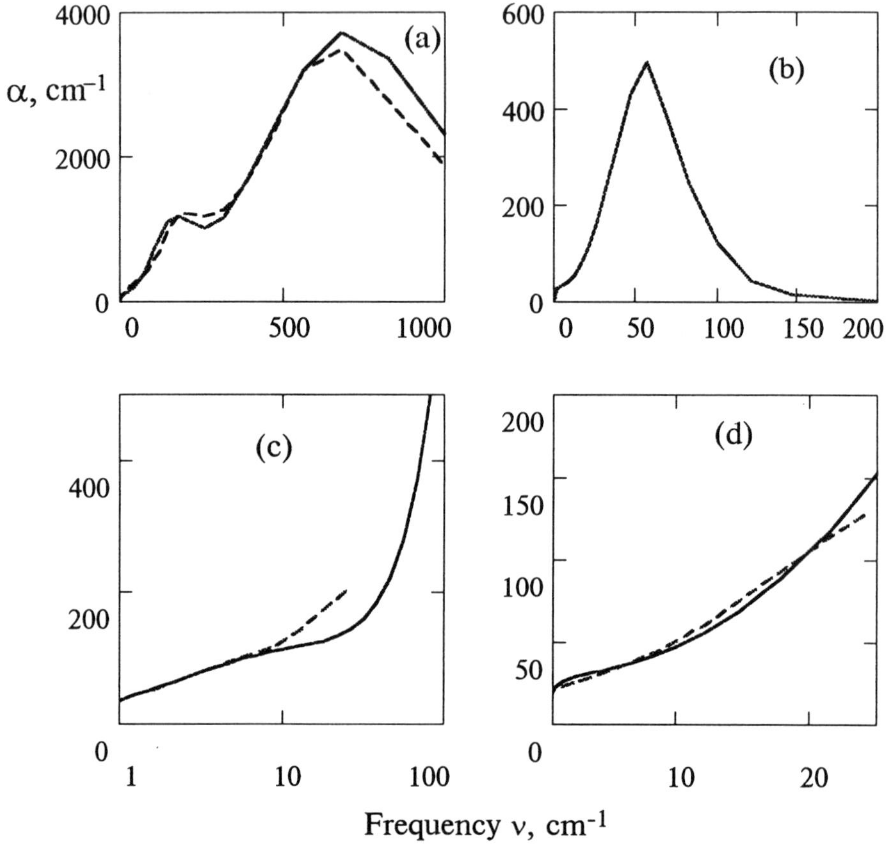

Fig. 15.11 Absorption of liquid water (a, c) and of DMSO (b, d) vs. frequency. Solid lines: calculated for the hybrid model, dashed lines: the observed data. Note: fitted parameters slightly differ from those, presented in Table. 15.16, viz: for water $\beta = 18.8°$, $u = 5.8$, $\tau = 0.42$ ps, $k_\mu = 1.19$; for DMSO $\beta = 24°$, $u = 7.9$, $\tau = 1.3$ ps, $k_\mu = 1.06$.

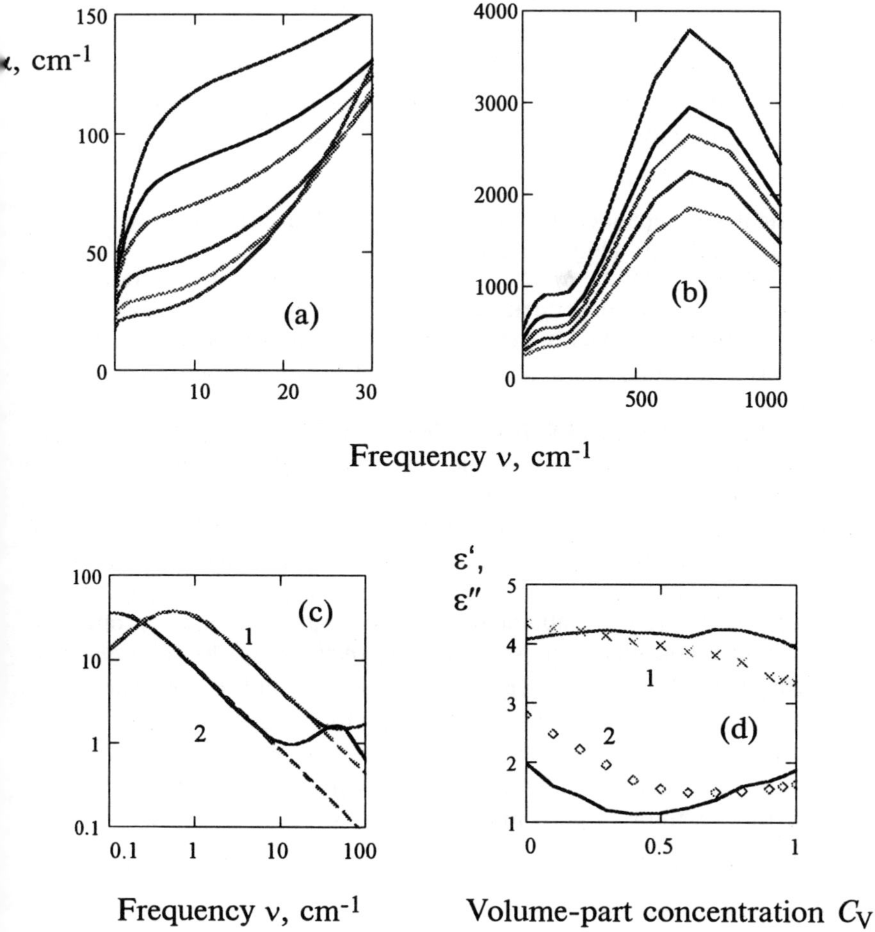

Fig. 15.12 Absorption (a, b) and loss (c) for DMSO-water solutions; real (curve 1) and imaginary (curve 2) parts of the permittivity vs. concentration of DMSO (d). Solid lines - calculation for two-fractional model, points - experimental data. **(a, b):** C_V = 0, 0.1, 0.2, 0.3, 0.4, 0.5 (from top to bottom). **(c):** curves 1 for C_V = 0 and curves 2 for C_V = 0.5; dashed lines - calculation for Debye formula with experimental parameters. **(d)**: wavelength λ = 0.41 mm.

Fig. 15.12a illustrates the "over-Debye" absorption in submillimeter range and Fig. 15.12b - transformation of the two-humped absorption spectrum, specific for water, to the spectral dependence $\alpha(\nu)$, having one maximum (cf. also Fig. 15.11a with Fig. 15.11b). In Fig. 15.12c we show, how the Debye loss curve $\varepsilon''(\nu)$ (dashed lines) transforms into "over-Debye" spectrum.

for water: $\beta = 18.8^o$, u=5.8, $\tau_L = \tau_R$ =0.42 ps, k_μ=1.19;

for DMSO: β=24°, u=7.9, τ =1.3 ps, k_μ=1.06.

Theoretical C_V dependencies of ε' and ε'' qualitatively agree with those, recorded by Dr. Apletalin (Fig. 15.12 d).

THE ABSORPTION DEFICIT. In Fig. 2.23 we see that theoretical and observed concentration dependencies of absorption agree qualitatively. Consequently, our theory yields description of the *absorption deficit,* i.e. the distinction between α and α_{add}, obliged to the decrease of rotational mobility in the solution.

CONCLUDING REMARKS. It is evident from Fig. 15.11c that the hybrid model yields *lower,* than it is found experimentally, *submillimeter-range absorption of water* (for $\nu \in$ [10, 50 cm^{-1}]). Indeed, theoretical $\alpha(\nu)$ curve exhibits signs of the Debye plateau, while the recorded absorption substantially rises with ν. Considerable discrepancy between calculated and observed absorption is seen, if one compares Fig. 2.23 a (theory) with Fig. 2.23 b (experiment). However, the elaborated model describes the main property of the solution under consideration, viz. the *sharp increase of the relaxation time* τ_D comparing with the observed τ_D-values of solvent and solute. The rise of τ_D with C_V is accompanied by alteration of structure / molecular dynamics in both fractions: *fitted lifetimes / intermolecular potentials are larger* in comparison with values of τ, u, fitted for individual liquids (water and DMSO).

This picture resembles that, gained for water, which is bound by a protein molecule (Sec. 15.3). However, in the last medium we observe *a number of relaxation regions,* characterized by different τ_{Dj}, while in DMSO-H_2O solution the loss curve $\varepsilon''(\nu)$ can be represented, in a first approximation, in the form of a single-Debye relaxation region, that is characterized by a single relaxation time τ_D.

CHAPTER 16
QUANTUM EFFECTS IN POLAR GASES AND GAS-LIKE LIQUIDS

16.1 The Quasi-classical Approach

16.1.1 Rotational Frequencies

Quantified energy E_J of linear molecules, frequencies Ω_J [rad/s] or ν_J [cm^{-1}] of *equidistant rotational spectrum* are functions of quantum numbers J (Townes and Schawlow 1975):

$$E_J = \frac{h^2}{8\pi^2 I} J(J+1), \tag{16.1}$$

$$\Omega_J = \frac{2\pi}{h}(E_{J+1} - E_{J+1}) = \frac{h}{2\pi I}(J+1), \tag{16.2}$$

$$\nu_J = \frac{\Omega_J}{2\pi c} = 2B\,(J+1); \tag{16.3}$$

$$B = \frac{h}{8\pi^2 cI} \tag{16.4}$$

being the rotational constant. It is convenient to employ dimensionless frequency (x), term of rotational energy (e_J), J th rotational frequency (p_J), normalized concentration (G):

$$x = \eta\omega;,\ e_J = \frac{E_J}{k_B T} = d^2 J(J+1),\ p_J = \frac{\nu_J d}{2B} = d(J+1);$$

$$\eta = \sqrt{\frac{I}{2k_BT}},\ d = \frac{\eta h}{2\pi I} = \frac{h}{2\pi\sqrt{2k_BTI}} = \sqrt{\frac{Bhc}{k_BT}} = p_0,\ G = \frac{\mu^2 N}{3k_BT}. \quad (16.5)$$

16.1.2 The Set of Lorentz Lines

THE COLLISIONLESS LIMIT. The loss susceptibility $\chi''(x)$, being an odd function, is expressed for a long lifetime limit as series of δ -functions, located at frequencies p_J (Van Vleck and Huber 1977). Let $\sum \equiv \sum_{J=0}^{\infty}$. Then

$$\chi''(x) = \frac{\pi G}{2Zd^2} \sum q_J p_J \left[\delta(p_J - x) - \delta(p_J + x)\right] \text{ for } \tau \to \infty; \quad (16.6)$$

$$q_J = \exp(-e_J) - \exp(-e_{J+1}) \quad (16.7)$$

are weight coefficients of rotational states ($\sum q_J = 1$). Rigorous and approximate expressions of statistic sum Z are:

$$Z = \sum (2J+1) \exp\left[-d^2 J(J+1)\right]; \quad (16.8a)$$

$$Zd^2 \approx 1 + \frac{d^2}{3} + \frac{d^4}{15} + \frac{4d^6}{316} + \frac{d^8}{315}. \quad (16.8b)$$

If T is very low, d is large, $Z \approx 1$, then

$$\lim_{T \to 0} \frac{G}{Zd^2} = \frac{8IN}{3}\left(\frac{\pi\mu}{h}\right)^2;$$

$G/(Zd^2)$ remains finite for all T-values. However, since a gas state may exist only at rather high temperature, usually $d << 1$ and Zd^2 is about 1.

In accordance with Eq. 16.6 and rule (5.172) the static susceptibility χ_s is given by

$$\chi_s = \frac{2}{\pi}\int_0^{\infty} \chi''(x)\frac{dx}{x} = \frac{G}{Zd^2}\sum q_J = \frac{G}{Zd^2} \quad (16.9)$$

The integrated absorption obeys the Gordon sum rule (5.145) and does not depend on T:

$$\int_{-\infty}^{\infty} \omega\chi''(\omega)\,d\omega = \frac{\pi G}{\eta^2 Z d^2}\sum q_J p_J^2 = \frac{2\pi\mu^2 N}{3I}\,\frac{S}{Z} \equiv \frac{2\pi\mu^2 N}{3I}; \quad (16.10)$$

$$S = \sum (J+1)^2 \left\{ \exp\left[-d^2 J(J+1) - \exp\left[-d^2 (J+1)(J+2)\right]\right]\right\} \equiv Z.$$

TRANSITION TO THE CASE OF FINITE LIFETIMES. If = $\eta/\tau > 0$, Eq.16.6 can be generalized to the set of *Lorentz lines*:

$$\chi^* = \overset{*}{\chi}_B(x) = \frac{G}{Zd^2} L(x); \; L(x) = \sum q_J L_J, \quad (16.11)$$

$$L_J = \frac{p_J^2}{p_J^2 - z_J^2} = \frac{p_J}{2}\left[\frac{p_J - x + iy_J}{(p_J - x)^2 + y_J^2} + \frac{p_J + x - iy_J}{(p_J + x)^2 + y_J^2}\right], \quad (16.12)$$

L and L_J are respectively *quantified spectral function* (SF) and Lorentzian; $z_J = x + iy_J$ and $y_J = \eta/\tau_J$ are dimensionless complex and collision frequencies, related to J th line; y_J is proportional to its half-width $\Delta\nu_J$. Eq. 16.11-16.12 follow from Eq.16.6, if, taking an usual representation of δ-function,

$$\delta(p_J - x) - \delta(p_J + x) = \frac{1}{\pi}\lim_{y_J \to 0}\left[\frac{y_J}{(p_J - x)^2 + y_J^2} - \frac{y_J}{(p_J + x)^2 + y_J^2}\right], (16.13)$$

we rewrite Eq.16.6 as follows:

$$\chi''(x) = \operatorname{Im}\left\{\lim_{y_J \to 0} \frac{G}{2Zd^2}\sum_{J=0}^{\infty} q_J p_J \left[\frac{1}{p_J - z_J} + \frac{1}{p_J + z_J}\right]\right\}. \quad (16.14)$$

Omitting symbols Im, lim and replacing χ'' by χ^*, we get Eq. 16.11.

In terms of our dynamic method the set (16.11) of Lorentz lines is a quantum-corrected susceptibility, calculated for the *Boltzmann induced distribution* F_B.

16.1.3 The Set of Gross Lines

More adequate description of quantum phenomena is obtained *by taking account of collisions in a scope of classical theory.* We apply the correlation-orientation collision model with induced distribution $F = F_K$ and make the following steps:

(A). We start from the usual expression for the complex susceptibility at F_K:

$$\chi^*(x) = \frac{gGzL(z)}{gx + iyL(z)}. \tag{16.15}$$

(B). The SF $L(z)$ is replaced by its quantum equivalent (16.11); complex / collision frequencies z, y are inserted *inside sums* just like it was made in Eq.16.11, 16.12:

$$z\,L(z) \to \sum z_J q_J L_J;\; y\,L(z) \to \sum y_J q_J L_J\,. \tag{16.16}$$

(C). We replace parameter G by its quantum analog,

$$G \to \frac{G}{Zd^2}, \text{so that } g \to \frac{\varepsilon_s - n_\infty^2}{4\pi \dfrac{G}{Zd^2}} \frac{2\varepsilon_s + n_\infty^2}{3\varepsilon_s}. \tag{16.17}$$

Finally we derive

$$\chi^*(x) = \frac{gG}{Zd^2} \sum L_{GJ};\; L_{GJ} = \frac{z_J q_J L_J}{gx + i\sum y_J q_J L_J}. \tag{16.18}$$

This Gross collision model describes *interference* of absorption lines, since *all terms affect due to denominator the dispersion of any J th* term L_{GJ}. On the contrary, each Lorentzian L_J contributes *independently* to χ^* in Eq. 16.11.

The single Lorentz line (16.12) describes resonance frequency dependence, specific for rare collisions ($y_J << 1$), while Eq. 16.11, 16.18 present a set of resonance lines. If collisions become frequent, these lines overlap, quantum specificity (the discrete spectrum) disappears. We shall show that if y_J is relatively large, Eq. 16.18 yields such rotational spectrum that coincides with that, described by the extended diffusion model (see Eq. 16.15). It is important, that this ED model is based on classical equations and does not involve, unlike quantified description (16.18), the rotational numbers J and Planck constant h.

Since the complex susceptibility, Eq. 16.18, is equivalent for rare collisions to rigorous quantum Eq. 16.6 and for frequent collisions - to classical expression (16.15), one may suggest that Eq.16.18 may be successfully applied to polar gases in a wide range of pressures.

16.2 Resonance Phenomena in Biatomic Gases

16.2.1 Molecular Parameters of a Gas State

It is convenient to express these parameters through rotational constant B (16.4):

$$I = \frac{h}{8\pi^2 cB}, \quad \eta = \frac{1}{4\pi}\sqrt{\frac{h}{ck_B TB}}, \quad d = \sqrt{\frac{hcB}{k_B T}}. \tag{16.19}$$

Let us identify τ with the mean time between collisions in an ideal gas, that obey Maxwellian velocity distribution. Following (Menzel 1957, p.258), we consider the gas mixture, consisting of spherical elastic molecules, characterized by diameters σ_a, σ_b and concentrations N_a, N_b. The mean collision frequencies ν_a, ν_b, lifetimes τ_a, τ_b, and the path l_a of a-molecules are given by:

$$\nu_a = \frac{2n_{aa} + n_{ab}}{N_a}, \quad \nu_b = \frac{2n_{bb} + n_{ab}}{N_b}, \quad \tau_a = \frac{1}{\nu_a}, \quad \tau_b = \frac{1}{\nu_b},$$

$$\frac{1}{\pi l_a} = \frac{N_a \sigma_a^2}{\sqrt{2}} + N_b \sigma_{ab}^2 \sqrt{1 + \frac{m_a}{m_b}}; \quad \sigma_{ab} = \frac{\sigma_a + \sigma_b}{2}, \tag{16.20}$$

$$n_{aa} = 2N_a^2 \sigma_a^2 \sqrt{\frac{\pi k_B T}{m_a}}; \quad n_{bb} = 2N_b^2 \sigma_b^2 \sqrt{\frac{\pi k_B T}{m_b}};$$

$$n_{ab} = 2N_a N_b \sigma_{ab}^2 \sqrt{\frac{2\pi(m_a + m_b)k_B T}{m_a m_b}}. \tag{16.21}$$

Path l and concentration N are given for *homogeneous gas* by

$$l = \frac{\sqrt{2}}{\pi N \sigma^2}, \quad N = L_0 \frac{P}{760}\frac{273}{T}, \quad L_0 = 2.687 \bullet 10^{19}\ \text{cm}^{-3}; \tag{16.22}$$

L_0 and P [mm Hg] being the Loschmidt number and pressure, respectively.

Concentration $N \equiv L_0$, corresponding to normal conditions (P = 760 mm Hg and T = 273 K), is termed 1 *amagat*.

16.2.2 Dielectric Behavior of Carbon Monoxide CO

The product $cB \bullet 10^{-9}$ yields rotational constant in GHz. For *pure CO gas* at $T = 293$ *K* we should take the following parameters:

$$\mu_0 = 0.118 \text{ D}, \ B = 57.898 \text{ GHz} = 1.944 \text{ cm}^{-1}, \ M = 28, \ \sigma = 2.44 \ \mathring{A},$$

$$d = 0.097, \ \eta = 1.338 \bullet 10^{-13} \text{ s}. \qquad (16.23)$$

We also consider the CO-N_2*gas mixture*; experimental half-widths $\Delta\nu_J$ are presented in Table 16.1 along with $\Delta\nu_J$ for gaseous HCl and DCl.

Now we regard concentration N_a of CO gas to be equal to N of pure CO gas for $P = 1$ atm, the rise of P in the mixture being obliged *to the increase of buffer-gas concentration* N_b. We calculate absorption α for $P = 10$ and 1000 atm, take $g = 1$, $n_\infty = 1$ and assume that $|n(\nu) - 1| \ll 1$. Simplified formulas for permittivity $\varepsilon*$ and absorption α hold:

$$\varepsilon^* \approx 4\pi\chi^* + 1, \ \alpha = \frac{\omega\,\varepsilon''}{cn} \approx 8\pi^2\nu\chi''. \qquad (16.24)$$

We calculate χ^* for the sets of Lorentz and Gross lines, described respectively by Eq. 16.11 and 16.18. If $P = 760$ mm Hg, then

$$N = N_a \approx L_0 \approx 2.5 \bullet 10^{19} \text{ cm}^{-1}, \ G \approx 2.87 \bullet 10^{-6}.$$

The distinction between G-values corresponding to gas or to liquid (in the latter G has order of 1) comprises 6 orders of magnitude. One may await that similar distinction should also concern the maximal absorption.

For arbitrary pressure P

$$y_J = 2\pi c\eta\,\Delta\nu_J\,10^{-3}\,P\,[\text{atm}]. \qquad (16.25)$$

For three lower lines in Table 16.1 we present (from top to bottom) slightly corrected values of $\Delta\nu_J$, that yield smooth falling dependence $\Delta\nu(J)$, y_J and $\tau_J = \eta/y_J$. We calculate absorption for $J \in [0, 50]$ by employing interpolation of the recorded $\Delta\nu(J)$ curve (see Fig. 2.2d). The calculated absorption spectrum of CO + N_2 mixture is plotted in Fig. 2.2 a, b, c.

For $P = 10$ atm discrete rotational lines are clearly seen (Fig. 2.2a), the distinction between the sets of Lorentz and Gross lines (i.e. between the induced distributions F_B and F_K) is substantial in the vicinity of the first line (see Fig. 2.2b) and on HF tail of the spectrum (see Fig. 2.2a).

Table 16.1 Experimental halfwidths $\Delta\nu_J$ of absorption lines for gaseous $CO+N_2$, HCl and DCl. T = 293 K. From (Gray and Van Kranedonk 1966). The three lower lines represent quantities $\Delta\nu_J$, $y_J \bullet 10^3$ [atm^{-1}] and τ_J [ms / atm], estimated by the author after smoothing of the upper line, corresponding to $\Delta\nu_J$ in $CO+N_2$ mixture

J	0	1	2	3	4	5	6	7	8	9	10	11
	$\Delta\nu_J$ [$cm^{-1}\bullet 10^3$ / atm]											
$CO+N_2$	82	79	77	74	72	70	66	64	64	61	63	59
HCl	268	270	295	293	261	232	170	161	131	112	97	85
DCl	229	246	263	276	278	267	249	228	202	180	161	141
	Corrected values $\Delta\nu_J$ [$cm^{-1}\bullet 10^3$ / atm]											
$CO+N_2$	82	79	77	74	72	70	67	65	63	61	60	59
	Corrected values $y_J \bullet 10^3$ [atm^{-1}]											
$CO+N_2$	2.1	2.0	1.9	1.9	1.8	1.8	1.7	1.6	1.6	1.5	1.5	1.5
	Estimated values of τ_J [ms / atm]											
$CO+N_2$	65	68	69	72	74	76	80	82	85	88	89	91

For P = 1000 atm individual rotational lines disappear because of their broadening; the distinction between collision models with F_B and F_K becomes substantial in all spectral region (see Fig 2.2c). At F_K maximum absorption is larger, than at F_B, its peak being shifted to lower frequencies; moreover, at F_K we have true expression for the static permittivity ε_S, in which the contribution to ε_S of rotational transitions does not depend on P (Fig.16.1, solid line). At F_B this contribution, found from Eq. 16.11 at ω = 0, depends on P (dashed line).

CLASSICAL DESCRIPTION. We employ the extended diffusion model with the induced distribution F_K, Eq. 16.15, where $z = 2\pi c\nu + i\langle y\rangle$; $L(z)$ is determined by Eq. 7.11. *The averaged collision frequency*, given by

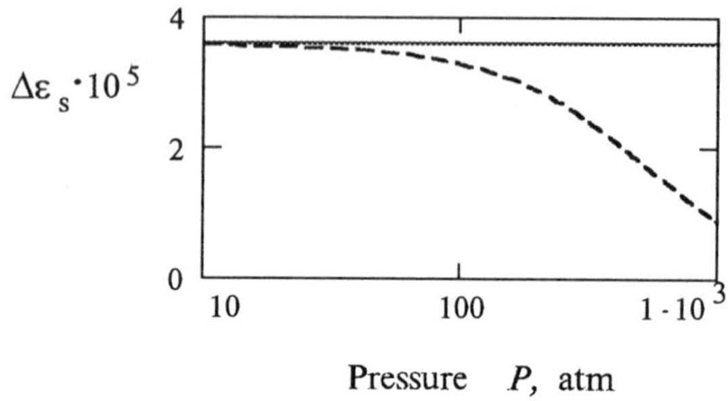

Fig.16.1 The contribution of rotational transitions to ε_S vs. P dependence: Sets of Gross (solid curve) and Lorentz (dashed curve) lines. Gas mixture CO-N_2, $T = 20^oC$.

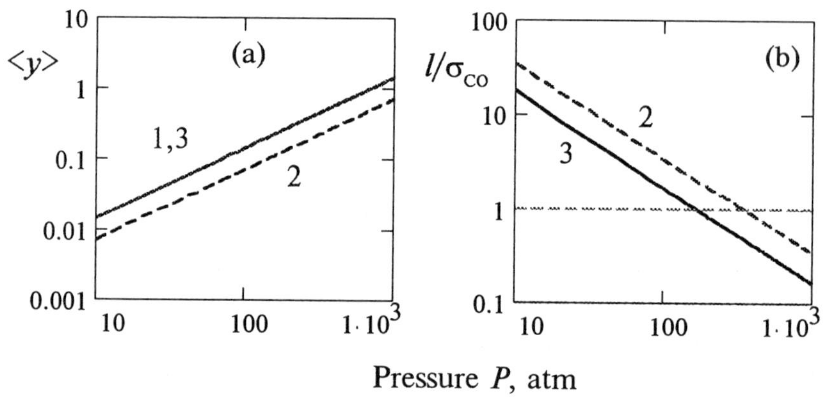

Fig.16.2 Mean normalized collision frequency (a) and path / diameter ratio (b) vs. pressure. CO component of CO-N_2 gas mixture: experimental linewidths (1), kinetic theory with σ(buffer) = 4.09 Å (2) or 7 Å (3).

$$\langle y \rangle \approx 2\pi c\eta\, 10^{-3}\, P \sum q_J\, \Delta\nu_J\,, \tag{16.26}$$

is the free parameter of the ED model, that can be estimated with use of Table 16.1.

The dependence of $\langle y \rangle$ on pressure is shown in Fig. 16.2a (solid line); $\langle y \rangle$ is very low for $P = 10$ atm, while for $P = 1000$ atm $\langle y \rangle$ is about 1, being commensurable with y-values in liquids. Comparison shows (cf. curves 1 and 3 in Fig. 2.2c), that if $P = 1000$ atm, classical and quantum descriptions coincide; we mentioned above this property.

Let us estimate the mean lifetime τ with use of classical kinetic theory of ideal gases. We find τ by assuming, that the same $y(P)$ dependence should be obtained for CO-N_2 mixture from the theory and experimental data. It follows from Eq. 16.21, 16.22, that if we put partial pressure $P_{CO} = 1$ atm, the buffer-gas concentration N_b will be related to P as follows:

$$N_b \approx \left(P\,[\mathrm{atm}] - 1\right)\frac{273}{T} L_0\, N_b, \tag{16.27}$$

the buffer diameter σ_b being a free parameter of the kinetic theory. If, following (Townes and Schawlow 1959, p.327), we take $\sigma_b = 4.09$ Å, then the $y(P)$ dependence, where $y = \eta/\tau_a$, becomes indeed close to $\langle y(P) \rangle$; it coincides with $\langle y \rangle$ for $\sigma = 7$ Å (see Fig. 16.2a).

Application of Eq. 16.22 shows (Fig. 16.2b), that for $P \in [10, 1000]$ atm the path l falls into the range [1, 0.1] Å.

Thus, quantum features disappear in CO-N_2 gas mixture at $P \approx 1000$ atm, the lifetime becomes commensurable with that, estimated from the kinetic theory; geometrical restrictions appear, since $l_{CO} < \sigma$, however the factor G and density ρ being still lesser, than in the corresponding liquid.

INDIVIDUAL ROTATIONAL LINE. Let us consider the case of low pressures (100 and 860 mm Hg). Now we apply the theory to a pure gas CO. In Fig. 16.3 we plot frequency dependencies of absorption $\alpha(\nu)$ and of the difference $\Delta\varepsilon'(\nu) \equiv [\varepsilon'(\nu) - 1]$. In Fig. 16.3a we take ν in the vicinity of the 9 $\rightarrow$ 10 transition (that corresponds to a maximum absorption) and in Fig. 16.3b - near frequency

$$\nu_{\min}(J) \approx \frac{p_{J-1} - p_J}{2 \cdot 2\pi c\nu},$$

to which the absorption minimum between two lines corresponds. We employ Eq. 16.22 for concentration $N(P)$ and the data of Table 16.1 for $\Delta\nu_J(P)$.

In Fig. 16.3a we see, that the *absorption peak practically does not depend nor on pressure* P nor on the collision model (we mean F_B or F_K induced distributions). The contribution to absorption of terms in sums with $J > 10$ is also small. If ν is near $\nu_{\min}$, then $\alpha_{\min}$ increases with P and function $\Delta\varepsilon'(\nu)$ goes through zero at frequency $\nu_{\min}$.

16.2.3 The Semi-Classical Approach in Terms of Dipolar ACF

ENSEMBLE OF LINEAR MOLECULES. In Eq. 5.41 we expressed the SF $L(z)$ through the ACF spectrum $\tilde{C}^0(\omega + i/\tau)$, that is found for periodic ("undamped") reorientation of a dipole in the prescribed potential $U(\vartheta)$.

Regarding free rotation, for which $U(\vartheta) \to 0$, we introduce here the *unnormalized* ACF spectrum $\tilde{C}^0(\omega^*)$ and rewrite Eq. 16.18 for the generalized Gross collision model in terms of normalized ACF $\tilde{c}^0(\omega^*)$ and static susceptibility $\chi_s = gG$:

$$L(z) = 1 + i\,\omega^*\,\tilde{c}^0(\omega^*); \quad \omega^* \equiv \omega + i/\tau; \tag{16.28}$$

$$\chi^* = \frac{gG\omega^* L}{g\omega + (i/\tau)L} = \chi_s \left[1 + \omega \frac{(1-g) + i\omega^*\tilde{c}^0(\omega^*)}{g\omega + \frac{i}{\tau} - \frac{\omega^*}{\tau}\tilde{c}^0(\omega^*)} \right]; \tag{16.29}$$

$$\chi_s = g\beta N C^0(0); \quad \tilde{c}^0(\omega^*) = \frac{\tilde{C}^0(\omega^*)}{C^0(0)}; \; \beta = \frac{1}{k_B T}, \tag{16.30}$$

where $C^0(0) \equiv C^0(t{=}0)$. We remark that both ACF spectra, unnormalized and normalized, $\tilde{C}^0(\omega^*)$ and $\tilde{c}^0(\omega^*)$, has dimension [s]; the normalized quantity $\tilde{c}^0(\omega^*)$ is the ratio of ACF *spectrum* to the *ACF itself.*

We may also rewrite Eq. 16.29 in terms of the unnormalized ACFs:

$$\chi^* = g\beta N\left[C^0(0) + i\omega\ \tilde{C}(\omega^*)\right]; \tag{16.31a}$$

$$\tilde{C}(\omega^*) \equiv \frac{1-g+i\omega^*\tilde{C}^0(\omega^*)}{g\omega+\frac{i}{\tau}-\frac{\omega^*}{\tau}\tilde{C}^0(\omega^*)}. \tag{16.31b}$$

In the quantum ensemble we employ the sets: of lifetimes τ_J, of complex frequencies $\omega_J^* = \omega + i/\tau_J$ and of "undamped" ACF $\tilde{C}_J^0(\omega^*)$ respectively instead of τ, ω^* and $\tilde{C}^0(\omega^*)$. Putting for simplicity the Kirkwood correlation factor $g = 1$ (this is justified for rarefied fluids or gas-like liquids), we generalize Eq. 16.31 by *formal transition to sums over quantum numbers J*:

$$\chi^* = \beta N\left[C^0(0) + i\omega\ \frac{\sum\omega_J^*\tilde{C}_J^0(\omega_J^*)}{\sum\omega_J^*\left[1-\frac{1}{\tau_J}\frac{\tilde{C}_J^0(\omega_J^*)}{C^0(0)}\right]}\right]. \tag{16.32}$$

Eq. 16.30 for χ_S remains valid (with $g = 1$), while $C^0(0)$ may slightly differ. The first of Eq. 16.31 holds also for the quantum ensemble. Consequently, the resultant ACF spectrum $\tilde{C}(\omega^*)$, in which collisions of molecules are taken into account (but without account of cross-correlation), is related to the ACF-terms as follows (Kalmykov and Titov 1984):

$$\tilde{C}(\omega^*) = \frac{\sum\omega_J^*\tilde{C}_J^0(\omega_J^*)}{\sum\omega_J^*\left[1-\frac{1}{\tau_J}\frac{\tilde{C}_J^0(\omega_J^*)}{C^0(0)}\right]} \tag{16.33}$$

This relation is based on combination of the *semi-classical* consideration of quantum rotator with the *classical* description of electromagnetic field / collisions. To elaborate this approach, first, a discrete set of rotational states is found. Second, the kinetic equation for the density matrix, including the collision integral, is solved.

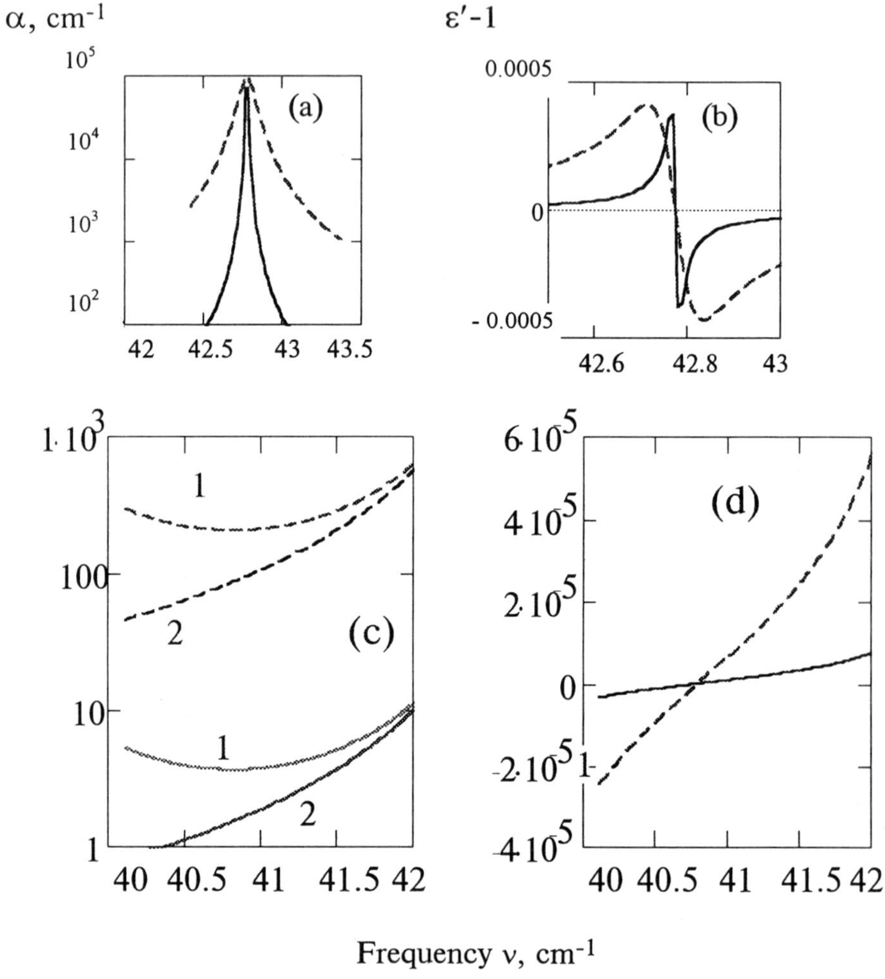

Fig. 16.3 Form of the rotational line, corresponding to $J = 9 \to 10$ transition (a, b) and form of the rotational spectrum in the vicinity of the transparency window between 9 th and 10 th peaks. Solid curves for $P = 100$ and dashed curves for $P = 760$ mm Hg. CO gas at 20°C. Calculated for the sets of Lorentz (curves 1) and Gross (curves 2) lines.

The relaxation time of the matrix is characterized by parameter τ. Complex susceptibility is related to the mean dipole moment of a polar medium. In its final form the problem is reduced to ensemble averaging of quantum states at equilibrium. One can regard this approach to be a realization of the Cubo linear-response theory for free quantum rotors.

ENSEMBLE OF SYMMETRIC-TOP MOLECULES. Expressions (16.32), (16.33) can be generalized (Kalmykov and Titov 1994) to this ensemble. Now additional K-set of quantum numbers appears, $\{J\} \to \{J, K\}$.

For *low concentration N* we have again the set of Lorentz-like rotational lines, each having the "dispersion" contour. The J th absorption peak is proportional to $N\tau$ and does not alter, when pressure P increases, since $N \propto P$, and $\tau \propto N^{-1}$. The half-widths $\Delta\nu_J$ *depend* on P, since $\Delta\nu_J \propto \tau^{-1} \propto N$.

In the low-frequency part of rotational spectrum (at microwaves), where $h\nu << 1$, the absorption coefficient *near the center of the J th rotational line*, where $\omega \approx \Omega_J$, is proportional to ω^3 (Strendberg 1956, p.82).

In the *non-resonant* low-frequency part of spectrum, where $\omega << \tau^{-1}$,

$$\alpha(\omega) \approx \frac{\omega\varepsilon''}{c} \approx \frac{4\pi\omega^2 N\mu^2}{3ck_BT}\frac{\tau_D^{-1}}{\omega^2+\tau_D^{-2}}. \tag{16.34a}$$

Hence, in the symmetric-top ensemble

$$\frac{\varepsilon''}{\omega} \approx \frac{4\pi N\mu^2}{3k_BT}\frac{\tau_D^{-1}}{\omega^2+\tau_D^{-2}}; \tag{16.34b}$$

the ratio ε''/ω *decreases* with the rise of ω.

Note: in the ensemble of linear molecules this ratio *increases* with ω, since

$$\frac{\varepsilon''}{\omega} \propto \frac{\alpha_J}{\omega^2} \propto \omega \qquad \text{for linear molecules.} \tag{16.34c}$$

The relaxation spectrum does *not* appear at all in this ensemble, since in distinction with (16.34b) the ratio (16.34b) tends to *constant* for $\omega \to 0$, while in Eq. 16.34c this ratio tends to *zero*.

The relaxation time τ_D was expressed for the Gross collision model by Kalmykov and Titov through the lifetime τ_0 of non-resonant transition:

$$\frac{1}{\tau_D} = \frac{1}{\tau_0}\left(1 - \frac{\sum_{JK} C^{\mathrm{o}}_{JK}}{C(0)}\right) \quad \text{for symmetric-top molecules.} \tag{16.35}$$

Another (completely quantum) feature is related to nuclear-spin statistics and consists in the *periodic alteration of weak and intense rotational lines.*

16.3 Classical Theory of Collisions

THE CLUE IDE . We have seen in Sec. 16.2.2, that the observed linewidths $2\Delta\nu_J$ depend on energy E_J of a molecule. Therefore, the lifetime τ_J of **J** th quantum state depends on angular frequency Ω_J of J th transition. *The lifetime* τ *of a rotating molecule may depend on its frequency* Ω: $\tau = \tau(\Omega)$.

On the other side, one can determine the lifetime $\tau_v(v)$ of a colliding particle in frames of classical collision theory, v being translational velocity of a particle corresponding to a prescribed potential U_v. Our aim is to estimate lifetimes τ_J of dipoles, rotating in polar gas-like liquids, where it is principally important to employ the $\tau(J)$ dependence.

If some correlation exists between rotational and translational velocities, Ω *and* v, *it is possible to recognize the relation of lifetime* τ *to rotational velocity* Ω *of a molecule or, in other words, to recognize the values* τ_J.

16.3.1 Collision Frequency of Binary Collisions

GENERAL RELATIONS. Let us consider the solution of the problem, considered in (Liboff 1969), where the kinetic equation method was applied to a steady-state molecular ensemble. It is possible to calculate the collision frequency $\left[\tau_v(v)\right]^{-1}$ of a polar particle. It is supposed that the translational velocity v (in laboratory coordinate system) of a particle is determined by the *repulsive* potential

$$U_v(r) = \Re r^{-n}, \qquad n = 2, 3 \dots . \tag{16.36}$$

Here r is the distance between two colliding particles, having respectively mass m (the probe particle) and mass m_1 (the particle of a solvent); n is here an index of interaction (the Reader should not mix it with the refraction index of the medium!). Dimension of U_v is [energy], dimension of $\Re$ should be [cm^n • energy]. When n increases from 2 to 8, $U_v(r)$ alters from the potential, specific for *dipolar interaction*, to potential, relevant to the *model of rigid spheres.* We assume that concentration N of probe particles is much lesser, than that (N_1) of a solvent, $N << N_1$, and put $\max r \approx \frac{1}{\sqrt[3]{N_1}}$.

Let

$$s = v\sqrt{\frac{m}{2k_BT}}$$

be normalized velocity of a probe particle 1. In view of Fig. 16.4 it flies to particle 2 from infinity and comes after scattering again to infinity. Its collision frequency

$$f_v(s) = \frac{1}{\tau_v(s)}$$

is given by

$$f_v(s) = C\int_0^{\psi_0} d\cos\psi \int_0^{\pi} d\cos\lambda \int_0^{\infty} \sigma^2 V s_1^2 \exp\left(-s_1^2\right)\cos\psi \, ds_1\,, \tag{16.37}$$

$$C = 4N_1\sqrt{\frac{\pi k_BT}{\overline{m}}};\; s_1 = v_1\sqrt{\frac{m_1}{2k_BT}};\; \theta = \pi - 2\psi,\; \overline{m} = \frac{m_1 m}{m_1 + m},$$

$$\sigma^2 = \left(\frac{\Re}{k_BT}\right)^{2/n} V^{-4/n}\,\frac{b\,db}{d\cos\theta};$$

$$V = \sqrt{s^2 + s_1^2 - 2ss_1\cos\lambda};\quad \psi(b) = \int_0^{\overline{\beta}} \frac{d\beta}{\sqrt{1-\beta^2\left(\frac{\beta}{b}\right)^n}}. \tag{16.38}$$

The upper limit $\overline{\beta}$ is a real positive root of the equation

$$1-\bar{\beta}^2-\left(\bar{\beta}/b\right)^n=0. \tag{16.39}$$

We may represent the b-parameter in the form $b \equiv \Im\left(\aleph/\Re\right)^{1/n}$, where $\Im$ is the sight-line and $\aleph$ is dimensionless energy of a *free* moving particle (for $U_0 \to 0$); V is normalized velocity of one particle relative another; λ is the angle between vectors-velocities of particles; σ^2 is the differential scattering cross-section, which determines the number of particles, that fall into a given sight angle for a given sight-line; $\beta = \Im/r$ is dimensionless inverse radius; $\bar{\beta}$ is the β-value in the turn point, in which the radial velocity vanishes and its kinetic energy becomes minimal.

To find f_V we should integrate $\sigma^2 V$ over all variables, in particular, over scattering angles ψ in the interval $(0, \psi_0)$, where ψ_0 can be found from physical considerations. The mean collision frequency is given by

$$\langle f_V \rangle = \frac{4}{\sqrt{\pi}} \int_0^\infty \tau_V^{-1}(s)\, s^2 \exp\left(-s^2\right) ds. \tag{16.40}$$

VELOCITY-DEPENDENT COLLISION FREQUENCY. Taking integrals over λ and b, we have

$$f_V(s) = \frac{C}{8}\left(\frac{\Re}{k_B T}\right)^{2/n} b^2(\psi_0)\,\Xi_n(s), \quad n = 2,3,\ldots; \tag{16.41}$$

$$\Xi_n(s) = \frac{1}{s}\int_{-\infty}^{\infty} t\,\exp\left(-t^2\right)\,|t+s|^{2-\frac{4}{n}}\,dt \tag{16.42a}$$

$$= \frac{\exp\left(-s^2\right)}{s}\int_0^\infty \sinh(2sx)\,\exp\left(-x^2\right)\,x^{2-\frac{4}{n}}\,dx. \tag{16.42b}$$

These formulas are derived in Sec. 16.3.2).

For $n = 2, 4, \infty$ the values $\Xi_n(0)$ and $\Xi_n(s)$ are expressed through the standard function (see Table 16.2).

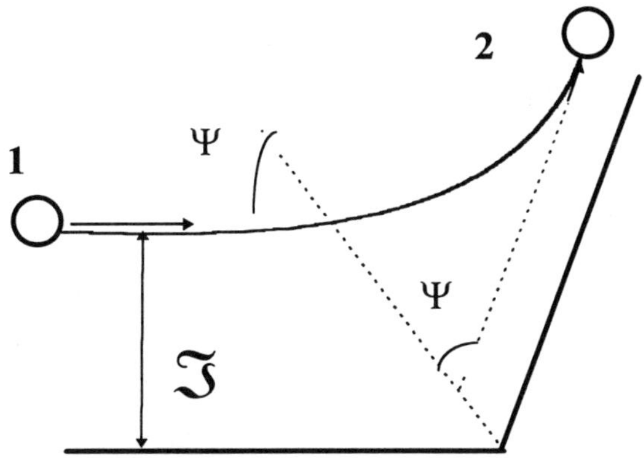

Fig. 16.4 A scheme, pertaining to a collision of two particles.

Note, the maximal scattering angle, max $\psi_0 = \pi/2$, is reached, if $b \to \infty$. We can avoid the divergence of max $\{b\}$, if we calculate the *normalized* quantity $\Xi_n(s)/\Xi_n(0)$:

$$\tau_0 f_v(s) \equiv \frac{\tau_0}{\tau(s)} = \frac{\Xi_n(s)}{\Xi_n(0)}; \ \tau_0 \equiv \frac{1}{f_v(0)} . \tag{16.43}$$

For a prescribed long-range interaction potential this ratio is shown in Fig. 16.5. The lifetime τ_0 will be estimated below from experimental data or from independent consideration.

The collision frequency $f_v(s)$ decreases with translational velocity s for the dipole-type interaction ($n = 2$ or $n = 3$) and increases with s for the rigid spheres-type intermolecular potential ($n > 4$).

THE MEAN PARAMETERS. Taking account of Eq.16.37,16.40, we find the *mean collision frequency* (details of derivation see below):

$$\langle f_v \rangle = C\left(\frac{\Re}{k_B T}\right)^{2/n} \frac{b^2(\psi_0)}{8} \frac{4}{\sqrt{\pi}} \mathbf{F}\left(\frac{1}{2} - \frac{2}{n}\right), \tag{16.44}$$

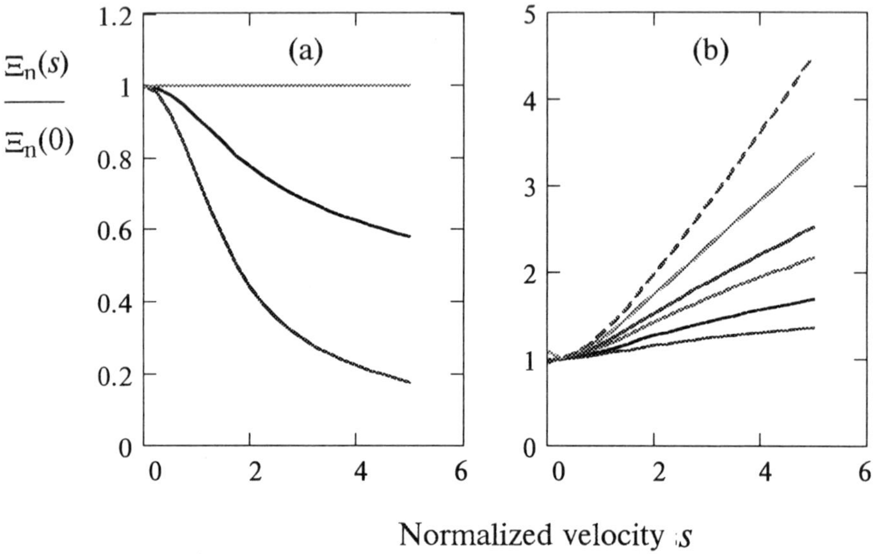

Fig.16.5 Normalized collision frequency $\Xi_n(s)$ / $\Xi_n(0)$ vs. normalized velocity s, calculated for n = 2, 3, 4 (a) and 5, 6, 8, 10, 20 (b) (solid lines, from beneath to top). Dashed curve relates to the model of rigid spheres.

$$\mathbf{F}(q) = \frac{\Gamma(q+3)}{8(q+1)} \int_{-\pi/2}^{\pi/2} \left[1 - 2^{q+1}\left(\sin\frac{\psi}{2}\right)^{2q+2}\right] \sin\psi \, d\psi \,. \qquad (16.45)$$

We also calculate the *mean scattering cross-section* by averaging σ^2 in Eq. 16.38 over s and s_1 (see Sec. 16.3.2):

$$\overline{\sigma^2} = \left(\frac{4}{\sqrt{\pi}}\right)^2 \int_0^{\pi} d\cos\lambda \int_0^{\psi_0} d\cos\psi \int_0^{\infty}\int_0^{\infty} \sigma^2 s^2 s_1^2 \exp\left(-s^2 - s_1^2\right) ds\, ds_1 \qquad (16.46)$$

$$= \left(\frac{\Re}{k_B T}\right)^{2/n} \frac{b^2(\psi_0)}{8} \left(\frac{4}{\sqrt{\pi}}\right)^2 \mathbf{F}\left(-\frac{2}{n}\right). \qquad (16.47)$$

If $n = 2$, $q = -\dfrac{2}{n} = -1$, then $\mathbf{F}(q)$ diverges due to multiplier $\dfrac{1}{1+q}$; *the mean cross-section tends to infinity for pure dipolar interaction.*

For $n = 3, 4, \ldots$ in view of Eq. 16.44, 16.46

$$\left(\frac{\Re}{k_B T}\right)^{2/n} \frac{b^2(\psi_0)}{8} = \frac{\pi\langle\sigma^2\rangle}{16\mathbf{F}\left(-\frac{2}{n}\right)};\ f_v(s) \approx \frac{C\pi\langle\sigma^2\rangle \Xi_n(s)}{16\mathbf{F}\left(-\frac{2}{n}\right)},$$

$$\langle f_v\rangle \approx \frac{C\sqrt{\pi}\langle\sigma^2\rangle \mathbf{F}\left(\frac{1}{2}-\frac{2}{n}\right)}{4\mathbf{F}\left(-\frac{2}{n}\right)};\ \frac{1}{\tau_0} \approx \frac{C\pi\langle\sigma^2\rangle \Xi_n(0)}{16\mathbf{F}\left(-\frac{2}{n}\right)}. \tag{16.48}$$

The last expression yields

$$\tau_0 \approx \frac{4}{\pi}\sqrt{\frac{\bar{m}}{\pi k_B T}} \frac{\Xi_n(0)}{\langle\sigma^2\rangle N_1 \mathbf{F}\left(-\frac{2}{n}\right)}. \tag{16.49}$$

Let

$$\gamma_n \equiv \frac{1}{\tau_v(0)} : \left\langle \frac{1}{\tau_v}\right\rangle \approx \frac{\sqrt{\pi}}{4} \frac{\Xi_n(0)}{\mathbf{F}\left(\frac{1}{2}-\frac{2}{n}\right)}. \tag{16.50}$$

The ratio $\gamma_n < 1$ for $n = 2, 3$ and $\gamma > 1$ for $n > 5$.

We define the effective diameter as follows:

$$\sigma_{ef} \equiv \sqrt{\overline{\sigma^2}}.$$

If we may estimate σ_{ef} *independently*, then the values $f_v(s)$, $\langle f_v(s)\rangle$ and τ_0^{-1} can be found together with σ_{ef}.

PATH l. We estimate l as follows:

$$l \equiv \langle \mathrm{v}(s)\,\tau(s) \rangle \approx \sqrt{\frac{2k_BT}{m}}\,\frac{4\,\mathbf{F}\left(-\frac{2}{n}\right)}{\pi\left\langle \sigma^2 \right\rangle N_1}\sqrt{\frac{\overline{m}}{\pi k_BT}}\left\langle \frac{s}{f_\mathrm{v}(s)} \right\rangle; \tag{16.51}$$

$$\left\langle \frac{s}{f_\mathrm{v}(s)} \right\rangle = \frac{4}{\sqrt{\pi}} \int_0^\infty \frac{s^3 \exp\left(-s^2\right) ds}{\Xi_n(s)}. \tag{16.52}$$

Table 16.2 Analytical representation of $\Xi_n(s)$.

n	$\Xi_n(s)$	$\Xi_n(0)$.	Model
2	$\frac{\sqrt{\pi}}{2}\frac{\mathrm{erf}(s)}{s}$	1	dipole potential
4	$\frac{\sqrt{\pi}}{2}$	$\frac{\sqrt{\pi}}{2}$	boundary case
∞	$\frac{1}{2}\left[\exp(-s^2) + \frac{\sqrt{\pi}}{2}\left(\frac{1}{s} + 2s\right)\mathrm{erf}(s)\right]$	1	rigid spheres
Note	$\Xi_n(0) = \int_0^\infty \exp(-t)\, t^{1-\frac{2}{n}}\, dt = \Gamma(2 - \frac{2}{n})$.		

Hence, for $n > 2$

$$l \approx \frac{16}{\pi^2}\sqrt{\frac{2\overline{m}}{m}}\,\frac{\mathbf{F}\left(-\frac{2}{n}\right)}{\left\langle \sigma^2 \right\rangle N_1} \int_0^\infty \frac{s^3 \exp\left(-s^2\right) ds}{\Xi_n(s)}. \tag{16.53}$$

From the dependence shown in Fig. 16.6 calculated for $m = m_1$ (so that $2\overline{m} = m$), we see, that the dimensionless path $lN_1\left\langle \sigma^2 \right\rangle$ varies with the rise of n between 3 and 4; $l(n)$ goes through a shallow maximum near $n = 5$.

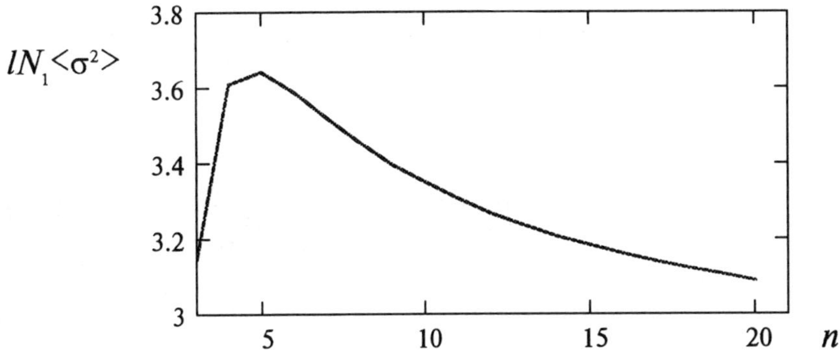

Fig.16.6 Normalized path vs. index n of the repulsive potential $U_V(r)$.

Various mean values, path l between them, weakly depend, unlike collision frequency, on index n.

16.3.2 Calculation of Collision Integrals

INTEGRAL REPRESENTATION OF $\Xi_n(s)$. We combine Eq. 16.37, 16.38, denote $t \equiv s_1$, $\cos\lambda \equiv x$, take account that $V > 0$, so that $V^q \equiv \left|t^2 + s^2 - 2ts\cos\lambda\right|^q$ and put $b^2(0) = 0$:

$$f_V(s) = C\left(\frac{\Re\, m_1}{\overline{m}\, k_B T}\right)^{2/n} \mathbf{S}_\psi \mathbf{S}_t, \tag{16.54}$$

$$\mathbf{S}_\psi \equiv -\int_0^{\psi_0} \frac{d\cos\psi}{d\cos(\pi - 2\psi)} \cos\psi\, b\,db = -\frac{1}{4}\int_0^{\psi_0} \frac{b\,db}{\sin\psi} \cos\psi \frac{d\cos\psi}{d\psi} = \frac{b^2(\psi_0)}{8},$$

$$\mathbf{S}_t = -\int_0^{\pi} d\cos\lambda \int_0^{\infty} V^{1-\frac{4}{n}}\, t^2 \exp\left(-t^2\right) dt \tag{16.55}$$

$$= -\int_0^{\pi} d\cos\lambda \int_0^{\infty}\left(t^2+s^2-2ts\cos\lambda\right)^{\frac{1}{2}-\frac{2}{n}} t^2\exp\left(-t^2\right)dt = \int_{-1}^{1} \mathbf{S}_x\, t^2\exp\left(-t^2\right)dt\,,$$

where

$$\mathbf{S}_x = \int_0^{\infty}\left(t^2+s^2-2ts\,x\right)^{\frac{1}{2}-\frac{2}{n}}dx = \frac{|t-s|^{3-\frac{4}{n}} - |t+s|^{3-\frac{4}{n}}}{\left(3-\frac{4}{n}\right)st}, \tag{16.56}$$

$$\mathbf{S}_t = \frac{1}{\left(3-\frac{4}{n}\right)s}\int_{-\infty}^{\infty}|t+s|^{3-\frac{4}{n}}\, t\exp\left(-t^2\right)dt \equiv \Xi_n(s). \tag{16.57}$$

Collecting expressions (16.54)-(16.57), we obtain finally Eq. 16.41. For derivation of analytic formulas for $\Xi_n(s)$, given in Table 16.2, it is convenient to transform integral (16.56). Then we get Eq. 16.42b:

$$\int_{-\infty}^{\infty}|t+s|^{3-\frac{4}{n}}\, t\exp\left(-t^2\right)dt = \frac{1}{2}\int_{-\infty}^{\infty}\exp\left(-t^2\right)\frac{d}{dt}|t+s|^{3-\frac{4}{n}}\,dt$$

$$= \frac{1}{2}\left(3-\frac{4}{n}\right)\int_0^{\infty} x^{2-\frac{4}{n}}\left[e^{-\left(x-s^2\right)} - e^{-\left(x+s^2\right)}\right]dx\,.$$

Putting $s = 0$ for arbitrary n or $n = 2, 4, 8$ for arbitrary s, we derive expressions, presented in Table 16.2. Indeed:

(a). If $s = 0$, we employ the definition of Γ-function:

$$\Xi_n(0) = 2\int_0^{\infty} x^{2-\frac{4}{n}}\exp\left(-x^2\right)x\,dx = \int_0^{\infty} t^{2-\frac{2}{n}-1}\exp(-t)\,dt = \Gamma\left(2-\frac{2}{n}\right).$$

(b). If $n = 2$,

$$\Xi_2(s) = \frac{1}{s}\exp\left(-s^2\right)\mathbf{F}(a),\ \mathbf{F}(a) \equiv \int_0^{\infty}\exp\left(-x^2\right)\sinh(ax)\,dx\,,$$

where $a = 2s$. Then we find that $\Xi_2(s) = \dfrac{\sqrt{\pi}}{2}\dfrac{\operatorname{erf}(s)}{s}$.

(c). If $n = 4$, then $\Xi_4 = \sqrt{\pi}/2$ and does not depend on s:

$$\Xi_4(s) = \frac{1}{s}\exp\left(-s^2\right)\mathbf{Q}(a);\ \mathbf{Q}(a) = \frac{a}{2}\int_0^\infty \exp\left(-x^2\right)\cosh(ax)\,dx = \frac{a\sqrt{\pi}}{4}\exp\left(\frac{a^2}{4}\right).$$

(d). If $n \to \infty$, we use the tabulated integral $\mathbf{F}(a)$:

$$\Xi_\infty(s) = \frac{1}{s}\exp\left(-s^2\right)\int_0^\infty \exp\left(-x^2\right)\sinh(2sx)\,x^2\,dx; \tag{16.58}$$

$$\mathbf{F}(a) = \int_0^\infty \exp\left(-x^2\right)\sinh(ax)\,dx = \frac{\sqrt{\pi}}{2}\exp\left(\frac{a^2}{4}\right)\operatorname{erf}\left(\frac{a}{2}\right). \tag{16.59}$$

Then we have, taking integral (16.103a) over parts:

$$\int_0^\infty \exp\left(-x^2\right)\sinh(2sx)\,x^2\,dx = \frac{1}{2}\left[\int_0^\infty \exp\left(-x^2\right)\sinh(2sx)\,dx\right.$$

$$\left.+\int_0^\infty 2xs\exp\left(-x^2\right)\cosh(2sx)\,dx\right] = \frac{1}{2}\left[\mathbf{F}(a) + a\frac{d\mathbf{F}(a)}{da}\right]$$

$$= \frac{\sqrt{\pi}}{4}\left[\exp\left(\frac{a^2}{4}\right)\left(1+\frac{a^2}{2}\right)\operatorname{erf}\left(\frac{a}{2}\right) + \frac{a}{\sqrt{\pi}}\right].$$

Since $a = 2s$, we get formula for $\Xi_\infty(s)$, given in Table 16.2.

CALCULATION OF INTEGRALS FOR AVERAGES $\langle\sigma^2\rangle$ *AND* $\langle f_v\rangle$.

Accounting for Eq. 16.40, 16.45, 16.54, 16.55, we derive expressions (16.44) and (16.46) for both averages, where

$$\mathbf{F}(q) = \int_{-1}^{1} dx \int_0^\infty\!\!\int_0^\infty \left|t^2 + s^2 - 2ts\,x\right|^q t^2 s^2 \exp\left(-t^2 - s^2\right)dt\,ds\,.$$

We recall the Reader, that variables t and s correspond to *different* particles. Now we introduce a sort of polar coordinates: $s = p\ \sin\varphi$, $t = p\ \cos\varphi$, so that $|D| = \left|\frac{\partial(t,s)}{\partial(p,\psi)}\right| = p$. Denoting $\psi = 2\varphi$, $x = \cos\lambda$, we have

$$\int_0^\infty\int_0^\infty (\bullet) t^2 s^2\ ds \equiv \int_0^\infty (\bullet) \exp\left(-p^2\right) p^3\, dp \int_0^{\pi/2} \sin^2\varphi\ \cos^2\varphi\ d\varphi$$

$$= \frac{1}{16}\int_0^\infty (\bullet) \exp(-h) h^2\, dh \int_0^\pi \sin^2\psi\ d\psi;\ \left|t^2 + s^2 - 2tsx\right|^q = h^q (1 - x\sin\psi)^q.$$

Taking integrals over x and h, we derive finally expressions (16.44)-(16.46) for $\langle f_V\rangle$ and $\langle\sigma^2\rangle$:

$$\frac{1}{1+q}\int_0^\pi \left[1 - (1 - \sin\psi)^{q+1}\right] \sin\psi\ d\psi$$

$$= \frac{2}{1+q}\int_{-\pi/2}^{\pi/2} \left[1 - 2^{q+1}\left(\sin\frac{\psi}{2}\right)^{2q+2}\right] \sin\psi\ d\psi;\ \int_0^\infty h^q \exp(-h) h^2\, dh = \Gamma(3+q);$$

$$\mathbf{F}(q) = \frac{\Gamma(3+q)}{8(1+q)} \int_{-\pi/2}^{\pi/2} \left[1 - 2^{q+1}\left(\sin\frac{\psi}{2}\right)^{2q+2}\right] \sin\psi\ d\psi\ . \qquad (16.60)$$

16.4 Gas-like Liquids

RELATION $\Delta\mathrm{v}(\Omega)$ *BETWEEN TRANSLATIONAL AND ANGULAR VELOCITIES*. Both velocities are assumed to be correlated in such a fashion that a molecule, rotating with angular velocity Ω, *as if rolls on the surface of other molecule*. Let $\Delta\mathrm{v}$ be velocity of a particle relative its center of mass. In view of Eq. 16.38 for V, denoting $s_1 \equiv t$, we find, by analogy with Eq. 16.37, the $\Delta\mathrm{v}(s)$ dependence after averaging V over t:

$$\Delta \text{v}(s) = \sqrt{\frac{2k_B T}{\overline{m}}}\,\langle V(s)\rangle, \tag{16.61a}$$

$$\langle V(s)\rangle = -\frac{4}{\sqrt{\pi}} \int_0^{\pi} d\cos\lambda \int_0^{\infty} V s_1^2 \exp\left(-s_1^2\right)\cos\psi \, ds_1 + \text{const}, \tag{16.61b}$$

where const is determined from the condition $\Delta\text{v}(0) = 0$, since we assume that $\Delta\text{v} \equiv 0$ *for a particle at rest.* Comparing (16.61b) with integrals (16.55) - (16.57) and accounting for formula for (16.58), we have:

$$\Delta \text{v}(s) = \sqrt{\frac{2k_B T}{\overline{m}}}\,\frac{4}{\sqrt{\pi}}\left(\Xi_\infty(s) - \Xi_\infty(0)\right), \tag{16.62a}$$

$$\Xi_\infty(s) - \Xi_\infty(0) \equiv \frac{1}{2}\left[\exp(-s^2) + \frac{\sqrt{\pi}}{2}\left(\frac{1}{s} + 2s\right)\operatorname{erf}(s) - 1\right]. \tag{16.62b}$$

Next, we relate Δv with angular velocity Ω of a "rolling" particle:

$$\left|\Delta \text{v}(s)\right| = \frac{1}{2}\xi\,\sigma_{\text{ef}}(m)\Omega, \tag{16.63}$$

$\sigma_{\text{ef}}(m)$ being an effective diameter of the particle and ξ - its *slip factor* (ξ is the free parameter of this model). Combining Eq. 16.61-16.63, we derive with account of Eq. 16.2:

$$\Xi_\infty(s_J) - \Xi_\infty(0) = \frac{\sqrt{\pi}}{4\cdot 2}\sqrt{\frac{\overline{m}}{2k_B T}}\,\xi\,\sigma_{\text{ef}}(m)\Omega_J; \tag{16.64}$$

$$\Xi_\infty(s_J) - \Xi_\infty(0) = DB(J+1), \quad D = \frac{\pi c}{2}\sqrt{\frac{\overline{m}\pi}{2k_B T}}\,\xi\,\sigma_{\text{ef}}(m). \tag{16.65}$$

Eq. 16.43 yields for $s = s_J$ and $\Delta\nu_J = (2\pi c\tau_J)^{-1}$:

$$\Delta\nu_J = \frac{1}{2\pi c\tau_0}\,\frac{\Xi_n(s_J)}{\Xi_n(0)}, \tag{16.66}$$

where $\Xi_n(0) = \Gamma\left(2-\frac{2}{n}\right)$, $\Xi_n(s)$ and $\Xi_\infty(s)-\Xi_\infty(0)$ are given by Eq. 16.42 and 16.62b, respectively.

One may regard τ_0 to be the free constant; τ_0 can also be estimated:

(1). From Eq. 16.49 for $n > 2$.

(2). From the *cell model of a liquid.* We may apply Eq. 16.53: $l = \frac{1}{\sqrt[3]{N_1}} - \sigma_{ef}$. Combining with Eq.16.49, 16.53, we derive:

$$\tau_0 = \frac{16\mathbf{F}(-\frac{2}{n})}{C\pi\sigma^2\,\Sigma_n(0)} \approx \frac{\left(\frac{1}{\sqrt[3]{N_1}} - \sigma_{ef}\right)\sqrt{\frac{\pi m}{2k_B T}}}{4\int\limits_0^\infty \frac{\Xi_n(0)}{\Xi_n(s)}\, s^3 \exp\left(-s^2\right) ds}. \quad (16.67)$$

ABSORPTION SPECTRA OF LIQUID HF-SF$_6$, HCl-SF$_6$, DCl-SF$_6$. These spectra were recorded by Birnbaum and Ho (1970), Birnbaum (1973). We introduce some critical number J_{min}. The most interesting features are:

(A). Existence of separate rotational lines (if $J \geq J_{min}$) for HF and HCl, located in non-polar SF_6 matrix, consisting of large molecules.

(B). Disappearance of few rotational lines for $J < J_{min}$.

(C). Disappearance of discrete rotational lines for DCl molecule having relatively small rotational constant B.

(D). *Descending* dependence of the half-width $\Delta\nu_J$ on rotational number J. This conclusion was made by Hernandez, Velasco and Mauricio (1986), Gaiduk and Titov (1989), Titov and Gaiduk (1990).

The $\Delta\nu(J)$ dependence agrees for HF-SF , HCl-SF solutions with that, calculated for dipolar potential ($n = 2$). For DCl-SF_6 this dependence is lesser enhanced and can be approximated by quadruple interaction ($n = 3$). Calculated and observed spectra agree (see Fig. 2.3) in main features. We employ half-widths, estimated (see Fig. 16.7) for parameters n, ξ, given in Table 16.3, where other molecular constants are also included. The sets of Lorentz and Gross lines appreciably differ for molecules with smaller constant B, viz. for HCl and DCl in our example.

Table 16.3 Molecular constants and fitted parameters used for liquid HF-SF_6, HCl-SF_6 and DCL-SF_6. $T = 273$ K.

Parameters	Polar solute				SF_6	Notes
	HF $n = 2$	HCl $n = 2$	HCl $n = 3$	DCl $n = 3$		
τ_D, ps	0.142	0.143	0.143	0.20		fitted
τ_0, ps	0.117	0.107	0.151	0.21		estimated
ξ	0.25	0.25	0.5	0.5		
τ, ps	0.392	0.449	0.326	0.5		
$\langle y \rangle$	0.104	0.177	0.129	0.160		
$N \bullet 10^{-18}$, cm^{-3}	1.22	8.4	8.4	9.65		fitted
$G \bullet 10^5$	3.33	8.07	8.07	11.1	7770	
β, cm^{-1}	20.94	10.44	10.44	5.39		
M	20	36	36	37	146	
μ, D	1.82	1.08	1.08	1.18		
σ_{eff}, Å	3.15	3.34	3.34	3.34	5.51	
ρ, $g \bullet cm^{-3}$					1.91	

Estimation of τ_0 for the cell model, Eq.16.67, yields values, that are rather close to the fitted τ_0-values (see Table 16.3). However, application for such estimate of Eq. 16.49 yields τ_0, that is about an order of magnitude *larger* than fitted τ_0. This discrepancy may mean *that colliding particles should overcome some potential barrier.*

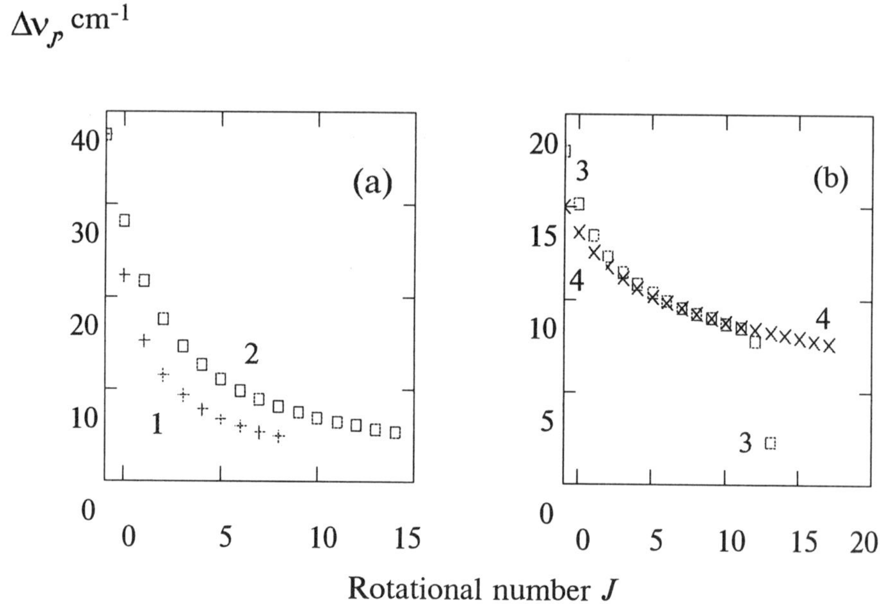

Fig. 16.7 Half-widths of rotational lines, fitted for HF-SF_6 (1), HCl-SF_6 (2, 3) and DCl-SF_6 (4) liquid mixtures. **a:** dipolar potential ($n = 2$), **b:** quadrupole potential ($n = 3$); the slip factor $\xi = 0.25$ (1, 2) and 0.5 (3, 4). $T = 293$ K.

We note, that similar assumption was already used in Sec. 14.9 for estimation of the self-diffusion coefficient in non-associated liquids. Thus, for a given index n in Eq. 16.36 *only two free constants (τ_0 and ξ) control the set of rotational lines.*

The first observed rotational line corresponds to $J = 2$ or 4, respectively for HF-SF_6 or DCl-SF_6 mixtures. For $J \leq J_{\min}$ absorption lines overlap (Fig. 2.3 a, b). If we take $\xi = 0.25$ for $n = 2$ or $\xi = 0.5$ for $n = 3$, then the only parameter (τ_0 in Eq. 16.43) will determine the set of half-widths $\Delta\nu_J$. The normalized concentration G has for gas-like liquids the same order of magnitude, as for a polar gas at $P \approx 10$ atm. That is why a discrete rotational spectrum appears in such liquids.

The recorded $\Delta\nu(J)$ dependence, given in Table 16.1 for gaseous CO + N_2 mixture, can easily be fitted, if one takes $n = 2$ or $n = 3$. However, for pure gaseous HCl or DCl the half-widths $\Delta\nu_J$ *increase* with the rise of J, if $J \leq J_{cr}$, and *decrease* with J, if J .> J_{cr}, $J_{cr} = 2$ for HCl and 4 for DCl. Such polar fluids behave *like a condensed medium* for small-energy molecules ($J \leq J_{cr}$) and like a rarefied gas for high-energy molecules ($J > J_{cr}$).

Appropriate $\Delta\nu(J)$ dependence probably can be found, if we introduce *corrections* into the classical theory of collisions. Namely, we may try to employ index $n(\mathrm{v},\mathrm{v}_1)$, *depending on velocities* of molecules.

THE AFTERWORD

Classical linear-reaction theory of orientational relaxation was presented in this book in the modification, termed the «dynamic method». The theory yields a satisfactory ground for consideration of various molecular models and for elaboration of new models. All discussed phenomena are actually based on *two effects* of alternating electric field $\boldsymbol{E}(t)$ on polar fluids. These effects are:

(1). *Resonance absorption* of electromagnetic energy by librating / rotating dipoles.

(2). *Orientation* (with a lag that leads to dielectric loss) of a dipole moment μ of a particle along time varying electric field $\boldsymbol{E}(t)$. This process is accompanied in a condensed medium by the *transition of a particle from one potential well to another via stochastic (Brownian) reorientation.*

The second mechanism **(2)**, dominating at low frequencies, is the cause of Debye-relaxation spectrum, while the first (resonance) one **(1)** stipulates in liquids the appearance of far-infrared absorption band. Both mechanisms are inherently related one to another and should be studied self consistently.

The resonance mechanism **(1)** *is studied in the book explicitly:*

(a) by solving equations of motion of a particle, reorienting inside a well at thermal equilibrium; the law of motion usually describes periodic oscillations of dipoles;

(b) by calculation of the spectral function (SF) $K(z)$ for an anisotropic polar medium [or $L(z)$ for an isotropic medium]. *The SF actually presents the spectrum of «undamped» dipolar ACF*; the SF depends on the potential profile and on the lifetime τ of a dipole in the well.

This procedure yields relatively simple analytic expressions (ordinary or double integrals) only for relatively simple profiles $U(\vartheta)$. For non-homogeneous profiles we have considered dielectric response in ensemble of *linear molecules.* For homogeneous profile, i.e. for $U = 0$, we have presented the dielectric-response theory of symmetric top molecules.

The Debye relaxation mechanism **(2)** *was studied implicitly:* stochastic equations of motion were not involved in the book. Instead, we used *«collision models», that determine induced distributions* F_B, F_D, F_G, F_K or F_T of dipoles in phase space. Employing these models, one can express free parameters of the spectral function through *experimental* Kirkwood correlation factor g. This simple procedure yields correct (but not rigorous!) description of low-frequency spectra.

However, for rather complex systems, such as *e.g.* dimethyl sulfoxide aqueous solution or a protein molecule in water surroundings, a problem arises: how many independent relaxation regions and Kirkwood factors g_m ($m = 1, 2, ...$) should one take into account?

Another problem, that is not practically touched in the book, concerns the description of polar fluids (such as concentrated water-electrolyte solutions, Sec. 15.2), which exhibit a strong distinction of the Cole-Cole diagram $\varepsilon''(\varepsilon')$ from the Debye semi-circle.

In the most important Part V we have demonstrated a number of applications of our theory. On the face of it, possible variety of molecular models is overwhelming! However, if one tries to fit tough criteria, related to peculiarities of observed spectra, it becomes obvious, that most models are bad and only few are more or less satisfactory. Here we mean the problem, how to reach the agreement of calculated positions and values of absorption / loss maxima, presence / absence of the Debye plateau, transition to the transparency on the HF edge of the orientational-relaxation region, *etc.* with the corresponding recorded values.

Simple and rather well description of dielectric spectra in liquids is obtained in frames of the *hybrid model.* The latter is characterized by the *rectangular potential well having flat edges.* If such well is deep, it transforms into the confined rotator model, that was widely used by the author with his collaborators and somewhere else, see e.g. (Grochulski and Gerschell 1992). This model partially «kills» the distinction between water and simple nonassociated liquids, such as liquid CH_3F. Indeed, high-energy dipoles appear in both fluids; an angular velocity of a fast particle keeps its sign during lifetime τ between strong collisions. However, a qualitative distinction between these fluids still remains, since in nonassociated liquids librations are damped, while in water a dipole performs about 10 librations during its lifetime τ.

ABOUT SOME NEW PROBLEMS

A. The version of the hybrid model, widely used in part V, is valid for small or moderate libration angles β, although the theory is *qualitatively* applicable for $\beta < \pi/2$. Application of the HM2 model, described in the end of Chap. 9, to concrete fluids will probably permit to avoid this restriction.

B. The hybrid → hat model transformation may stimulate solution of new problems, *viz.*:

(α). Consistent theory of an *isotopic effect in liquid water.* In Sec. 15.1.6 we proved that this effect may be explained by contribution of high-energy *hindered* rotators into complex susceptibility.

(β). *Contraction of theoretical absorption band in liquid water.* This band is wider, than recorded one, Sec. 15.2. The approach, given in Sec. 13.6, shows that absorption band may contract for a potential profile that declines from the rectangular one.

(γ). Elaboration of a molecular model, capable to describe *reorientation in ice Ih and its dielectric response.* Absorption spectra, observed in liquid water and ice Ih, are similar (Fig. 2.13). In view of the previous item **(β)** we suggest that one can reach the agreement of theoretical R- and L-absorption bands with those, observed experimentally after appropriate correction of a potential profile. Taking relatively high potential U / long lifetime τ it is also principally possible to obtain in new theory a very long relaxation time τ_D, that is typical for ice.

(C). It is interesting to improve our study of electrolytes, given in Sec.15.2, by taking into account the electrostatic ion-water interaction. Estimations of angular frequencies of H_2O dipoles, given in GT, p.318, could be useful in this respect.

A fundamental problem concerns the frequency dependence $\sigma(\nu)$ of conductivity due to hydrated ions, translating in some near-order potential. In the above calculations, Sec.15.2, we regard σ to be independent on frequency ν. For typical electrolytes σ should appreciably alter in the millimeter wave range, because due to inertia ions are actually in rest for so high frequencies of electromagnetic field. It would also be very interesting to elaborate the molecular theory of the static conductivity $\sigma(0)$, e.g. in terms of the rectangular potential well, in which ions perform multiple oscillations during the ionic lifetime τ_i, see the preliminary investigation in (Gaiduk, Tseitlin, and Liberman 1999).

ADDENDUM II
ON THE EVOLUTION OF MOLECULAR MODELS OF ORIENTATIONAL RELAXATION

Here we mean models elaborated by the author with his collaborators.

THE CONFINED ROTATOR (CR) MODEL. We have already mentioned in Addendum I that within the frames of the dynamic method the planar CR model by (Gaiduk, Kalmykov, and Tseitlin 1979) was the first one, applied to polar fluids. The Debye collision model (i.e. with $F = F_D$) was envisaged, that is characterized by the relaxation time τ_D. It should be mentioned that the formula for the susceptibility was derived earlier in (Higasi, Minami, Takahashi, and Ohmo 1973), (Minami and Ohmo 1973) by considering ACF of dipoles, librating in the rectangular potential well. Unfortunately, the theory, elaborated in the aforecited works, is not applicable in the FIR region.

Starting elaboration of the CR model, the author imagined, that a dipole performs *many* librations during its lifetime τ. How ever, the very first calculations showed, that the fitted lifetime τ is actually so *short,* that during τ a dipole scarcely may perform one oscillation in the potential box. On the other side, in (Gaiduk, Kalmykov, and Tseitlin 1979) only one absorption maximum was described *for water,* though two FIR absorption peaks are observed experimentally. Further applications of the CR model were related:

(A). To the improvement of the theory with the aim to obtain better description of dielectric relaxation in nonassociated liquids.

(B). To the complication of the model in order to seek a correct explanation of dielectric spectra in water and in aqueous systems. Briefly, **A** and **B** trends consist in the following:

A1. Generalization of the theory with the purpose to describe *spatial* librations of confined rotators. An *approximate* theory was given in (Gaiduk 1986). *Rigorous* cone-confined rotator (CCR) theory was published in GT, p.231 and is described at great length in Sec. 8.5.

A2. Generalization of the collision model. In GT, p.185 other induced distributions were found: Gross (F_G), correlation-orientation (F_K), isothermal (F_T). The estimated relaxation time τ_D was related to the free parameters of the CR model.

A3. Only *qualitative* agreement of theoretical and recorded dielectric spectra was arrived at in first publications, related to the CR model, see *e.g.* GK, (Novskova, Kukebaev, and Gaiduk 1986), (Brekhovkikh and Gaiduk, 1989). Better agreement was reached, when it was recognized [see, e.g. (Gaiduk, Novskova, and Brekhovkikh 1991) and GT] that the *exact* description of the static permittivity should be obtained with use of Eq. 1.12 and that it is useful to employ the dipole correction factor k_μ in order to obtain better description of the maximal FIR absorption.

A4. Using temperature dependence of the fitted parameters, an attempt was undertaken in (Gaiduk, Novskova, and Brekhovskikh 1991) and in GK to picture specific features of molecular dynamics in liquids. The «rotational» cell model and calculation scheme for the self-diffusion coefficient *D(T)* were suggested. One of new results concerns to the prediction of the critical temperature of liquids, see Chap. 14.

B1. The idea appeared (see GK) to introduce the R-fraction of molecules with greater rotational mobility in addition to the main L-fraction of dipoles, librating in the rectangular potential well. In GK, p. 167 it was suggested that R-molecules rotate freely ($\tau \to \infty$); correspondingly their contribution to the dielectric susceptibility of water was described on the basis of the extended rotational diffusion model with $y = 0$. Later, in (Gaiduk, Gus'kova, and Novskova 1988), (Gaiduk, Novskova, and Brekhovskikh 1993) the lifetime τ was considered to be finite. It was evident that librating dipoles constitute the network of H-bonds. However, physical origin of the R-fraction was not recognized that time. Thus, the CR/ED model has appeared; it was based on the consideration of two *independent* groups of dipoles, the proportion r of R-particles being the free parameter of the model.

B2. A number of aqueous systems was studied, in which dielectric response of water was described with use of two fractional CR/ED model. We mean hemoglobin-water solution (Gaiduk and Novskova, 1987), dimethyl sulfoxide-water mixtures (Novskova, Gaiduk, Kudryashova, and Khurgin 1990), aqueous solutions of electrolytes (Lyashchenko, Novskova, Lileev, Zasetski, and Gaiduk 1993) and humidified collagen (Novskova and Gaiduk, 1989), the latter being a dielectric model of a skin; see also review (Novskova and Gaiduk 1996). A number of elementary «units» were employed in the theory of such fluids, where each unit represents a fraction, whose dielectric behavior is described on the ground of a simple model, such as CR/ED (for water), CR(for bound water) or ED (for a protein).

FIELD MODELS. Following (Kalmykov 1986, 1987a, 1987b), where the kinetic equation method was used, we employed the dynamic method to describe dielectric response of polar molecules, librating / rotating in a plane under effect of the cosine or cosine squared conservative potential $U(\vartheta)$. The approximate and rigorous solutions of these problems were found (Gaiduk, Tseitlin, Novskova, and McConnell 1993), (Gaiduk, Gaiduk, Novskova, Tseitlin, and McConnell 1995), GT, (Novskova and Gaiduk 1996). It was shown, that these potential profiles are characterized by *too narrow* absorption band, if the potential welldepth U_0 is commensurable with $k_B T$ and / or if collisions are not very frequent ($y = \eta/\tau < 1$). We have applied both field models: **(a)** to the subensemble of hindered rotors in liquid water (Gaiduk, Gaiduk and McConnell, 1995) and **(b)** to ice (Gaiduk, Gaiduk, Novskova, Tseitlin and McConnell, 1995), the potential well U_0 being about 9 $k_B T$.

The dynamic method is generally characterized by *instantaneous collisions*, since we regard that a dipole, librating in the well U, moves during mean time τ between strong collisions and then suddenly disappears as a particle, related to some point of the phase space. Therefore, formally one may regard that the *collision time* τ_c is zero. On the other side, considering field models one can envisage that the time τ_c in a liquid is commensurable with the mean half-period $\langle T_L / 2 \rangle$ of librations, see (Gaiduk, Tseitlin, Novskova, and McConnell 1993). This approach presents some alternative to that, developed by (Burshtein and McConnell 1989), (Kalmykov and McConnell 1993), where, unlike our approach, *the effective potential is not introduced, since it is replaced by some modified collision model.* Our idea (it was described in the book) consists in the following. Assuming that τ_c is ca. $\langle T_L / 2 \rangle$, we can reveal the specificity of the liquid comparing with the gas state. In the former the lifetime τ is generally lesser, than τ_c, while in the gas usually $\tau >> \tau_c$.

THE HAT MODEL. Applicability of the aforementioned «field models», having smooth potential profile, turned out to be very restricted, comparing with the applicability of the rectangular potential box. The author suggested (Gaiduk 1994), that an adequate intermolecular potential well should have (at least, in the cases of water / ice) the form of the box with curved edges. In other words, an adequate potential U(J) should have the form of a hat with curved brims. A thorough description of dielectric response was given in Chap. 13 for such model. The theory was elaborated by the author in collaboration with B.M. Tseitlin but it was not yet applied for the study of concrete fluids.

THE HYBRID MODEL. Let the form of the potential well be the rectangular box with flat edges. Such well represents a simplest particular case of the hat-like potential. The similar molecular model was considered by (Gaiduk, Novskova, and Brekhovskikh 1991), see also GT, p.342. However, this approach occurred to be too simplified. Later, more adequate theory was elaborated by B. M. Tseitlin and the author [see (Gaiduk and Gaiduk 1997), (Gaiduk, Novskova, and Tseitlin 1998)]. This corrected approach is widely presented in the book (Chap. 9). The corrected hybrid model is applicable (within frames of its validity) for nonassociated liquids, for water and aqueous systems, see also (Gaiduk and Liberman 1997), (Novskova, Lyashchenko, and Gaiduk 1996), as well this book, Sec. 14.6-14.9 and Chap. 15.

THE EXTENDED ROTATIONAL DIFFUSION MODEL. This «old» model, being hardly applicable to liquids, is useful:

(1) for comparison with «modern» models, operating with some intermolecular potential well $U(\vartheta)$;

(2) as a model, used for description of dielectric response of some molecular fractions, characterized by large rotational mobility, see e.g. (Gaiduk and Novskova, 1997) and Sec. 15.3;

(3) for description of dielectric response of polar gases after quantum correction is introduced into formulas for susceptibility, see (Titov and Gaiduk 1990) and Chap. 16 of this book.

REFERENCES

Abramowitz, M. and Stegun, A. 1964 (eds.). *Handbook of mathematical functions* (National Bureau of Standarts, Applied Mathematics Series).

Akhadov, Ya. Yu. 1977. *Dielectric properties of binary solutions* (Nauka, Moscow) (in Russian).

Belaya, M. L. and Levadny, V. G. 1987. *Molecular structure of water* (Znaniye, Moscow) (in Russian).

Bennowna, M., Cachet, H., Lestrade, J. C., and Birch, J. R. 1981. *Chem. Phys.* **62**, 439.

Berens, P. H. and Wilson, K. R. 1981. *J. Chem. Phys.,* **71**, 4872.

Bernal, J. D. and Fowler, R. H. 1933. *J. Chem. Phys.*, **1**, 515.

Birnbaum, G. and Ho, W. 1970. *Chem. Phys. Lett.*, **5**, 334.

Birnbaum, G. 1973. *Mol. Phys.,* **25**, 241.

Bone, S. and Pethig, R. 1985. *J. Mol. Biol.,* **181**, 323.

Brekhovskikh, V. V. and Gaiduk, V. I. 1989. *Khim.. Phiz.*, **8**, 1257 (in Russian).

Brekhovskikh, V. V. 1993. *Prediction of dielectrric properties of homogeneous polar media and stratificated systems.* Thesis (Moscow Inst. Phys. & Technology.) (in Russian).

Brot, C. 1975. In: *Dielectric and Related Molecular Processes* (Chem. Soc. London, Vol. 2), Chap. 1.

Brown, W. F. 1956. *Encyclopedia of physics, Vol. 17, Dielectrics* (Ed. S. Flugge, Springer-Verlag, Berlin).

Burshtein, A. I. and McConnell, J. 1989. *Physica* **A, 157**, 933.

Chantry, G. W. 1976. *Submillimetre spectroscopy* (Academic Press. London).

Coffey, W., Evans, M., and Grigolini, P. 1984. *Molecular diffusion and spectra* (Wiley, New-York).

Cole, K. S and Cole, R. H. 1941. *J. Chem. Phys.*, **9**, 341.

Davis, P. K. 1972. *J. Chem. Phys.*, **57**, 517.

Debye, P. 1929. *Polar molecules* (Chemical Catalog Company, New York).

Downing, H. D. and Williams, D. 1975. *J. Geophys. Res.*, **80**, 1656.

Einstein, A. 1905. *Annal. Phys.*, **17**, 549.

Evans, M., Evans, G. J., Coffey, W. T., and Grigolini, P. 1982. *Molecular dynamics and theory of broad band spectroscopy* (Wiley. New. York).

Filippov, N. N., Vakoulin, A. A., and Orlova, N. D. 1986. *Opt. spektr.* **61**, 52 (in Russian).

Frohlich, H. 1958. *Theory of dielectrics, dielectric constant and dielectric loss* (Second edition, Clarendon Press, Oxford).

Gaiduk, V. I. 1960a. *Radiotekh. Elektron*, **5**, 239 (in Russian).

Gaiduk, V. I. 1960b. *Doklady Acad. Nauk. SSSR,* **188**, 760 (in Russian).

Gaiduk, V. I., Palatov, K. I., and Petrov, D. M. 1971. *Physical principles of microwave electronics* (Soviet. Radio, Moscow) (in Russian).

Gaiduk, V. I., Loskutov, V. S., and Sekistov, V. N. 1976. *Radiotekh. Elektron*, **21**, 2270 (in Russian).

Gaiduk, V. I., Kalmykov, Yu. P., and Tseitlin, B. M. 1979. *Radiotekh. Elektron*, **24**, 1170 (in Russian).

Gaiduk, V. I. and Tseitlin, B. M. 1980. *Radiotekh. Elektron*, **25**, 352 (in Russian).

Gaiduk, V. I. and Kalmykov, Yu. P. 1981. *J. Chem. Soc., Faraday Trans. 2,* **77**, 929.

Gaiduk, V. I. 1986. *Izv. Vyssh. Uchebn. Zaved., Radiofiz.,* **28**, 1366 (in Russian).

Gaiduk, V. I. and Kalmykov, Yu. P. 1987. *J. Mol. Liquids*, **34**, 1.

Gaiduk, V. I. and Novskova, T. A. 1987. *Biofiz.,* **32**, 579 (in Russian).

Gaiduk, V. I., Gus'kova, V. D., and Novskova, T. A. 1988. *Izvest. Vyssh. Uchebn. Zaved., Radiofiz.,* **31**, 799 (in Russian).

Gaiduk, V. I. 1991. *Water, radiation, life* (Znaniye, Moscow) (in Russian).

Gaiduk, V. I., Novskova, T. A., and Brekhovskikh, V. V. 1991, *J. Chem. Soc. Faraday Transactions*, **87**, 559.

Gaiduk, V. I., Novskova, T. A., and Brekhovskikh, V. V. 1993. *J. Chem. Soc. Faraday Transactions*, **89**, 1975.

Gaiduk, V. I., Tseitlin, B. M., Novskova, T. A., and McConnell, J. 1993, *Physica* **A, 197**, 75.

Gaiduk, V. I. 1994. *Mendeleev Communic.*, No. 2, 15.

Gaiduk, V. I. and Tseitlin, B. M. 1994. *Adv. Chem. Phys.,* **87,** 125 (referred throughout as GT).

Gaiduk, V. I., Gaiduk, V. V., and McConnell, J. 1995. Physica **A, 222,** 46.

Gaiduk, V. I., Gaiduk, V. V., Novskova, T. A., Tseitlin, B. M., and McConnell, J. 1995. *J. Chem. Phys.,* **103**, 5246.

Gaiduk, V. I. and Novskova, T. A. 1996. *Mendeleev Communic.*, No. 2, 75.

Gaiduk, V. I. and Gaiduk, V. V. 1997. *Mendeleev Communic.,* No. 2, 76.

Gaiduk, V. I., Novskova, T. A., and Tseitlin, B. M. 1998, *Khim. Fiz.,* **17,** 50 (in Russian).

Gaiduk, V. I. and Liberman, B. M. 1998, *Russian J. Chem. Phys.*, **71,** 1619.

Gaiduk, V. I., Liberman, B. M., Apletalin, V. N., and Meriacri, V. V. 1999. *Radiotekhn. Elektron.*, **71**, 1619 (in press) (in Russian).

Gaiduk, V. I., Tseitlin, B. M., and Liberman, B. M. 1999. *Radiotekhn. Elektron.*, **71**, No.2 (in press) (in Russian).

Galiyarova, N. M. and Shakhparonov, M. I. 1976. *Vestnik MGU, Khim.* (Deposited by VINITI, No. 3612-76) (in Russian).

Gerschel, A., Dimicoli, I., Jaffre, A., and Riou, A. 1976, *Mol. Phys.,* **32**, 679.

Gradshtein, I. S. and Rydzyk, I. M. 1963. *Tables of integrals, sums, series, and products* (Ed. 4, Fizmatgiz, Moscow) (in Russian).

Grant, E. H., Sheppard, R. G., and South, G. P. 1978. *Dielectric behaviour of biological molecules in solution* (University Press, Oxford, Clarendon).

Gray, C. C. 1974. *J. Chem. Phys.* **61**, 418.

Gray, C. C. and Van Kranendonk, J. 1966. *Canad. J. Phys.*, **44**, 2411.

Grochulski, T. and Gerschell, A. 1992. *Mol. Phys.,* **77**, 539.

Hernandez, A. C., Velasco, S., and Mauricio, F. 1986. *Phys. Rev.* **A., 33**, 750.

Higasi, K., Minami, R., Takahashi, H., and Ohmo, A. 1973. *J. Chem. Soc. Faraday Trans. II*, **64**, 1579.

Hill, N. E. 1963. *Proc. Phys. Soc.*, **82**, 723.

Hippel, A. R. 1954. *Dielectrics and waves.* (Russian translation: Izdat. Inostr. Literat., Moscow, 1960).

Kalmykov, Yu. P. 1986. *Radiotekh. elektron.* **31**, 1358 (in Russian).

Kalmykov, Yu. P. 1987a. *Khim. Fiz.*, **6**, 592 (in Rissian).

Kalmykov, Yu. P. 1987b. *Khim. Fiz.*, **6**, 745 (in Rissian).

Kalmykov, Yu. P. 1988. *Khim. Fiz.*, **7**, 1341 (in Rissian).

Kalmykov, Yu. P. and Limonova, S. V. 1989. *J. Mol. Liquids*, **43**, 71.

Kalmykov, Yu. P. and McConnell, J. 1993. *Physica* **A, 193**, 394.

Kalmykov, Yu. P. and Titov S. V. 1994. *Adv. Chem. Phys.* **87**, 31.

Khurgin, Yu. I. 1976, *J. Mendeleev Chem. Soc.*, 684 (in Russian).

Khurgin, Yu. I., Kudryashova, V. A., Zavizion, V. A., and Betski, O. V. 1994. *Adv. Chem. Phys.* **87**, 483.

Landau, L. D. and Lifshits, E. M. 1973. *Theoretical Physics, Vol.1. Mechanics* (3th ed., Nauka, Moscow) (in Russian).

Lane, J. A. and Saxton, J. A. 1952. *Proc. Roy. Soc.* 214 **A**, 531.

Lassier, B. and Brot, C. 1968. *Chem. Phys. Lett.*, **1**, 581.

Liboff, R. 1969. *Introduction to the theory of kinetic equations.* (Wiley, New-York).

Liebe, H. J., Hufford, J. A., and Manabe, T. 1991. *Intern. J. of Infrared and Millimeter Waves*, **12**, 659.

Lyashchenko, A. K., Novskova, T. A., Lileev, A. S., Zasetski, A. Yu., and Gaiduk, V. I. 1993. *J. Chem. Soc. Faraday Transactions*, **89,** 1985.

Mashimo, S., Kuwabara, S., Yagihara S., and Higasi, K. 1987. *J. Phys. Chem.,* **89**, 6337.

McConnell, J. R. 1980. *Rotational Brownian motion and dielectric theory* (Academic Press, London).

Menzel, D. H. 1957 (ed.). *Fundamental formulas of physics* (Russian translation: Izdat. Inostr. Literat, Moscow).

Minami, R. and Ohmo, A. 1973. *J. Phys. Soc. Japan*, **35**, 1730.

Novskova, T. A., Kukebaev, A. M., and Gaiduk, V. I. 1986. *Izvest. Vuzov, Radiofizika*, **29**, 41 (in Russian).

Novskova, T. A. and Gaiduk, V. I. 1989. In: *Millimetre waves in medicine and biology*. (Ed. N. D. Dev'yatkov, IREE, USSR Academy of Sciences, Moscow), p. 276 (in Russian).

Novskova, T. A., Gaiduk, V. I., Kudryashova, V. A., and Khurgin, Yu. I. 1990. *Khim. Fiz.*, **9**, 969 (in Russian).

Novskova, T. A. and Gaiduk, V. I. 1996. *J. Mol. Liquids*, **69,** 143.

Novskova, T. A. and Gaiduk, V. I. 1996. *Biofiz.*, **41,** 565 (in Russian).

Novskova, T. A., Lyashchenko, A. K., and Gaiduk, V. I. 1996. *Millimeter waves in Biology and Medicine, Institute of Radio Engineering & Electronics RAS*, No. 8, 38 (in Russian).

Pinkley, L. W., Sethna, P. P., and Williams, D. 1976. *J. Opt. Soc. Am.*, **67**, 494.

Poley, J. P. 1955. *Appl. Sci. Research*, **413**, 337.

Prudnikov, A. P., Brychkov, Yu. A., and Marychev, O. I 1981. *Integrals and series, Elementary functions* (Nauka, Moscow) (in Russian).

Samoilov, O. Ya. 1957. *Structure of aqueous solutions of electrolytes and hydration of ions*. (Izdat. Akademii Nauk SSSR, Moscow) (In Russian).

Scaife, B. K. P. 1989. *Principles of dielectrics* (Clarendon Press, Oxford).

Sethna, P. P., Palmer, K. F., and Williams, D. 1978. *J. Opt. Soc. Am.*, **68,** 815.

Shakhparonov, M. I. and Galiyarova, N. M 1980. In: *Phisics and physical chemistry of liquids. Vyp. 4* (eds. L. P. Filippov and M. I. Shakhparonov, MGU Press), p. 75 (in Russian).

Strendberg, M. 1956. *Radiospectroskopie* (Russian translation: Inostr. Literat., Moscow) (in Russian).

Titov, S. V. and Gaiduk, V. I. 1990. *Opt. Spektr.* **68**, 1043 (in Russian).

Townes, Ch. H. and Shawlow, A. L. 1975. *Microwave spectroscopy* (Dover Publications, New-York).

Van Vleck, J. H. and Huber, D. L. 1977. *Rev. Mod. Phys.*, **49**, 939.

Water. A comprehensive treaties. 1972. Vol. 1 (Ed. F. Franks), Chap. 5.

Wyllie, G. 1971. *J. Phys. C. Solid St. Phys.*, **4**, 564.

Zatsepina, G. N. 1974. *Properties and structure of water* (Moscow University) (in Russian).

SUBJECT INDEX

SERIES IN CONTEMPORARY CHEMICAL PHYSICS

Vol. 1 The Photon's Magnetic Field — Optical NMR Spectroscopy
by M. W. Evans and F. Fahari

Vol. 2 Beltrami Fields in Chiral Media
by A. Lakhtakia

Vol. 3 Quantum Mechanical Irreversibility and Measurement
by P. Grigolini

Vol. 4 The Photomagneton and Quantum Field Theory: Quantum Chemistry, Vol. 1
by A. A. Hasanein and M. W. Evans

Vol. 5 Computational Methods in Quantum Chemistry: Quantum Chemistry, Vol. 2
by A. A. Hasanein and M. W. Evans

Vol. 6 Transport Theory of Inhomogeneous Fluids
by L. A. Pozhar

Vol. 7 Dynamic Kerr Effect: The Use and Limits of the Smoluchowski Equation and Nonlinear Inertial Responses
by J.-L. Dejardin

Vol. 8 Dielectric Relaxation and Dynamics of Polar Molecules
by V. I. Gaiduk

Vol. 9 Water in Biology, Chemistry and Physics: Experimental Overviews and Computational Methodologies
by G. W. Robinson, S. B. Zhu, S. Singh and M. W. Evans

Vol. 10 The Langevin Equation: With Applications in Physics, Chemistry and Electrical Engineering
by W. T. Coffey, Yu P. Kalmykov and J. T. Waldron

Vol. 11 Structure and Properties in Organised Polymeric Materials
eds. E. Chiellini, M. Giordano and D. Leporini

Vol. 12 Proceedings of the Euroconference on Non-Equilibrium Phenomena in Supercooled Fluids, Glasses and Amorphous Materials
eds. M. Giordano, D. Leporini and M. P. Tosi

Vol. 13 Electronic Structure and Chemical Bonding
by J. R. Lalanne

Vol. 14 Dialogues on Modern Physics
by M. Sachs

Vol. 15 Phase in Optics
by V. Perinova, A. Luks and J. Perina

Vol. 16 Extended Electromagnetic Theory: Space-Charge in Vacuo and the Rest Mass of the Photon
by S. Roy and B. Lehnert